Aircraft Structures
Second Edition

Aircraft Structures
Second Edition

G. Lakshmi Narasaiah

Formerly Professor
Dept. of Aeronautical Engineering
Institute of Aeronautical Engineering
Dundigal, Hyderabad.

BS Publications

A unit of **BSP Books Pvt. Ltd.**
4-4-309/316, Giriraj Lane,
Sultan Bazar, Hyderabad - 500 095

Aircraft Structures, *Second Edition*
by G. Lakshmi Narasaiah

© 2021, 2010 *by Author*
Second Edition, 2021; *First Edition*, 2010
All rights reserved

Disclaimer: The authors and the publishers have taken due care to provide the authentic, reliable and up to date information related to the subject. However, neither the authors nor the publisher shall be responsible for any liability for any damage caused as a result of use of this book. The respective user must check the accuracy from other sources too.

Published by:

 **BS Publications**

A unit of **BSP Books Pvt. Ltd.**
4-4-309/316, Giriraj Lane, Sultan Bazar,
Hyderabad – 500 095.
Phone : 040 – 23445688
e-mail:info@bspbooks.net
www.bspbooks.net

Printed at

Adithya Offset Process (I) Pvt. Ltd.
Hyderabad.

Price : Rs. 695.00
ISBN : 978-93-89974-92-8 (Paperback)

To my wife *Sandhya Rani*
whose support and encouragement
were instrumental in my successful career

Preface to Second Edition

A few simple mistakes were noticed in the first printed version of the book, in symbols and equations. They have all been corrected. To avoid confusion in reading equations correctly, their format is improved.

To meet the needs of students, as per the syllabus of some universities, new topics have been added. A more detailed presentation is included in some topics like Analysis of redundant trusses, Analysis of Isotropic Flat Plates in Tension, Plate Subjected to Lateral Load and In-plane Load, Energy method for analysis of thin plates in bending, Bending of plates having a small initial curvature, Structural idealization for wing and fuselage sections etc.

G. Lakshmi Narasaiah

Ph : 91-9989860244

e-mail: *gogineni_ln@yahoo.co.in*

15.07.2020

Preface to First Edition

Many books are available for the analysis of structures. The methods explained in most of these books are suitable for civil and mechanical engineering applications. Not many books on aerospace structures are written, for two reasons – firstly, institutes offering aeronautical engineering are very few in the world and writing a book for meeting their specific needs is not economically viable and secondly, analysis of aircraft structures is largely supported by experimental data, most of which is company specific and confidential.

The words Airplane, Aeroplane and Aircraft are used synonymously in different books. The word Aircraft was coined when the aeroplane in early stages was more of craftsmanship (primarily with wood) rather than engineering. Today an aeroplane is an engineering marvel and its design has propelled development of Finite Element Method (FEM), Computational Fluid Dynamics (CFD), Fracture Mechanics, Super alloys and Composite materials – to name a few. The technology of aeroplane is now extended to missiles and space vehicles. Aeronautical Engineering is thus renamed as Aerospace Engineering in many universities / Institutes.

Since 70s, India has made significant strides in aerospace engineering with the development of - light combat aircraft (LCA) and light combat helicopter (LCH); Polar satellite launch vehicles (PSLV); and supersonic missiles. With the open-sky policy, air transportation and overhauling facilities of aeroplanes have increased many folds in India. To cater to the growing demand in HAL, ISRO, DRDL, Civil air transportation etc., many engineering colleges are now offering undergraduate and postgraduate courses in aerospace engineering.

Aerospace vehicle structures have special features - in terms of nature of loads, type of construction, redundancy to be provided, materials used and safe operation. Construction features like ring frames in fuselage, box beams in wings, tension field beams, stiffened panels and stiffeners of different shapes need special methods of analysis. Loading conditions contribute to bending, shear and torsion in wing, fuselage and control surfaces. These slender components are also susceptible to buckling, warping and torsion-bending. In addition, aero-elastic behaviour such as flutter is specific to aerospace vehicle structures.

An aerospace engineer should, therefore, be exposed to all these special aspects. The engineer has to estimate the loads and responses as accurately as possible so that the structure can be designed for minimum weight to achieve best operational efficiency of the aeroplane.

The subject is vast and is, therefore, split into two parts in many universities and covered in two semesters. Topics like Mechanical properties of materials, properties of sections (covering centroid and moment of inertia) etc., are covered as parts of other subjects and form prerequisites to this course. These topics are briefly included here so that students can refer to the same, whenever necessary. Finite element method is usually offered as a separate subject later and is included here briefly. Other related topics like aeroelasticity and materials of aeroplane construction, which do not form an essential part of this subject, are included for a better understanding of the subject.

Wings and fuselage, main parts of aircraft structure, are made of beams and shear panels. Analysis of determinate and indeterminate beams, which is covered in many universities as 'Mechanics of solids' or 'Strength of materials', is included in detail here for necessary extension to other special cases. More emphasis is given for stiffened panels and shear panels in bending and torsion as well as buckling of columns and panels, which are more specific to aeroplane structures.

It is considered helpful for the students and graduate engineers to have a comprehensive reading material, including all related topics. The material for this book was compiled from my class notes as a student at I.I.T., Kanpur and from books by David J Peery, Paul Kuhn, Ernest E Sechler & Louis G Dunn, A C Kermode and Robert M Rivello. Selected topics from Engineering mechanics, Strength of materials. Theory of elasticity, Machine design, Material science and FEM formed additional material.

In spite of my best efforts, some numerical mistakes and typing mistakes might have crept in. I will be grateful for your comments and suggestions for improving the presentation and content in the subsequent editions.

G. Lakshmi Narasaiah
Ph : 91-40-66881678
e-mail:gogineni_ln@yahoo.co.in

15.07.2010

Acknowledgements

Discussions with colleagues and aeronautical students in engineering colleges, have highlighted the need for a comprehensive book on this subject. The author wishes to thank all those, who have directly or indirectly contributed to the presentation of this subject in this form.

The author also wishes to thank the management of MNR College of Engineering & Technology, Sangareddy, for giving me an opportunity to teach this subject in two parts in two semesters, as a Professor in the Department of Aeronautical Engineering.

I thank Prof (Retd) P N Murthy, IIT, Kanpur for laying the foundation for this subject as a teacher and consolidating the theoretical concepts as my project guide.

I thank M/s B S Publications for publishing this book, even though it caters to a small section of students and, therefore, has limited market.

-Author

Contents

Chapter 1

Introduction   1-4

PART - I : BASICS OF STRUCTURAL ANALYSIS

Chapter 2

Mechanical Properties of Materials   7-18

Chapter 6

Analysis of Trusses   **81-96**

Chapter 7

Analysis of Continuum Structures   **97-130**

PART – II : DESIGN AND ANALYSIS OF AEROPLANE STRUCTURES

Chapter 8

Design Aspects of Aeroplane Structure   133-154

Chapter 9

Basics of Aerodynamics   155-160

Chapter 10

Loads on Aeroplane Structure   161-176

Chapter 11

Analysis of Determinate and Indeterminate Beams   177-234

Chapter 12

Analysis of Shafts   235-254

Chapter 13

Buckling of Columns   255-278

Chapter 14

Bending and Buckling of Thin Plates   279-322

Chapter 15

Shear Flow, Shear Center and Shear Lag   323-368

Chapter 16

Analysis of Aeroplane Fuselage and Wing   369-412

Chapter 17

Fatigue Analysis   413-422

PART - III OTHER RELEVANT TOPICS

Chapter 18

Materials of Aeroplane Construction   425-430

Chapter 20

Aeroelasticity 481-500

C H A P T E R **1**

INTRODUCTION

Many structures such as bridges, cranes, ships, aeroplanes are commonly noticed. Each part of the structure has a definite purpose to fulfill as an individual member of that structure. In some of these structures like bridges and cranes, individual members are seen even in the completed structure. In ships and aeroplanes, the structure is covered by an outer skin and, therefore, complexity associated with the structure is not easily noticed. In the earlier planes, outer skin served twin purposes of providing smooth air flow over it and protecting the structure and payload from external environment (temperature, rain, dust,..). In modern aeroplanes, even the outer metal skin is designed to resist some external load in addition to maintaining proper shape and is to be treated as a part of the structure. Each member of the structure is designed for a particular purpose and for withstanding certain types of loads (bending, shear, torsion, axial).

Mechanical design is the design of a component for optimum size, shape, etc., ***against failure*** under the application of operational loads. A good design should also minimise the cost of material and cost of production. Failures that are commonly associated with mechanical components are broadly classified as:

    (a) Failure by breaking of brittle materials and fatigue failure (when subjected to repetitive loads) of ductile materials

    (b) Failure by yielding of ductile materials, subjected to non-repetitive loads

    (c) Failure by elastic deformation

The last two modes cause change of shape or size of the component rendering it useless and, therefore, refer to functional or ***operational failure***. Most of the design problems refer to one of these two types of failures. Designing, thus, involves estimation of stresses of the components at different critical points of a component for the specified loads and boundary conditions, so as to satisfy operational constraints.

Design is associated with the calculation of dimensions of a component to withstand the applied loads and perform the desired function. ***Analysis*** is associated with the estimation of stresses in a component of assumed dimensions so that adequacy of assumed dimensions is validated. ***Optimum design*** is obtained by many iterations of modifying dimensions of the component based on the calculated values of stresses *vis-à-vis* permitted values and re-analysis.

An analytic method is applied to a model problem rather than to an ***actual physical problem*** (Ref Fig 1.1). Even many laboratory experiments use models. A ***geometric model*** (Ref Fig 1.2) for analysis can be devised after the physical nature of the problem has been understood. A model excludes superfluous details such as bolts, nuts, rivets, but includes all essential features, so that analysis of the model is not unnecessarily complicated and yet provides results that describe the actual problem with sufficient accuracy. A geometric model becomes a ***mathematical model*** when its behaviour is described or approximated by incorporating restrictions such as homogeneity, isotropy, constancy of material properties and mathematical simplifications applicable for small magnitudes of strains and rotations.

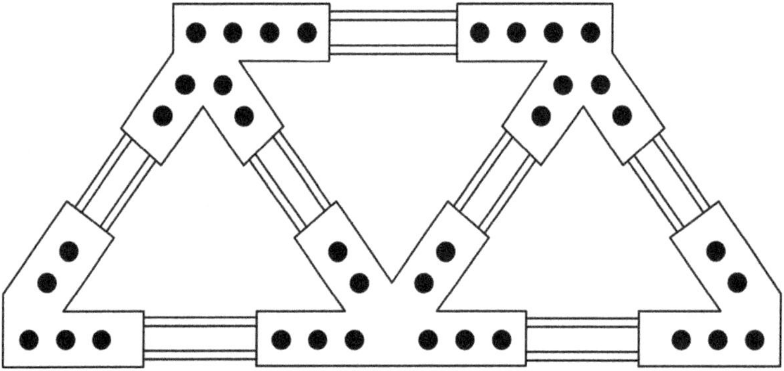

FIGURE 1.1 Physical structure of a plane truss

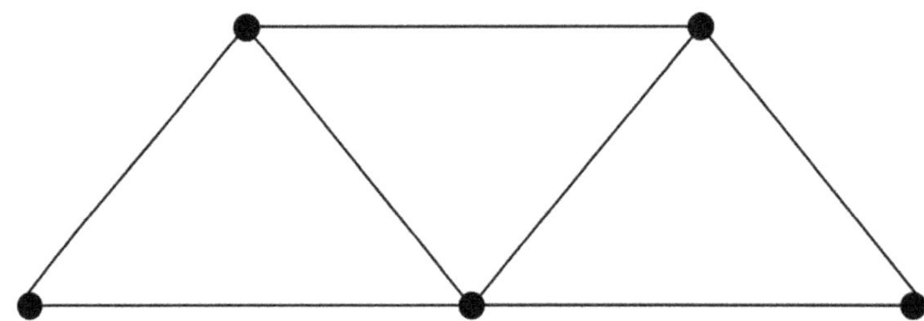

FIGURE 1.2 Geometric model of a plane truss

A review of main parts of an aeroplane and their functions are briefly covered for better appreciation of the subject in this book. Main parts of an aeroplane can be broadly classified as structure, aerodynamic and control surfaces, engine or power plant to generate lift and forward motion and avionics for effective flight control & navigation.

Type of loads and design of structural members are discussed in detail in the subsequent chapters.

1.1   SCOPE OF THIS BOOK

The subject is covered in this book in three parts. Topics, which are related to this subject and taught earlier, are briefly covered to serve as reference material to the students, whenever a need arises. Thus, basics of structural analysis covering mechanical properties of materials (Chapter-2), properties of sections (Chapter-3), mechanics of rigid bodies (Chapter-4), energy principles (Chapter-5), analysis of trusses (Chapter-6) and analysis of continuum structures (Chapter-7) are included in Part-1. These topics are covered in many universities before a student is eligible to study 'aircraft structures'.

Aircraft structures essentially consist of stiffened panels and box beams. Design and analysis of these structures is covered in sufficient detail as course material in Part-2. It starts with Design aspects of Aeroplane structure (Chapter-8) and includes Basics of aerodynamics (Chapter-9), Loads on Aeroplane structure (Chapter-10), Analysis of determinate and indeterminate beams (Chapter-11), Analysis of shafts (Chapter-12), Buckling of columns (Chapter-13), Bending and buckling of thin plates covering tension field beams and stiffened panels (Chapter-14), Shear flow, shear center and shear lag (Chapter-15), Analysis of aeroplane fuselage and wing (Chapter-16) and Fatigue Analysis (Chapter-17).

Matrix method of structural analysis was initially developed for analysing discrete structures, whose element stiffness matrices are obtained from strength of materials approach. This method has been generalised into Finite Element Method to analyse continuum structures as well as combinations of different types of elements. Element stiffness matrices are evaluated using the principle of minimum potential energy and variational principle. Analysis of any structure by this method leads to a large set of linear algebraic simultaneous equations, which can only be solved with the help of a computer. All the basic steps in the analysis of a structure by matrix method and finite element method are identical. In this background, matrix method of structural analysis is excluded from this book. Instead, finite element method is covered in sufficient detail in Chapter-19 of Part-3.

Some related topics, like Materials of aircraft construction (Chapter-18) and Aeroelasticity (Chapter-20), which are usually covered in detail as separate subjects, are also briefly covered in Part-3 for a reasonable exposure to the students.

Discussion on engine and avionics is outside the scope of this book.

Part - I
Basics of Structural Analysis

Chapter **2**

MECHANICAL PROPERTIES OF MATERIALS

Every engineer is vitally concerned with the materials available to him. He must have an intimate knowledge of the properties and behavioural characteristics of the materials he proposes to use. In making his choice, the engineer must take into account such properties as strength, electrical and/or thermal conductivity, density, etc. Further, he must consider the behaviour of the material during processing and use (where formability, machinability, electrical stability, chemical durability and radiation behaviour are important) as well as cost and availability. Many improved designs depend on the development of completely new materials. Since it is obviously impossible for the engineer to have detailed knowledge of the many thousands of materials already available, as well as to keep abreast of new developments, he must have a firm grip of the underlying principles that govern the properties of all materials.

Important mechanical properties of a structural material are elasticity, ductility, creep, hardness and toughness. In addition, the material of a component may have to withstand different types of loads such as normal force, bending moment, shear force, torsion etc. A single material may not be suitable for all sorts of applications and all types of loads. To provide a working basis for making comparisons between structural properties and the effects of

7

in-service behaviour on those properties, we shall first define some of the more commonly used engineering terms.

2.1 STRESS

Stress is defined as the internal reaction in a component for an applied force. It is expressed as force per unit area, defined at a point on the six faces of an imaginary small cube around that point, and is measured in units of N/sq.m or Pascal (Pa). Since the force vector can have any arbitrary orientation, the force on any surface is resolved into two components - normal and parallel to the surface. **Normal stress** (σ) is defined perpendicular to a surface or cross section of the component. **Shear stress** (τ) is defined along a surface and is produced by torque, friction etc. A force (or component) normal to a section produces normal stress while a force (or component) parallel to the surface produces shear stress.

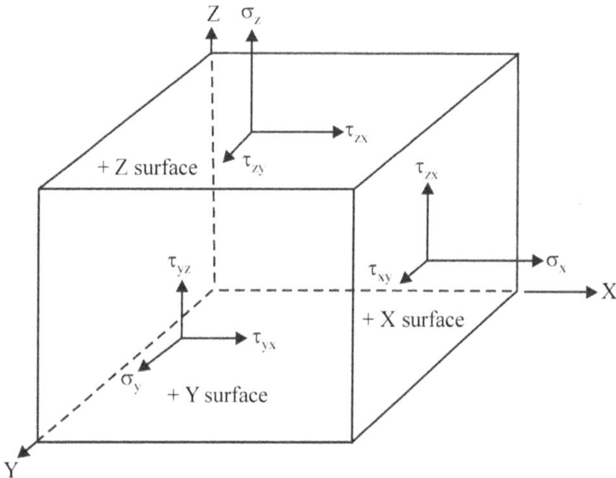

FIGURE 2.1 Stress at a point

Stress on a plane inclined at an angle θ to a surface (Ref. Fig. 2.2) will have normal stress and shear stress, as functions of angle θ, given by the force equilibrium equations along and perpendicular to BC.

$$\sigma_n \times BC = (\sigma_X \times AC) \times \cos\theta + (\sigma_Y \times AB) \times \sin\theta$$
$$+ (\tau_{XY} \times AC) \times \sin\theta + (\tau_{XY} \times AB) \times \cos\theta$$

and $\tau \times BC = (\sigma_X \times AC) \times \sin\theta + (\sigma_Y \times AB) \times \cos\theta$
$$+ (\tau_{XY} \times AC) \times \cos\theta + (\tau_{XY} \times AB) \times \sin\theta$$

Substituting BC = AC / cos θ = AB / sin θ, these equations are simplified as

$$\sigma_n = \sigma_X \cos^2\theta + \sigma_Y \sin^2\theta + \tau_{XY} \sin 2\theta \qquad \dots(2.1)$$

and

$$\tau = [(\sigma_X - \sigma_Y)/2] \sin 2\theta - \tau_{XY} \cos 2\theta \qquad \dots(2.2)$$

It can be seen that τ decreases, due to the negative term, as a function of θ while σ_n increases. Minimum value of τ is zero corresponding to

$$\tan 2\theta = 2\tau_{XY} / (\sigma_X - \sigma_Y) \qquad \dots(2.3)$$

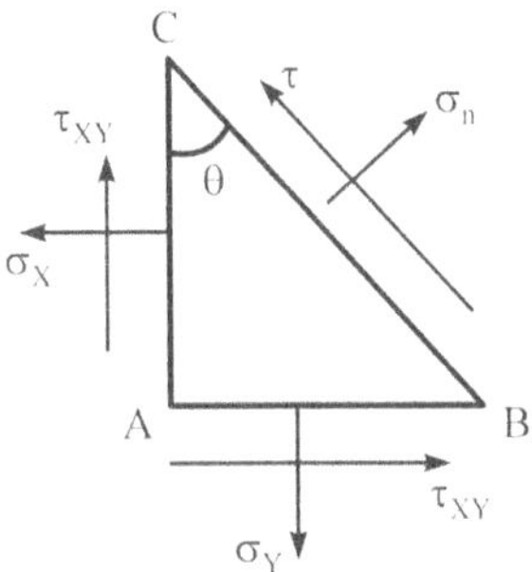

FIGURE 2.2 Stresses on an inclined plane

Substituting this angle in the equation for σ_n gives maximum and minimum normal stress values (σ_1 and σ_2)

$$\sigma_1, \sigma_2 = (\sigma_X + \sigma_Y)/2 \pm \sqrt{[(\sigma_X - \sigma_Y)/2]^2 + \tau_{XY}^2} \qquad \dots(2.4)$$

The maximum shear stress corresponds to $2\theta = 90°$ and is given by

$$\tau_{max} = (\sigma_1 - \sigma_2)/2 \qquad \dots(2.5)$$

These maximum (σ_1) and minimum (σ_2) values on two perpendicular planes are called **Principal normal stresses**. These values are important for safe design of any component. The planes inclined at the angle θ, corresponding to the principal normal stresses, are called the **Principal planes**.

These can be graphically obtained from Mohr's circle (Ref. Fig. 2.3), drawn with normal stresses along X-axis and shear stress along Y-axis. Radius of the circle gives the maximum shear stress while the extreme values along the X-axis indicate maximum and minimum principal stresses.

Angle θ between the principal planes and the actual plane of stresses is obtained from angle 2θ between the line AB and X-axis in the Mohr's circle.

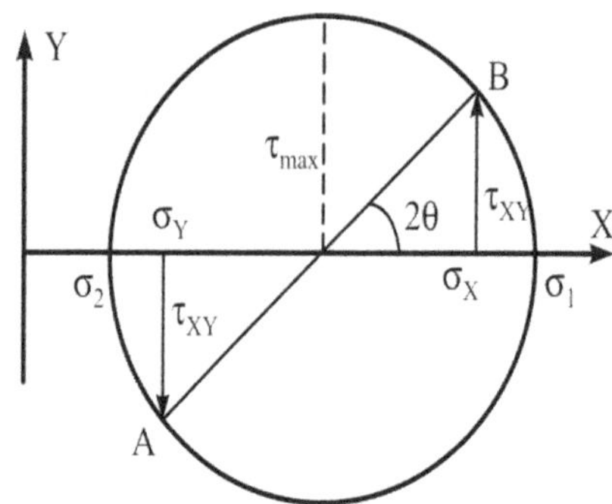

FIGURE 2.3 Mohr's circle of stresses on a plane

2.2   STRAIN

Strain is the deformation of a material due to applied load, in proportion to the original dimension. ***Normal strain*** (ε) is defined as the ratio of change in the dimension of the component (deformation) in a particular direction to the original dimension and, therefore, is dimensionless. Since the deformation in most engineering structures is very small, strain is usually defined as a percent or in micro-strain units (micro-meter deformation per meter length). It is related to normal stress at the point. ***Shear strain*** (γ) is defined as the change of angle between two adjacent sides of a section or component and is dimensionless (Ref Fig 2.4). It is related to shear stress. If u and v are displacements (changes in lengths) of a small element of size dx × dy, along X and Y directions, then

$$\gamma_{XY} = \frac{\partial u}{\partial y} + \frac{\partial v}{\partial x}$$

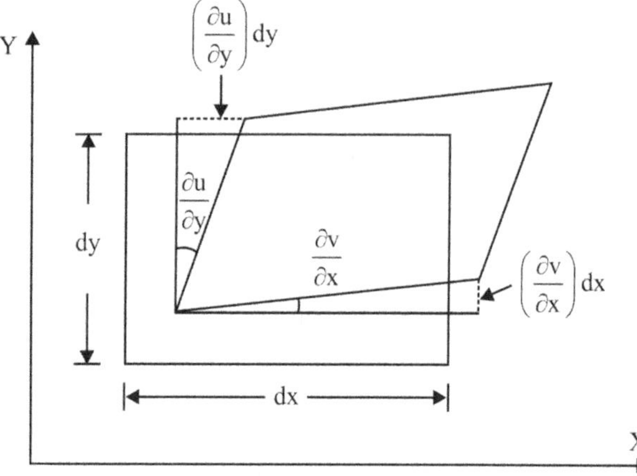

FIGURE 2.4 Shear strain at a point

Temperature change of a component can produce normal strain only, since thermal strain ($\alpha \times \Delta T$, where α is the coefficient of linear thermal expansion) is same in all the directions and does not alter the included angles of the geometry of the component. So, *shear strain is not produced, due to temperature changes*. However, it can produce bending stresses if there is a temperature gradient across thickness, such as in a steam turbine casing.

Elastic strain is reversible strain; it disappears after the stress or load is removed. During elastic strain the same atomic neighbours are retained. *Plastic strain* is the strain permanently given to a material by stresses, which exceed the elastic limit. Plastic strain is the result of a permanent displacement of the atoms inside the material.

2.3   HOOKE'S LAW, MODULUS OF ELASTICITY AND MODULUS OF RIGIDITY

Elastic strain is proportional to the amount of applied stress for most engineering materials, up to a limit, called *limit of proportionality*. This proportionality is called *Hooke's law*.

For normal stress (σ) and normal strain (ε), $\sigma \propto \varepsilon$ or $\sigma = E \times \varepsilon$, where E is the proportionality constant, popularly known as Modulus of elasticity or Young's modulus. Similarly, for shear stress (τ) and shear strain (γ), $\tau \propto \gamma$ or $\tau = G \times \gamma$, where G is the proportionality constant, popularly known as Modulus of rigidity. Since strains are dimensionless, modulus of elasticity and modulus of rigidity have the same units as the stress (N/sq.m or Pascal).

An exception to the Hooke's law is thermal strain, which produces thermal stress only when the thermal strain is constrained. *Dimensions of an unconstrained body, subjected to temperature increase/decrease, are increased/reduced in all directions producing thermal strain without any thermal stress.* If a component is partially constrained, then thermal stress is proportional to the constrained part of strain.

For example stress in a uniform bar subjected to temperature rise by ΔT, is dependent on the end condition as shown below. Let α be the coefficient of linear thermal expansion. Total elongation of a bar of length L due to increase in its temperature by ΔT is L $\propto \Delta T$. Then, stress in the bar depends on the constraint (boundary condition) for its expansion, as shown Fig. 2.5. In each case,

Stress, $\sigma = E \times \varepsilon = \dfrac{E}{L} \times$ Restrained part of expansion.

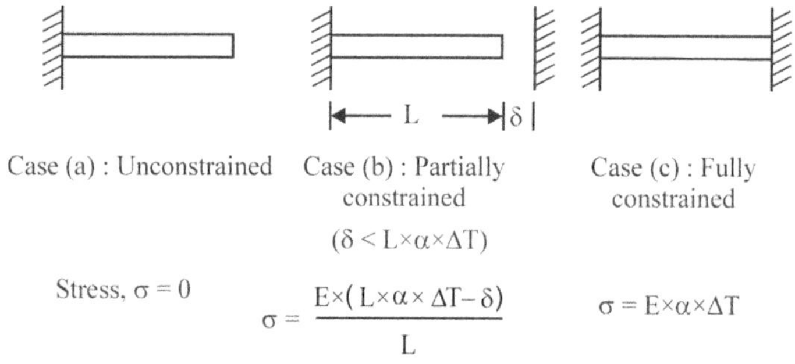

Case (a) : Unconstrained   Case (b) : Partially   Case (c) : Fully
constrained   constrained

$(\delta < L \times \alpha \times \Delta T)$

Stress, $\sigma = 0$ $\sigma = \dfrac{E \times (L \times \alpha \times \Delta T - \delta)}{L}$ $\sigma = E \times \alpha \times \Delta T$

FIGURE 2.5

2.4   DUCTILITY

It is the amount of plastic deformation at the breaking point. Thus, its value may be expressed as elongation. A second measure of ductility is the percentage reduction in area, compared to the original area, at the point of fracture.

A *ductile* material has an elastic limit (or limit of proportionality) beyond which yield or significant plastic deformation occurs before fracture. A *brittle* material experiences very small plastic deformation and causes sudden failure of the component, which may damage the equipment or endanger the lives of people working around it. *Ductile materials are used in most of the engineering applications, since the highly stressed component can be identified from its plastic deformation, and replaced even before the component physically breaks.*

2.5   POISSON'S RATIO

When a component elongates in one (longitudinal) direction due to the applied load, its dimensions in the other orthogonal (lateral) directions reduce and vice versa, in order to minimise change in the volume of the component, as shown in Fig. 2.6. This is called *Poisson's effect* and is quantified by *Poisson's ratio* (ν or $1/m$), which is defined as ratio of lateral strain (perpendicular to the direction of applied load) to the longitudinal strain (in the direction of applied load). Thus, Poisson's ratio, $\nu = $ Lateral strain / Longitudinal strain

i.e., $\nu = \varepsilon_Y / \varepsilon_X$ or $\varepsilon_Z / \varepsilon_X$ for load applied in X-direction   .....(2.6)

It is found to be a constant for each material. If tensile normal strain is taken as positive and compressive normal strain as negative, Poisson's ratio is always negative. However, the value is normally defined as positive and its negative effect is considered in the relevant equations. Its value ranges theoretically from 0 to ½. However, *for most engineering materials, it ranges from 1/3 to 1/4*. For a perfectly plastic material, $v = ½$.

Deformation due to axial compressive load   Deformation due to axial tensile load

FIGURE 2.6 Deformation of a long bar due to compressive and tensile loads

This phenomenon is not relevant for shear loads and, hence, *this ratio is not defined for shear strains*.

The two elastic modulii are related, in terms of Poisson's ratio as

$$G = E / 2(1 + v) \qquad\qquad .....(2.7)$$

An exception to the Poisson's effect is thermal strain. Due to temperature rise (fall), dimensions of the component increase (decrease) in all the directions in the same proportion. Volume of the material changes but increase (decrease) of dimension in one direction does not reduce (increase) dimensions of the component in the other (lateral) directions.

2.6 STRENGTH

The maximum stress that a material can resist without resulting in yielding or plastic deformation is called the *yield strength*. It has, obviously, the same units as stress. It may include a small part of non-linear elastic stress (beyond limit of proportionality).

Tensile or ultimate strength of a material is calculated by dividing the maximum tensile load applied on a long slender specimen by the original cross sectional area (before necking or yielding of a ductile material). It also has the same units as stress.

In some materials, the yield strength is marked by a definite *yield point*. In other materials, where it is not so clearly visible in stress-strain diagram of uni-axial tensile test, it is common to define the *yield stress* as that stress which gives 0.2% plastic strain on release of load.

2.7 STRESS-STRAIN CURVE

Properties (such as yield strength, ultimate or tensile strength, modulus of elasticity, ductility, toughness, etc..) of ductile metals are usually *determined from a uniaxial tensile test on a standard test specimen*. This curve is given in Fig. 2.7. for two different types of materials – ductile and brittle. For a ductile

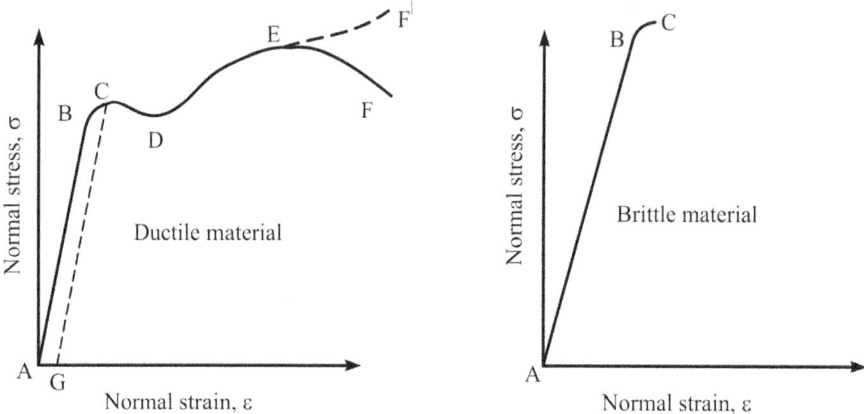

FIGURE 2.7 Stress-strain curve of a tensile test specimen

material, the curve consists of 4 regions – linear elastic portion AB up to the proportionality limit B; non-linear elastic portion BC up to the upper yield point or yield stress C; yielding portion C-D-E up to the necking point E and yielding followed by necking up to the breaking point F (on engineer's curve).

The actual yield line follows true stress line $E - F'$. The point E on the engineer's stress-strain curve is the max stress or ultimate tensile stress (UTS). In some materials, the lower yield point D may not be clearly seen. In some materials, upper yield point C is not clearly seen and is taken as the point corresponding to 0.2% plastic strain (AG) along GC (drawn parallel to AB). Slope of the line AB gives modulus of elasticity, E.

True stress (σ') is the force divided by instantaneous area of cross section ($\sigma' = P/A$). If the area is assumed to remain constant and equal to original area, the stress is called engineering stress (σ) and is given by $\sigma = P/A_0$

True strain (ε') is the integral of (incremental change in length divided by instantaneous length) over the limits of length. It is given by

$$\varepsilon' = \int_{L_o}^{L} \frac{dL'}{L'} = \log_e \frac{L}{L_o}$$

Engineering strain (ε) is the change in length divided by original length. In the elastic range, original length can be assumed constant and so,

$$\varepsilon = \int_{L_0}^{L} \frac{dL}{L_0} = \frac{L - L_0}{L_0} = \frac{\delta}{L_0}$$

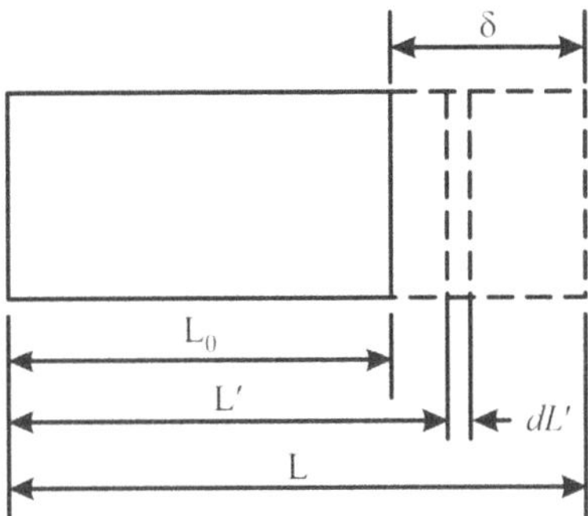

Most engineering structures are designed to operate within the elastic range, in which case true and engineering values are very close.

In plastic range, they are not close but can be correlated, based on the assumption of constant volume ($A_0 \times L_0 = A \times L$)

$$\varepsilon' = \log_e(L/L_0) = \log_e[(L_0 + \delta)/L_0] = \log_e[1 + (\delta/L_0)]$$

$$= \log_e(1 + \varepsilon)$$

and $$\sigma' = P/A = P \times L/(A_0 \times L_0) = (P/A_0) \times (L/L_0) = (P/A_0) \times [\,(L_0+\delta) / L_0\,]$$

$$= \sigma \times (1 + \varepsilon)$$

Once the material starts yielding, its length increases while its cross section reduces significantly. It is difficult to measure the *true* or physical stress in the material, based on actual area at that load, continuously. Hence, *engineering stress* based on the original area of cross section is used by the engineer

Stress-strain curve of a ductile material indicates that beyond ultimate strength, local necking and instability starts, resulting in a *decrease of engineering stress with increasing strain* till fracture of test specimen (an apparent anomaly that fracture takes place at a stress lower than ultimate strength), while in fact *true stress increases with increasing strain* due to necking.

Stress-strain curve of brittle materials follows the curve ABC (linear elastic line AB and non-linear elastic line BC) without significant yielding.

2.8   FACTOR OF SAFETY

To account for scatter of these results for various test specimens of the same material, variations in the operating conditions (magnitude and nature of load,..) compared to the design conditions and deterioration of properties with age and weather conditions, mechanical strength properties are divided by a *factor of safety* (>1 and specified in the design data books or design codes for various operating conditions) to obtain *allowable stress values*, which are used in design calculations.

2.9 ENDURANCE STRENGTH

Components made of ductile materials and subjected to n number of cyclic or fluctuating loads can fail by brittle fracture at a stress much lower than the yield strength (Ref Fig. 2.8). This phenomenon is called *fatigue*. The stress below which a material will not fail, even when subjected to a large number of stress cycles (usually $>10^6$) is called *endurance limit* or *endurance strength* (σ_E).

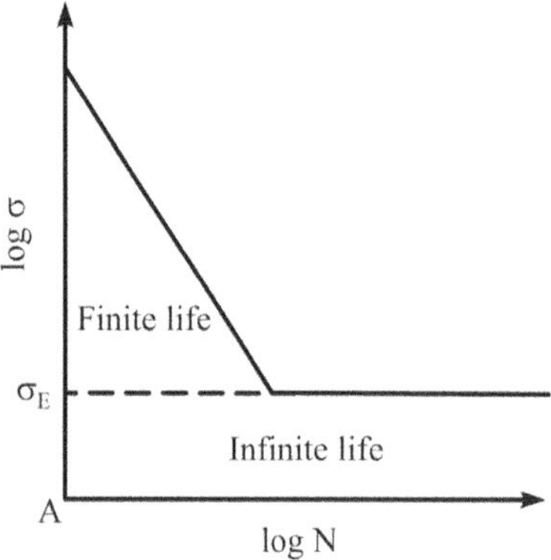

FIGURE 2.8 Fatigue curve

Rotating components of equipment (such as turbines, generators, motors, pumps, compressors, fans etc) are subjected to large number ($>10^6$) of stress cycles and are, therefore, designed for infinite life (based on endurance strength). Other components such as turbine casing, subjected to a smaller number of thermal and/or pressure stress cycles, are designed for finite life (or low cycle fatigue).

2.10 RUPTURE TIME

Components loaded for a long duration (time, t) at a stress (σ) even below elastic limit, may yield and undergo plastic deformation (Ref Fig. 2.9). This phenomenon is called *creep*. It is more prominent at high operating temperatures. The duration of time (corresponding to a particular stress) after which creep deformation is observed, is called ***rupture time.***

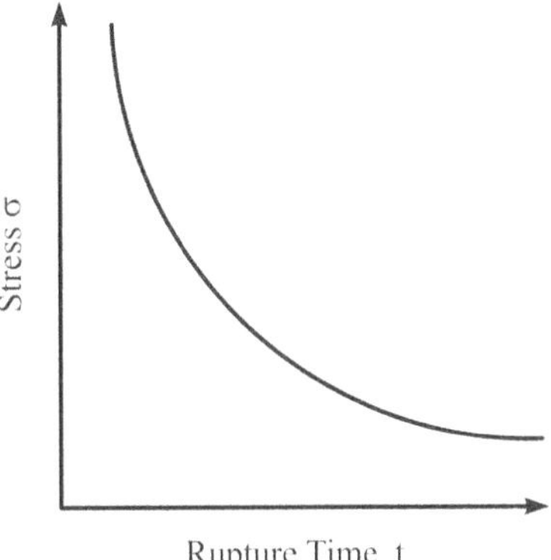

FIGURE 2.9 Creep curve

2.11 VARIATION OF PROPERTIES WITH TEMPERATURE

Most of these properties vary significantly for many metals, with significant changes in the temperature of the component. Since atmospheric temperature reduces by about 7^0C per km of altitude, aeroplane structural components are exposed to large temperature changes, ranging from a maximum of about $+60^0$C at ground level during summer to a minimum of about -55^0C at very high altitudes (about 10km). Properties like modulus of elasticity, yield strength, fatigue strength, etc. at the appropriate temperatures should, therefore, be used in the design calculations.

2.12 HARDNESS

It is defined as the resistance of a material to penetration of its surface. Hardness and strength of a material are closely related. The ***Brinnel hardness number*** (BHN) is a hardness index calculated from the *area of penetration* by a large standard indenter. The ***Rockwell hardness number*** (RHN) is measured from the *depth of penetration* by a small standard indenter.

2.13 TOUGHNESS

It is a measure of the energy required to break a material. A ductile material with the same tensile strength as a brittle material will require more energy for breaking and be tougher. It is the measure of area under the load-deflection curve A-B-C-D-E-F of ductile material or A-B-C of brittle material, up to the breaking point, as shown in Fig. 2.10.

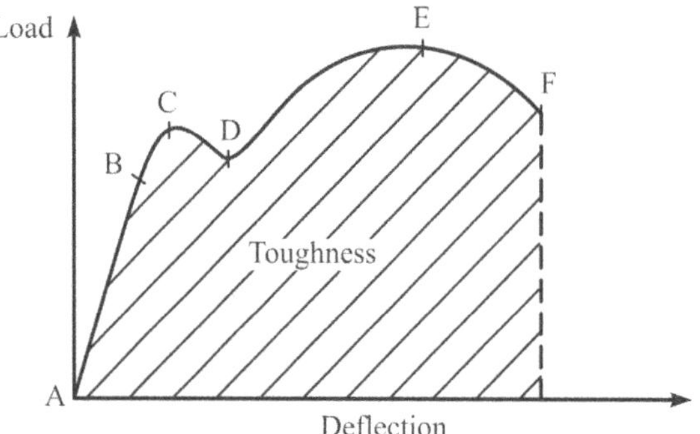

FIGURE 2.10 Toughness of a material

2.14   RESILIENCE

It is the strain energy stored in a body when loaded to a particular stress and strain. ***Proof resilience*** is the strain energy stored in a body when loaded up to its elastic limit. It is the measure of area under the load deflection curve AB, as shown in Fig. 2.11. ***Modulus of resilience*** is the proof resilience per unit volume of the material.

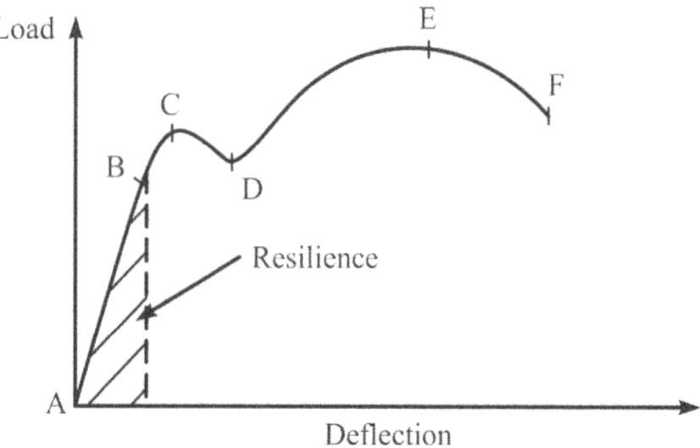

FIGURE 2.11 Resilience of a material

CHAPTER 3

PROPERTIES OF SECTIONS

3.1 CENTROID

Centroid is defined as the central point where the entire physical property is assumed to be concentrated, to give the same first moment as that obtained by considering all elements of the body. These central points for length, area and volume are usually called *centroids* while the central points for the distributed masses and gravitational forces are called *center of mass* and *center of gravity* respectively.

Location of center of gravity plays a crucial part in the flight of an aeroplane and depends on the intended role – civil transport aeroplane for passengers & cargo or military aeroplane with ammunition. Even a small change in the location of center of gravity due to movement of persons, cargo or consumption of fuel can create an upward or downward tilt of the aeroplane unless the aerodynamic force is adjusted accordingly

3.1.1 CENTROIDS OF SIMPLE SHAPES/BODIES

Position vector of centroid C of a line element is given by

$$r_C = (\int r \times dL) / (\int dL)$$

It can also be expressed in terms of Cartesian coordinates in the form

$$X_C = (\int x \times dL) / (\int dL)$$

$$Y_C = (\int y \times dL) / (\int dL)$$

$$Z_C = (\int z \times dL) / (\int dL) \qquad \qquad .....(3.1)$$

Similarly, the position vectors of area and volume elements are given by

$$r_C = (\int r \times dA) / (\int dA)$$

$$\text{or} \qquad X_C = (\textstyle\int x \times dA)/(\textstyle\int dA)$$

$$Y_C = (\textstyle\int y \times dA)/(\textstyle\int dA)$$

$$Z_C = (\textstyle\int z \times dA)/(\textstyle\int dA) \qquad\qquad .....(3.2)$$

$$\text{and} \qquad r_C = \textstyle\int (r \times dV)/(\textstyle\int dV)$$

$$\text{or} \qquad X_C = (\textstyle\int x \times dV)/(\textstyle\int dV)$$

$$Y_C = (\textstyle\int y \times dV)/(\textstyle\int dV)$$

$$Z_C = (\textstyle\int z \times dV)/(\textstyle\int dV) \qquad\qquad .....(3.3)$$

The centroid may or may not lie on the element itself. This method of locating centroid of an element is called ***moment method***. It is used here to locate centroid of some simple lines and areas.

An important property of centroids, applicable to all line and area elements, is -

Distance of centroid from X-axis, $Y_C = 0$

when the element is *symmetric about X-axis*

and

Distance of centroid from Y-axis, $X_C = 0$

when the element is *symmetric about Y-axis*

For volume elements,

Distance of centroid from X-axis, $Y_C = Z_C = 0$

when the element is *symmetric about X-axis*

Similarly, $Z_C = X_C = 0$ when the element is *symmetric about Y-axis*

and $X_C = Y_C = 0$ when the element is *symmetric about* Z-axis

It is thus obvious that for solids of revolution *about X-axis*, $Y_C = Z_C = 0$ and so on.

(i) Centroid of Circular Arcs

Let us consider a circular arc of radius 'R' and subtending angle $-\alpha$ to $+\alpha$

Consider a small segment of the arc at an angle 'θ' and subtending an angle 'dθ' at the origin.

Length of this segment, $dL = R \times d\theta$

Center of this segment 'C' is located at 'R sin θ' and 'R cos θ' from X and Y axes

Then, $X_C = (\int x \times dL) / (\int dL) = (\int R\cos\theta \times R\, d\theta) / (\int R\, d\theta)$

$$= (R^2 \sin\theta)/(R\,\theta)$$

$$= R\sin\theta / \theta \qquad \text{for} \quad -\alpha \le \theta \le +\alpha \qquad \dots(3.4)$$

Similarly, $Y_C = (\int y \times dL) / (\int dL) = (\int R\sin\theta \times R\, d\theta) / (\int R\, d\theta)$

$$= -(R^2 \cos\theta) / (R\,\theta)$$

$$= -R\cos\theta / \theta \qquad \text{for} \quad -\alpha \le \theta \le +\alpha \qquad \dots(3.5)$$

For a semi-circular arc with limits of 'θ' equal to $-\pi/2$ and $+\pi/2$,

$$X_C = [R\sin\theta / \theta] = R\,[\sin(\pi/2) - \sin(-\pi/2)] / [(\pi/2) - (-\pi/2)]$$

$$= 2R/\pi$$

and $Y_C = [-R\cos\theta/\theta] = -R\,[\cos(\pi/2) - \cos(-\pi/2)] / [(\pi/2) - (-\pi/2)] = 0$

These values are tabulated here for a few simple cases.

Case	Line	Limits of θ	X_C	Y_C
1	Full circle (Sym about X & Y axes)	0 to $+2\pi$	0	0
2	Semi circle (Sym about X-axis)	$-\pi/2$ to $+\pi/2$	$2R/\pi$	0
3	Semi circle (Sym about Y-axis)	0 to $+\pi$	0	$2R/\pi$
4	Quarter circle (Sym about X-axis)	$-\pi/4$ to $+\pi/4$	$2\sqrt{2}\,R/\pi$	0
5	Quarter circle	0 to $+\pi/2$	$2R/\pi$	$2R/\pi$

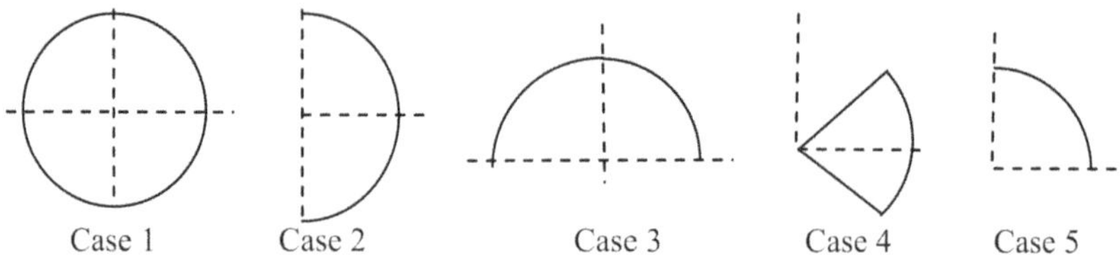

Case 1   Case 2   Case 3   Case 4   Case 5

(ii) Centroid of Plane Areas (assuming uniform thickness and density)

    (a) *Centroid of a sector of a circular plate:* Let us consider a circular plate of radius 'R' and subtending angle $-\alpha$ to $+\alpha$

Let us consider a small segment of the plate at mean radius 'r', width 'dr', mean angle 'θ' and subtending an angle 'dθ' at the origin.

    Area of this segment, $dA = (r\,d\theta) \times dr$

    Center of this segment 'C' is located at 'r sin θ' and 'r cos θ' from X and Y axes

Then, $X_C = (\int x \times dA) / (\int dA)$

$$= (\iint r\cos\theta \times r\,d\theta \times dr) / (\iint r\,d\theta \times dr)$$

$$\text{for } 0 \le r \le R \quad \text{and} \quad -\alpha \le \theta \le +\alpha$$

$$= (R^3 \sin\theta / 3) / (R^2 \theta / 2)$$

$$= 2R \sin\theta / 3\theta \qquad \text{for } -\alpha \le \theta \le +\alpha \qquad\qquad .....(3.6)$$

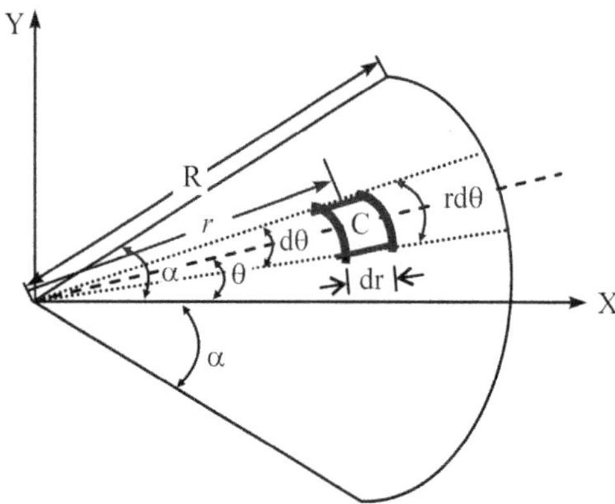

Similarly, $Y_C = (\int y \times dA) / (\int dA)$

$$= (\int r\sin\theta \times r\,d\theta \times dr) / (\int r\,d\theta \times dr)$$

$$= -(R^3 \times \cos\theta / 3) / (R^2 \times \theta / 2)$$

$$= -2R \times \cos\theta / 3\theta \qquad\qquad .....(3.7)$$

For a semi-circular plate with limits of 'θ' equal to $-\pi/2$ and $+\pi/2$,

$$X_C = [2R \sin \theta / 3\theta] \qquad \text{for} \quad -\pi/2 \le \theta \le +\pi/2$$

$$= 2R[\sin(\pi/2) - \sin(-\pi/2)] / 3[(\pi/2) - (-\pi/2)] = 4R/3\pi$$

and

$$Y_C = [-2R \cos \theta / 3\theta] \qquad \text{for} \quad -\pi/2 \le \theta \le +\pi/2$$

$$= -2R[\cos(\pi/2) - \cos(-\pi/2)] / 3[(\pi/2) - (-\pi/2)] = 0$$

These values are tabulated here for a few simple cases.

Case	Plate	Limits of θ	X_C	Y_C
1	Full circle (Sym about X & Y axes)	0 to $+2\pi$	0	0
2	Semi circle (Sym about X-axis)	$-\pi/2$ to $+\pi/2$	$4R/3\pi$	0
3	Semi circle (Sym about Y-axis)	0 to $+\pi$	0	$4R/3\pi$
4	Quarter circle (Sym about X-axis)	$-\pi/4$ to $+\pi/4$	$4\sqrt{2}R/3\pi$	0
5	Quarter circle	0 to $+\pi/2$	$4R/3\pi$	$4R/3\pi$

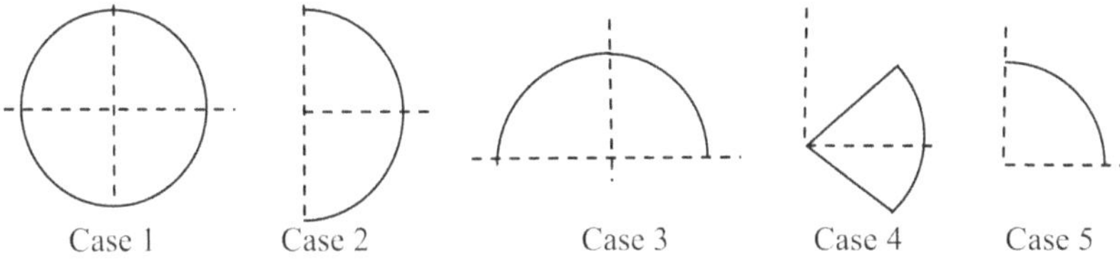

(iii) Centroid of other plane areas

(a) *Triangle:* Let us consider a triangle of base 'b' and height 'h'. Its centroid can be calculated by two approaches.

Method - 1

Let us consider a small elemental area of 'dA', identified by EF at a distance of 'y' from the base. Let BC = b; AD = h; GD = y   and EF = x_1, which is function of y

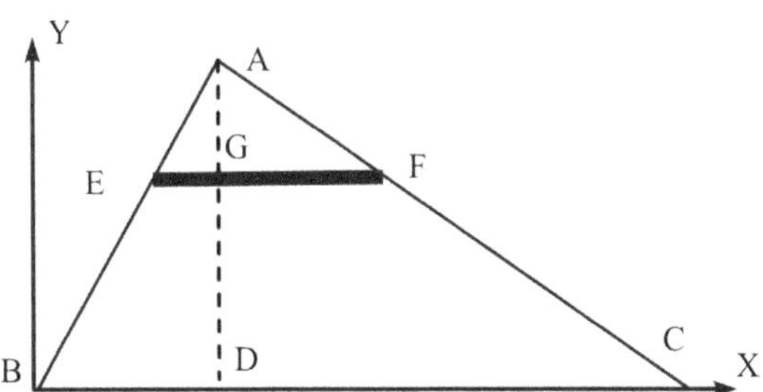

From similar Δs AEF and ABC, EF/BC = AG/AD or $x_1 / b = (h - y)/h$

and   Area of the segment, $dA = x_1 \times dy = [\, b \times (h - y)/h \,]\, dy$

Therefore $Y_C = (\int y \times dA) / (\int dA)$

$$= (\int y \times [\, b \times (h - y)/h \,]\, dy) / (\int [\, b \times (h - y)/h \,]\, dy)$$

$$= [hy^2/2 - y^3/3] / [\, hy - y^2/2 \,]\quad \text{for } 0 \le y \le h$$

$$= (h^3/6) / (h^2/2) = \textbf{\textit{h/3}} \qquad\qquad .....(3.8)$$

Method - 2

Let us consider a small rectangular area dA, identified by its sides dx and dy and located at a distance of 'y' and 'x' from X and Y axes. Here, limits of x are functions of y

$$Y_C = (\int y \times dA\,) / (\int dA\,)$$

$$= (\int \int y \times dx \times dy\,) / (\int \int dx \times dy)$$

$$= [\int (\int dx\,) \times y\, dy\,] / [\int (\int dx)\, dy]$$

$$= [\int (\,x\,) \times y\, dy\,] / [\int (\,x\,)\, dy\,]$$

within the limits of x, identified by the lines AB and AC as

$$y = m_1 x + c_1 \quad \text{and} \quad y = m_2 x + c_2$$

and is more involved for a general case.

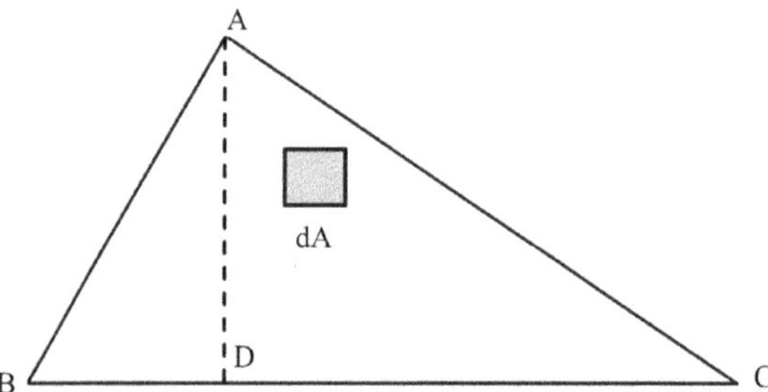

For the particular case of a right angled triangle with AB perpendicular to BC, the limits of x are 0   and $b \times (h - y)/h$

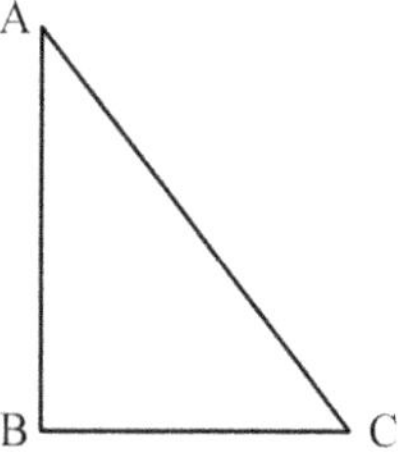

$$Y_C = [\int \{\, (b\,(h - y) / h) - 0\} \times y\, dy\,] /$$

$$[\int \{(b\,(h - y) / h\,) - 0\}\, dy\,]$$

$$= [hy^2/2 - y^3/3] / [hy - y^2/2\,]\quad \text{for } 0 \le y \le h$$

$$= (h^3/6) / (h^2/2) = \textbf{h/3}$$

(b) *Parabola:* Let us consider a parabola, defined by $y^2 = k \times x$, and bounded by the X-axis and $x = b$ line (parallel to Y-axis). Let us consider a small elemental area of 'dA', identified by EF at a distance of 'y' from the base.

Let $BC = b$; $AC = h$; $FC = y$ and $EF = x_1$

Then, at point A on the parabola,

$$b = h^2/k$$

and $x_1 = b - x_E = b - (y^2/k) = (h^2 - y^2)/k$

Area of the segment,

$$dA = x_1 \times dy$$

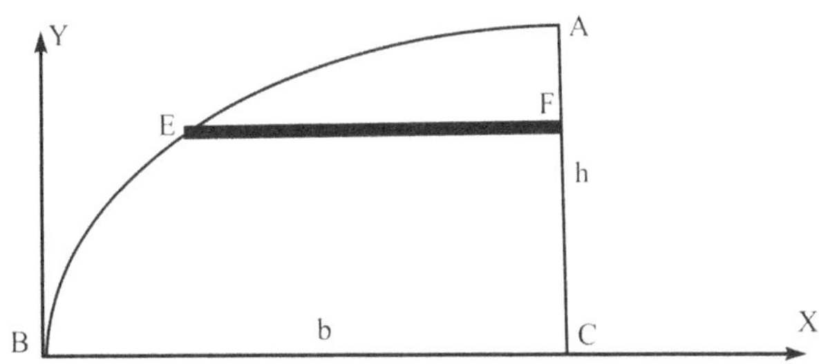

$$Y_C = (\textstyle\int y \times dA) / (\textstyle\int dA) = (\textstyle\int y \times x_1 \times dy) / (\textstyle\int x_1 \times dy)$$
$$= (\textstyle\int y \times [(h^2 - y^2)/k] \, dy) / (\textstyle\int [(h^2 - y^2)/k] \, dy)$$
$$= [h^2 y^2/2 - y^4/4] / [h^2 y - y^3/3] \qquad \text{for } 0 \le y \le h$$
$$= (h^4/4) / (2h^3/3) = \mathbf{3h/8}$$
$$X_C = (\textstyle\int x_2 \times dA) / (\textstyle\int dA) \quad \text{where, } x_2 = (y^2/k) + x_1/2 = (h^2 + y^2)/2k$$
$$= (\textstyle\int [(h^2 + y^2)/2k] \times [(h^2 - y^2)/k] \, dy) / (\textstyle\int [(h^2 - y^2)/k] \, dy)$$
$$= [(h^4 y - y^5/5)/2k^2] / [(h^2 y - y^3/3)/k] \qquad \text{for } 0 \le y \le h$$
$$= (4h^5/10 k^2) / (2h^3/3k) = 3h^2/5k$$
$$= \mathbf{3b/5}$$

3.1.2 CENTROIDS OF COMPOSITE SECTIONS

There are many situations in which a particular component can be treated as a combination of different simple elements. Centroid of such combinations can be calculated by using the known locations of the simpler elements, which together make up the desired component. The combination may include addition and/or deletion of simpler elements. Integrals used in the case of continuously distributed property are replaced by the algebraic addition/deletion of the property of individual elements. A few examples will illustrate practical application of this approach.

(i)   *Trapezium:* Calculate centroid of a trapezium of height 'h' and parallel sides 'a' and 'b'

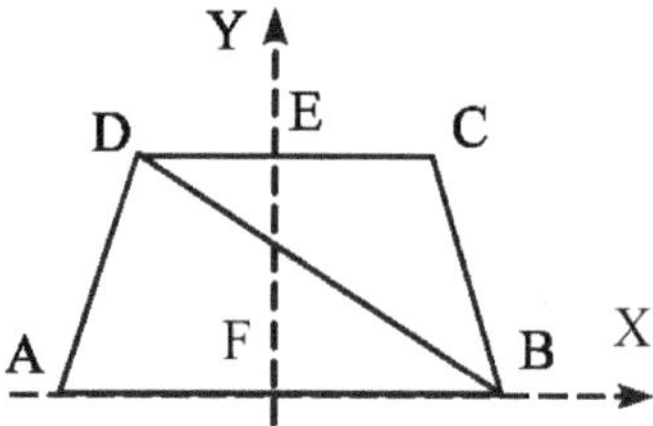

Solution

Let AB = a ; DC = b and EF = h

The trapezium can be considered as a combination of two triangles ABD and BCD. Since the trapezium is symmetric about Y-axis, $X_C = 0$

For the triangle ABD, $A_1 = a \times h/2$; $Y_1 = h/3$

For the triangle BCD, $A_2 = b \times h/2$; $Y_2 = 2h/3$

Then,

$Y_C = (A_1 Y_1 + A_2 Y_2)/(A_1 + A_2)$

$= [(a \times h/2) \times (h/3) + (b \times h/2) \times (2h/3)] / [(a \times h/2) + (b \times h/2)]$

$\mathbf{= h(a + 2b) / 3(a + b)}$ from AB   .....(3.9)

Distance of centroid from CD = $h - Y_C = \mathbf{h(2a + b) / 3(a + b)}$

(ii)   **I-*Section:*** Calculate the position of the centroid of an I-section of base width 200 mm, top width 100 mm, height 140 mm and thickness of flanges and web 20 mm.

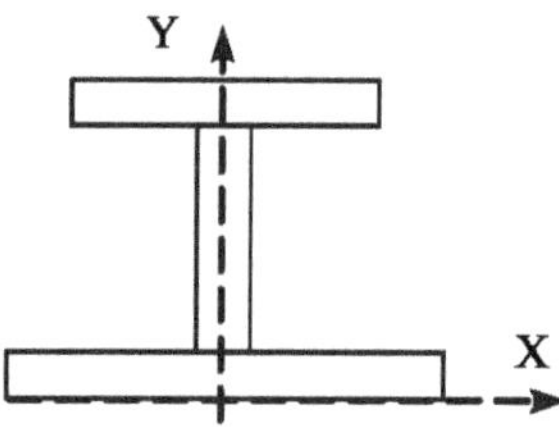

Solution

The I-section can be considered as a combination of three rectangles i.e., bottom flange (1), web (2) and top flange (3). Because of symmetry of the composite area about Y-axis, $X_C = 0$

$A_1 = 200 \times 20 = 4000 \text{ mm}^2$; $y_1 = 20/2 = 10$ mm

$A_2 = 20 \times 100 = 2000 \text{ mm}^2$; $y_2 = 20 + (100/2) = 70$ mm

$A_3 = 100 \times 20 = 2000 \text{ mm}^2$; $y_3 = 120 + 20/2 = 130$ mm

$$Y_C = [A_1Y_1 + A_2 Y_2 + A_3 Y_3] / (A_1 + A_2 + A_3)$$

$$= (4000 \times 10 + 2000 \times 70 + 2000 \times 130)/(4000 + 2000 + 2000) = 55 \text{ mm}$$

3.1.3 CENTROIDS OF SOLIDS OF REVOLUTION

An arc rotated about an axis generates a hollow body while an area rotated about an axis generates a solid body. While arcs and areas have two coordinates (X_C and Y_C) for their centroids, a solid body has three coordinates (X_C, Y_C and Z_C). *A body generated by rotating an area about Y-axis will have its centroid on the axis of rotation* i.e., $X_C = 0$ and $Z_C = 0$. Centroid of a solid body of revolution is calculated by integration of an elemental volume property, similar to the method used for calculating centroids of arcs or areas.

(i) *Cone:* Find centroid of a right circular cone of radius 'R' and height 'h'.

Solution

The cone is generated by rotating a right angle triangle of sides 'R' and 'h' about the edge of length 'h'. Since the cone is symmetric about Y-axis, its centroid lies along the axis of revolution. Hence, X and Z coordinates of centroid are zero. Considering an elemental volume dV of a circular disc of radius r and thickness 'dy' at 'y' from the origin,

$$dV = \pi r^2 \, dy \quad \text{and from similar } \Delta s \text{ ABC \& ADE, } r / R = (h - y) / h$$

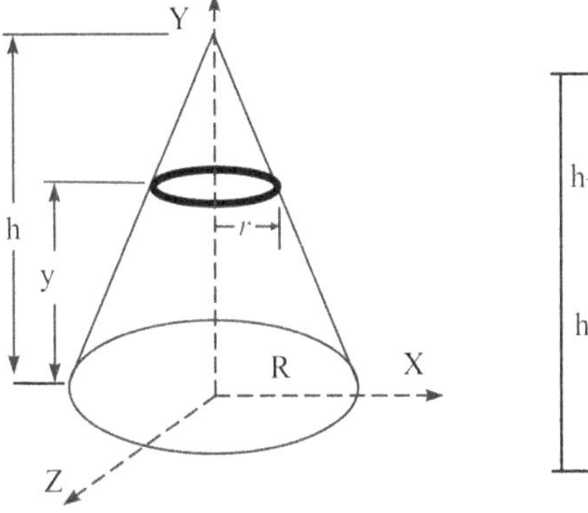
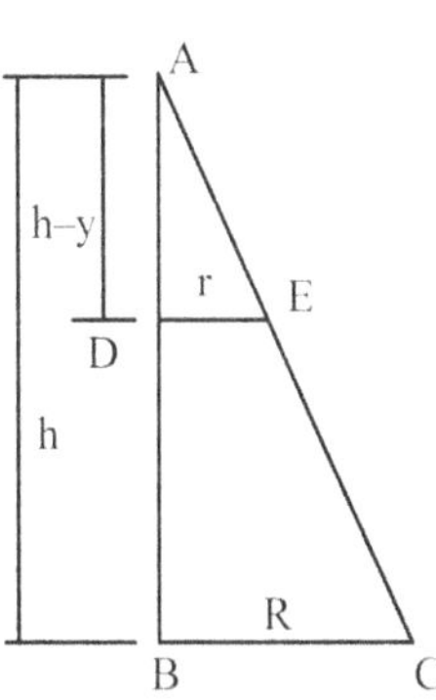

Then,

$$Y_C = \int y \times dV / \int dV = [\int y \times (\pi r^2 \, dy)] / [\int \pi r^2 \, dy]$$

$$= [\int y \times (h - y)^2 \, dy] / [\int (h - y)^2 \, dy] \quad \text{for} \quad 0 \leq y \leq h$$

$$= [h^2y^2/2 - 2hy^3/3 + y^4/4] / [h^2y - 2hy^2/2 + y^3/3]$$

$$= \boldsymbol{h/4} \qquad\qquad\qquad .....(3.10)$$

(ii) *Hemisphere:* Find centroid of a hemisphere of radius 'R'.

Solution

The hemisphere is generated by rotating a quarter circle of radius 'R' about its vertical edge. Since the hemisphere is symmetric about Y-axis, its centroid lies along the axis of revolution. Hence, X and Z coordinates of centroid are zero. Considering an elemental volume dV of radius 'r' and thickness 'dy' at a distance of 'y' from the origin, $dV = \pi r^2 dy$. Here, r and y are related by $r^2 = R^2 - y^2$ from the triangle shown, since any point A on the periphery is at a distance of 'R' from the center of base or origin 'O'.

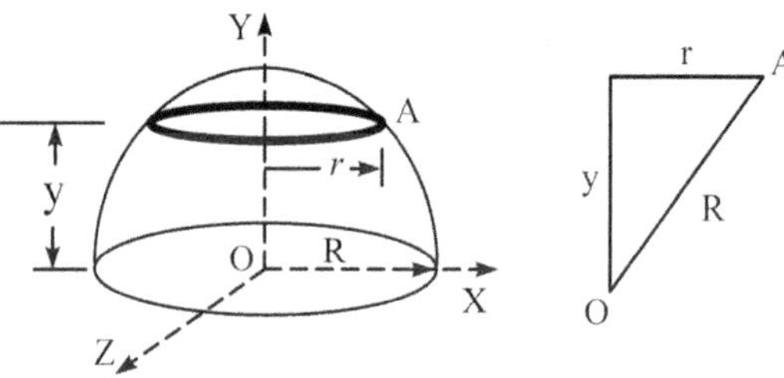

Then, $Y_C = \int y \times dV \ / \int dV = [\int y \times (\pi r^2 dy)] / [\int \pi r^2 dy]$

$= [\int y \times (R^2 - y^2) \times dy] / [\int (R^2 - y^2) dy]$ for $0 \le y \le R$

$= [(R^2 y^2/2) - (y^4/4)] / [R^2 y - (y^3/3)] = \mathbf{3\,R\,/\,8}$ .....(3.11)

(iii) *Paraboloid*: Find centroid of a paraboloid of base radius 'R' and height h, if the parabola is defined by $h - y = k \times x^2$ (since the origin is at the base of the hyperboloid)

Solution

The paraboloid is generated by rotating a parabola $h - y = k \times x^2$ about Y-axis. Since the paraboloid is symmetric about Y-axis, X and Z coordinates of centroid are zero.

Considering an elemental volume dV of a circular disc of radius r and thickness dy at y from the origin, $dV = \pi r^2 dy$ and $r^2 = x^2 = (h - y)/k$

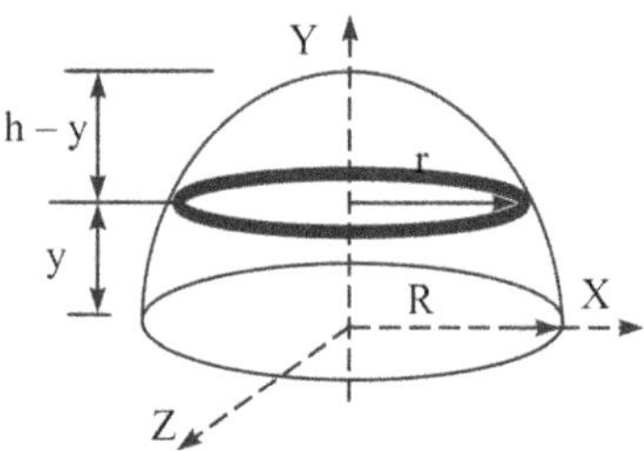

$$Y_C = \int y \times dV / \int dV = [\, y \times \int \pi \{(h-y)/k\}\, dy\,] / [\int \pi \{(h-y)/k\}\, dy]$$

$$= [\int y \times (h-y)\, dy] / [\int (h-y)\, dy] \qquad \text{in the limit } y = 0 \text{ to } h$$

$$= \mathbf{h/3}$$

3.1.4 CENTROIDS OF COMPOSITE SOLIDS OF REVOLUTION

Centroids of composite solids of revolution are calculated following the same procedure used for composite areas, as a combination of different simple solids.

Example 3.1

A homogeneous body consists of a circular cylindrical portion of radius 'R' and height 'h', attached to hemispherical portion of the same radius. Find 'h' such that centroid of the composite body lies at the center of the circular plane of the hemisphere.

Solution

Since the composite body is symmetric about Y-axis, its centroid lies along the axis of revolution. Hence, $X_C = Z_C = 0$. Considering hemisphere as part-1 and cylinder as part-2, volume and distance of centroid from the origin are :

$$V_1 = 2\pi R^3/3 \; ; \quad Y_1 = h + 3R/8 \qquad \text{and} \qquad V_2 = \pi R^2 h \; ; \quad Y_2 = h/2$$

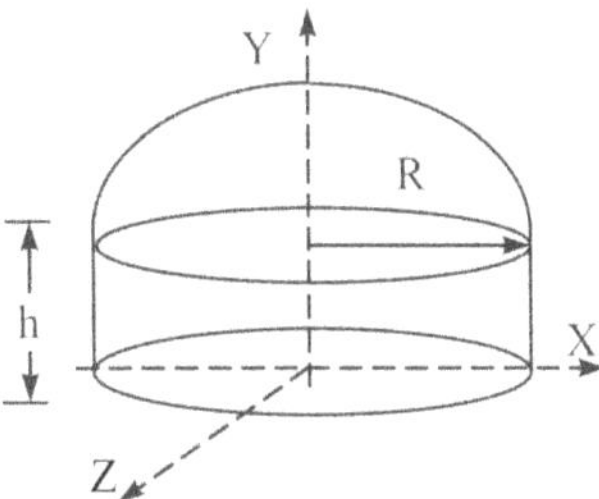

Then, $Y_C = (V_1 \times Y_1 + V_2 \times Y_2) / (V_1 + V_2)$

or $h = [\,(2\pi R^3/3) \times (h + 3R/8) + (\pi R^2 h) \times (h/2)\,] / [\,(2\pi R^3/3) + (\pi R^2 h)\,]$

$$= R/\sqrt{2}$$

3.1.5 PAPPU'S THEOREMS

Theorem – 1

The area of surface generated by revolving a plane curve of length 'L' about a non-intersecting axis in its plane is equal to the product of length of the curve and the distance traveled by the centroid of the curve during the generation of the surface.

If a surface is generated by revolving a curve about X-axis, then

$$\text{Surface area} = L \times (2\pi\, Y_C) \qquad .....(3.12)$$

Similarly, if a surface is generated by revolving a curve about Y-axis, then

$$\text{Surface area} = L \times (2\pi\, X_C) \qquad .....(3.13)$$

Theorem – 2   (Also referred to as **Guldinus theorem**)

The volume of a body generated by revolving a plane of area 'A' about a non-intersecting axis in its plane is equal to the product of area of the plane and the distance traveled by the centroid of the area during the generation of the volume.

If a volume is generated by revolving an area 'A' about X-axis, then

$$\text{Body volume} = A \times (2\pi\, Y_C) \qquad .....(3.14)$$

$$\text{where, } Y_C \text{ is the distance of centroid of the area from X-axis}$$

If a volume is generated by revolving an area 'A' about Y-axis, then

$$\text{Body volume} = A \times (2\pi\, X_C) \qquad .....(3.15)$$

$$\text{where, } X_C \text{ is the distance of centroid of the area from Y-axis}$$

These laws can also be used to calculate location of the centroid of a given curve or area, if the area or volume of the object is known by other means.

Example 3.2

(a)   Volume of a cone of base radius 'R' and height 'h',

$$V = (\text{Area of the triangle of base 'R' and height 'h'}) \times (2\pi\, X_C)$$

$$\text{where, } X_C \text{ is the distance of centroid of the triangle from Y-axis}$$

$$= (R \times h/2) \times [2\,\pi \times (R/3)] = \boldsymbol{\pi\, R^2 h\, /\, 3}$$

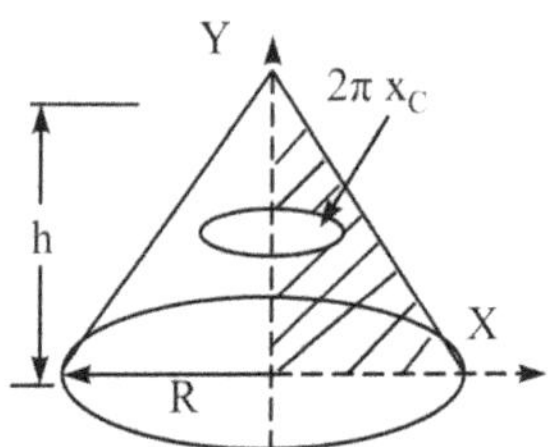

(b)   Volume of a hemisphere of radius 'R',

$$V = (\text{Area of the quarter circle of radius 'R'}) \times (2\pi\, X_C)$$

$$\text{where, } X_C \text{ is the distance of centroid of the quarter circle from Y-axis}$$

$$= (\pi R^2/4) \times [2\,\pi \times (4R/3\pi)] = \boldsymbol{(2/3)\,\pi\, R^3}$$

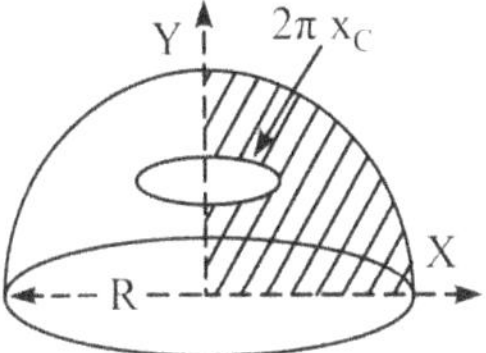

Similarly, Volume of a sphere, $V = (\pi R^2/2) \times [2\pi \times (4R/3\pi)]$

$$= (4/3)\,\pi\,R^3$$

(c) Surface area of annular torus (for example, cycle tyre) of mean radius 'R', with hollow circular cross section of radius 'r',

A = Perimeter of the hollow circular cross section $\times (2\pi\,Y_C)$

where, Y_C is the distance of centroid of the cross section from the axis of revolution

$$= (2\pi\,r) \times (2\pi\,R) = \mathbf{4\pi^2\ r\ R}$$

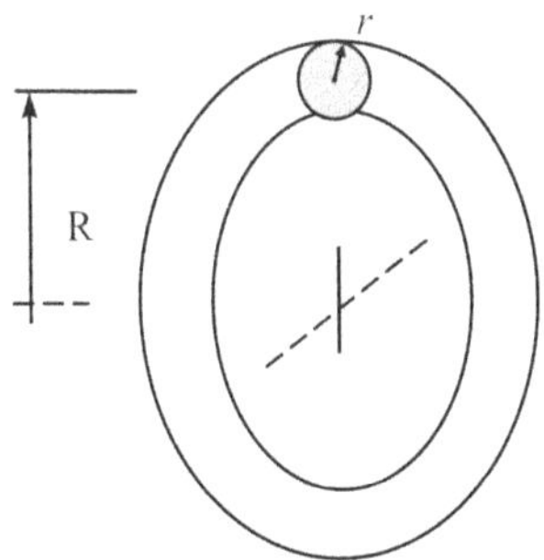

(d) Calculate centroid of a semi-circular plate, using Pappu's theorem, if the volume of the generated sphere is known.

$$V = (\pi\,R^2/2\,) \times (2\pi\,X_C) = (4/3)\,\pi\,R^3$$

Therefore, $\mathbf{X_C = 4\,R\,/\,3\,\pi}$

3.2   MOMENT OF INERTIA

Moment of inertia (also called 2nd moment of area) is a property of cross sectional area of a member, used in the calculation of bending stresses in beams subjected to loads perpendicular to its axis. Moment of inertia about any axis is defined as the area of the cross section multiplied by square of the distance of the area from the axis. Hence, for the same area, moment of inertia varies with the particular axis about which it is calculated. Since area is always positive and distance square is also positive, moment of inertia of any area is always positive about any axis.

For a small area 'A' at a distance of x from Y-axis and y from X-axis,

$$I_{XX} = A \times y^2 \quad ; \quad I_{YY} = A \times x^2$$

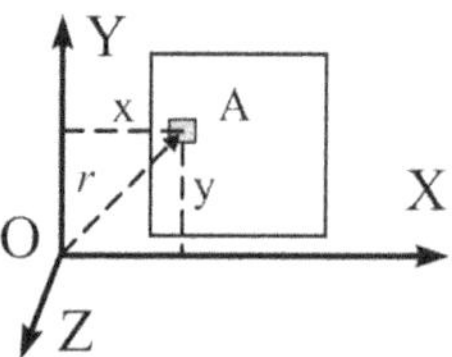

Moment of inertia of a large area can be obtained by integration as

$$I_{XX} = \int (dA \times y^2) \quad ; \quad I_{YY} = \int (dA \times x^2) \qquad .....(3.16)$$

Moment of inertia about Z-axis (also called **Polar moment of inertia**), perpendicular to the plane of cross section,

$$I_{ZZ} = \int (dA \times r^2) = \int (dA \times y^2) + \int (dA \times x^2)$$

or $\quad I_{ZZ} = I_{XX} + I_{YY} \qquad .....(3.17)$

This is also called **perpendicular axis theorem**. I_{ZZ} is a property used in members subjected to torsional moment or moment about the axis (Z-direction) of the beam or shaft.

3.2.1   MOMENT OF INERTIA OF SOME SIMPLE SHAPES

(a) *Rectangular section of width 'b' and depth 'd'*

Considering a small area dA of sides 'ds' and 'dt' at a distance of 't' from Y-axis and 's' from X-axis,

Moment of inertia about X-axis, passing through O, is

$$\mathbf{I_{XX}} = \int dA \times t^2 = \int \int ds \times dt \, t^2 \quad \text{for } 0 \le s \le b \text{ and } 0 \le t \le d$$

$$= \int [b] \, t^2 \, dt = b \, [t^3 / 3] = \mathbf{b \, d^3 / 3} \qquad .....(3.18)$$

Similarly, $\quad I_{YY} = b^3 \, d / 3$

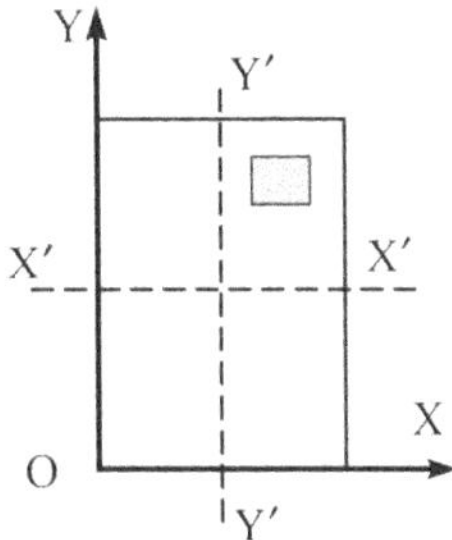

Moment of inertia about $\left(X'-X'\right)$ axis passing through the centroid, can be obtained in the same way, by using appropriate limits for 't', as

$$\mathbf{I}_{X'X'} = \int dA \times t^2 = \int \int (ds \times dt)\, t^2$$

$$\text{for }\ 0 \le s \le b \ \text{ and } -d/2 \le t \le d/2$$

$$= \int [b]\, t^2\, dt = b\,[\,t^3/3\,] = \mathbf{b\,d^3/12} \qquad .....(3.19)$$

Similarly, $I_{Y'Y'} = b^3\, d/12$

This method of integration w.r.t. y alone is possible because width is constant for different values of y (unlike in the case of a triangle, circle, etc.), throughout the depth.

Alternative method

Considering a small area 'dA' of width 'b' and thickness 'dy' at a height of 'y' from the axis A – A, moment of inertia,

$$I_{AA} = \int dA \times y^2 = \int (b \times dy) \times y^2 \qquad \text{for }\ 0 \le y \le d$$

$$= b\, d^3/3$$

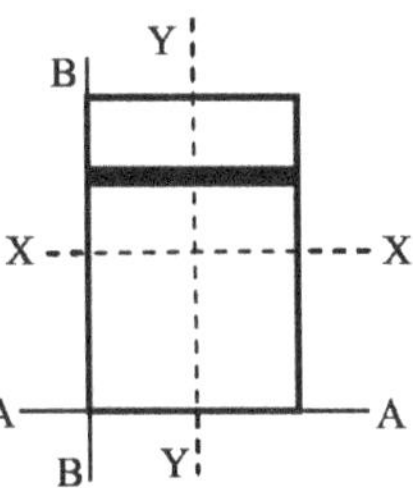

Similarly, moment of inertia about BB, $I_{BB} = b^3 d/3$

(b) *Triangular section of width 'b' and depth 'd'*

Considering a small area dA of width 's' and thickness 'dt' at a distance of 't' from Y-axis,

Moment of inertia about X-axis, passing through O, is

$$\mathbf{I_{XX}} = \int dA \times t^2 = \int (s \times dt) \times t^2$$

$$\text{with}\quad s/b = (d-t)/d \quad \text{and} \quad 0 \le t \le d$$

$$= \int [b \times (d-t)/d]\, t^2\, dt = (b/d) \times [(dt^3/3) - (t^4/4)]$$

$$= \mathbf{b\, d^3 / 12} \qquad\qquad\qquad \dots(3.20)$$

Similarly, $I_{YY} = b^3\, d / 12$

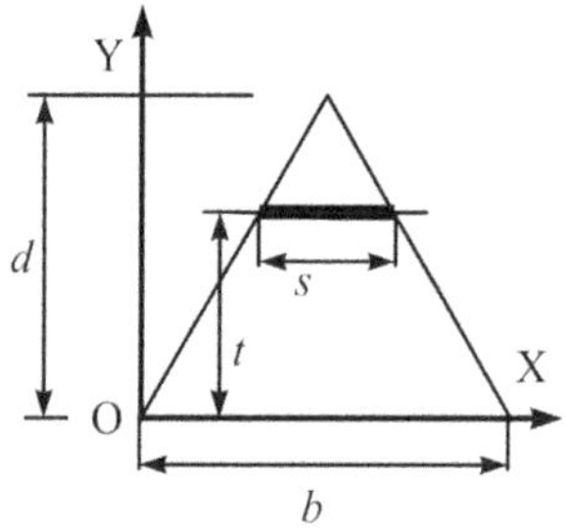

(c) *Circular section of radius 'r'*

Consider a small elemental area $(r\, d\theta) \times dr$ at a distance of 'y' $(= r \sin \theta)$ from X-axis through its centroid,

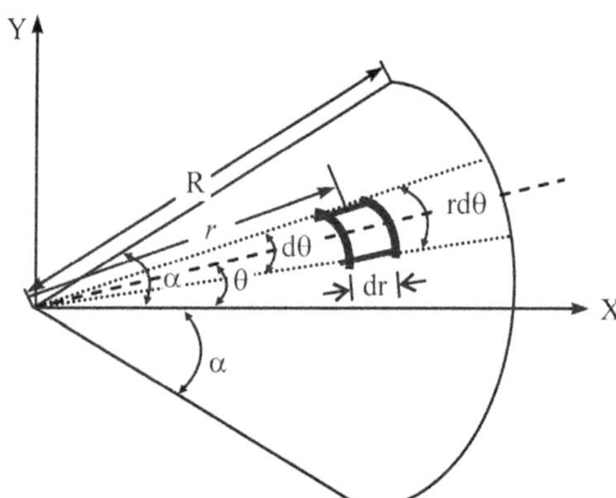

$$\mathbf{I_{XX}} = \int dA \times y^2 = \int\int [(r\, d\theta) \times dr] \times (r \sin \theta)^2$$

$$\text{for}\quad 0 \le r \le R \quad \text{and} \quad -\alpha \le \theta \le +\alpha$$

$$= \int [r^4/4] \sin^2\theta\, d\theta = (R^4/4) \int [(1 - \cos 2\theta)/2]\, d\theta$$

$$= (R^4/8) \times [\theta + \cos 2\theta] = (R^4/8) \times [2\alpha + 2 \sin 2\alpha]$$

For a circular plate, $\alpha = \pi$ and, therefore,

$$\mathbf{I_{XX} = \pi\, R^4 / 4} \quad \text{or} \quad \pi\, D^4 / 64$$

Since circular plate is symmetric about any diametral axis,

$$\mathbf{I_{YY} = I_{XX} = \pi\, R^4 / 4} \qquad \text{or} \qquad \pi\, D^4 / 64 \qquad \ldots\text{(3.21)}$$

Polar moment of inertia of a circular plate,

$$\mathbf{I_{ZZ} = I_{XX} + I_{YY} = \pi\, R^4 / 2} \quad \text{or} \quad \pi\, D^4 / 32 \quad \ldots\text{(3.22)}$$

Following the same procedure, with appropriate change of limits of θ,

Moment of inertia of a semi-circular plate (symmetrical about Y-axis), $\mathbf{I_{XX} = \pi\, R^4 / 8}$

and Moment of inertia of a quarter circular plate,

$$\mathbf{I_{XX} = I_{YY} = \pi\, R^4 / 16}$$

(d) *Thin circular ring of mean radius 'r' and thickness 't'*

Consider a small elemental area $(R\, d\theta) \times t$ at a distance of 'R' from the center of the ring.

$$\mathbf{I_{ZZ}} = \int dA \times R^2 = \int [(R\, d\theta) \times t] \times R^2 \quad \text{for} \quad 0 \le \theta \le 2\pi$$

$$= \mathbf{2\pi\, R^3 t} \quad \text{or} \quad \pi\, D^3 t / 4$$

Since the ring is symmetric about any diametral axis,

$$\mathbf{I_{XX} = I_{YY} = I_{ZZ} / 2 = \pi\, R^3 t} \qquad \ldots\text{(3.23)}$$

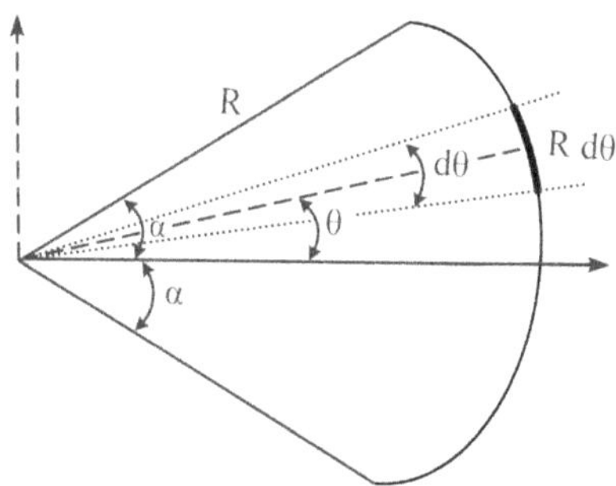

In the same way, polar moment of inertia of a **semi-circular ring**,

$$I_{ZZ} = \pi\, R^3 t$$

and   polar moment of inertia of a *quarter* **circular ring**,

$$I_{ZZ} = \pi\, R^3 t / 2$$

(e) *Ellipse:* Calculate moment of inertia of an ellipse of semi-major axis 'a' and semi-minor axis 'b'.

Let us consider a small elemental area 'dA' of height '2y' and width 'dx', at a distance 'x' from the origin of the coordinate system. Equation of ellipse is $x^2/a^2 + y^2/b^2 = 1$

$$I_{YY} = \int x^2 \times dA = \int x^2 \times (2y \times dx)$$

$$= \int x^2 \times 2b\sqrt{1-(x^2/a^2)} \times dx \quad \text{for} \quad -a \leq x \leq +a$$

This integral can be evaluated by substituting, $x = a \sin\theta$

Then, $dx = a \cos\theta \times d\theta$ with $\theta = -\pi/2$ to $+\pi/2$

corresponding to $x = -a$ to $+a$ and $\sqrt{\left[1-(x^2/a^2)\right]} = \cos\theta$

Therefore, $I_{YY} = \int (a \sin\theta)^2 \times 2b \times \cos\theta \times a \cos\theta \times d\theta$

$$= 2\pi a^3 b \times \{(\pi/8) - [0-0]/32\}$$

$$\mathbf{= \pi\, a^3\, b\, /\, 4} \qquad\qquad \text{.....(3.24)}$$

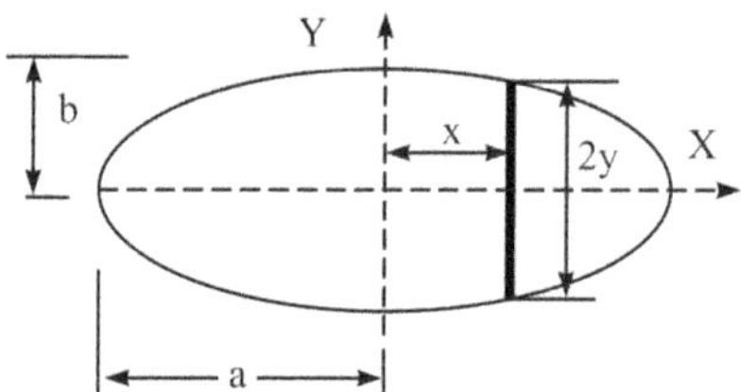

Similarly, $\mathbf{I_{XX} = \pi\, a\, b^3\, /\, 4}$

and $\mathbf{I_{ZZ} = I_{XX} + I_{YY} = \pi\, a\, b\, (a^2 + b^2)\, /\, 4}$

3.2.2 PARALLEL AXIS THEOREM

Moment of inertia calculated about an axis (X-X) through its centroid can be transferred to a parallel axis using the relation,

$$I_{AA} = I_{XX} + A \times y^2 \qquad\qquad \text{.....(3.25)}$$

where, A is the area of the plane and 'y' is the distance between the two parallel axes X-X and A – A. This is also called ***Transfer of axis theorem.***

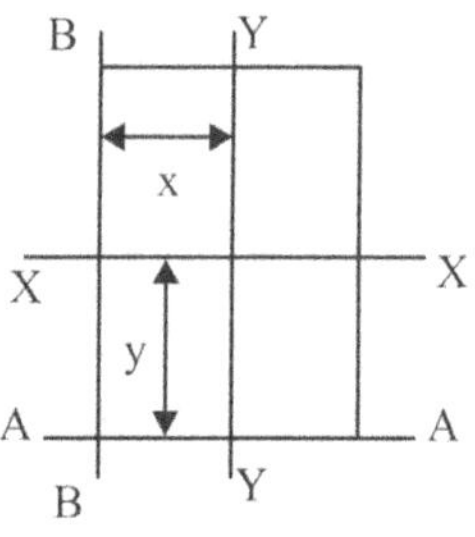

Similarly, $I_{BB} = I_{YY} + A \times x^2 \qquad\qquad \text{....(3.26)}$

Example 3.3

Moment of inertia of a rectangle through its centroidal axes can be obtained from the values about one of their edges, using parallel axes theorem, as shown here.

$$I_{XX} = I_{AA} - A \times y^2 = (bd^3/3) - (b \times d) \times (d/2)^2 = b\,d^3\,/\,12$$

Similarly,

$$I_{YY} = I_{BB} - A \times x^2 = (b^3d/3) - (b \times d) \times (b/2)^2 = b^3d\,/\,12$$

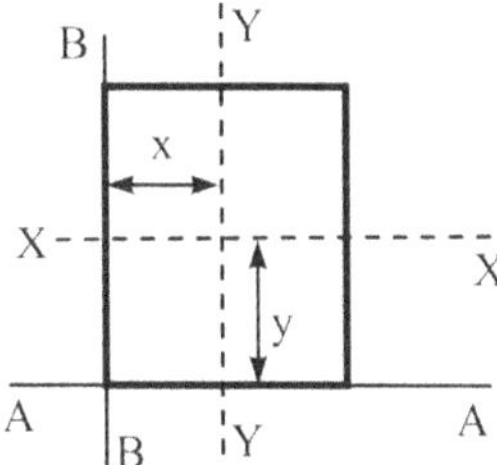

Example 3.4

Moment of inertia of a triangle through its centroidal axis X-X can be obtained from the moment of inertia value about its edge A-A, using parallel axes theorem, as

$$I_{XX} = I_{AA} - A \times y^2 = (bd^3/12) - (b \times d/2) \times (d/3)^2 = b\,d^3\,/\,36$$

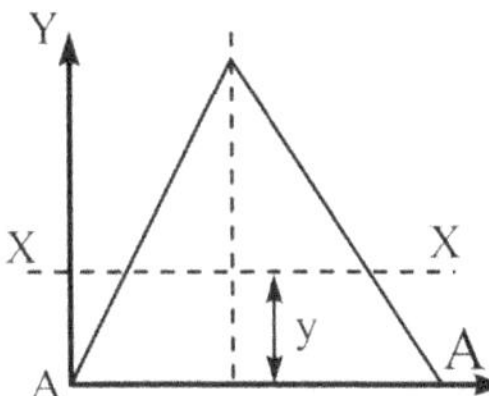

Example 3.5

Moment of inertia of a semi-circular plate through its centroidal axis X-X can be obtained from the moment of inertia value about its edge A-A, using parallel axes theorem, as

$$I_{XX} = I_{AA} - A \times y^2 = (\pi\,r^4/8) - (\pi\,r^2/2) \times (4\,r/3\pi)^2 = 0.11\,r^4$$

Due to symmetry, centroid lies on Y-axis. So, $I_{YY} = \pi\,r^4/8$

Example 3.6

Moment of inertia of a quarter circle through its centroidal axis X-X can be obtained from the moment of inertia value about its edge A-A, using parallel axes theorem, as

$$I_{XX} = I_{YY} = I_{AA} - A \times y^2 = (\pi r^4/16) - (\pi r^2/4) \times (4r/3\pi)^2$$

$$= \mathbf{0.055\ r^4}$$

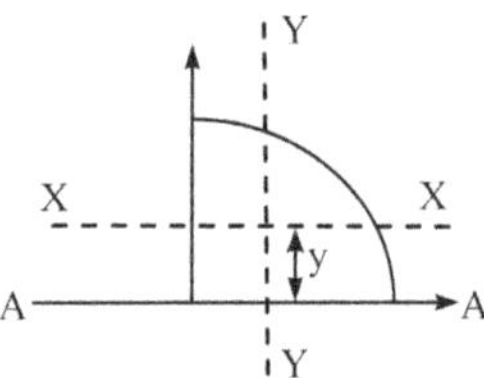

3.2.3 MOMENT OF INERTIA OF COMPOSITE SECTIONS

Moment of inertia of a composite section is the algebraic sum of moments of inertia of its simpler components.

$$I_{XX} = \sum [\ I_{xx}\ ]_i \ ; \quad I_{YY} = \sum [\ I_{yy}\ ]_i \quad \text{for } i = 1, n \qquad ....(3.27)$$

Moments of inertia of all individual sections should be added only after transferring all of them to a common axis.

Example 3.7

(a) *Hollow rectangular section* of outer dimensions 'B' and 'D' and inner dimensions 'b' and 'd' :

$$I_{XX} = [\ B \times D^3 - b \times d^3\ ] / 12$$

$$I_{YY} = [\ B^3 \times D - b^3 \times d\ ] / 12$$

(b) *Hollow circular section* of outer radius 'R' and inner radius 'r' :

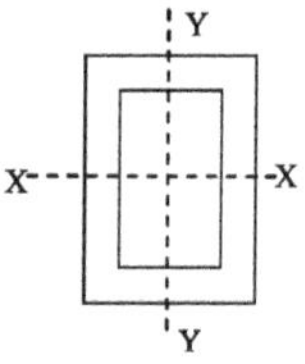

$$I_{XX} = I_{YY} = \pi (\ D^4 - d^4\ ) / 64$$

A general formula for the other cases is not possible.

A few specific cases are given here.

(c) *T-section:* Width 20cm; Height 30 cm;

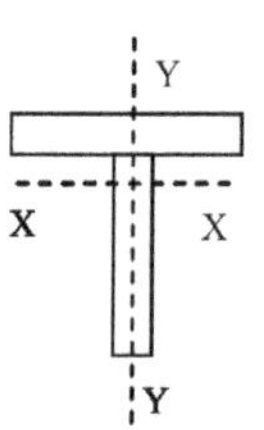

Thickness of vertical web and horizontal flange 4 cm

Section is symmetric about Y-axis.   Therefore, $X_C = 0$

Taking horizontal flange as area-1 and vertical web as area-2 and taking all measurements w.r.t. the bottom edge,

$A_1 = 20 \times 4$; $A_2 = (30–4) \times 4$; $Y_1 = 30 – (4/2) = 28$; $Y_2 = (30 – 4)/2 = 13$

$Y_C = [A_1 \times Y_1 + A_2 \times Y_2] / [A_1 + A_2]$

$\quad = [\,(20 \times 4) \times 28 + (26 \times 4) \times 13\,] / [\,(20 \times 4) + (26 \times 4)\,] = 19.52$ cm

$I_{XX} = [\,(b_1 \times d_1^3/12) + (b_1 \times d_1) \times (Y_1 – Y_C)^2\,]$

$\qquad\qquad + [\,(b_2 \times d_2^3/12) + (b_2 \times d_2) \times (Y_C – Y_2)^2\,]$

$\quad = [\,(20 \times 4^3/12) + (4 \times 20) \times (28 – 19.52)^2\,]$

$\qquad\qquad\qquad + [\,(26^3 \times 4 / 12) + (4 \times 26 \times (19.52 – 13)^2\,]$

$\quad = 16139.3$ cm^4

$I_{YY} = (b_1^3 \times d_1/12) + (b_2^3 \times d_2/12) = (20^3 \times 4 / 12) + (26^3 \times 4 / 12)$

$\quad = 8525.4$ cm^4

(d)   ***I-section:*** Width 20cm at top & 30cm at bottom; Height 40 cm; Thickness of flanges and web 4 cm

Section is symmetric about Y-axis. Therefore, $X_C = 0$

Taking horizontal flange as area-1, vertical web as area-2 and bottom flange as area-3 and taking all measurements w.r.t. the bottom edge,

$Y_1 = 40 – (4/2) = 38$; $Y_2 = (40 – 4 – 4)/2 + 4 = 20$; $Y_3 = 4/2 = 2$

$Y_C = [A_1 \times Y_1 + A_2 \times Y_2 + A_3 \times Y_3] / [A_1 + A_2 + A_3]$

$\quad = [\,(20 \times 4) \times 38 + (4 \times 32) \times 20 + (30 \times 4) \times 2\,]$

$\qquad\qquad\qquad\qquad / [\,(20 \times 4) + (4 \times 32) + (30 \times 4)\,]$

$\quad = 17.8$ cm

$I_{XX} = [(b_1 d_1^3/12) + (b_1 \times d_1) \times (Y_1 – Y_C)^2]$

$\qquad + [(b_2 \times d_2^3/12) + (b_2 \times d_2) \times (Y_C – Y_2)^2]$

$\qquad + [(b_3 \times d_3^3/12) + (b_3 \times d_3) \times (Y_C – Y_3)^2]$

$\quad = [\,(20 \times 4^3 / 12) + (20 \times 4) \times (38 – 17.8)^2]$

$\qquad\qquad + [(4 \times 32^3 / 12) + (4 \times 32) \times (17.8–20)^2]$

$\qquad\qquad + [(30 \times 4^3/ 12) + (30 \times 4) \times (17.8 – 2)^2]$

$\quad = 73794.5$ cm^4

$I_{YY} = (b_1^3 \times d_1/12) + (b_2^3 \times d_2/12) + (b_3^3 \times d_3/12)$

$\quad = (20^3 \times 4 / 12) + (4^3 \times 32 / 12) + (30^3 \times 4 / 12) = 11837.4$ cm^4

(e) *L – section:*

Width 20cm; Height 30 cm; Thickness of web and flange 4 cm

Taking measurements for the horizontal rectangle (area-1) and vertical rectangle (area-2), w.r.t. the bottom left corner,

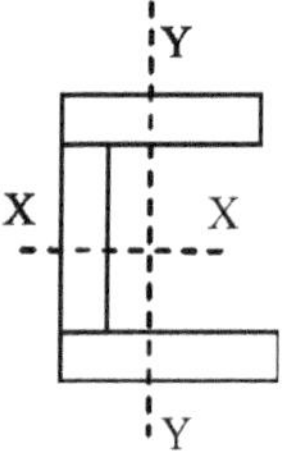

$X_1 = 20/2 = 10;$ $Y_1 = 4/2 = 2$;

$X_2 = 4/2 = 2;$ $Y_2 = (30 – 4)/2 + 4 = 17$

$X_C = [A_1 \times X_1 + A_2 \times X_2] / [A_1 + A_2]$

$\quad = [(20 \times 4) \times 10 + (26 \times 4) \times 2] / [(20 \times 4) + (26 \times 4)]$

$\quad = 5.48$ cm

$Y_C = [A_1 \times Y_1 + A_2 \times Y_2] / [A_1 + A_2]$

$\quad = [(20 \times 4) \times 2 + (26 \times 4) \times 17] / [(20 \times 4) + (26 \times 4)] = 10.48$ cm

$I_{XX} = [(b_1 \times d_1^{3}/12) + (b_1 \times d_1) \times (Y_C – Y_1)^2]$
$$+ [(b_2 \times d_2^{3}/12) + (b_2 \times d_2) \times (Y_C – Y_2)^2]$$

$\quad = [(20 \times 4^3 / 12) + (4 \times 20) \times (10.48 – 2)^2]$
$$+ [(4 \times 26^3 / 12) + (4 \times 26) \times (10.48 – 17)^2]$$

$\quad = 16139.3$ cm^4

$I_{YY} = [(b_1^{3} \times d_1/12) + (b_1 \times d_1) \times (X_1 – X_C)^2]$
$$+ [(b_2^{3} \times d_2/12) + (b_2 \times d_2) \times (X_C – X_1)^2]$$

$\quad = [(20^3 \times 4 / 12) + (20 \times 4) \times (15 – 10.48)^2]$
$$+ [(26^3 \times 4 / 12) + (4 \times 26) \times (5.48 – 2)^2]$$

$\quad = 11419.3$ cm^4

(f) *C - section:*

Width 20cm; Height 30 cm; Thickness of web and flanges 4 cm

Section is symmetric about X-axis. Therefore, $Y_C = 0$

Taking top flange as area-1, vertical web as area-2, bottom flange as area-3 and taking all measurements w.r.t. the left edge, $A_1 = A_3 = 20 \times 4 = 80$

$A_2 = (30 – 2 \times 4) \times 4 = 88$

$X_1 = X_3 = 20/2 = 10;$ $X_2 = 4/2 = 2$

$Y_1 = Y_3 = (30 – 4)/2 = 13;$ $Y_2 = 0$

$X_C = [A_1 \times X_1 + A_2 \times X_2 + A_3 \times X_3] / [A_1 + A_2 + A_3]$

$\quad = [2 \times (20 \times 4) \times 10 + (4 \times 22) \times 2] / [2 \times (20 \times 4) + (4 \times 22)] = 7.16$ cm

$$I_{XX} = 2\,[(b_1 \times d_1^3/12) + (b_1 \times d_1) \times (Y_C - Y_1)^2]$$
$$+\,[(b_2 \times d_2^3/12) + (b_2 \times d_2) \times (Y_C - Y_2)^2]$$
$$= 2\,[(20 \times 4^3 / 12) + (20 \times 4) \times (0 - 13)^2)]$$
$$+\,[(4 \times 22^3 / 12) + (4 \times 22) \times (0-0)^2]$$
$$= 30802 \text{ cm}^4$$

$$I_{YY} = 2\,[(b_1^3 \times d_1/12) + (b_1 \times d_1) \times (X_C - X_1)^2]$$
$$+\,[(b_2^3 \times d_2/12) + (b_2 \times d_2) \times (X_C - X_2)^2]$$
$$= 2\,[(20^3 \times 4 / 12) + (20 \times 4) \times (7.16 - 10)^2]$$
$$+\,[(4^3 \times 22 / 12) + (4 \times 22) \times (7.16 - 2)^2]$$
$$= 9084.4 \text{ cm}^4$$

Example 3.8

A quarter circular area of radius 'a' is removed from a square area of side 'a'. Calculate moment of inertia of the remaining section about the three axes.

Solution

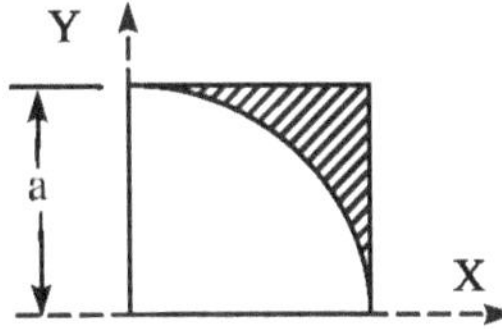

$$I_{XX} = (I_{XX})_{Square} - (I_{XX})_{Quadrant}$$
$$= (a^4/3) - (\pi\,a^4/16) = a^4(16 - 3\pi) / 48$$
$$= 0.137a^4$$

Similarly, $I_{YY} = 0.137\ a^4$

and $\quad I_{ZZ} = I_{XX} + I_{YY} = 0.274\ a^4$

3.2.4   PRODUCT OF INERTIA

This property does not have much of practical use, except for finding out principal planes. For an area A, it is defined as $\quad I_{XY} = A \times (x \times y)$

In general, for any plane area, it is defined as $\quad I_{XY} = \int dA \times (x \times y)$

$$\dots\dots(3.28)$$

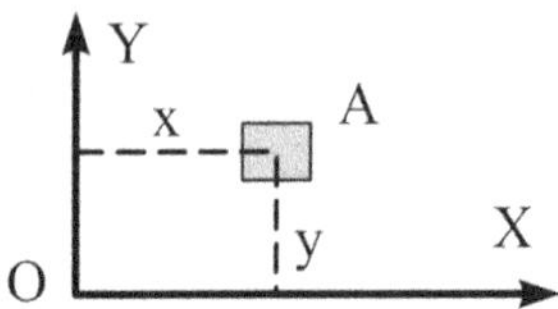

Depending on the nature of cross section and coordinate axes, I_{XY} can take +ve or –ve values. As the coordinate axes through centroid are rotated, the product of inertia value changes from +ve to –ve or vice versa. Therefore, with reference to some particular axes, the product of inertia value becomes zero. These axes are called **Principal axes**. These axes have significance in terms of maximum stress in a cross section, as can be seen in detail in 'Mechanics of solids'. *Obviously, the product of inertia about any geometrically symmetric axis is zero.*

Thus, product of inertia for circular, semi-circular, elliptic, parabolic, rectangular sections, about any line of symmetry is zero. Also, *the product of inertia of a given area w.r.t. the principal axes at the centroid is zero. i.e.,* **line of symmetry is a principal axis.**

Parallel axis theorem is also applicable to product of inertia with appropriate change

$$\text{i.e., } I_{X'Y'} = I_{XY} + A \times (a \times b) \qquad \qquad ....(3.29)$$

where I_{XY} is the product of inertia about centroidal X and Y axes and a and b are the distances between the two sets of axes.

(a) Rectangular section of width 'B' and height 'D', with axes coinciding adjacent sides

$$I_{XY} = (I_{XY})_C + (B \times D) \times [(B/2) \times (D/2)] = 0 + B^2 D^2 / 4 = B^2 D^2 / 4$$

(b) Quadrant of a circle of radius 'R' considering a small element of length x (parallel to X- axis) and thickness dy at a distance y from X-axis

$$I_{XY} = \int (x/2) \times y \times (x \times dy)$$
$$= (1/2) \int x^2 \times y \times dy = R^4/8$$
$$= (1/2) \int (R^2 - y^2) \times y \times dy$$
$$= (1/2) [R^2 y^2/2 - y^4/4] \quad \text{for} \ \ 0 \le y \le R$$

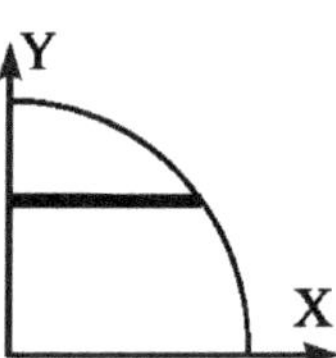

3.2.5 ROTATION OF AXES

Let us consider a small area dA. With respect to axes X-Y,

$$I_{XX} = \int y^2 \times (dA); \quad I_{YY} = \int x^2 \times (dA); \quad I_{XY} = \int x \times y \times (dA)$$

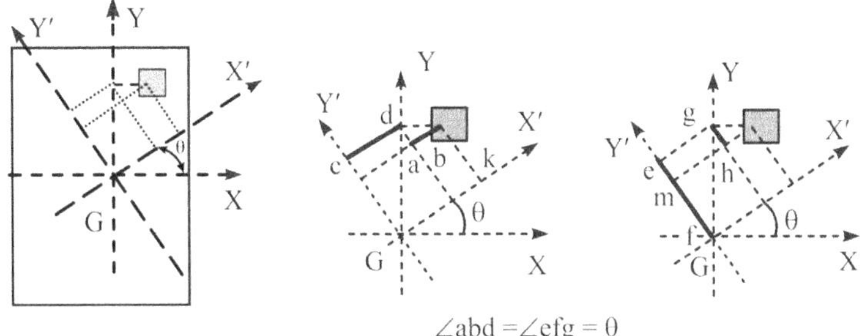

If X-Y axes are rotated through an angle θ to X′–Y′ axes, from the above fig,

$$x' = Gk = ab + cd = x \cos\theta + y \sin\theta; \quad y' = fm = ef - gh = y \cos\theta - x \sin\theta$$

then, $I_{X'X'} = \int (y')^2\, dA = \int (y \cos\theta - x \sin\theta)^2\, dA$

$$= I_{XX} \cos^2\theta + I_{YY} \sin^2\theta - 2\, I_{XY} \sin\theta \cos\theta$$

$$= (I_{XX} + I_{YY})/2 + (I_{XX} - I_{YY}) \times \cos 2\theta /2 - I_{XY} \times \sin 2\theta \quad \ldots(3.30)$$

and $\quad I_{Y'Y'} = \int (x')^2\, dA = \int (x \cos\theta + y \sin\theta)^2\, dA$

$$= I_{XX} \sin^2\theta + I_{YY} \cos^2\theta + 2\, I_{XY} \sin\theta \cos\theta$$

$$= (I_{XX} + I_{YY})/2 - (I_{XX} - I_{YY}) \times \cos 2\theta /2 + I_{XY} \times \sin 2\theta \quad \ldots(3.31)$$

Adding and subtracting these two equations, we get,

$$I_{X'X'} + I_{Y'Y'} = I_{XX} + I_{YY} = I_{ZZ}$$

and $\quad I_{X'X'} - I_{Y'Y'} = (I_{XX} - I_{YY}) \times \cos 2\theta - 2\, I_{XY} \times \sin 2\theta$

Product of inertia w.r.t. the new axes,

$$I_{X'X'} = (I_{XX} - I_{YY}) \times \sin 2\theta /2 + I_{XY} \times \cos 2\theta$$

If X′-Y′ are principal axes, $I_{X'Y'} = 0$ or $\tan 2\theta = 2\, I_{XY} / (I_{YY} - I_{XX})$ ..(3.32)

Maximum and minimum moments of inertia of any section are referred to the principal axes and are given by

$$I_{X'X'}, I_{Y'Y'} = \left(\frac{I_{XX} + I_{YY}}{2} \right) \pm \sqrt{ \left(\frac{I_{XX} - I_{YY}}{2} \right)^2 + I_{XY}^2 } \qquad \ldots(3.33)$$

3.2.6 MOHR'S CIRCLE OF INERTIA

This is used to find the principal planes geometrically, by plotting I_{XX} and I_{YY} along X-axis and I_{XY} along Y-axis. Mark E and G such that $OE = I_{XX}$; $OG = I_{YY}$

Mark D and H such that $ED = -HG = I_{XY}$

Then, draw a circle with C, mid-point of GE, as the center and CD or CH as the radius.

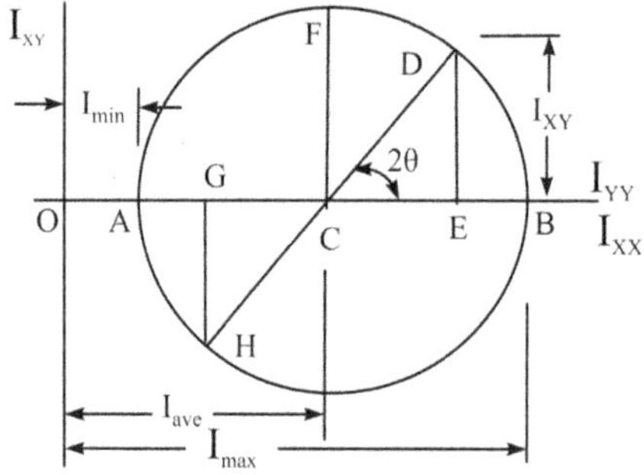

Corresponding to $I_{XY} = 0$, $\quad I_{max} = OB = OC + CB$

$$\text{and} \qquad\qquad I_{min} = OA = OC - CA$$

$$\text{where,} \quad OC = I_{Ave} = (I_{XX} + I_{YY}) / 2$$

$$\text{and} \quad CB = CA = CF = (I_{XY})_{max} = \sqrt{\left[(I_{XX} - I_{YY})/2\right]^2 + I_{XY}^2}$$

Inclination θ of the principal axes w.r.t X-Y axes is obtained from the angle 2θ between HD and AB in the Mohr's circle.

Since the moments of inertia I_{XX} and I_{YY} are always positive, Mohr's circle of inertia will always be in the first and fourth quadrants.

Example 3.9

A square hole of side 'r' is cut centrally from a circular cross section of radius 'r'. Calculate moment of inertia about a diagonal of the square hole.

Solution

Because of symmetry about X and Y axes, centroid lies at the center of the square. Moment of inertia about X-Y can be obtained by rotation of the geometry from X'- Y' axes by an angle of $\theta = 45^0$.

$$I_{X'X'} = I_{Y'Y'} = [I_{X'X'}]_{Circle} - [I_{X'X'}]_{SqHole} = (\pi\, r^4/4) - (r^4/12)$$

$$= r^4 (3\pi - 1) / 12 = 0.702\, r^4$$

$$\text{and} \quad I_{X'Y'} = 0$$

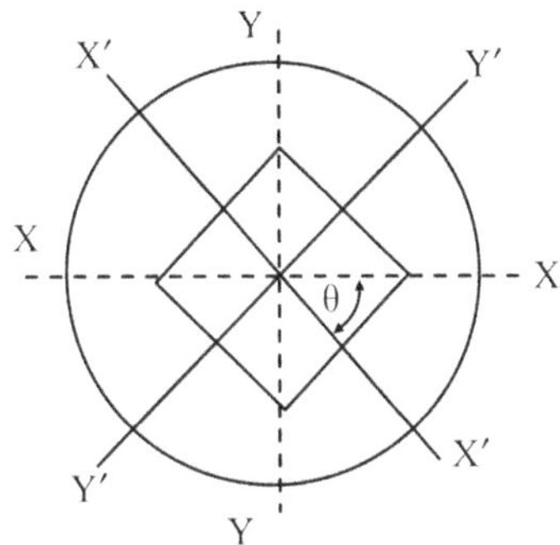

Therefore, rotating coordinate axes by $\theta = 45^0$,

$$I_{XX} = I_{YY} = [I_{X'X'} + I_{Y'Y'}]/2 + [I_{X'X'} - I_{Y'Y'}]\cos 2\theta / 2 - I_{X'Y'}\sin 2\theta$$

$$= [I_{X'X'} + I_{Y'Y'}]/2 + [I_{X'X'} - I_{Y'Y'}] \times (0)/2 - (0) \times (1)$$

$$= [0.702\ r^4 + 0.702\ r^4]/2 = 0.702\ r^4$$

Example 3.10

A circular hole of radius 'r' is cut centrally from a square cross section of side '4r'. Calculate moment of inertia about a diagonal of the square.

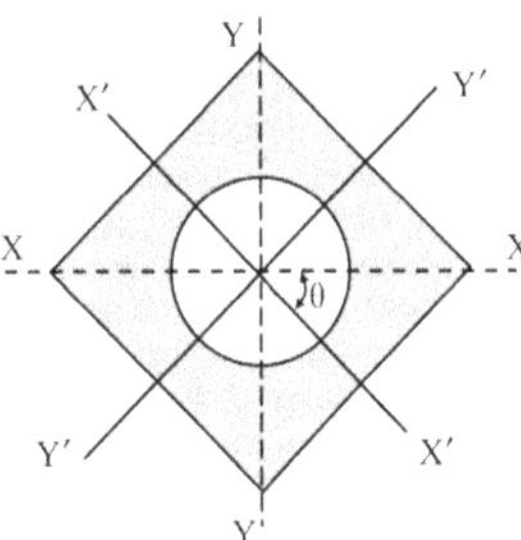

Solution

Because of symmetry about X and Y axes, centroid lies at the center of the square. Moment of inertia about X-Y axes can be obtained by rotation of the geometry from X' - Y' axes by an angle of $\theta = 45^0$.

$$I_{X'X'} = I_{Y'Y'} = [I_{X'X'}]_{Square} - [I_{X'X'}]_{Circle} = [(4r)^4/12] - (\pi r^4/4)$$

$$= r^4(256 - 3\pi)/12 = 20.548\ r^4$$

and $\qquad I_{X'Y'} = 0$

Therefore, $\qquad I_{XX} = I_{YY} = [I_{X'X'} + I_{Y'Y'}]/2 + [I_{X'X'} - I_{Y'Y'}]\cos 2\theta/2 - I_{X'Y'}\sin 2\theta$

$$= [I_{X'X'} + I_{Y'Y'}]/2 + [I_{X'X'} - I_{Y'Y'}] \times (0)/2 - (0) \times (1)$$

$$= [20.548\ r^4 + 20.548\ r^4]/2$$

$$= 20.548\ r^4$$

Example 3.11

Compute moment of inertia of 10cm × 15cm rectangle about X-X axis, passing through corner A of the rectangle and inclined at $\alpha = \sin^{-1}(4/5)$ to the longer edge.

Solution

Method - 1

The rectangle is divided into *four triangles ABE, AEF, FCE and FCD*, as shown, such that FC is parallel to AE (X-axis).

$$\angle BAE = \angle DCF = \angle FDJ = \theta = 90 - \alpha$$

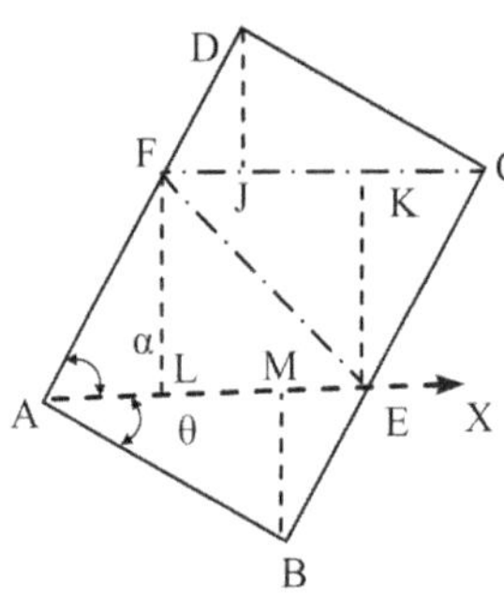

$$BM = AB \sin(90 - \alpha) = 6\text{cm} ; \quad FC = AE = AB/\cos(90 - \alpha) = 12.5\text{cm}$$

$$DJ = CD \sin(90 - \alpha) = 6\text{cm} ; \quad FD = DJ/\sin\alpha = 7.5\text{cm}$$

$$EK = FL = (AD - FD) \sin\alpha = 6\text{cm}$$

$$\begin{aligned}
I_{XX} &= (I_{XX})_{ABE} + (I_{XX})_{AEF} + (I_{XX})_{FCE} + (I_{XX})_{FCD} \\
&= (12.5 \times 6^3 / 12) + (12.5 \times 6^3 / 12) \\
&\quad + [(12.5 \times 6^3 / 36) + (12.5 \times 6 / 2) \times (2 \times 6 / 3)^2] \\
&\quad + [(12.5 \times 6^3 / 36) + (12.5 \times 6 / 2) \times \{6 + (6 / 3)\}^2] \\
&= 3600 \text{ cm}^4
\end{aligned}$$

Method - 2

This problem can also be solved by using *rotation of axes*. Considering AB as X′-axis and AD as Y′-axis,

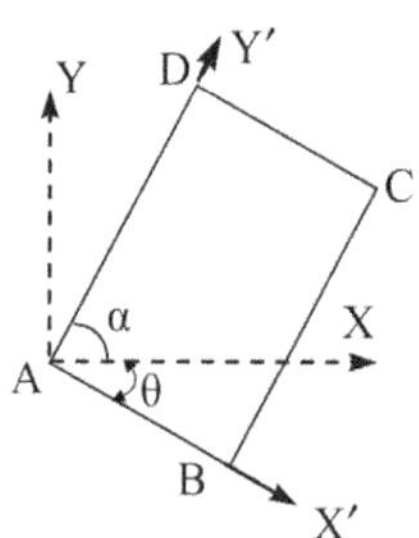

$$I_{X'X'} = 10 \times 15^3/3 = 11250 \text{ cm}^4$$

$$I_{Y'Y'} = 10^3 \times 15/3 = 5000 \text{ cm}^4$$

and $\quad I_{X'Y'} = 10^2 \times 15^2/4 = 5625 \text{ cm}^4$

The angle between X-axis and X′ axis, $\theta = 90 - \alpha$

So, $\quad \sin \theta = 3/5; \; \cos \theta = 4/5$

and $\quad \cos 2\theta = \cos^2\theta - \sin^2\theta = 7/25; \; \sin 2\theta = 2 \sin \theta \times \cos\theta = 24/25$

Using the relation for rotation of axes,

$$I_{XX} = [I_{X'X'} + I_{Y'Y'}]/2 + [I_{X'X'} - I_{Y'Y'}] \cos 2\theta / 2 - I_{X'Y'} \sin 2\theta$$

$$= [11250 + 5000]/2 + [11250 - 5000] \times (7/25)/2$$

$$- 5625 \times (24/25)$$

$$= 8125 + 875 - 5400 = 3600 \text{ cm}^4$$

Method - 3

This problem can also be solved by using another combination of triangles with triangle GCK deleted from triangle AKD and the remaining area added to triangle ABG

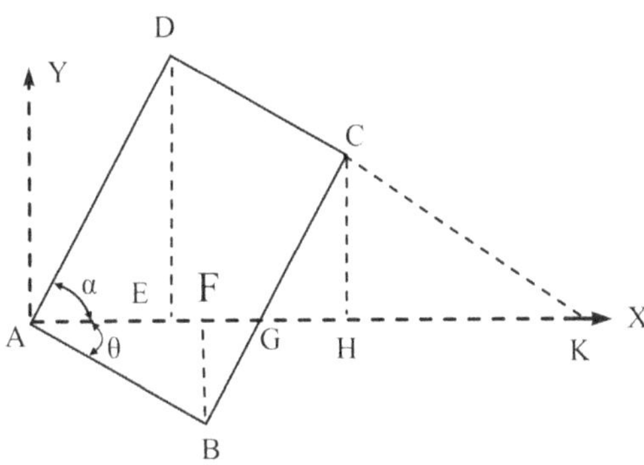

$\sin \alpha = \cos \theta = 4/5 \; ; \; \cos \alpha = \sin \theta = 3/5$

AG = AB / cos θ = 12.5 cm ; $\quad$ BF = AB sin θ = 6 cm

AK = AD/cos α = 25 cm $\quad$; $\quad$ GK = AK – AG = 12.5 cm

DE = AD sin α = 12 cm $\quad$; $\quad$ BG = AB tan θ = 7.5 cm

GC = BC – BG = 7.5 cm $\quad$; $\quad$ CH = GC sin α = 6 cm

$$I_{XX} = (I_{XX})_{ABG} + (I_{XX})_{AKD} - (I_{XX})_{GCK}$$

$$= [AG \times BF^3 + AK \times DE^3 - GK \times CH^3]/12$$

$$= [12.5 \times 6^3 + 25 \times 12^3 - 12.5 \times 6^3]/12 = 3600 \text{ cm}^4$$

3.2.7   RADIUS OF GYRATION

It is a function of moment of inertia and cross sectional area and is an important property of rotating components. Corresponding to the three moments of inertia, there are three radii of gyration. It has the dimensions of length. It can be interpreted as the distance 'k' at which the complete area of cross section 'A' is squeezed and kept as a thin rectangle, parallel to the axis such that there is no change in its moment of inertia.

Then, $I = A \times k^2$

or
$$k_{XX} = \sqrt{I_{XX}/A} \; ; \quad k_{YY} = \sqrt{I_{YY}/A} \; ; \quad k_{ZZ} = \sqrt{I_{ZZ}/A}$$

Radius of gyration is not defined for product of inertia

These values are given below for a few simple sections

(i)   Rectangular section of breadth 'b' and depth 'd', through its centroid :

$$k_{XX} = \sqrt{\frac{b \times d^3/12}{b \times d}} = \sqrt{d^2/12} = d/\sqrt{12}$$

$$k_{YY} = \sqrt{\frac{d \times b^3/12}{b \times d}} = \sqrt{b^2/12} = b/\sqrt{12}$$

$$k_{ZZ} = \sqrt{\frac{b \times d \times (b^2 + d^2)/12}{b \times d}} = \sqrt{(b^2 + d^2)/12}$$

(ii)   Circular cross section of diameter 'd', axis through diameter :

$$k_{XX} = k_{YY} = \sqrt{\frac{\pi d^4/64}{\pi d^2/4}} = \sqrt{d^2/16} = d/4$$

$$k_{ZZ} = \sqrt{[\pi d^4/32]/[\pi d^2/4]} = \sqrt{d^2/8} = d/\sqrt{8}$$

(iii)   Semi-circular cross section of diameter 'd', axis through diameter

$$k_{XX} = k_{YY} = \sqrt{\frac{\pi d^4/32}{\pi d^2/2}} = \sqrt{d^2/16} = d/4$$

$$k_{ZZ} = \sqrt{\frac{\pi d^4/16}{\pi d^2/2}} = \sqrt{d^2/8} = d/\sqrt{8}$$

(iv)   Hollow circle of outer diameter 'D' and inner diameter 'd', axis through its diameter :

$$k_{XX} = k_{YY} = \sqrt{\frac{\pi(D^4 - d^4)/64}{\pi(D^2 - d^2)/4}} = \sqrt{(D^2 + d^2)/16}$$

$$k_{ZZ} = \sqrt{\frac{\pi(D^4 - d^4)/32}{\pi(D^2 - d^2)/4}} = \sqrt{(D^2 + d^2)/8}$$

Example 3.12

Two semi-circular areas of radius 'a' are cut from a square area of side '4a', as shown. Find polar moment of inertia and polar radii of gyration through the centroid of the remaining area.

Solution

Because of symmetry about X and Y axes, centroid lies at the center of the square. Using suffix 'SQ' for square and 'SC' for semi-circle,

$$A = (4a) \times (4a) - 2\,[\,\pi \times a^2/2\,] = 12.86\ a^2$$

$$I_{XX} = [I_{XX}]_{SQ} - 2[I_{XX}]_{SC} = (4a)^4/12 - 2\{\,0.11a^4 + (\pi a^2/2)\,[2a - (4a/3\pi)]^2\,\}$$

$$= 13.31\ a^4$$

$$I_{YY} = [I_{YY}]_{SQ} - 2[I_{YY}]_{SC} = (4a)^4/12 - 2[\pi a^4/8] = 20.54\ a^4$$

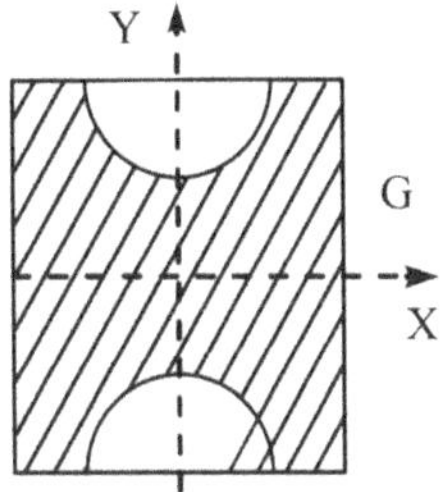

Polar M.I., $I_{ZZ} = I_{XX} + I_{YY} = 13.31\ a^4 + 20.54\ a^4 = 33.85\ a^4$

Radius of gyration about X-axis, $k_{XX} = \sqrt{I_{XX}/A} = \sqrt{13.31\ a^4/12.86\ a^2}$

$$= 1.017\ a$$

Radius of gyration about Y-axis, $k_{YY} = \sqrt{I_{YY}/A} = \sqrt{20.54\ a^4/12.86\ a^2}$

$$= 1.264\ a$$

Radius of gyration about Z-axis, $k_{ZZ} = \sqrt{I_{ZZ}/A} = \sqrt{33.85\ a^4/12.86\ a^2}$

$$= 1.622\ a$$

CHAPTER 4

MECHANICS OF RIGID BODIES

4.1   FORCE, REACTION AND MOMENT

Design of any component or structure is based on calculation of forces (or stresses) at various points of the component as well as calculation of resulting displacements (or deformations). As a first step, every component is assumed as a rigid body and the effect of applied loads at different points on the component is estimated, based on equations of static equilibrium. This approach is called 'Mechanics of rigid bodies'. We summarise here concepts of forces, reactions and equations of equilibrium.

Any force acting on a body has magnitude as well as direction and hence is a vector, identified by '**F**'. Thus, a force may be identified by its magnitude and the direction of force w.r.t. the chosen coordinate system. For example, a force **F** in X-Y plane is defined by its magnitude '**f**' (in units of Newtons) and its inclination '**θ**' with respect to the positive X-direction. It is more convenient to express each force by its vector components '**f cos θ** or F_X' along X-direction and '**f sin θ** or F_Y' along Y-direction as shown in Fig. 4.1. The effect of a force on a body is dependent only on these two quantities and is independent of the point of application of the force.

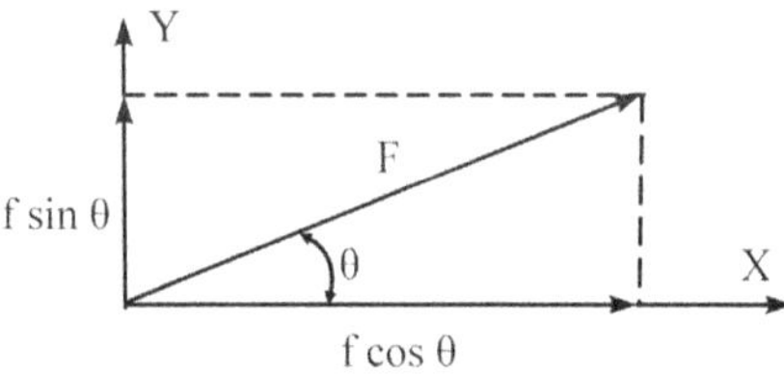

FIGURE 4.1 Force & its components in a plane

When forces are applied on a component, reactions '**R**' of appropriate magnitude and direction are generated at the points of support, as per Newton's 3rd law of motion, so as to balance applied forces. This produces a net force of 'zero' and keeps the component in static equilibrium. For example, a body of weight '**W**' produces an equal and opposite reaction '**R**' at the support location (Fig. 4.2a) and a concentrated load '**P**' acting at a distance 'a' from the left end of a simply supported beam of length 'L' produces reactions '**R₁**' and '**R₂**' of magnitudes 'P × (L – a)/L' and 'P × a/L' respectively at the two end supports (Fig. 4.2b). For the particular case of a = L/2, we get $P_1 = P_2 = P/2$

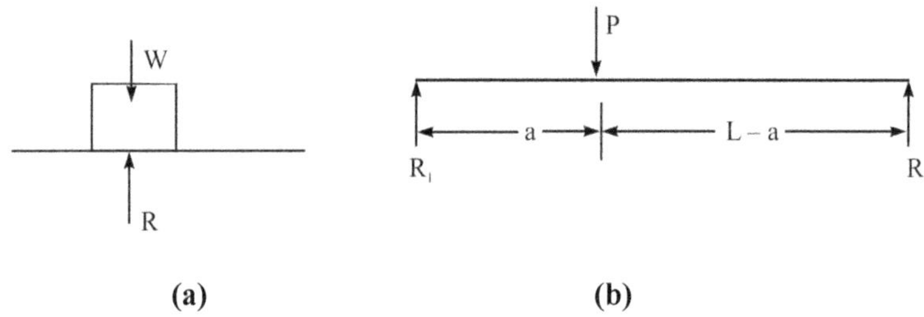

(a)   (b)

FIGURE 4.2 Reactions in a body

Force systems acting on a component (including the reactions developed) are broadly classified, based on their orientation as *co-planar forces* (forces lying in a single plane) and *spatial forces* (forces not confined to a single plane). They can also be classified as *concurrent forces* (lines of action of all forces meeting at a single point) and *non-concurrent forces* (lines of action of all forces not meeting at a single point). *Collinear forces* (forces acting along a single line of action, in positive direction and /or negative direction) form one particular category of co-planar forces. Thus, various combinations of these categories form different types of force systems as

1. Co-planar concurrent forces
2. Co-planar non-concurrent forces
3. Spatial concurrent forces
4. Spatial non-concurrent forces

The shape and size of a body have no effect in Mechanics of Rigid bodies. The equations of equilibrium are applied conveniently on a *free body diagram*, which is a simplified line sketch showing the position and direction of external forces and reactions at the support locations.

A force **F** applied at point A will produce a *moment* about B (tendency to rotate the body). The direction of moment **M** is defined as the direction of rotation of the body it produces at the point B, counter-clockwise (+ve) or clockwise (–ve). A pair of two non-collinear forces, equal in magnitude and opposite in direction, is called a ***couple***. Moment of a couple is the product of the force and the normal distance between the two forces.

A pair of equal and opposite forces **F** and − **F** applied at point B has no consequential effect on the force system. Then, force **F** at A is equivalent to applying force **F** at B and a clockwise moment $M_B = F \times d$ of the remaining couple. By the same logic, a single force **F** at A is equivalent to a force **F** and a counter-clockwise moment $M_C = F \times x$ about some other point C (Ref.Fig.4.3).

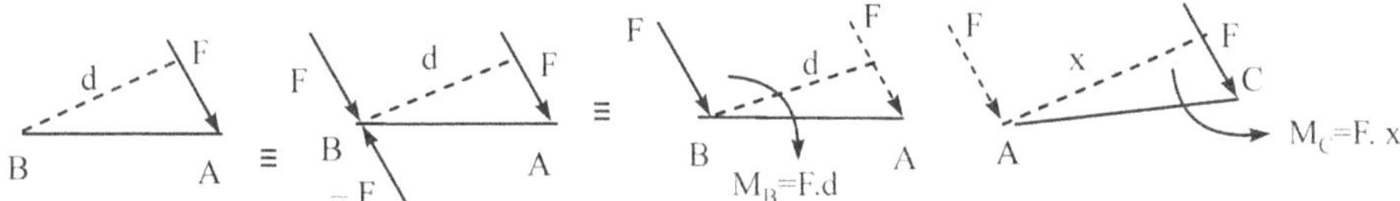

FIGURE 4.3 Equivalent forces at a point

The equations of static equilibrium are based on Newton's 1ˢᵗ law of motion, which states that "every body continues to be in its state of rest or of uniform motion unless compelled by an external force". It, therefore, means that a static body continues to be stationary when there is no net force (or moment) acting on it or the vector sum of applied forces ($\sum$**F**) is completely balanced by the vector sum of reactions ($\sum$**R**). It can be mathematically stated as $\sum F + \sum R = 0$

In addition, for non-concurrent forces, $\sum M_F + \sum M_R = \sum M = 0$

where, M_F is the moment due to the applied forces and M_R is the moment due to the reaction forces. Addition of force vectors follows parallelogram law or triangle law for every two vectors (different from algebraic sum, unless the forces are collinear).

4.2   RESOLUTION OF A FORCE

Resolution of a force into two components is carried out such that the two components produce the same effect on the body as produced by the applied force, as shown in Fig. 4.4

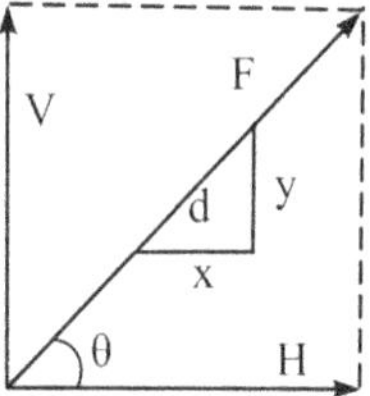

FIGURE 4.4 Resolution of a force

Resolution into two perpendicular components (usually along X and Y axes) is the most common application. Thus,

$$H = F \times \cos\theta \;;\; V = F \times \sin\theta \;;\; \tan\theta = V/H \;\text{ and }\; F^2 = H^2 + V^2 \qquad \ldots..(4.1)$$

If slope is defined by lengths x and y,

$$H/x = V/y = F/d \qquad \text{where,} \qquad d^2 = x^2 + y^2$$

4.3   RESULTANT OF FORCES

Parallelogram law : The *resultant of two concurrent forces*, meeting at a point, is the diagonal of the parallelogram whose sides represent the force vectors to some scale (Ref Fig. 4.5a). It can not be proved but can only be demonstrated by experiment.

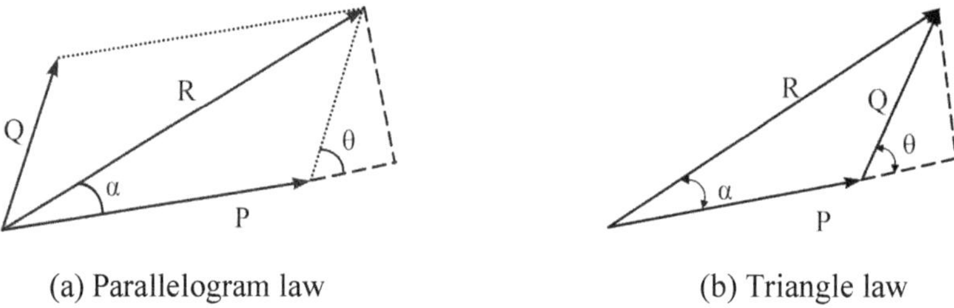

(a) Parallelogram law   (b) Triangle law

FIGURE 4.5 Resultant of two forces

From trigonometry, we know $R = \sqrt{(P + Q\cos\theta)^2 + (Q\sin\theta)^2}$

$$= \sqrt{P^2 + Q^2 + 2\,P\,Q\cos\theta} \qquad \ldots..(4.2)$$

and $\qquad \tan\alpha = Q \times \sin\theta \;/\;[\,P + Q \times \cos\theta\,]$ $\qquad \ldots..(4.3)$

Triangle law (Corollary of parallelogram law) **:** If two forces are represented to some scale by their vectors placed tip-to-tail, their resultant is the vector directed from the tail of the first vector to the tip of the second vector. (Ref Fig. 4.5 b)

The resultant of two collinear vectors (angle between them is 0^0 or 180^0) is their algebraic sum, in the direction of the larger vector.

Resultant of three or more coplanar concurrent forces

1. *Graphical method:* Considering two force vectors at a time, resultant force can be obtained using parallelogram law. This operation can be repeated with the resultant of the earlier forces and one more force. This method is explained for a set of three forces F_1, F_2 and F_3 in Fig.4.6a. Alternatively, a **force polygon** can be constructed using triangle law successively, as shown in Fig.4.6 (b).

In this example, resultant R_1 of forces F_1 and F_2 is first obtained by parallelogram law (in Fig.4.6a) and by triangle law (in Fig.4.6b). Then, the resultant of forces F_1, F_2 and F_3 is obtained as the resultant of forces R_1 and F_3 by parallelogram law or triangle law. This procedure can be repeated for any number of forces acting at a point.

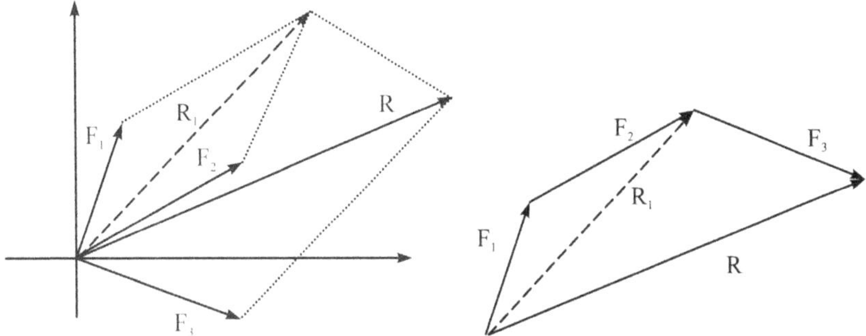

(a) Resultant by Parallelogram Law   (b) Resultant by Triangle Law

FIGURE 4.6 Resultant of multiple forces

2. ***Analytical method or Components of forces method****:* It is not convenient to deal with ***a system of coplanar forces*** by repeated application of parallelogram law or triangle law. A more convenient alternative is to express these vector equations as algebraic equations, in terms of components of these vectors along mutually perpendicular axes. Since X-components of all forces are collinear, their vector sum is equal to their algebraic sum. Similarly, Y-components of all forces are collinear and hence their vector sum is their algebraic sum. Thus, the vector sum of forces can be written as a set of algebraic equations, of the form

$$R_X = \sum H ; \qquad R_Y = \sum V$$

$$R = \sqrt{\left(R_X^2 + R_Y^2\right)} = \sqrt{\left(\sum H\right)^2 + \left(\sum V\right)^2} \qquad .....(4.4)$$

Inclination θ of the resultant, with the X-axis, is given by

$$\tan \theta = R_Y / R_X \qquad .....(4.5)$$

4.4 EQUATIONS OF EQUILIBRIUM

In general, ***for a system of spatial, non-concurrent forces*** acting on a body and keeping it in equilibrium, we can write Newton's 1st law of motion as equations of equilibrium so that the body has no net force or moment acting on it and remains static. Thus,

$$\sum F_X = 0 \quad ; \quad \sum F_Y = 0 \qquad \text{and} \quad \sum F_Z = 0$$
$$\text{Similarly,} \quad \sum M_X = 0 \quad ; \quad \sum M_Y = 0 \qquad \text{and} \quad \sum M_Z = 0 \qquad .....(4.6)$$

All these equations may not be relevant for each force system. For coplanar force systems in X-Y plane, $\sum F_Z = 0$; $\sum M_X = 0$ and $\sum M_Y = 0$ are not relevant. For concurrent force systems, which produce no moments, $\sum M_X = 0$; $\sum M_Y = 0$ and $\sum M_Z = 0$ are not relevant.

If the system of forces acting at a point is in equilibrium, then the resultant is zero and, hence, the force polygon is a closed polygon. If three forces acting at a point (including reactions, if any) are in equilibrium, then their force polygon is a closed triangle (Ref.Fig.4.7). Sine rule of triangle is applicable to the lengths of the triangle (which represent the three forces to the same scale). Sine rule applied to a force triangle is called Lami's theorem.

Lami's theorem *or* ***Sine rule*** - If three forces P, Q and R acting at a point are in equilibrium, then according to sine rule of triangle,

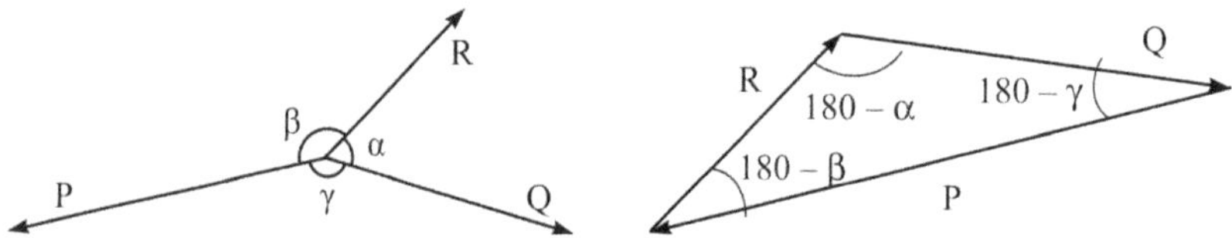

FIGURE 4.7 Application of Lami's theorem

$$P / \sin \alpha = Q / \sin \beta = R / \sin \gamma = R / \sin [180 - (\alpha + \beta)] = R / \sin (\alpha + \beta)$$
$$\text{or } P / \sin (180 - \alpha) = Q / \sin (180 - \beta) = R / \sin (180 - \gamma) \qquad ......(4.7)$$

This idea can also be used to find ***equilibrant*** of any two concurrent forces acting on a body, which is defined as the force required to keep the body in equilibrium. It is equal in magnitude and opposite in direction to the resultant of the two forces acting at any point on a body.

The following rules of force systems can be logically understood, from the basic concepts

- If three forces, acting on a body, are keeping the body in equilibrium, then the three forces are either concurrent or parallel

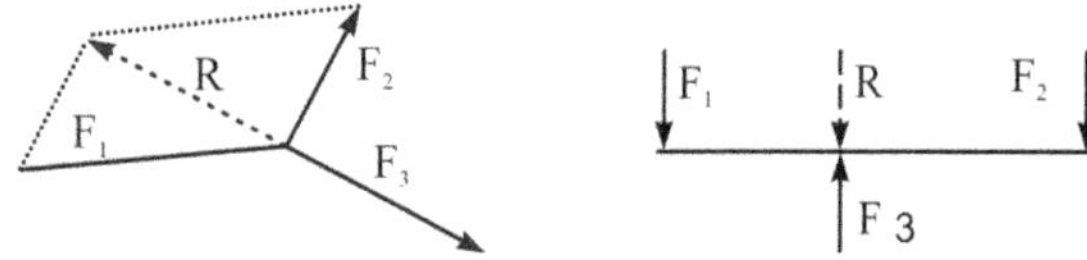

The two situations are obvious from the figures shown here. The three forces can remain in equilibrium only if the resultant (R) of any two forces (F_1 and F_2) is equal and opposite to the third force (F_3).

- In a system of 'n' concurrent forces acting on a body, every force is the equilibrant of the remaining (n – 1) forces.

4.5   DETERMINATE AND INDETERMINATE STRUCTURES

If a structure has as many unknown reactions as the number of relevant equations of equilibrium, all the support reactions can be evaluated from the equations of static equilibrium. Such a structure is called a *'determinate structure'*. A structure, in which the number of reactions is more than the number of equations of equilibrium, is called an *'indeterminate structure'*. Such a structure requires additional conditions of known displacements 'δ' and slopes 'θ', also called **compatibility equations** or kinematic conditions, to evaluate the reactions. A few examples of both these structures are given in Fig.4.8 with the unknown reactions in table, assuming the structures in X-Y plane.

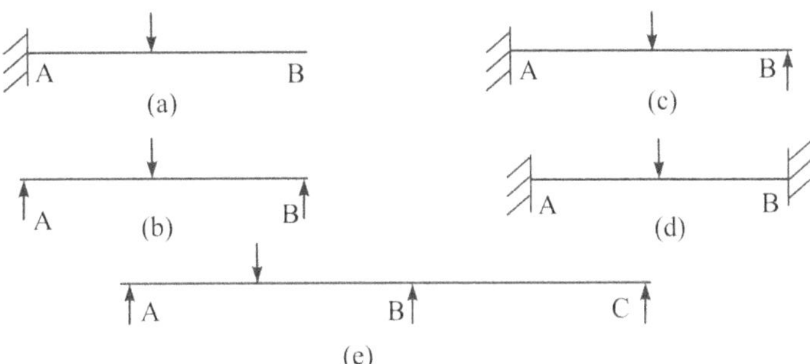

FIGURE 4.8 Determinate and indeterminate structures

In all these cases, equations of static equilibrium are only two i.e., $\sum F_Y = 0$, $\sum M_Z = 0$, in the absence of applied horizontal loads.

Case	Unknown reactions	Number of		Type of structure	Displacement end conditions
		equations	unknowns		
a	R_A, M_A	2	2	Determinate	-
b	R_A, R_B	2	2	Determinate	-
c	R_A, M_A, R_B	2	3	Indeterminate	$\delta_A=0$, $\delta_B=0$, $\theta_A=0$
d	R_A, M_A, R_B, M_B	2	4	Indeterminate	$\delta_A=0$, $\delta_B=0$, $\theta_A=0$, $\theta_B=0$
e	R_A, R_B, R_C	2	3	Indeterminate	$\delta_A=0$, $\delta_B=0$, $\delta_C=0$

A few problems of determinate structures are included here to understand '**mechanics of rigid bodies**' better. Books on 'Engineering Mechanics' or 'Mechanics of solids' can be referred, for detailed solutions of indeterminate structures.

Example 4.1

An electric light fixture weighing 20N hangs from a point C by two strings AC and BC, inclined at 30^0 and 45^0 to the vertical respectively. Determine forces in the strings AC and BC.

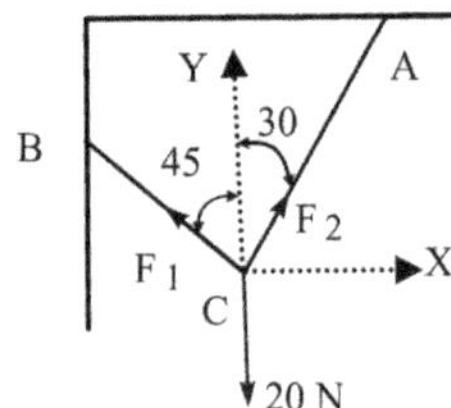

Solution

$$\Sigma F_X = 0 \Rightarrow -F_1 \times \sin 45 + F_2 \times \sin 30 = 0 \quad \text{or} \quad F_1 = \sqrt{2}\ F_2$$
$$\Sigma F_Y = 0 \Rightarrow F_1 \times \cos 45 + F_2 \times \cos 30 = 20 \quad \text{or} \quad F_2\,(2 + \sqrt{3}\,) = 2 \times 20$$
$$F_2 = 10.354\ \text{N} \quad \text{and} \quad F_1 = 14.641\ \text{N}$$

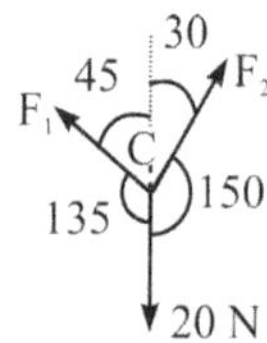

Alternatively, using Lami's theorem for equilibrium at C, on the free body diagram

$$F_1 / \sin 150 = F_2 / \sin 135 = 20 / \sin 75$$

or $\quad F_1 = 20 \times \sin 150 / \sin 75 = 14.641\ \text{N}$

and $\quad F_2 = 20 \times \sin 135 / \sin 75 = 10.354\ \text{N}$

Example 4.2

Forces 10 N, 20 N, 30 N, 40 N and 50 N are acting on one of the angular points of a regular hexagon, towards the other five angular points taken in order. Find the direction and magnitude of the resultant.

Solution

Many possible interpretations with 10 N force from any one side of hexagon and other forces in clockwise or counter-clockwise sequence are possible. For the system shown in figure, resolving forces along X and Y directions,

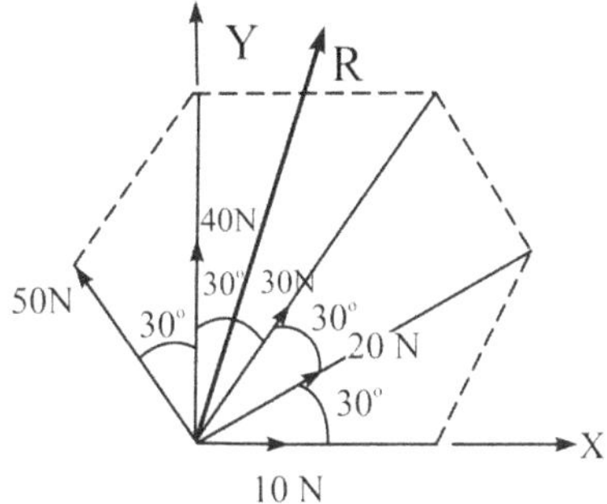

$$R_X = \sum F_X = 10 \times \cos 0 + 20 \times \cos 30 + 30 \times \cos 60 + 40 \times \cos 90 + 50 \times \cos 120$$

$$= 17.32N$$

$$R_Y = \sum F_Y = 10 \times \sin 0 + 20 \times \sin 30 + 30 \times \sin 60 + 40 \times \sin 90 + 50 \times \sin 120$$

$$= 119.28 \text{ N}$$

Resultant, $R = \sqrt{R_X^2 + R_Y^2} = 120.53$ N

Angle between R and X, $\theta = \tan^{-1}(R_Y/R_X) = 81^0 44'$

Example 4.3

A derrick crane carries a load of 10 kN as shown. If AB, BC and CA are 4m, 4m and 2m respectively, calculate forces in BC and CA.

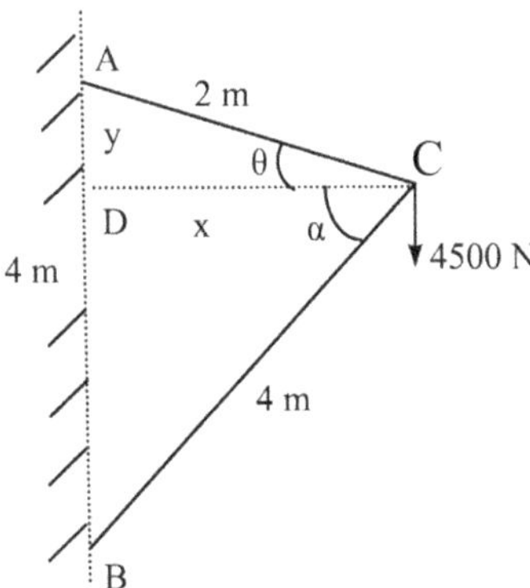

Solution

Let $CD = x$ and $AD = y$

Then, $x^2 + y^2 = 2^2$ and $x^2 + (4 - y)^2 = 4^2$

or $y = 0.5$ m   and $x = 1.9365$ m

Alternatively,

Area of $ABC = \sqrt{s(s-a)(s-b)(s-c)} = (4 \times)/2 = 2x$

where, $s = (a + b + c)/2 = (2 + 4 + 4)/2 = 5m$

Then, $\sqrt{5(5-4)(5-4)(5-2)} = 2x$

or $x = \sqrt{15/2} = 1.9365$ m ; $y = \sqrt{\left[2^2 - 1.9365^2\right]} = 0.5$ m

Also, $\sin\theta = y/2 = 1/4$ or $\theta = 14.48^0$

and $\sin\alpha = (4 - y)/4 = 7/8$ or $\alpha = 61.05^0$

Using Lami's theorem,

$$10/\text{Sin}(180 - \theta - \alpha) = F_{BC}/\text{Sin}(90 + \theta) = F_{AC}/\text{Sin}(90 + \alpha)$$

⇨ $F_{BC} = -10$ kN (Compressive)   and $F_{AC} = 5$ kN (tensile)

Example 4.4

A rope supported at A and B carries a load of W at D and 25 kN at C, as shown.
Find the value of W so that CD remains horizontal.

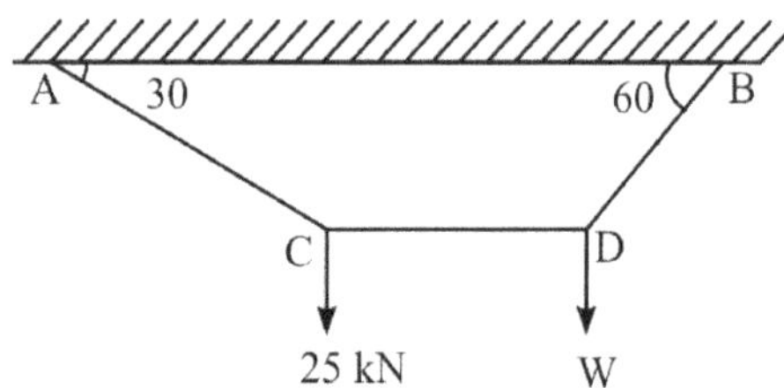

Solution

Using Lami's theorem at C,

$$25/\sin 150 = F_{AC}/\sin 90 = F_{CD}/\sin 120   \text{or}   F_{CD} = 43.3 \text{ kN}$$

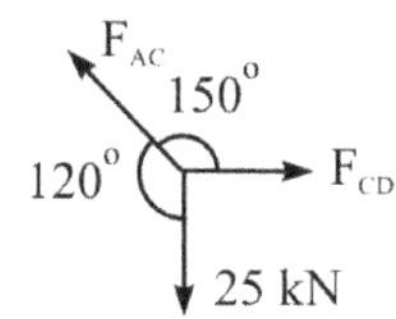

Using Lami's theorem at D,

$$W / \sin 120 = F_{BD} / \sin 90 = F_{CD} / \sin 150 \quad \text{or} \quad W = 75 \text{ kN}$$

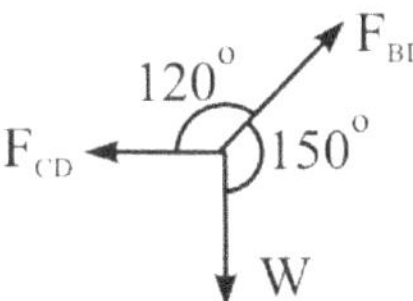

Example 4.5

Two smooth spheres, each of radius 10 cm and weight 100 N, rest in a 30 cm wide horizontal channel having vertical walls, as shown. Find the reactions on the wall and the floor, at the points of contact.

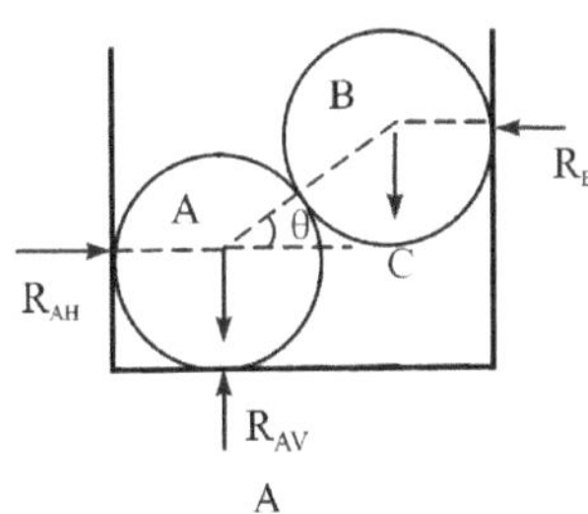

Solution

$$\cos \theta = AC/AB = (30 - 10 - 10) / 20 = 10 / 20 \quad \text{or} \quad \theta = 60$$

Using Lami's theorem at center of B, on the free body diagram

$$R_B / \sin (90 + \theta) = R_{AB} / \sin 90 = 100 / \sin (180 - \theta)$$

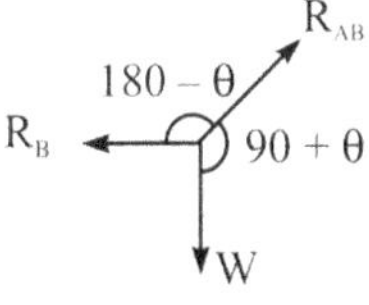

or $R_B = 100 \times \cos \theta / \sin \theta = 57.75$ N

$R_{AB} = 100/\sin \theta = 115.5$ N

Considering equilibrium of sphere A,

$$R_{AH} = R_{AB} \times \cos \theta = 115.5 \times \cos 60 = 57.75 \text{ N}$$

$$R_{AV} = W + R_{AB} \times \sin \theta$$

$$= 100 + 115.5 \times \sin 60 = 200 \text{ N}$$

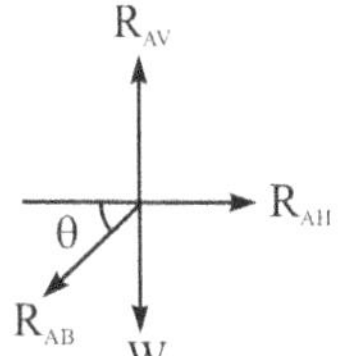

Example 4.6

A vertical bar AB of length 50cm, fixed at both ends, is subjected to downward axial load P of 120 N at a point C, 20 cm from the top end. (a) Calculate the reactions at the two ends. (b) What is the effect of change in area of cross section or modulus of elasticity between the two parts AC and CB.

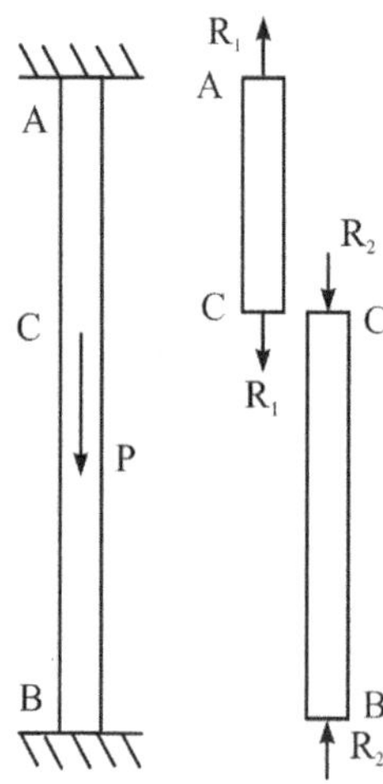

Solution

The elongation of a bar due to axial load P is given by

$$\delta = \varepsilon \times L = (\sigma/E) \times L = \sigma \times (L/E) = (P/A) \times (L/E) = (P \times L)/(A \times E)$$

Case - a

Considering free body diagrams of portions AC and CB,

Let R_1 be the reaction at end A and R_2 be the reaction at end B

Then, equation of equilibrium, $\sum F_Y = R_1 + R_2 - P = 0$

$\Rightarrow \qquad R_1 + R_2 = P$ .....(4.8)

The portion AC of the bar is subjected to axial tensile force of R_1 and the portion CB is subjected to axial compressive force of R_2. The two unknown reactions R_1 and R_2 can not be evaluated from the single equation of equilibrium, unless the kinematic condition that the length of the bar remains unchanged is used.

i.e., $\delta_{AC} + \delta_{CB} = 0$, taking elongation as +ve and compression as –ve.

or $\quad (R_1 \times L_{AC})/(A_{AC} \times E_{AC}) - (R_2 \times L_{CB})/(A_{CB} \times E_{CB}) = 0$ .....(4.9)

If area of cross section (A) and modulus of elasticity (E) are same for the two parts AC and CB, equation 4.9 reduces to

$$R_1 \times L_{AC} = R_2 \times L_{CB} \qquad\qquad .....(4.10)$$

From equations 4.8 and 4.10, $R_1 = P / [1 + (L_{AC}/L_{CB})] = 72N$

and $\qquad\qquad\qquad\qquad\qquad R_2 = P / [1 + (L_{CB}/L_{AC})] = 48N$

In this example, the values of area of cross section (A) and modulus of elasticity (E), being equal for the two parts, are redundant in the evaluation of reactions.

Case - b

If the two portions AC and CB are of different areas of cross sections and different modulii of elasticity, then the end reactions R_1 and R_2 depend on these values.

Let $A_{AC} = 8$ sq.cm; $A_{CB} = 12$ sq.cm ; $E_{AC} = 200$ GPa and $E_{CB} = 70$ GPa

Then, solving equations 4.8 and 4.9, $R_1 = 88.9$ N   and $R_2 = 31.1$ N

Example 4.7

Copper bar AB and aluminium bar CD, each of 1cm diameter, are fixed rigidly as shown, with initial gap of 5mm between them. If the temperature of both the bars is increased by 250^0C, calculate the reactions, if any. Take coefficient of linear thermal expansion (α) and modulus of elasticity (E) as $17 \times 10^{-6} /^0$C and 110 GPa for copper and $23 \times 10^{-6} /^0$C and 70 GPa for aluminium respectively.

Solution

Reactions are developed at the two ends only when thermal expansion of the bars is prevented. i.e., if the total expansion of the two bars exceeds the initial gap provided. Let δ_{AB} be the thermal expansion of bar AB and δ_{CD} be the thermal expansion of bar CD. If $\delta_{AB} + \delta_{CD} > 5$mm, then a force (P) is developed at the junction of the two bars, constraining the total expansion to 5mm. Let u_{AB} be the contraction of bar AB and u_{CD} the contraction of bar CD, due to the force at the junction, given by P/EA. Then,

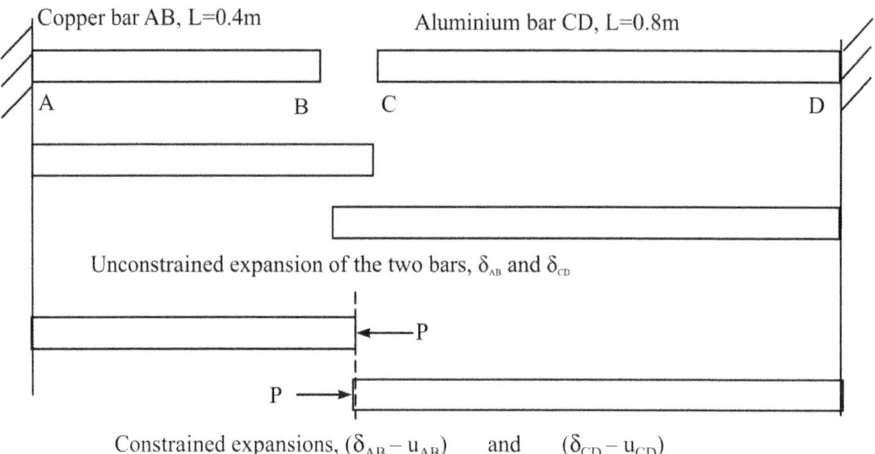

Net expansion of bar AB = $\delta_{AB} - u_{AB}$

and   Net expansion of bar CD = $\delta_{CD} - u_{CD}$,

such that $(\delta_{AB} - u_{AB}) + (\delta_{CD} - u_{CD}) = 5$ mm

Here,

$$\delta_{AB} = L_{AB} \times \alpha_{AB} \times \Delta T = 400 \times 17 \times 10^{-6} \times 250 = 1.7 \text{ mm}$$

$$u_{AB} = P \times L_{AB} / (E_{AB} \times A) = P \times 400 / (110 \times 10^3 \times \pi \times 10^2 / 4) = 4.628 \times 10^{-5} P$$

$$\delta_{CD} = L_{CD} \times \alpha_{CD} \times \Delta T = 800 \times 23 \times 10^{-6} \times 250 = 4.6 \text{ mm}$$

$$u_{CD} = P \times L_{CD} / (E_{CD} \times A) = P \times 800 / (70 \times 10^3 \times \pi \times 10^2 / 4) = 14.54 \times 10^{-5} P$$

Therefore,

$$1.7 - 4.628 \times 10^{-5} P + 4.6 - 14.54 \times 10^{-5} P = 5$$

$$P = (1.7 + 4.6 - 5) / (14.54 \times 10^{-5} + 4.628 \times 10^{-5})$$

$$= 6780 \text{ N}$$

Stress in each bar, $\sigma = P/A = 86.3$ MPa

Net expansion of copper bar AB $= \delta_{AB} - u_{AB} = 1.7 - 6780 \times (4.628 \times 10^{-5})$
$$= 1.387 \text{ mm}$$

Net expansion of aluminium bar CD $= \delta_{CD} - u_{CD} = 4.6 - 6780 \times (14.54 \times 10^{-5})$
$$= 3.613 \text{ mm}$$

4.6   DYNAMIC ANALYSIS

Dynamics is a special branch of mechanics where inertia of accelerating masses must be considered in the force-deflection relationships. In order to describe motion of the mass system, a component with distributed mass is approximated by a finite number of mass points.

Every structure is associated with certain frequencies and mode shapes of free vibration (without continuous application of load), based on the distribution of mass and stiffness in the structure. Any time-dependent external load acting on the structure, whose frequency matches with the natural frequencies of the structure, causes resonance and produces large displacements leading to failure of the structure. Calculation of natural frequencies and mode shapes is therefore very important.

4.6.1   D'ALEMBERT'S PRINCIPLE

Consider i^{th} mass m_i of a system of connected rigid bodies and the force components F_j $(j = 1,2,..6)$ acting upon it in three-dimensional space. If the mass m_i is in equilibrium at rest, then $\Sigma F_i = 0$.

If mass m_j is not in equilibrium, it will accelerate in accordance with Newton's second law i.e., $F_j = m_i \ddot{u}_j$

The force $(- m_i \times \ddot{u}_j)$ is called the reversed effective force or inertia force. According to D'Alembert's principle, the net external force and the inertia force together keep the body in a state of *'fictitious equilibrium'*.

i.e., $\Sigma \left(F_j - m \ddot{u}_j \right) = 0$

If the displacement of the mass m_i is represented by δu_j $(j = 1,2,..6)$, then the virtual work done by these force components on the mass m_i in equilibrium is given by $\delta W_i = \Sigma F_j \times \delta u_j = 0$

D'Alembert's principle rewritten in the form,

$$\delta W_i = \Sigma F_j \times \delta u_j - \Sigma (m \ddot{u}_j) \times \delta u_j = 0 \qquad \qquad .....(4.11)$$

is a statement of **virtual work for a system in motion**.

4.6.2 NATURAL FREQUENCIES AND MODE SHAPES

For a simple spring of stiffness 'k' and a lumped mass 'm' under steady state undamped condition of oscillation without external force, the force equilibrium condition of the system is given by

$$k \times u(t) + m \times \ddot{u}(t) = 0, \qquad \qquad .....(4.12)$$

where, $F_i = -k \times u(t)$ is the reactive elastic force applied to the mass

Displacement in vibration is assumed to be a simple harmonic motion and can be represented by a sinusoidal function of time as

$$u(t) = u \sin \omega t \qquad \qquad .....(4.13)$$

where, ω is the frequency of vibration in radians/sec

It is more often expressed in 'f' cycles/sec or Hertz (Hz) where $\omega = 2\pi f$

Then, velocity $\qquad \qquad \dot{u}(t) = -\omega u \cos \omega t$

and acceleration $\qquad \quad \ddot{u}(t) = -\omega^2 u \sin \omega t = -\omega^2 u(t)$

$$\therefore \qquad \qquad k \times u(t) + m \times \ddot{u}(t) = (k - \omega^2 m) \times u(t) = 0$$

In general, for a system with 'n' degrees of freedom, stiffness 'k' and mass 'm' are represented by stiffness matrix [K] and mass matrix [M] respectively.

Then, $\qquad \qquad ([K] - \omega^2 [M]) \{u\} = \{0\} \qquad \qquad .....(4.14)$

or $\qquad \qquad ([M]^{-1}[K] - \omega^2 [I]) \{u\} = \{0\}$

Here, [M] is the mass matrix of the entire structure and is of the same order, say $n \times n$, as the stiffness matrix [K].

This is a typical eigenvalue problem, with ω^2 as eigenvalues and $\{u\}$ as eigenvectors. A structure with 'n' degrees of freedom (or displacements defining the structure) will therefore have 'n' eigenvalues and 'n' eigenvectors. Some eigenvalues may be repeated and some eigenvalues may be complex, in pairs.

The equation can be represented in the standard form, $[A]\{x\}_i = \lambda_i \{x\}_i$. In dynamic analysis, ω_i indicates i^{th} natural frequency and $\{x\}_i$ indicates i^{th} natural mode of vibration. A natural mode is a ___*qualitative*___ plot of nodal displacements. In every natural mode of vibration, all the points on the component will reach their maximum values at the same time and will pass through zero displacements at the same time. Thus, in a particular mode, all the points of a component will vibrate with the same frequency and their relative displacements are indicated by the components of the corresponding eigenvector. These relative (or proportional) displacements at different points on structure remain same at every time instant for undamped free vibration (Ref. Fig. 4.9). Hence, without loss of generality, $\{u(t)\}$ can be written as $\{u\}$.

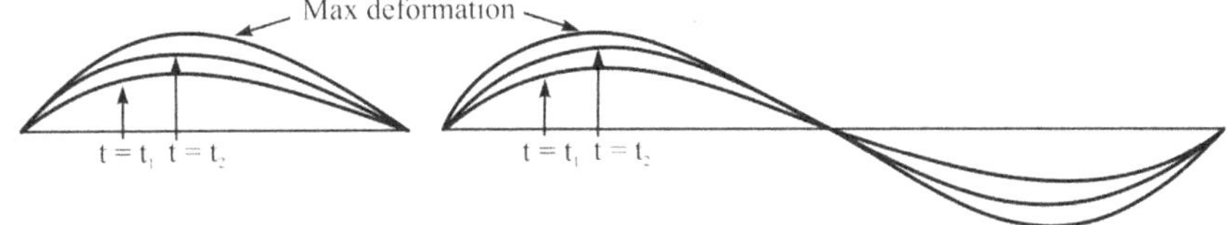

FIGURE 4.9 Mode shapes

Since $\{u\} = \{0\}$ forms a trivial solution, the homogeneous system of equations $([A] - \lambda[I])\{u\} = \{0\}$ gives a non-trivial solution only when

$$([A] - \lambda[I]) = \{0\},$$

which implies $\text{Det}([A] - \lambda[I]) = 0$ .....(4.15)

This expression, called ***characteristic equation***, results in n^{th} order polynomial in λ and will therefore have n roots. For each λ_i, the corresponding eigenvector $\{u\}_i$ can be obtained from the n homogeneous equations represented by $([K] - \lambda[M])\{u\} = \{0\}$. The mode shape represented by $\{u(t)\}$ gives relatives values of displacements in various degrees of freedom. It can also be represented as

$$[A][X] = [X][\Lambda]   .....(4.16)$$

where, $[A] = [M]^{-1}[K]$

$[X]$ is called the ***modal matrix***, whose i^{th} column represents i^{th} eigenvector $\{x\}_i$ and $[\Lambda]$ is called the ***spectral matrix*** with i^{th} diagonal element representing one eigenvalue, corresponding to i^{th} eigenvector and off-diagonal elements are all equal to zero.

CHAPTER 5

ENERGY PRINCIPLES

5.1 STRAIN ENERGY

Whenever a component is subjected to some external force(s), the component retains energy which is a function of stress and strain developed in the component. This is called **strain energy (U)**. This energy per unit volume is called **strain energy density** (U^*). Graphically, the area under the stress-strain curve, upto the maximum stress developed in the component due to the applied load is the strain energy density.

Thus, for a gradually applied load,

$$U = \int (1/2) \times (\sigma \times \varepsilon)\, dV \qquad \text{(Ref Fig.5.1 a)}$$

and for a suddenly applied load,

$$U = \int (\sigma \times \varepsilon)\, dV \qquad \text{(Ref Fig.5.1 b)}$$

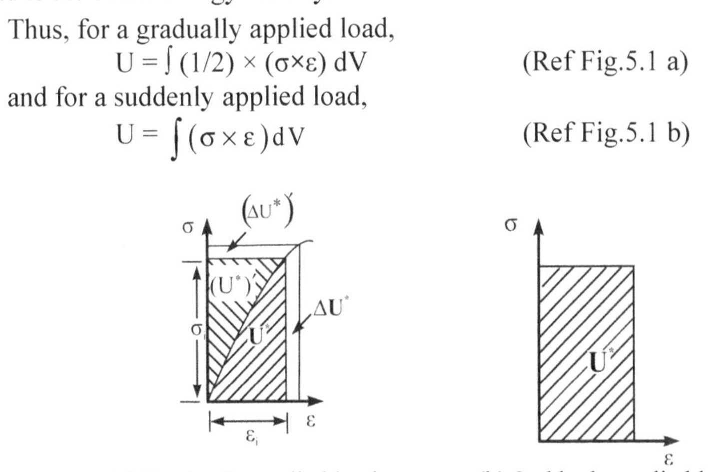

(a) Gradually applied load   (b) Suddenly applied load

FIGURE 5.1 Strain energy density

The strain energy density, corresponding to the stress at limit of proportionality, is called **modulus of resilience.** It is a measure of the maximum energy that a component can withstand without any plastic deformation after unloading. The strain energy stored in a component upto the breaking point is called **toughness.** It is the measure of maximum energy that a component can store before failure.

For a linear elastic bar on which, at any section 'x' along its length, internal loads are axial force F_X, bending moment M_y, bending moment M_z, shear force F_Y, shear force F_z and torque T. If we assume only one axial stress is present in the bar, strain energy due to axial stress is given by,

$$U = (1/2) \times \int (\sigma \times \varepsilon)\, dV = (1/2) \times \int (\sigma^2/E)\, dV$$

where, $E = \sigma/\varepsilon$ is the modulus of elasticity

Axial stress may result from direct axial load or bending moment. Substituting $\sigma_X = F_X/A$ due to axial load, $\sigma_X = M_Y \times z/I_{YY}$ due to bending moment M_Y and $\sigma_X = M_Z \times y/I_{ZZ}$ due to bending moment M_Z, strain energy can be written as

$$\begin{aligned}
U_{axial} &= (1/2) \times \int \{F_X^2/(E \times A^2)\}\, dV + (1/2) \times \int \{M_Y^2 \times z^2/(E \times I_{YY}^2)\}\, dV \\
&\qquad + (1/2) \times \int \{M_Z^2 \times y^2/(E \times I_{ZZ}^2)\}\, dV \\
&= (1/2) \times [\,\int\!\int \{F_X^2 / (E \times A^2)\}\, (dA \times dx) + \int\!\int \{M_Y^2 \times z^2 / (E \times I_{YY}^2)\}\, (dA \times dx) \\
&\qquad + \int\!\int \{M_Z^2 \times y^2 / (E \times I_{ZZ}^2)\}\, (dA \times dx)\,] \\
&= (1/2) \times [\,\int \{\{F_X^2/(E \times A^2)\}\, dx \times \int dA + \int \{\{M_Y^2/(E \times I_{YY}^2)\}\, dx \times \int z^2\, dA \\
&\qquad + \int \{\{M_Z^2/(E \times I_{ZZ}^2)\}\, dx \times \int y^2\, dA\,] \\
&= (1/2) \times [\,\int\{F_X^2/(E \times A)\}\, dx + \int\{M_Y^2/(E \times I_{YY})\}\, dx + \int\{M_Z^2/(E \times I_{ZZ})\}\, dx\,] \\
&\qquad\qquad\qquad\qquad\qquad\qquad\qquad\qquad\qquad\qquad\qquad\qquad\qquad .....(5.1)
\end{aligned}$$

$$\text{since } A = \int dA \,,\ I_{YY} = \int z^2\, dA \quad \text{and} \quad I_{ZZ} = \int y^2\, dA$$

Strain energy due to transverse shear is given by

$$\begin{aligned}
U_{shear} &= (1/2) \times [\,\int (\tau_{YZ} \times \gamma_{YZ})\, dV + \int (\tau_{ZX} \times \gamma_{ZX})\, dV] \\
&= (1/2) \times [\int (\tau_{YZ}^2/G)\, dV + \int (\tau_{YZ}^2/G)\, dV] \\
&= (1/2) \times [\,\int \{(F_Y/A)^2/G\}\, (dA \times dx) + \int \{(F_Z/A)^2/G\}\, (dA \times dx)\,] \\
&= (1/2) \times [\,\int \{F_Y^2/(G \times A)\}\, dx + [\,\int\{F_Z^2/(G \times A)\}\, dx\,] \qquad\qquad .....(5.2)
\end{aligned}$$

$$\text{where, } G \text{ is the shear modulus}$$

Similarly, strain energy due to torque is given by $U_{torsion} = (1/2)\int \{T^2/(G \times J)\}\, dx$

If a structure consists of 'm' structural bar elements, strain energy due to the three components of force and three components of moments is given by

$$U = U_{axial} + U_{shear} + U_{torsion}$$

i.e., $$U = (1/2) \times \sum \int [\,\{F_X^2/(E \times A)\} + \{F_Y^2/(G \times A)\} + \{F_Z^2/(G \times A)\} + \{T^2/(G \times J)\}$$
$$+ \{M_Y^2/(E \times I_{YY})\} + \{M_Z^2/(E \times I_{ZZ})\}\,]\, dx \quad \text{for } I = 1,m \ ......(5.3)$$

5.2   COMPLEMENTARY STRAIN ENERGY

For a particular stress component (σ_i) and corresponding strain component (ε_i) *strain energy* or energy stored in the structure per unit volume is given by, $U^* = (\sigma_i \times \varepsilon_i)$. It is represented by the area below the stress-strain curve shown in Fig 5.1. Total strain energy stored in the volume of the structure is given by $U = U^* \times V$, where V is the volume of the structure. The area above this curve $(U^*)'$ is called *complementary strain energy* per unit volume.

Similarly, *work* done by the gradually applied force (P_i) when it produces displacement (u_i) is given by $W_i = (1/2) \sum (P_i \times u_i)$ and is represented by the area below the load-displacement curve shown in Fig. 5.2. The area above this curve W_i' is called *complementary work*.

It can be seen that the work done is equal to the strain energy stored in the volume of the structure for gradually applied load,

$$W_i = \tfrac{1}{2} P_i \times u_i = \tfrac{1}{2} (\sigma_i \times A) \times (\varepsilon_i \times L) = \tfrac{1}{2} (\sigma_i \times \varepsilon_i) \times (A \times L) = U^* \times V = U$$

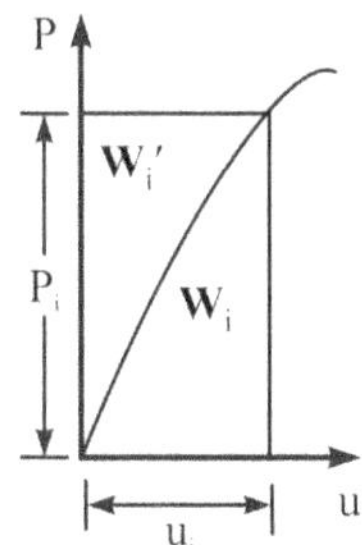

FIGURE 5.2 Work done

Similarly, complementary work (W′) is equal to the complementary strain energy U′

For linear elastic material (linear stress-strain relation),

$$U = U' \quad \text{and} \quad W = W'$$

5.3   PRINCIPLE OF VIRTUAL DISPLACEMENTS

If the structure is given an infinitesimal (or virtual) displacement (δu_i) at the same load, the change in work done (or virtual work) is given by $\delta W_i = P_i \times \delta u_i$ and the corresponding change in strain energy (virtual strain energy) is given by $\delta U^* = \sigma_i \times \delta \varepsilon_i$. If we consider all the components of forces acting on the structure and corresponding displacements, the total virtual work done is obtained by summation as

$$\delta W = \sum (P_i \times \delta u_i) = [\delta u] \{P\} \qquad \qquad .....(5.4)$$

where, $[\delta u]$ represents row vector of all displacement components

and $\{P\}$ represents column vector of all force components

Similarly, corresponding variation in strain energy density can be expressed as

$$\delta U^* = \sum(\sigma_i \times \delta\varepsilon_i) = [\delta\varepsilon]\,\{\sigma\} \quad \text{and} \quad \delta U = \int \delta U^* \, dV$$

Every component subjected to external applied loads reaches stable equilibrium, when its potential energy or the difference between work done by external forces and internal strain energy due to stresses developed is zero. It can also be expressed as - "During any arbitrary kinematically consistent virtual displacement from the equilibrium state, satisfying constraints prescribed for the body, potential energy equals zero or the work done by the external forces equals the increment in strain energy". This is called *Principle of Minimum potential energy* i.e., $\delta I = \delta W - \delta U = 0$

The same principle is also expressed as *Principle of virtual displacements,* which states that "an elastic deformable structure is in a state of equilibrium if the virtual work (δW) done by forces is equal to the virtual strain energy (δU) for every arbitrary virtual displacement consistent with the constraints of the structure". i.e., $\delta W = \delta U$ or $[\delta u]\{P\} = \delta U$

Using Taylor series expansion,

$$\delta U = \sum_i (\partial U / \partial u_i) \times \delta u_i + (1/2) \sum_i \sum_j [\partial^2 U/(\partial u_i \, \partial u_j)] \times \delta u_i \times \delta u_j + ..$$

$$= [\delta u]\,\{\partial U / \partial u\} + (1/2)\,[\delta u]\,[S]\,\{\delta u\} + ...$$

where, $[S]$ is the stiffness matrix such that $S_{ij} = \partial^2 U/(\partial u_i \, \partial u_j)$

If higher order terms in δu are neglected, $[\delta u]\,\{P_i\} = [\delta u]\,\{\partial U / \partial u\}$

or $[\delta u]\,(\{P\} - \{\partial U / \partial u\}) = 0$

For a non-trivial solution, $\{P\} - \{\partial U / \partial u\} = 0$ or $\{P\} = \{\partial U / \partial u\}$ .....(5.5)

$P_i = \partial U / \partial u_i$ is **Castigliano's first theorem**

5.4   PRINCIPLE OF VIRTUAL FORCES

If an additional infinitesimal (or virtual) force (δP_i) is applied at the same displacement, the change in complementary work done (or virtual complementary work) is given by $\delta W' = \delta P_i \times u_i$ and the corresponding change in complementary strain energy (virtual complementary strain energy) density is given by

$$(dU^*)' = \delta\sigma_i \times \varepsilon_i$$

If we consider all the components of forces acting on the structure and corresponding displacements, the total virtual complementary work done is obtained by summation as $\delta W' = \sum (\delta P_i \times u_i) = [\delta P] \{u\}$

where, $[\delta P]$ represents row vector of all force components

and $\{u\}$ represents column vector of all displacement components

Similarly, corresponding variation in complementary strain energy density can be expressed as

$$(dU^*)' = \sum(\delta\sigma_i \times \varepsilon_i) = [\delta\sigma] \{\varepsilon\}$$

and $dU' = \int (dU^*)' \, dV$

Principle of virtual forces (also called ***Principle of virtual work***) states that an elastic deformable structure is in a state of equilibrium if the complementary virtual work $\delta W'$ done by forces is equal to the virtual complementary strain energy $\delta U'$ for every arbitrary virtual force consistent with the constraints of the structure.

i.e., $\delta U' = \delta W' = [\delta P] \{u\}$

Using Taylor series expansion,

$$\delta U' = \sum_i (\partial U'/\partial P_i) \times \delta P_i + (1/2) \sum_i \sum_j [\partial^2 U'/(\partial P_i \, \partial P_j)] \times \delta P_i \times \delta P_j +..$$

$$= [\delta P] \{\partial U'/\partial P\} + (1/2) [\delta P] [F] \{\delta P\} + \ldots$$

where, $[F]$ is the flexibility matrix such that $F_{ij} = \partial^2 U'/(\partial P_i \, \partial P_j)$

If higher order terms in δP are neglected, $[\delta P] \{u\} = [\delta P] \{\partial U'/\partial P\}$

or $[\delta P] (\{u\} - \{\partial U'/ \partial P\}) = 0$

For a non-trivial solution, $\{u\} - \{\partial U'/\partial P\} = 0$ or $\{u\} = \{\partial U'/\partial P\}$ .....(5.6)

$$\mathbf{u_i = \partial U' /\partial P_i} \quad \text{is } \textbf{Castigliano's second theorem}$$

For linear elastic systems, common to most of the engineering applications, $U = U'$ and, so, this theorem can also be expressed as $\mathbf{u_i = \partial U/\partial P_i}$

In order to apply this theorem to obtain displacement (slope) *at a point in a particular direction*, a force (moment) must be applied *at that point in that direction*. If a force / moment is not applied there, a virtual force / moment need to be considered in the calculation, as shown in the examples

Example 5.1

In the 3-member structure shown, with fixed end at 1 and hinged connections at 2, 3 and 4 and all three members of same material and same cross section, calculate vertical deflection at 3.

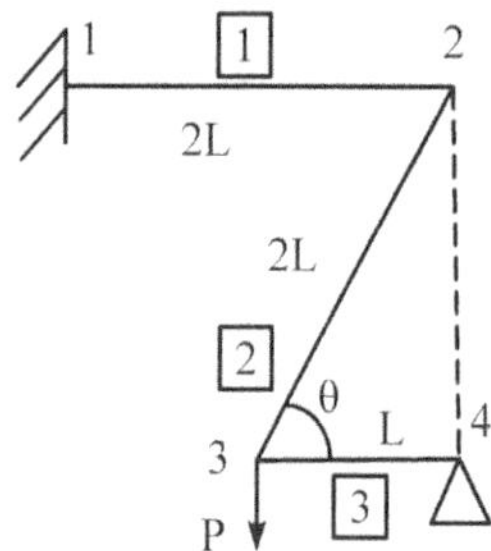

Solution

Due to the hinged connections at 3 and 4, member-3 tends to rotate about 4 without any bending. The applied load P is balanced by forces in member-2 and member-3. From elementary mechanics, equations of force equilibrium at 3 give

$$\sum F_X = F_2 \times \cos\theta - F_3 = 0 \quad \text{and} \quad \sum F_Y = F_2 \times \sin\theta - P = 0$$

In this structure, $\cos\theta = L / 2L = 1/2, \ \sin\theta = \sqrt{3}/2$

Therefore, $F_2 = P/\sin\theta = 2P/\sqrt{3}$ and $F_3 = F_2\cos\theta = (2P/\sqrt{3}) \times (1/2) = P/\sqrt{3}$

Member-2, with both ends 2 and 3 hinged, is subjected to only axial force F_2. Member-1 with end 1 fixed and end 2 hinged behaves like a cantilever beam. Force F_2 in member-2 can be resolved into two components – $(F_2)_X$ along 1-2 and $(F_2)_Y$ perpendicular to 1-2. $(F_2)_X$ produces axial deformation in member-1 while $(F_2)_Y$ produces constant shear deformation from 2 to 1 and bending deformation due to bending moment varying linearly from zero at 2 to $(F_2)_Y \times 2L$ at 1.

$$F_1 = (F_2)_X = F_2 \times \cos\theta = P/\sqrt{3}; \quad (F_2)_Y = F_2 \times \sin\theta = P \quad \text{and} \quad M = (F_2)_Y \times (2L - x)$$

$$\text{for } \ 0 \le x \le 2L$$

Total strain energy stored in the structure, $U = U_1 + U_2 + U_3$

$$U = (1/2)\int_0^L \left(F_3^2 / EA\right) dx_{3-4} + (1/2)\int_0^{2L} \left(F_2^2 / EA\right) dx_{3-2}$$

$$+ (1/2)\int_0^{2L}\left[\left(F_1^2/EA\right) + (F_2)_Y^2/GA + \left(M^2/EI\right)\right] dx_{1-2}$$

$$= (1/2) \times [\{F_3^2 \times L/(E \times A)\} + \{F_2^2 \times (2L)/(E \times A)\}$$

$$+ \{F_1^2 \times (2L)/(E \times A)\} + \{(F_2)_Y^2 \times (2L)/(G \times A)\}$$

$$+ \int\{(F_2)_Y^2 \times (2L - x)^2/(E \times I)\} dx \ ]$$

since F_3, F_2, F_1 and $(F_2)_Y$ are constant throughout the lengths of members and $(\frac{1}{2}) \times \int \{(F_2)_Y^2 \times (2L - x)^2/(E \times I)\} dx = \{(P^2/(2E \times I)\} \times \int (4L^2 - 4Lx + x^2)\, dx$

$$= 4P^2 \times L^3/(3E \times I),$$

$$U = \{P^2 \times L/(2A \times E)\}\, [1/3 + 8/3 + 2/3]$$
$$+ \{P^2 \times L/(G \times A)\} + \{4\,P^2 \times L^3/(3E \times I)\}$$

Vertical deflection at 3, using Castigliano's first theorem, is

$$(\delta_Y)_3 = \partial U/\partial P = [\{11L/(3A \times E)\} + \{2L/(G \times A)\} + \{8L^3/(3E \times I)\}\,]\, P$$

Neglecting shear deformation, $(\delta_Y)_3 = [\{11L/(3A \times E)\} + \{8L^3/(3E \times I)\}\,]\, P$

Example 5.2

Find the slope and vertical deflection at the free end of a stepped cantilever, shown in figure, subjected to a vertical load P at the free end.

Solution

In order to apply Castigliano's theorem to obtain slope at the free end, a fictitious moment M must be applied at the free end. Then, $M_X = M + P \times x$

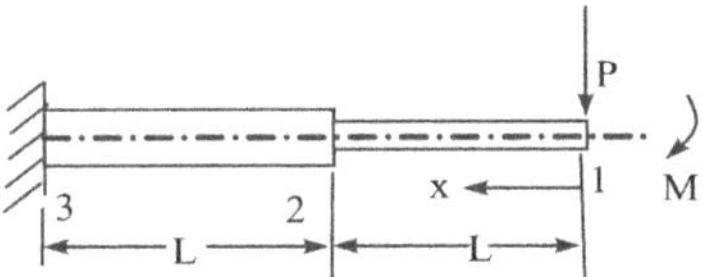

Total strain energy,

$$U = (1/2)\int (M_X^2/EI)dx = (1/2)\left[\int_0^L (M_X^2/EI_{1\text{-}2})dx + \int_L^{2L} (M_X^2/EI_{2\text{-}3})dx\right]$$

$$= \{1/(2E \times I_{1-2})\}\,[M^2 \times L + 2M \times P \times (L^2/2) + P^2 \times (L^3/3)]$$

$$+ \{1/(2E \times I_{2-3})\}\,[M \times {}^2L + 2M \times P \times (3L^2/2) + P^2 \times (7L^3/3)]$$

Slope at the free end, $\theta_1 = \partial U/\partial M = \{1/(2E \times I_{1-2})\}\,[2ML + 2P \times (L^2/2) + 0]$

$$+ \{1/(2E \times I_{2-3})\}\,[2ML + 2P \times (3L^2/2) + 0]$$

Since Moment at the free end is fictitious, substituting $M = 0$ in the above equation,

$$\theta_1 = \{1/(2E \times I_{1-2})\}\,[2P \times (L^2/2)] + \{1/2E \times I_{2-3})\,[2P \times (3L^2/2)]$$

$$= \{P \times L^2/(2E \times I_{1-2})\} + \{3P \times L^2/(2E \times I_{2-3})\}$$

Vertical deflection at the free end,

$$(\delta_Y)_3 = \partial U/\partial P = \{1/(2E \times I_{1-2})\} \times [0 + 2M.(L^2/2) + 2P \times (L^3/3)]$$

$$+ \{1/(2E \times I_{2-3})\} \times [0 + 2M \times (3L^2/2) + 2P \times (7L^3/3)]$$

Since Moment at the free end is fictitious, substituting $M = 0$ in the above equation,

$$(\delta_Y)_3 = \{1/(2E \times I_{1-2})\} [0 + 0 + 2P \times (L^3/3)]$$

$$+ \{1/(2E \times I_{2-3})\} [0 + 0 + 2P \times (7L^3/3)]$$

$$= \{P \times L^3/(3E \times I_{1-2})\} + \{7P \times L^3/(3E \times I_{2-3})\}$$

If $I_{2-3} = 2I_{1-2}$, $\theta_1 = 5P \times L^2 / (4E \times I)$; $(\delta_Y)_3 = 3P \times L^3 / (2E \times I)$

If $I_{2-3} = I_{1-2}$ (uniform section), $\theta_1 = 2P \times L^2/(E \times I)$; $(\delta_Y)_3 = 8P \times L^3/(3E \times I)$

Example 5.3

Find the vertical deflection at 1 and angular twist at 2 of the structure shown in figure, neglecting shear deformation, due to a vertical load P at its free end.

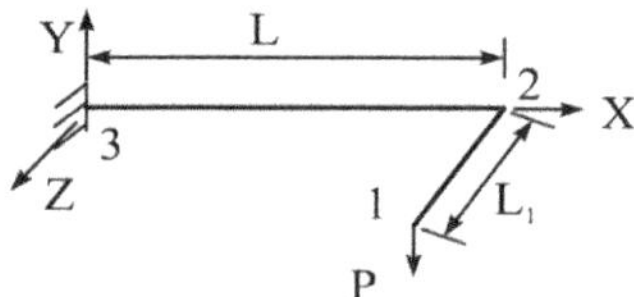

Solution

In order to find angular twist at 2, fictitious torque 'T' has to be applied at 2.

Then, total strain energy stored in the structure is

$$U = (1/2) \int \{M_{1-2}^2/(E \times I)\} \, dz + (1/2) \int \{M_{2-3}^2/(E \times I)\} \, dx + (1/2) \int \{T_{2-3}^2/(G \times J)\} \, dx$$

Substituting $M_{1-2} = P \times z$; $M_{2-3} = P \times x$ and $T_{2-3} = T + P \times L_1$

$$U = (1/2) [\{P^2/(EI)\} \int z^2 \, dz + \{P^2/(EI)\} \int x^2 \, dx + \{(T + P \times L_1)^2 / 2(G \times J)\} \int dx]$$

$$= (1/2)[\{P^2/(E \times I)\} \times (L_1^3/3) + \{P^2/(E \times I)\} \times (L^3/3)$$

$$+ \{(T^2 + 2T \times P \times L_1 + P^2 \times L_1^2)/(2G \times J)\} \times L]$$

Vertical deflection at the free end,

$$(\delta_Y)_1 = \partial U/\partial P = (1/2) [\{2P \times L_1^3/(3E \times I)\} + \{2P \times L^3/(3E \times I)\}$$

$$+ (0 + 2T \times L_1 + 2P \times L_1^2) \times L/(2G \times J)]$$

$$= \{P/(3E \times I)\} \times (L_1^3 + L^3) + P \times L_1^2 \times L/(2G \times J)$$

(substituting fictitious torque, $T = 0$)

Angular twist at 2,

$$\alpha_2 = \partial U/\partial T = (1/2) [0 + 0 + (2T + 2P \times L_1 + 0) \times (L/(2G \times J)]$$

$$= P \times L \times L_1/(2G \times J)$$

If $L = L_1$, $(\delta_Y)_1 = P \times L^3 \times [\{2/(3E \times I)\} + \{1/(G \times J)\}]$

and $\alpha_2 = P \times L^2/(2G \times J)$

5.5   PRINCIPLE OF SUPERPOSITION

This principle is of significant help in analysing redundant structures. If a bar of length L is subjected to an axial tensile load of $(P_1 + P_2)$, its displacement

$$\delta = (P_1 + P_2) \times L / (A \times E) = (P_1 \times L) / (A \times E) + (P_2 \times L) / (A \times E) = \delta_1 + \delta_2 \quad ..(5.7)$$

where,   A is the area of cross section of the bar

and   E is the modulus of elasticity of the material of the bar

This principle is applicable as long as the load-displacement (or stress-strain) relationship is linear. This principle also implies that the strain energy due to different loads can be algebraically added to get the total strain energy for linearly elastic behaviour.

5.6   MAXWELL'S RECIPROCAL THEOREM

If the displacement at a point 'A' in the direction of load 'P_A' due to load 'P_B' at point 'B' is designated by δ_{AB} and that at point 'B' in the direction of load 'P_B' due to load 'P_A' at point 'A' is designated by δ_{BA}, then this theorem states that

$$P_A \times \delta_{AB} = P_B \times \delta_{BA} \qquad ....(5.8)$$

To prove the theorem, let us consider a simple bar AC of length L, fixed at end A. Let B and C be the points at distances of L/2 and L from the fixed end C. Due to axial tensile load of P_B at B,

Displacement at C, $\delta_{CB} = P_B \times (L/2)/(A \times E)$

Similarly, due to axial tensile load of P_C at C,

Displacement at B, $\delta_{BC} = P_C \times (L/2)/(A \times E)$

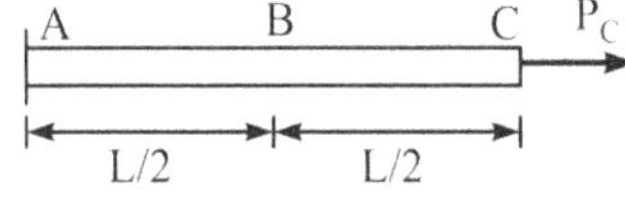

Then, $P_B \times \delta_{BC} = P_B \times [\ P_C \times (L/2) / (A \times E)\ ] = P_C \times [\ P_B \times (L/2) / (A \times E)\ ] = P_C \times \delta_{CB}$

In a particular case, if $P_B = P$, $P_C = 2P$,

$$P_B \times \delta_{BC} = P \times [2P \times (L/2) / (A \times E)] = 2P \times [P \times (L/2) / (A \times E)] = P_C \times \delta_{CB}$$

and if $P_B = P_C = P$, $\delta_{BC} = \delta_{CB} = P \times L / (2A \times E)$

Stiffness coefficient k_{ij} is defined as the load at 'i' for unit displacement at 'j' or P_i/δ_j. It follows from eq (5.8) that $P_i/\delta_j = P_j/\delta_i$ or $k_{ij} = k_{ji}$. Thus, this theorem forms the basis for stiffness matrix [K] to be symmetric, as we see in Chapter 19 - Finite Element Method.

5.7 UNIT LOAD METHOD

From the principle of virtual forces, we have seen that $(du^*)' = [\delta\sigma]\,\{\varepsilon\}$

The virtual complementary work, $\delta W' = \int (dU^*)'\,dV = \int [\delta\sigma]\,\{\varepsilon\}\,dV$

or $[\delta P]\,\{u\} = \int [\delta P]\,[C]\,\{\varepsilon\}\,dV$

where, $[\delta\sigma] = [C]\,[\delta P]$ are the equilibrium conditions

If r^{th} virtual force is given a unit value and all other virtual forces are set equal to zero, then $u_r = \int [C]\,\{\varepsilon\}\,dV$ This is called ***unit-load method***

From $[\delta\sigma] = [C]\,[\delta P]$, for this condition, we get $C_r = \delta\sigma_r$

The principle of stationary total complementary potential is given by

$$\delta(W' + U') = \delta V' = 0$$

where, $\delta W' = \sum u_i \times \delta P_i$

This is equivalent to $\partial V'/\partial P_i = 0$ for $i = 1, 2, \ldots n$

which represents the set of compatibility conditions at all coordinates.

ADDITIONAL PROBLEMS FOR PRACTICE

1. Calculate horizontal and vertical displacements of end C of the frame ABC, due to vertical load 'W' at C. Length of AB is 'L' while length of BC is 'L/2'. Compare the vertical displacement at C with the displacement when the member AB doesn't exist and end B is fixed.

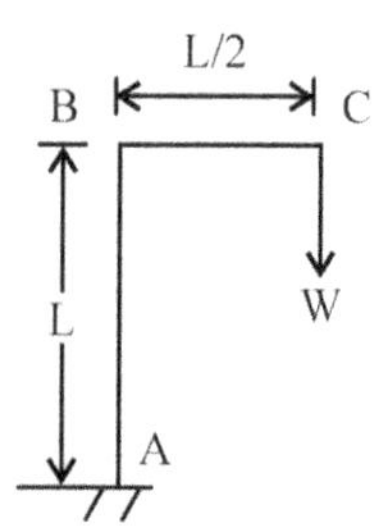

$$[\text{Ans} : \delta_{CV} = (7/24) \times W \times L^3/(A \times E)\,;\ \delta_{CH} = W \times L^3/(4E \times I);$$

$$\delta_{CV} = (1/24) \times W \times L^3/(A \times E)\ \text{when AB does not exist and B is fixed}\,]$$

2. Calculate vertical displacement at A in the truss shown due to a vertical load of 15kN at A. Assume $A = 600\text{mm}^2$ and E = 200GPa for all the members.

$$[\text{Ans} : \delta_{AV} = 64.3/(A \times E)\,]$$

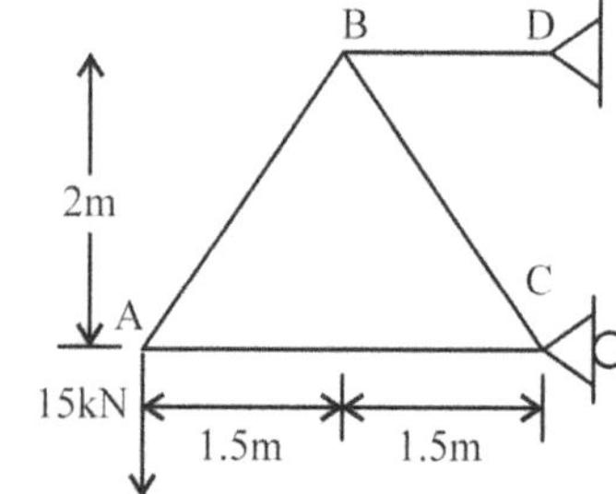

3. Calculate vertical displacement at the mid point C of a bar AB of length 1.8m, when it is suspended at A and B by springs of stiffness 250kN/m and 160kN/m respectively and a vertical load of 8kN is applied at C.

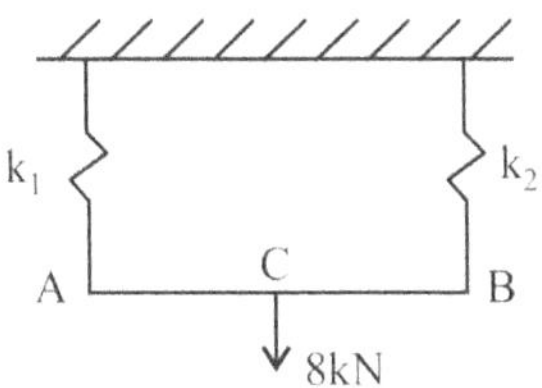

[Ans : δ_{CV} = 25mm]

4. Calculate vertical and horizontal displacements of A in the truss shown due to a vertical load 'W' at A. Assume area of each tensile member is 'A' and area of each compression member is '2A'. E is same for all the members.

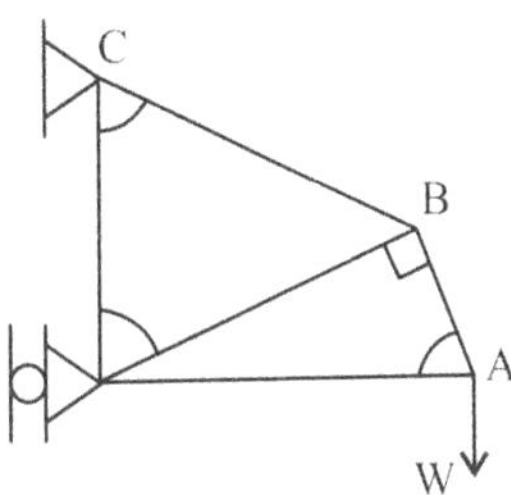

[Ans : δ_{AV} = 2.57 × W × L/(A × E); δ_{AH} = 0.29 × W × L/(A × E)]

5. Calculate horizontal displacement of C in the truss shown due to a horizontal load of 80kN at B. Area of members AB, BC and CD is 800mm² and area of members AD and CD is 1600mm². E = 200GPa for all the members.

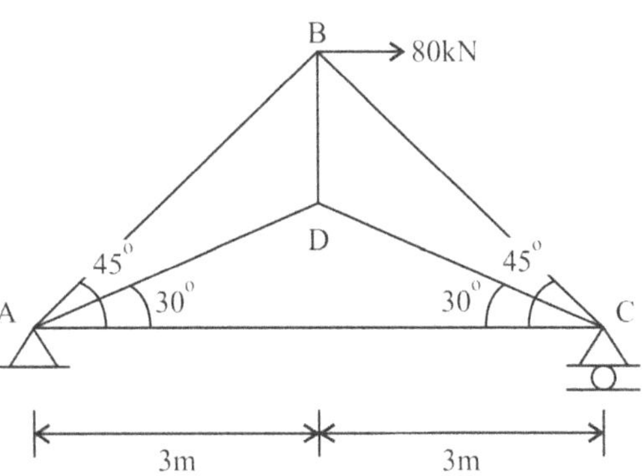

[Ans : δ_{CH} = 16.35mm]

6. A square truss with a horizontal diagonal member, as shown, is supported from one corner and a vertical load of 10kN is applied at the opposite corner. Calculate vertical displacement of the loaded end

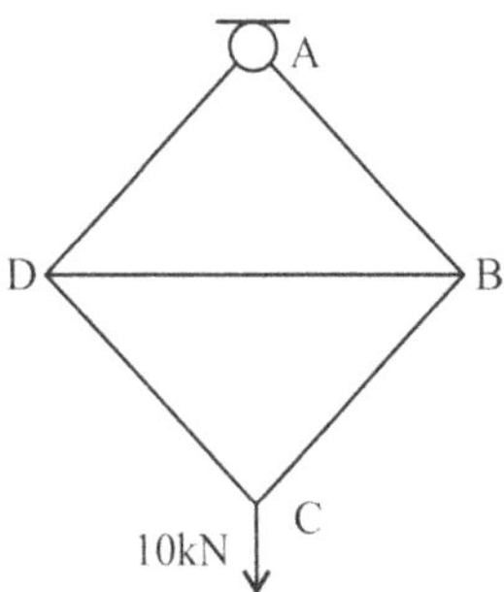

CHAPTER 6

ANALYSIS OF TRUSSES

Structures such as trusses and frames, which have many identifiable members, connected only at their end points or nodes are called ***discrete structures***. Each member of the structure is considered as a one-dimensional (1-D) element along its length, identified by its end point coordinates. Their lateral dimensions are reflected in element properties like area of cross section in trusses and moment of inertia and depth of section in beams.

A truss consists of axially loaded members, called spars, hinged together at joints. Naturally, all the loads are applied at joints (neglecting self weight of the member distributed along its length) and the truss is also supported at the joints. Trusses are broadly classified as planar trusses (if all the members are oriented in a single plane) or spatial trusses. If they have just adequate number of members so that they resist applied loads effectively, (less number of members behave like a mechanism producing gross rigid body motion of members and more number of members restrain the hinged joints for effective realignment of members due to applied loads) are called '***perfect frames***'(Ref.Fig.6.1 a). They can be analysed for member forces by repeated application of equations of static equilibrium at different joints or across different sections. The system of loads (including applied loads, member forces and reactions at the supports) on a truss is non-concurrent, while the system of forces at a joint is always concurrent, since all the member forces act along the member axes and pass through the joint. The number of members 'm' is related to the number of joints 'j', in a perfect frame or determinate truss, by the relation $m = 2j - 3$ (plane truss) or $m = 3j - 6$ (space truss)

A perfect frame, with minimum members, can always be designed for a particular load set. But, it may not be suitable for a different load set and, hence, more members are added. Such a structure with $m > 2j - 3$ (plane truss) or $m > 3j - 6$ (space truss) is called '***imperfect frame***' or internally redundant or indeterminate truss (Ref.Fig.6.1 b).

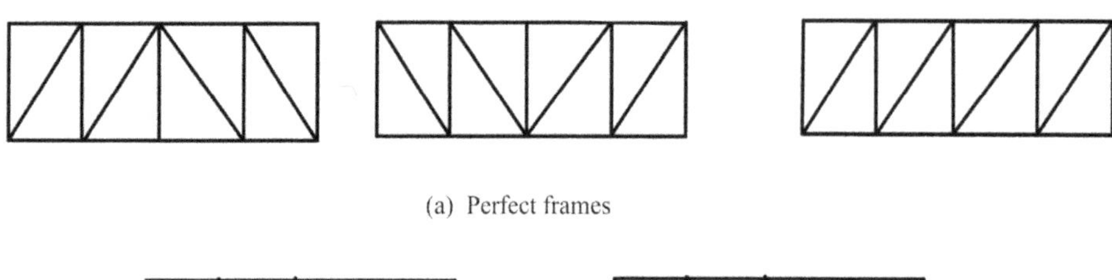

(a) Perfect frames

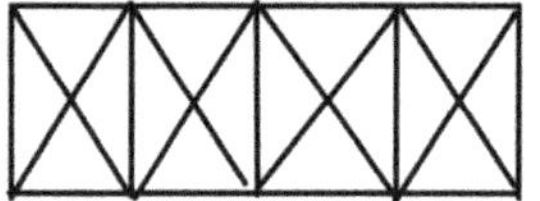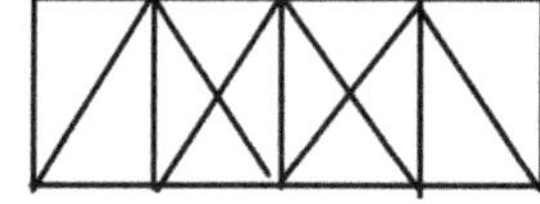

(b) Imperfect frames

FIGURE 6.1 Types of frames

Reactions are calculated by considering the entire truss and using the equations of equilibrium, applicable to the system of non-concurrent forces

$$\Sigma F_X = 0, \ \Sigma F_Y = 0 \ \text{ and } \ \Sigma M_Z = 0 \qquad \text{for a plane truss}$$

and $\ \Sigma F_X = 0, \Sigma F_Y = 0, \Sigma F_Z = 0, \Sigma M_X = 0, \ \Sigma M_Y = 0 \ \text{ and } \ \Sigma M_Z = 0 \ $ for a space truss

Trusses are fixed at one support and free to slide at other supports, so that the members are free to adjust to their new positions, when loaded, without causing additional forces in the members due to support restraints. Thus, a plane truss will have a maximum of three reactions – two perpendicular (horizontal and vertical) components at the fixed support (A) and one normal (vertical) reaction at the sliding or roller support (B).

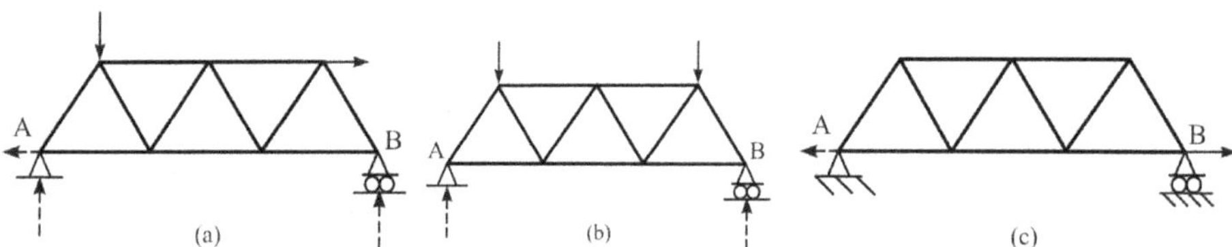

FIGURE 6.2 Support reactions in trusses

Let us consider all forces along horizontal (X) direction and vertical (Y) direction. It is possible for a plane truss to have one or two reactions only for some specific cases of applied loads. In Fig. 6.2 a, three reactions (R_{AX}, R_{AY} and R_{BY}) exist which can be evaluated from $\Sigma F_X = 0$, $\Sigma F_Y = 0$ and $\Sigma M_Z = 0$. In Fig. 6.2 b, $\Sigma F_X = 0$ is redundant since, there are no forces in X-direction. So, only two reactions (R_{AY} and R_{BY}) exist, which can be evaluated from $\Sigma F_Y = 0$ and $\Sigma M_Z = 0$. In Fig. 6.2 c, $\Sigma F_Y = 0$ is redundant since, there are no forces in Y-direction. Also, $\Sigma M_Z = 0$ is redundant, since the applied force passes through the support points and does not produce a moment at either support. So, only one reaction (R_{AX}) exists, which can be evaluated from $\Sigma F_X = 0$.

The rigid body mechanics approach of truss analysis assumes that the change in the lengths of members is so small that orientations of the truss members remain practically unchanged. This assumption does not alter the calculated member forces. Stress in each member is obtained by dividing force in the member with the corresponding area of cross section. Stress is assumed to be uniform across the cross section in each member.

But, no member can be stressed without a proportional strain, according to Hooke's law. The finite element method of analysis based on energy principles, with stress and proportional strain always co-existing in a member, calculates displacements of joints as primary unknowns and stresses in members as secondary unknowns from stress-displacement relation using modulus of elasticity of material and area of cross section.

Each member subjected to compressive load has to be checked separately for buckling.

6.1   METHOD OF JOINTS

In this method, the analysis for calculation of member forces is carried out by applying equations of static equilibrium for concurrent forces at each joint of the truss, choosing the joints in a sequence such that the system of equations matches with the number of unknowns in the set of equations. Thus, in a plane truss in X-Y plane, the equations of equilibrium ($\Sigma F_X = 0$ and $\Sigma F_Y = 0$) can be used at each joint to evaluate two unknown member forces. Similarly, in a space truss, the equations of equilibrium ($\Sigma F_X = 0$, $\Sigma F_Y = 0$ and $\Sigma F_Z = 0$) can be used at each joint to evaluate three unknown member forces. The following example illustrates practical application of this method

Example 6.1

Find forces in all the members of the plane truss, shown below, for the vertical loads of 20 kN applied at joints C and D. DE = EF = FG = 3 m ; AG = 4 m

Solution

Let $\angle CDE = \theta$. Considering equilibrium of each joint, two unknown forces can be evaluated using the two equations of equilibrium

$$\Sigma F_X = 0 \text{ and } \Sigma F_Y = 0$$

$$\tan \theta = AG/DG = AG/(DE + EF + FG) = 4/(3 + 3 + 3) = 4/9$$

$$\Rightarrow \quad \sin \theta = 4/\sqrt{97} ; \quad \cos \theta = 9/\sqrt{97}$$

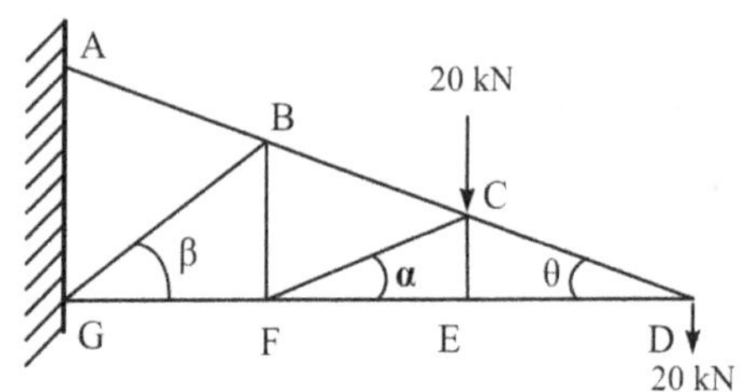

At D, $\sum F_Y = F_{CD} \times \sin\theta - 20 = 0$

$\Rightarrow$ $F_{CD} = 20/\sin\theta = 20/\left(4/\sqrt{97}\right) = 49$ kN

$\sum F_X = F_{ED} - F_{CD} \times \cos\theta = 0$

$\Rightarrow$ $F_{ED} = F_{CD} \times \cos\theta = 49 \times \left(9/\sqrt{97}\right) = 45$ kN

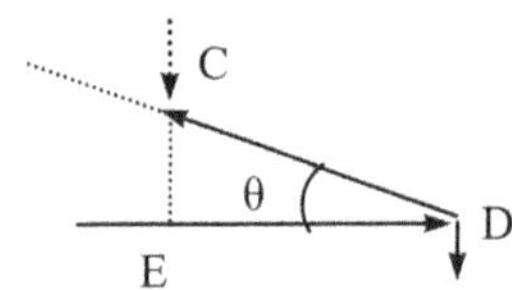

At E, $\sum F_X = F_{ED} - F_{EF} = 0$

$\Rightarrow$ $F_{EF} = F_{ED} = 45$kN

and $\sum F_Y = F_{CE} = 0$

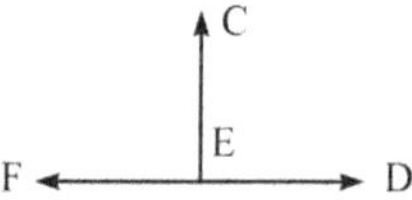

At C, $\sum F_Y = -F_{CD} \times \sin\theta + F_{FC} \times \sin\alpha + F_{BC} \times \sin\theta - 20 = 0$

where, $\tan\alpha = CE/FE = (4/3)/3 = 4/9$

$\sin\alpha = 4/\sqrt{97}$; $\cos\alpha = 9/\sqrt{97}$

and $\sum F_X = -F_{BC} \times \cos\theta + F_{CD} \times \cos\theta + F_{FC} \times \cos\alpha = 0$

Solving these two equations F_{BC} and F_{FC} can be obtained as

$F_{BC} = 73.87$ kN

and $F_{FC} = 24.62$ kN

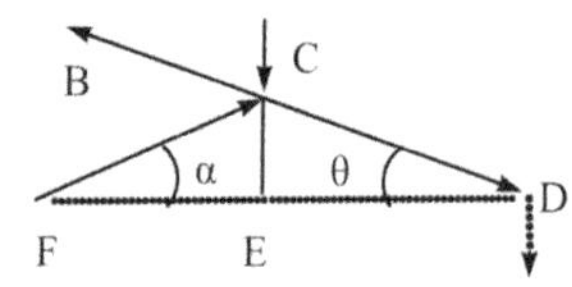

At F, $\qquad \sum F_X = -F_{FC} \times \cos \alpha - F_{FE} + F_{GF} = 0$

$\qquad\qquad\quad \sum F_Y = -F_{FC} \times \sin \alpha + F_{FB} = 0$

from which F_{FB} and F_{GF} can be obtained as

$$F_{FB} = F_{FC} \times \sin \alpha = 24.62 \times \left(4/\sqrt{97}\right) = 10 \text{ kN}$$

and $\qquad F_{GF} = F_{FC} \times \cos \alpha + F_{FE} = 24.62 \times \left(9/\sqrt{97}\right) + 45 = 67.5 \text{ kN}$

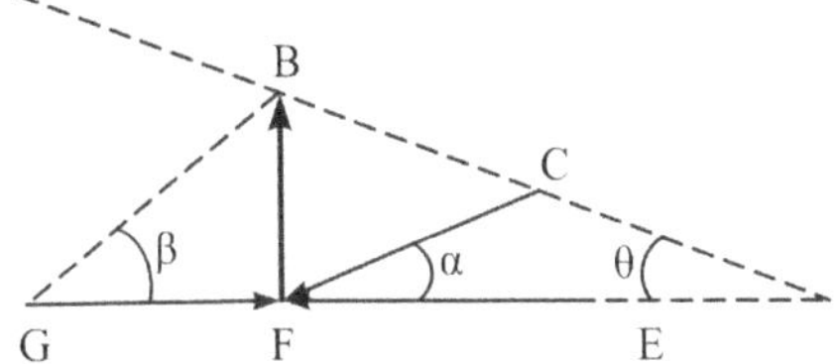

At B, $\qquad \sum F_Y = -F_{BC} \times \sin \theta + F_{AB} \times \sin \theta + F_{GB} \times \sin \beta - F_{FB} = 0$

and $\qquad \sum F_X = F_{BC} \times \cos \theta - F_{AB} \times \cos \theta + F_{GB} \times \cos \beta = 0$

where, $\qquad \tan \beta = BF/GF = (8/3)/3 = 8/9$

$\Rightarrow \qquad \sin \beta = 8/\sqrt{145} \;;\; \cos \beta = 9/\sqrt{145}$

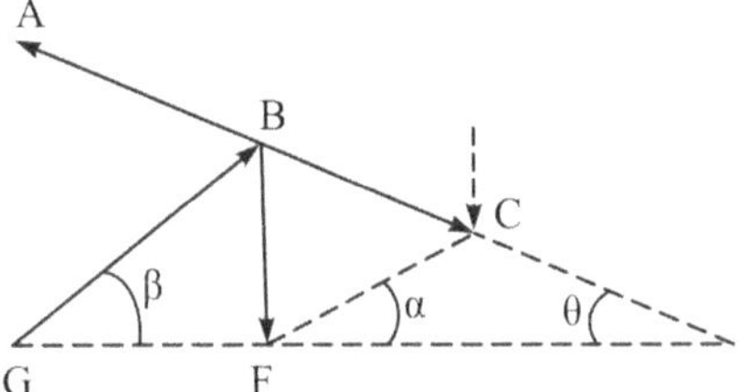

Solving these two equations, F_{GB} and F_{AB} can be obtained as

$$F_{GB} = 82 \text{ kN} \qquad \text{and} \qquad F_{AB} = 10 \text{ kN}$$

*Please **note** that*

- Equations of equilibrium are written for the assumed directions of member forces, (indicating tensile or compressive forces) as shown in the figures. If the assumed direction is wrong, the calculated value of member force will be negative. It is assumed in the above equations that forces directed towards +ve X-axis (to the right, in the horizontal direction) are +ve and those directed towards –ve X-axis are –ve. Similarly, forces directed towards +ve Y-axis (upwards, in the vertical direction) are +ve and those directed along –ve Y-axis are –ve.

- Directions of member forces at its two ends are opposite. A member in tension is represented by inward arrow at each joint while a member in compression is represented by outward arrow at each joint. The force directions in each member are internal reactions and, hence, the tensile force directions appear to produce compression in the member and compressive force directions appear to produce elongation of the member. Thus, if a member I-J has its internal force (shown in figure by solid arrow) oriented towards J, while analysing joint I, then internal force is

taken oriented towards I while analysing joint J. Similarly, if the member I-J has its internal force oriented towards I while analysing joint I, then the internal force is taken oriented towards J while analysing joint J.

- A member with zero force, in this example CE, is called a ***null member***

- At the ends A and G, the member force and reaction form a system of collinear forces along member axis and, hence, application of equations of equilibrium will give reactions which are equal and opposite to the member forces

- The sequence of joints is so chosen as to have only two unknowns in the equations of equilibrium at each joint. For example, in the above problem, if joint C is considered after joint D, with unknown member forces in CD, CE and BC then the two equations of equilibrium are not adequate to obtain a solution. Similarly, if joint B is considered after joint E, with unknown member forces in BF, BG and BA, then also the two equations are not adequate to obtain a solution

6.2   METHOD OF SECTIONS

In this method, the analysis for calculation of member forces is carried out by applying equations of static equilibrium repeatedly considering that the truss is cut by imaginary section planes (straight or curved) and the truss is in stable equilibrium due to the contribution of member forces acting on the free body diagram of one part of the cut truss. The forces on the free body diagram (applied loads, reactions and forces in the cut members) are usually non-concurrent and, hence, applicable equations of equilibrium are considered.

The same example is solved here to illustrate practical application of this method.

Example 6.2

Find forces in all the members of the plane truss, shown below, for the vertical loads of 20kN applied at joints C and D. DE = EF = FG = 3m ; AG = 4m

Solution

Let $\angle CDE = \theta$. Considering equilibrium of each joint, two unknown forces can be evaluated using the two equations of equilibrium.

$$\Sigma F_X = 0 \quad \text{and} \quad \Sigma F_Y = 0$$

$$\tan \theta = AG/GD = AG / (GF + FE + ED) = 4 / (3 + 3 + 3) = 4/9$$

$$\Rightarrow \quad \sin \theta = 4 / \sqrt{97} ; \quad \cos \theta = 9 / \sqrt{97}$$

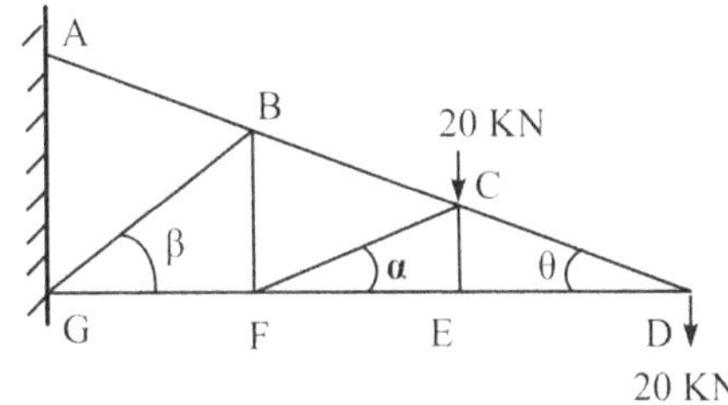

If the truss is cut by an imaginary section plane-1, cutting members CD and DE as shown, for the equilibrium of the free body, we get

$$\Sigma F_Y = F_{CD} \times \sin \theta - 20 = 0$$

or $\quad F_{CD} = 20/\sin \theta = 20 / \left(4 / \sqrt{97} \right) = 49 \text{ kN}$

$$\Sigma F_X = F_{ED} - F_{CD} \times \cos \theta = 0$$

$$\Rightarrow \quad F_{ED} = F_{CD} \times \cos \theta = 49 \times \left(9 / \sqrt{97} \right) = 45 \text{ kN}$$

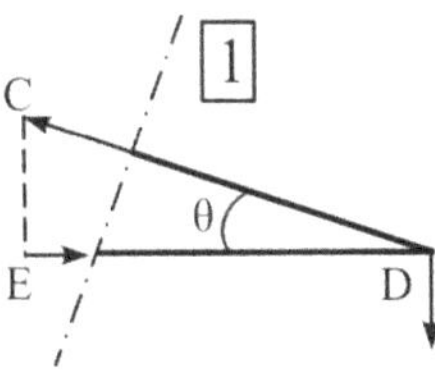

If the truss is cut by an imaginary section plane-2, cutting members CD, CE and FE as shown, for the equilibrium of the free body, we get

$$\Sigma F_Y = F_{CD} \times \sin \theta + F_{EC} - 20 = 0$$

$$\Rightarrow \qquad F_{EC} = 20 - F_{CD} \times \sin \theta = 20 - 49 \times \left(4/\sqrt{97}\right) = 0$$

and $\qquad \sum F_X = F_{EF} - F_{CD} \times \cos \theta = 0$

$$\Rightarrow \qquad F_{EF} = F_{CD} \times \cos \theta = 49 \times \left(9/\sqrt{97}\right) = 45 \text{ kN}$$

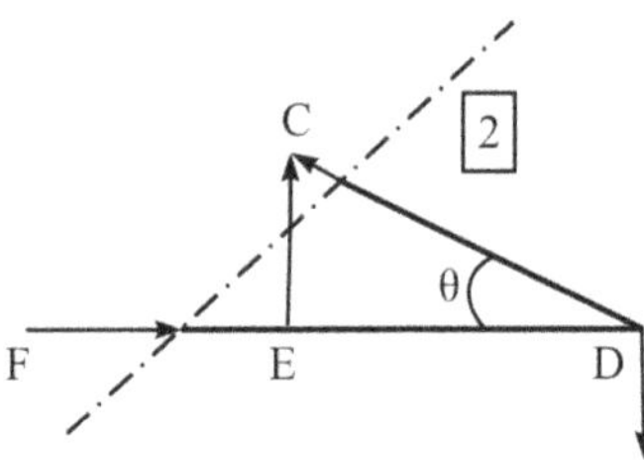

If the truss is cut by an imaginary section plane-3, cutting members FE, FC and BC as shown, for the equilibrium of the free body, we get

$$\sum F_Y = F_{FC} \times \sin \alpha + F_{BC} \times \sin \theta - 20 - 20 = 0$$

where, $\quad \tan \alpha = CE/FE = (4/3)/3 = 4/9$

$$\Rightarrow \qquad \sin \alpha = 4/\sqrt{97} \ ; \cos \alpha = 9/\sqrt{97}$$

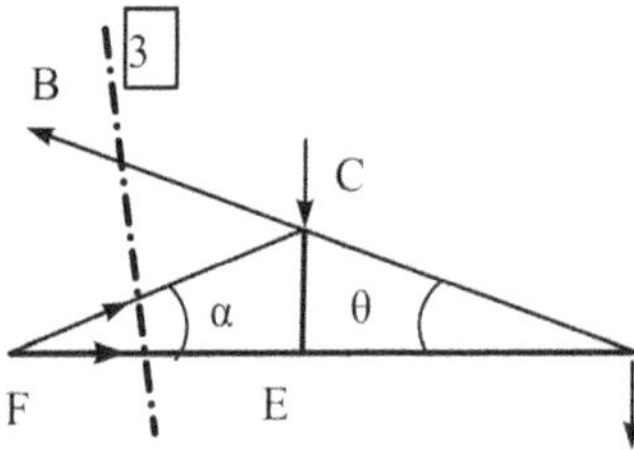

and $\qquad \sum F_X = -F_{BC} \times \cos \theta + F_{FC} \times \cos \alpha + F_{FE} = 0$

Solving the two conditions, F_{BC} and F_{FC} can be obtained as

$$F_{BC} = 73.87 \text{ kN}$$

and $\qquad F_{FC} = 24.62 \text{ kN}$

If the truss is cut by an imaginary section plane-4, cutting members GF, BF and BC as shown, for the equilibrium of the free body, we get

$$\sum F_X = F_{GF} - F_{BC} \times \cos \theta = 0$$

and $\qquad \sum F_Y = F_{FB} + F_{BC} \times \sin \theta - 20 - 20 = 0$

from which F_{FB} and F_{GF} can be obtained as

$$F_{GF} = F_{BC} \times \cos \alpha = 73.87 \times \left(9/\sqrt{97}\right) = 67.5 \text{ kN}$$

and $\quad F_B = 20 + 20 - F_{BC} \times \sin\theta = 40 - 73.87 \times \left(4/\sqrt{97}\right) = 10$ kN

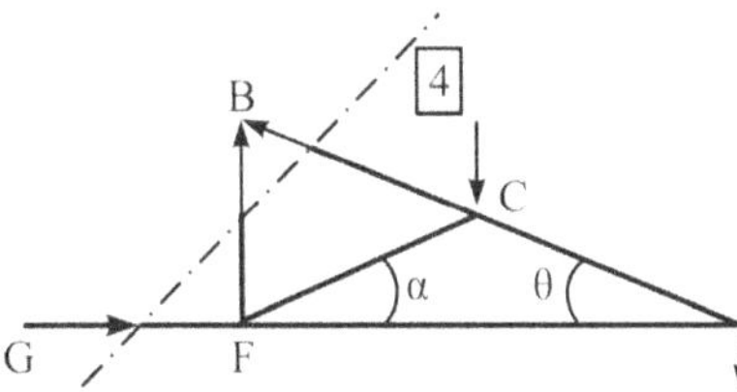

If the truss is cut by an imaginary section plane-5, cutting members GF, GB and AB as shown, for the equilibrium of the free body, we get

$$\sum F_Y = F_{AB} \times \sin\theta + F_{GB} \times \sin\beta - 20 - 20 = 0$$

and $\quad \sum F_X = F_{GF} - F_{AB} \times \cos\theta + F_{GB} \times \cos\beta = 0$

where, $\quad \tan\beta = BF/GF = (8/3)/3 = 8/9$

$$\sin\beta = 8/\sqrt{145} \ ; \ \cos\beta = 9/\sqrt{145}$$

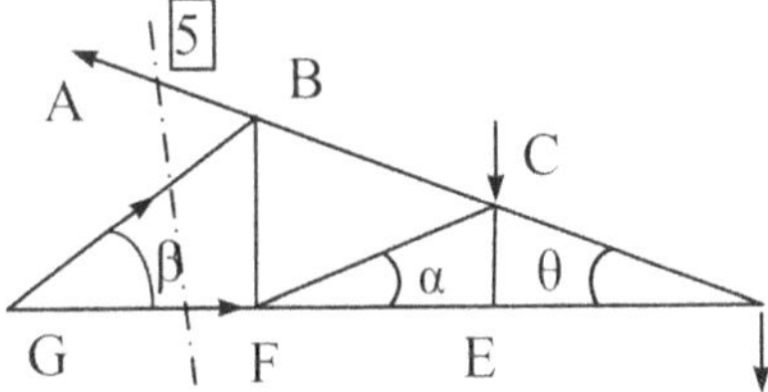

Solving these two equations, F_{GB} and F_{AB} can be obtained as

$$F_{GB} = 82 \text{ kN} \ \text{ and } \ F_{AB} = 10 \text{ kN}$$

*Please **note** that*

- Equations of equilibrium are written for the assumed directions of member forces, as shown in the figures. If the assumed direction is wrong, the calculated value of member force will be negative. It is assumed in the above equations that forces directed towards +ve X-axis (to the right, in the horizontal direction) are +ve and those directed towards –ve X-axis are –ve. Similarly, forces directed towards +ve Y-axis (upwards, in the vertical direction) are +ve and those directed along –ve Y-axis are –ve.

- The free body diagram of the truss obtained by cutting with the section plane-1 is acted upon by a system of concurrent forces; whereas the parts of truss obtained by cutting with the section planes 2 to 5 are acted upon by non-concurrent forces. It is therefore possible to choose different section planes so that three unknown member forces can be obtained from the three equations of equilibrium (for example, free body diagram with section plane-2 can be used to evaluate forces in three members BC, FC and EC; similarly free body diagram with section plane-4 can be used to evaluate forces in AB, GB and FB). There is *no unique set of section planes* for this analysis.

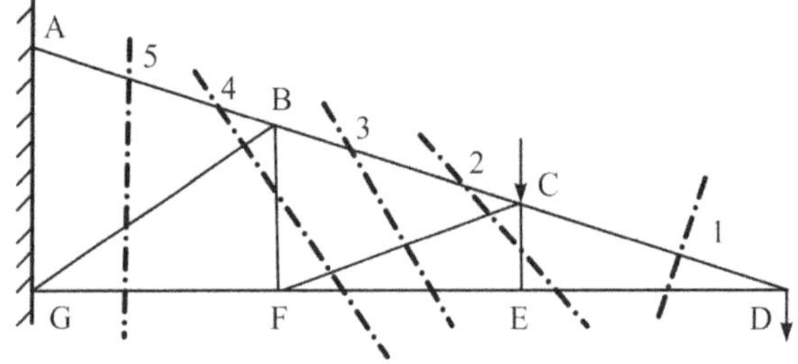

- It is possible to consider any one of the two parts of the truss, obtained by cutting with the section plane, for evaluating unknown member forces so long as the number of equilibrium equations match with the number of unknown member forces.

6.3   WORK-STRAIN ENERGY METHOD

Method of joints and method of sections give member forces only. The strain energy method can be used to calculate displacements in the members, based on the assumption that work done by external forces (W) is equal to the strain energy (U) stored in the member i.e., W=U. This is true for any *conservative system* where, energy is neither created by phenomenon like heating nor lost due to phenomenon like friction.

Example 6.3

A 4500 N load is attached to a pin at C as shown. Determine the forces acting in members AC and BC. Also calculate vertical displacement at C

Solution

$$\tan \theta = AB/AC = 900/1200 = 3/4$$

Therefore, $\sin \theta = 3/5$; $\cos \theta = 4/5$

Applying equation of equilibrium at C,

$$\sum F_X = 0 \Rightarrow F_{AC} + F_{BC} \times \cos \theta = 0$$

$$\sum F_Y = 0 \Rightarrow F_{BC} \times \sin \theta = +4500 \text{ N}$$

$$\Rightarrow \quad F_{BC} = 7500 \text{ N (tensile)}$$

$$F_{AC} = -6000 \text{ N (Compressive)}$$

Alternatively, using Lami's theorem, on free body diagram at C

$$F_{AC} / Sin\,(90 + \theta) = F_{BC} / Sin90 = 4500 / Sin\,(180 - \theta)$$

or $\qquad$ $F_{AC} = 4500 \times Sin\,(90 + \theta) / Sin(180 - \theta) = 6000\ N$

and $\qquad$ $F_{BC} = 4500 \times Sin\,90 / Sin\,(180 - \theta) = 7500\ N$

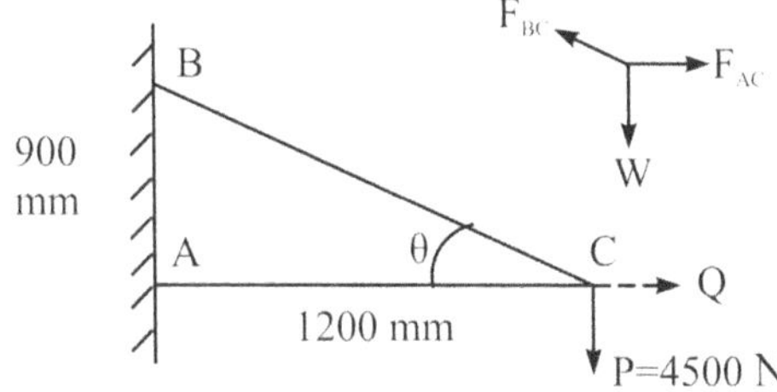

Forces F_{AC} and F_{BC} depend only the direction of members and the applied load. However, calculation of vertical displacement (δ_V) at the load point C depends on section property (A) as well as material property (E) of the members.

Work done by gradually applied load,

$W = (1/2) \times$ Force $\times$ displacement $= (1/2) \times 4500 \times \delta_V$

Strain energy, U for the system is equal to the sum of strain energies stored in the two members AC and BC.

Strain energy stored in each member,

$U_i = (1/2) \times F \times \delta = (1/2) \times F \times (F \times L) / (A \times E) = (1/2) \times (F^2 \times L) / (A \times E)$

Thus, Strain energy stored in the system,

$$U = U_{AC} + U_{BC} = (1/2) \times F_{AC} \times \delta_{AC} + (1/2) \times F_{BC} \times \delta_{BC}$$
$$= (1/2) \times (F_{AC}^2 \times L_{AC}) / (A \times E) + (1/2) \times (F_{BC}^2 \times L_{BC}) / (A \times E)$$
$$= 6000^2 \times 1200 / (2A \times E) + 7500^2 \times 1500 / (2A \times E)$$
$$= 432 \times 10^8 / (2A \times E) + 843.75 \times 10^8 / (2A \times E) = 637.875 \times 10^8 / (A \times E)$$

Since $\quad W = \dfrac{1}{2} \times 4500 \times \delta_v = U$, we get

$$\delta_V = 2U / 4500 = 28.35 \times 10^6 / (A \times E)$$

Horizontal component of displacement at C can be obtained from Castigliano's theorem, using an imaginary horizontal load at C, in the next section.

6.4   USING CASTIGLIANO'S THEOREM

Castigliano's theorem states that if total strain energy of a structure is expressed as a function of applied loads and corresponding displacements, its partial derivative w.r.t. each applied load will give displacement along that load at the point of application of load .i.e., $\delta_P = \partial U / \partial P$

In example 6.3, $F_{BC} = P/\sin\theta$ and $F_{AC} = P/\tan\theta - Q$

Then, total strain energy,

$$U = U_{AC} + U_{BC} = (1/2) \times (F_{AC}^2 \times L_{AC}) / (A \times E) + (1/2) \times (F_{BC}^2 \times L_{BC}) / (A \times E)$$

$$= (1/2) \times (P/\tan\theta - Q)^2 \times 1200 / (A \times E) + (1/2) \times (P/\sin\theta)^2 \times 1500 / (A \times E)$$

Displacements at C,

$$\delta_H = \partial U / \partial Q = -(P/\tan\theta - Q) \times 1200/(A \times E) + 0$$

$$\delta_V = \partial U / \partial P = (P/\tan\theta - Q) \times 1200 / (A \times E \times \tan\theta)$$

$$+ (P/\sin\theta) \times 1500 / (A \times E \times \sin\theta)$$

Substituting $Q = 0$, $\delta_H = -(P/\tan\theta) \times 1200/(A \times E)$

$$= -(4500 \times 4/3) \times 1200 / (A \times E) = -7.2 \times 10^6 / (A \times E)$$

and $\delta_V = (P/\tan\theta) \times 1200 / (A \times E \times \tan\theta) + (P/\sin\theta) \times 1500 / (A \times E \times \sin\theta)$

$$= (4500 \times 4/3) \times 1200/(A \times E \times 3/4) + (4500 \times 5/3) \times 1500/(A \times E \times 3/5)$$

$$= 28.35 \times 10^6 / (A \times E)$$

δ_V value obtained from Castigliano's theorem is same as the displacement value obtained by equating work done with total strain energy. In addition, δ_H value could also be calculated using Castigliano's theorem.

Please note that displacement sign corresponds to the force direction. Thus, –ve sign for δ_H indicates x-displacement at C along $-Q$ direction (to the left) while +ve sign for δ_V indicates y-displacement at C along $+P$ direction (downward).

6.5 ANALYSIS OF REDUNDANT TRUSSES

A redundant truss with more than the minimum required spar members can be analysed using method of superposition. The structure is first analysed as a non-redundant truss by removing redundant members. Unit loads are applied along one redundant member at a time and corresponding forces in all the members are evaluated. Actual forces in the redundant members are calculated for zero displacement in the redundant members. Actual forces in all the spars are then obtained by superposition of all these cases for the particular set of forces in the redundant members. The method is explained through the following example of a truss with two redundant members, requiring superposition of three cases.

Let us consider the frame shown here, having 6 joints and 11 members. A perfect frame with 6 joints should have m = 2j – 3 = 9 members. Thus, the given structure is redundant with two additional members. A vertical load at joint E and a horizontal load at joint F are applied. The frame is hinged at joint A and supported on rollers at joint C.

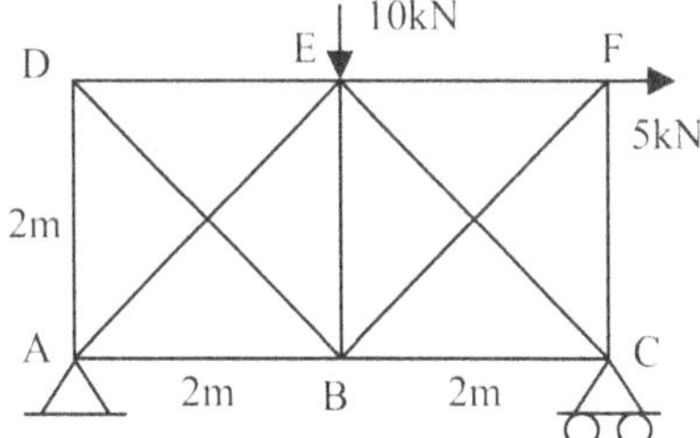

It can be analysed for the applied loads considering the frame as a perfect frame with 9 members as shown, by deleting members BD and CE. It can also be considered as a perfect frame by deleting some other members (a) AE and BF or (b) BD and BF or (c) AE and EC etc.

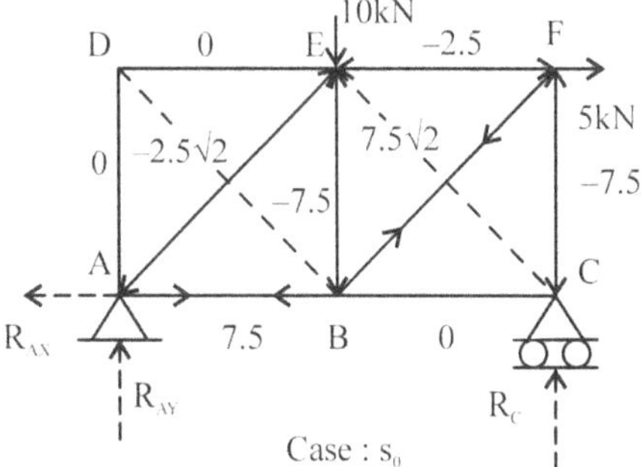

Support reactions are obtained from the three equations of static equilibrium, as

$$R_C = R_{CY} = (10 \times 2 + 5 \times 2)/4 = 7.5 \text{ kN}; \quad R_{AY} = 10 - 7.5 = 2.5 \text{ kN}; \quad R_{AX} = 5 \text{ kN}$$

Member forces (F_0 in kN) in the perfect frame obtained using method of joints are indicated in the figure, along with their type i.e., tensile (+ve) and compressive (–ve)

The effect of unit load in each of the redundant members is separately analysed. In these cases, reactions at the supports are non-existent as there are no unbalanced external forces in the frame. Member forces in the perfect frame corresponding to these unit loads (F_1 and F_2) are shown in the figures below.

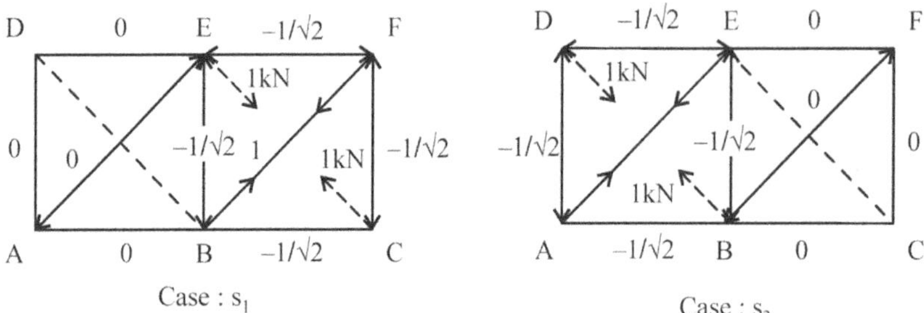

Lengths of diagonal members AE, BD, BF and CE are $2\sqrt{2}$ m while all other members are of length 2m.

Member forces in the actual redundant structure are now obtained from,

$$F = F_0 + a \times F_1 + b \times F_2$$

where, 'a' and 'b' are unknown constants (multiplication factors for the member forces), which are calculated from the two kinematic conditions (displacements along the two redundant members are equal to zero).

Using Castigliano's theorem, these conditions can be written as

$$\partial U/\partial P_1 = \partial U/\partial a = (\partial/\partial a) \times [F^2 L /(2A \times E)] = \{ L/(2A \times E) \} \times (\partial F^2/\partial a) = 0$$

If '$A \times E$' is same for all the members of the truss,

$$\{ L/(2A \times E) \} \times [\partial(F_0 + a \times F_1 + b \times F_2)^2 /\partial a] = 0$$

$$\Rightarrow \quad \{ L/(2A \times E) \} \times [\sum (2F_0 \times F_1) + 2a \times \sum F_1^2 + 2b \times \sum (F_1 \times F_2)] = 0$$

$$\Rightarrow \quad a \times \sum F_1^2 + b \times \sum (F_1 \times F_2) = - \sum (F_0 \times F_1) \qquad .....(6.1)$$

and $\quad \partial U/\partial P_2 = \partial U/\partial b = (\partial/\partial b) \{F^2 L /(2A \times E)\} = \{L/(2A \times E)\} \times (\partial F^2/\partial a) = 0$

If 'AE' is same for all the members of the truss,

$$\{L/(2A \times E)\} \times [\partial(F_0 + a \times F_1 + b \times F_2)^2 /\partial b] = 0$$

$$\Rightarrow \quad \{L/(2A \times E)\} \times [\sum (2F_0 \times F_2) + 2b \sum F_2^2 + 2a \sum (F_1 \times F_2)] = 0$$

$$\Rightarrow \quad a \times \sum (F_1 \times F_2) + b \times \sum F_2^2 = - \sum (F_0 \times F_2) \qquad .....(6.2)$$

By solving the two simultaneous equations (6.1) and (6.2), the constants a and b can be calculated.

In this example, assuming '$A \times E$' is same for all members, these equations reduce to

$$+ 9.656\, a + b = - 54.745 \quad \text{and} \quad a + 9.656\, b = 10$$

Solving them, we get $\quad a = -5.839 \quad$ and $\quad b = 1.636$

Member forces (in kN) in the actual redundant structure are now obtained from,

$$F = F_0 + a \times F_1 + b \times F_2 \quad \text{and are shown in the figure below.}$$

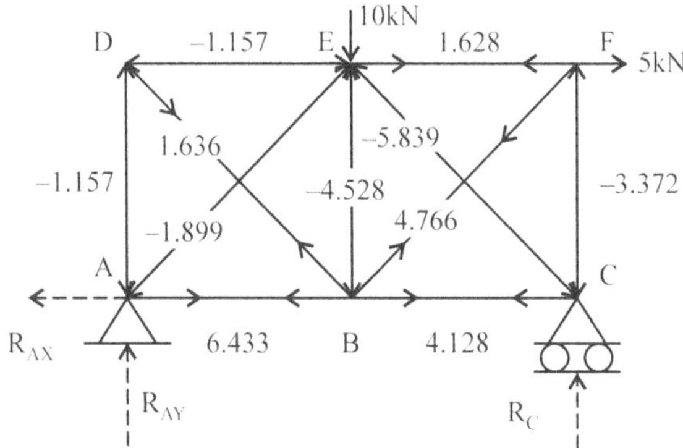

It may be noted that the reactions at the supports for the redundant frame and the perfect frame are same, being dependent only on the overall geometry of the structure and are independent of the member forces and number of members.

In general, if there are 'm' members in the truss and 'n' redundant members, the constants 'a_i' are calculated from

$$\partial U/\partial P_i = \partial U/\partial a_i = (\partial/\partial a_i) \left[\sum (F_j^2 \times L)/(2A \times E) \right] = 0 \text{ for } j = 1, m \text{ and } i = 1, n$$

$$\Rightarrow \qquad (\partial/\partial a_i) \left[\left\{ \sum F_0 + \sum (a_i \times F_i) \right\}^2 \times L/(2A \times E) \right] = 0$$

Actual forces in all the 'm' members can be obtained from

$$F_j = (F_0)_j + \left(\sum a_i \times F_i \right)_j \qquad \text{for} \quad i = 1, n \text{ and } j = 1, m$$

ADDITIONAL PROBLEMS FOR PRACTICE

1. A square truss with diagonal members is supported from one corner and a vertical load of 10kN is applied at the opposite corner. Calculate forces in all the members.

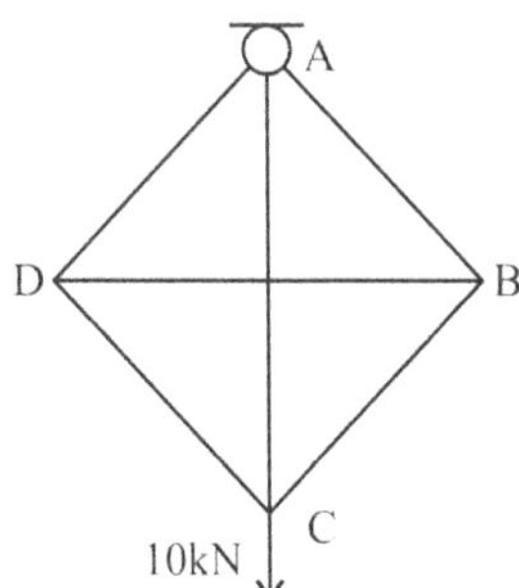

2. Determine forces in the members of the frame shown. Area of cross section of members AB and AC is '2A' and that of all other members is 'A'.

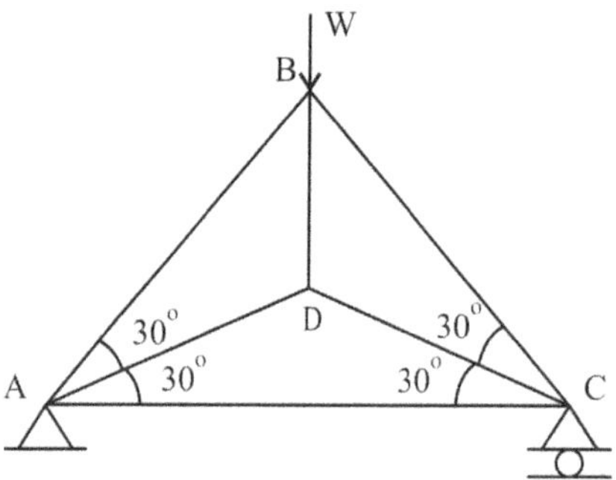

3. A braced cantilever is loaded as shown. Calculate forces in all the members.

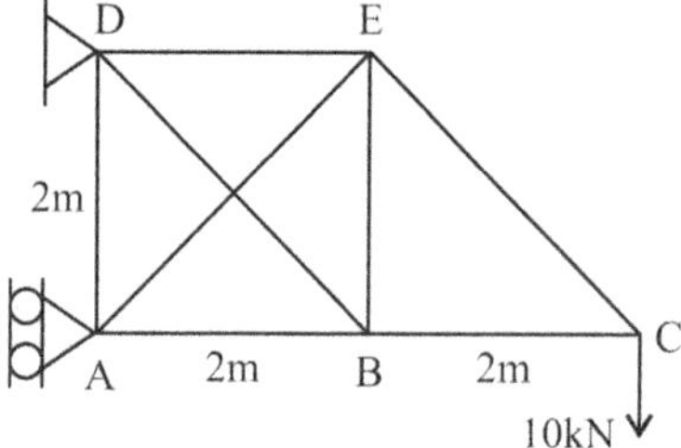

CHAPTER 7

ANALYSIS OF CONTINUUM STRUCTURES

Every body subjected to external loads deforms, however small the deformation may be, based on the material and nature of applied forces. A brief review of theory of elasticity is presented here for a clear understanding of the '***Mechanics of deformable solids***', in the subsequent chapters. For a more detailed explanation, the reader may refer to other standard books on this subject.

Every physical component is a three-dimensional solid. However, based on the relative dimensions along three coordinate directions and nature of applied loads / boundary conditions, they are approximated as 1-D, 2-D or 3-D components, neglecting variation of response in some directions. This approximation helps in analysing the component quickly and at a lower cost.

7.1 DEGREES OF FREEDOM (DOF)

The direction in which a point in a structure is free to move is defined as its degree of freedom (DOF). In general, any point in a component can move along an arbitrary direction in space and rotate about an arbitrary direction, depending upon the loads applied on the component. Since specifying this arbitrary direction through angles is tedious, the movement and rotation at any point are identified by their components in the chosen coordinate system. Thus, in Cartesian coordinate system, a point can at best have translation identified by its three components *along* the three coordinate directions X, Y and Z; and rotation identified by its three components *about* the three coordinate directions X, Y and Z. Depending on the way a member is assembled in a structure, some or all of its DOF at a point may be fixed or free. Truss members are designed for only axial loads and rotational DOFs are not relevant whereas beam is designed for bending loads which result in displacement at each point normal to the axis. Its derivative or slope is independently constrained and hence is considered as an independent DOF. Similarly, in plates subjected to in-plane loads, rotational DOFs are not significant and need not be considered as independent DOFs whereas in plates and shells subjected to bending loads, rotational DOF are significant and independently constrained. Hence, they are treated as independent DOFs.

7.2   RIGID BODY MOTION

It is the motion of entire component or part under the influence of external applied loads. Such a rigid body motion, with no relative deformation between any two points in the component, can not induce stresses or strains in a component. Since design or analysis of a component involves calculation of stresses and strains in a component, any static part with 'n' degrees of freedom can not be solved unless it is restrained, from moving as a rigid body by constraining it at least at one point along each DOF. The number of DOFs to be constrained depends on the type of component. For example, a truss has to be constrained for rigid body motion along X and Y axes; a plane frame (in X-Y plane) has to be constrained along X and Y axes as well as for rotation about Z axis; a thick shell has to be constrained along X, Y and Z axes as well as for rotation about X, Y and Z axes. In the conventional analysis by closed form solution to the differential equation, the rigid body motion is constrained by using relevant boundary conditions for evaluating constants of integration.

7.3   DISCRETE AND CONTINUUM STRUCTURES

Structures such as trusses and frames, which have many identifiable members, connected only at their end points or nodes are called *discrete structures*. Each member of the structure is considered as a one-dimensional (1-D) element along its length, identified by its end point coordinates. Their lateral dimensions are reflected in element properties like area of cross section in trusses and moment of inertia and depth of section in beams.

Structures such as plates, thin shells, thick shells, solids, etc which do not have distinctly identifiable members are called *continuum structures*. These are used along with discrete members in many aircraft structures, such as fuselage structure consisting of thin shell strengthened by ring frames and longitudinal stringers; shear webs consisting of plate in shear surrounded by axial loaded spars in wings. In such structures, each member has a distinct and different purpose.

7.4   MATERIAL TYPES

Many structures may consist of members made up of different materials with distinctly varying material properties, such as Modulus of elasticity (E), Modulus of rigidity (G), coefficient of linear thermal expansion. These properties may be

(a)   constant (linear stress-strain relationship) or variable (non-linear stress-strain relationship) over the range of load

(b)   same in all directions (isotropic) or vary in different directions (anisotropic or orthotropic)

(c)   constant over the temperature range or vary with temperature, particularly when the temperature range over the component is large

These variations may be inherent in the material or induced by the manufacturing processes like rolling, casting, welding etc. Treatment of non-homogeneous material, with varying properties at different locations of a component, is difficult and is also very unusual. In most cases, variation of material properties with the direction and with temperature may not be significant and hence neglected. So, an isotropic, homogeneous material is most often used in the analysis of a component.

7.5   LINEAR ANALYSIS

It is based on linear stress-strain relationship (Hooke's law) and is usually applicable to stress below the elastic limit (or yield stress), at any point in the component. In this analysis, ***linear superposition of results*** obtained for individual loads on a component is valid in order to obtain stresses due to any combination of these loads acting simultaneously. In some designs, it is necessary to check for many combinations of loads during the flight. In such a case, analysing for unit load of varying nature such as fuel weight; multiplying the stress results with the actual load corresponding to that particular time of the transient and adding to the stresses due to other constant loads will be economical.

7.6   NON-LINEAR ANALYSIS

In many cases where stress-strain relationship of a material or load displacement relationship is non-linear, the mathematical formulations are based on small deflection theory. A component with large deflections due to loads, such as aircraft wing, comes under the category of '*geometric non-linearity*'. In some aerospace applications, where the component is designed for single use, stress level above yield point may be permitted. In some other cases involving non-metallic components, material may exhibit non-linear stress-strain behaviour in the operating load range. These two cases come under the category of '*material non-linearity*'. In both these cases, analysis is carried out by applying the load in small steps and modifying geometry or material properties after each load step.

7.6.1   GEOMETRIC NON-LINEARITY

In these problems, geometry of the component is redefined after every load step by adding the displacements at various nodes to the nodal coordinates for defining the true geometry to be used for the next load step.

7.6.2   MATERIAL NON-LINEARITY

In these problems, total load on the component is applied in small steps and non-linear stress-strain relationship in the material, usually represented by the value of Young's Modulii of elasticity (normal stress / normal strain) E_X, E_Y and E_Z in different directions, is considered as linear in each load step (Ref.Fig.7.1). Here, normal stress and normal strain can be tensile (+ve) or compressive (-ve) and E has the same units as stress, since strain is non-dimensional. These values are suitably modified after each load step, till the entire load range is covered.

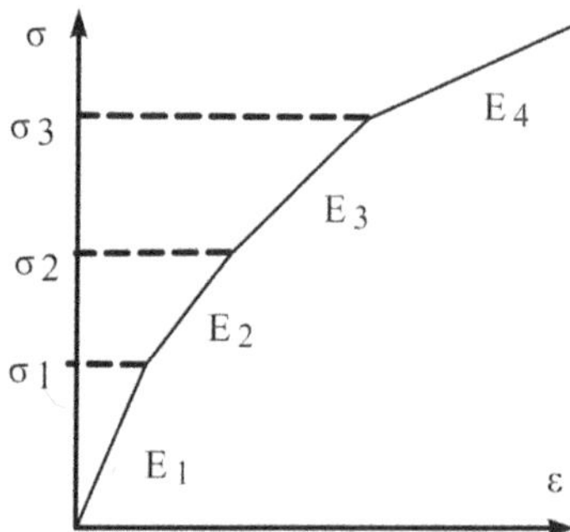

FIGURE 7.1 Stress-strain diagram of a non-linear material

7.7   STIFFNESS AND FLEXIBILITY

Loads and displacements in an element are related through stiffness and flexibility coefficients. Stiffness coefficient (K) is the force required to produce unit displacement, while Flexibility coefficient (F) is the displacement produced by a unit force. They are usually defined for 1-D elements such as truss elements or spring elements by

$$P = K \times u$$

where, $K = \dfrac{A \times E}{L}$ is the stiffness coefficient

$$\text{or} \qquad u = F \times P$$

where, $\qquad F = \dfrac{L}{A \times E}$ is the flexibility coefficient

since $\qquad \sigma = \dfrac{P}{A} = E\,\varepsilon = E\left(\dfrac{u}{L}\right)$

where, u is the change in length and, therefore, $P = \left(\dfrac{A \times E}{L}\right) \times u$

However, a more general definition of stiffness coefficient k_{ij} is the force required at node 'i' to produce unit displacement at node 'j'. i.e., $P_i = k_{ij} \times u_j$

An important feature of the **stiffness coefficient** is that it **has different units with reference to different loads and different displacements**. For example, units of k_{ij} relating load at 'i' to displacement at 'j' in a truss element are in 'N/mm' whereas units of k_{ij} relating moment at 'i' to slope at 'j' in a beam element is in 'N mm'

Stiffness coefficients k_{ij} connecting loads at 'n' degrees of freedom with displacements at these 'n' degrees of freedom thus form a square matrix of order $n \times n$, represented by [K]. In structures having linear force-deflection relationship, the stiffness coefficients have the property $k_{ij} = k_{ji}$. This is called **Maxwell's reciprocity relationship**. This makes the stiffness matrix symmetric.

7.8   STRESS AND STRAIN AT A POINT

Stress is the internal reaction in a component subjected to external forces. Thus, stress exists only when external force is applied on a component and varies from point to point. Stress at any point in a component is defined as a tensor with three components on each of the six faces of an infinitesimal cube around that point, as shown in Fig.7.2

Stress components are identified by two subscripts, the first subscript representing the direction of outward normal of the plane or face of the cube and the second subscript representing the direction of stress on that plane. Thus σ_{XX} (or σ_X) represents normal stress (along X-direction) on the plane whose outward normal is along +ve X-axis (identified as +X face) while σ_{XY} (or τ_{XY}) and σ_{XZ} (or τ_{XZ}) represent shear stresses along Y and Z directions on the same plane (+X face). Stresses are similarly defined on the other five faces of the cube.

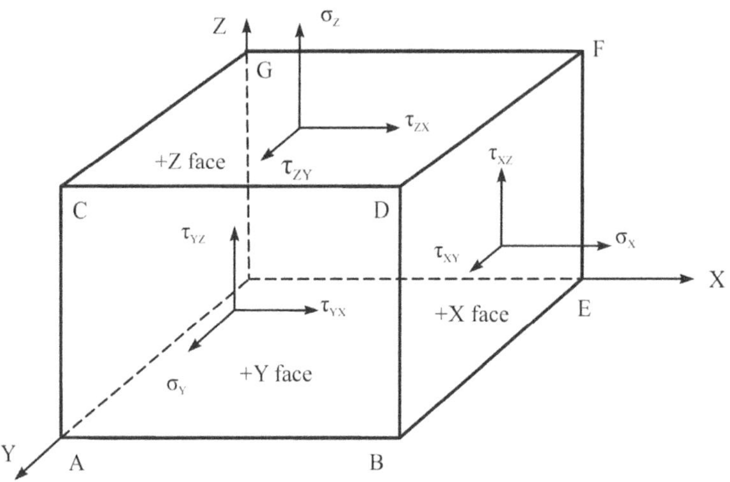

FIGURE 7.2 Stress at a point

For the equilibrium of this small element along X, Y and Z directions, it is seen that σ_X, τ_{XY} and τ_{XZ} on the +X face are equal to σ_X, τ_{XY} and τ_{XZ} on the $-$X face. Similar conditions hold good on the Y and Z faces. So, out of the eighteen stress components on the six faces of the cube, only the following nine stress components are considered independent. These are represented as

$$\sigma_X \quad \tau_{XY} \quad \tau_{XZ} \qquad \tau_{YX} \quad \sigma_Y \quad \tau_{YZ} \qquad \tau_{ZX} \quad \tau_{ZY} \quad \sigma_Z$$

Also, for the equilibrium of the element for moments about X, Y and Z directions, $\tau_{XY} = \tau_{YX}$, $\tau_{XZ} = \tau_{ZX}$ and $\tau_{YZ} = \tau_{ZY}$ being the complementary shear stresses on two perpendicular faces. Thus, only three normal stress components σ_X, σ_Y, σ_Z and three shear stress components τ_{XY}, τ_{YZ} and τ_{ZX} are identified at each point in a component. In the books on Finite Element Method (FEM), these components are written as a stress vector $\{\sigma\}$ for convenience of matrix operations.

Normal stress on any plane whose normal is **N** (with components N_X, N_Y and N_Z along Cartesian X, Y and Z directions) can be obtained from the stress components available in Cartesian coordinate system by the following relationship.

$$\sigma_n = T_X \times N_X + T_Y \times N_Y + T_Z \times N_Z$$

where,

$$T_X = \sigma_X \times N_X + \tau_{XY} \times N_Y + \tau_{XZ} \times N_Z$$
$$T_Y = \tau_{XY} \times N_X + \sigma_Y \times N_Y + \tau_{YZ} \times N_Z$$
$$T_Z = \tau_{XZ} \times N_X + \tau_{YZ} \times N_Y + \sigma_Z \times N_Z$$

In 2-D case, $\quad \tau_{XZ} = \tau_{YZ} = \sigma_Z = 0$; $N_X = \cos\theta$; $N_Y = \sin\theta$

Therefore, $\quad \sigma_n = \sigma_X \times N_X^2 + \sigma_Y \times N_Y^2 + 2\,\tau_{XY} \times N_X \times N_Y$

$$= \sigma_X \times \cos^2\theta + \sigma_Y \times \sin^2\theta + 2\,\tau_{XY} \times \cos\theta \times \sin\theta$$

$$= \sigma_X \times \cos^2\theta + \sigma_Y \times \sin^2\theta + \tau_{XY} \times \sin 2\theta$$

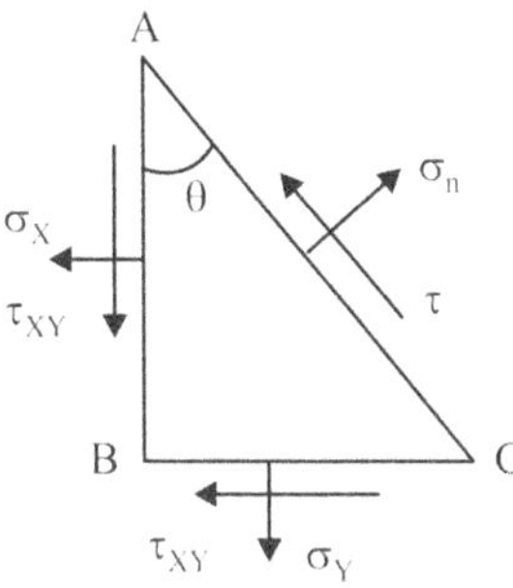

It can also be seen from force equilibrium condition, by resolving forces normal to AC and along AC,

$$\sigma_n \times AC = (\sigma_X \times AB) \times \cos\theta + (\sigma_Y \times BC) \times \sin\theta$$

$$+ (\tau_{XY} \times AB) \times \sin\theta + (\tau_{XY} \times BC) \times \cos\theta$$

$$= (\sigma_X \times AC \times \cos\theta) \times \cos\theta + (\sigma_Y \times AC \times \sin\theta) \times \sin\theta$$

$$+ (\tau_{XY} \times AC \times \cos\theta) \times \sin\theta + (\tau_{XY} \times AC \times \sin\theta) \times \cos\theta$$

Dividing throughout by AC, $\sigma_n = \sigma_X \cos^2\theta + \sigma_Y \sin^2\theta + \tau_{XY} \sin 2\theta$

Similarly, $\quad \tau \times AC = (\sigma_X \times AB) \times \sin\theta - (\sigma_Y \times BC) \times \cos\theta$

$$- (\tau_{XY} \times AB) \times \cos\theta + (\tau_{XY} \times BC) \times \sin\theta$$

$$= (\sigma_X \times AC \times \cos\theta) \times \sin\theta - (\sigma_Y \times AC \times \sin\theta) \times \cos\theta$$

$$- (\tau_{XY} \times AC \times \cos\theta) \times \cos\theta + (\tau_{XY} \times AC \times \sin\theta) \times \sin\theta$$

Dividing throughout by AC, $\quad \tau = [(\sigma_X - \sigma_Y)/2] \times \sin 2\theta - \tau_{XY} \times \cos 2\theta$

In a similar way, strain at a point is defined as a tensor of eighteen components, out of which six components consisting of normal strains ε_X, ε_Y and ε_Z and shear strains γ_{XY}, γ_{YZ} and γ_{ZX} are considered independent. Normal strain along a direction is defined as the change in length per unit length along that direction while shear strain ($\gamma_{XY} = \partial u/\partial y + \partial v/\partial x,..$) is defined as the change in the included angle as shown in Fig.7.3. In the books on FEM, these components are written as a strain vector $\{\varepsilon\}$ for convenience of matrix operations.

In the case of 2-D and 3-D elements, general relations between displacements and strains and between strains and stresses, as obtained in the theory of elasticity, are used for calculating element stiffness matrices. These relations are given later with the following notation.

Notation: u, v, w are the displacements along X, Y and Z directions

ε_X, ε_Y, ε_Z are the normal strains

γ_{XY}, γ_{YZ}, γ_{ZX} are the shear strains

σ_X, σ_Y, σ_Z are the normal stresses

τ_{XY}, τ_{YZ}, τ_{ZX} are the shear stresses

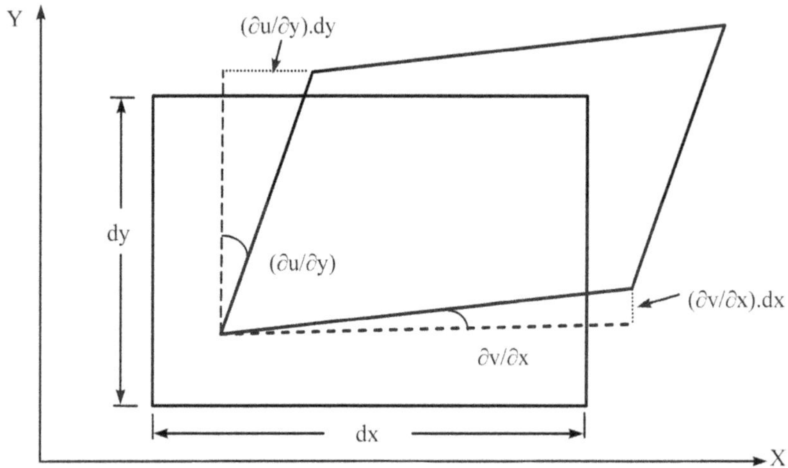

FIGURE 7.3 Shear strain at a point

7.9   PRINCIPAL STRESSES

Depending on the type of structure and the load applied, three normal stresses and three shear stresses can exist at any point in the structure. These stress components are calculated with reference to the coordinate system used. However, there exists a plane along which shear stress is zero and the corresponding normal stresses (maximum and minimum) are called principal normal stresses or principal stresses. Along some direction, inclined to the two principal stresses, there exists maximum or principal shear stress. These are of interest to any designer as the component has to be designed to limit these stresses to the allowable limits of the material. They can be calculated, for a two-dimensional stress state, from the condition

$$\partial\sigma_n/\partial\theta = -\,2\,\sigma_X \times \cos\theta \times \sin\theta + 2\,\sigma_Y \times \sin\theta \times \cos\theta + 2\,\tau_{XY} \times \cos 2\theta = 0$$

$$\Rightarrow \quad -\,(\sigma_X - \sigma_Y) \times \sin 2\theta + 2\,\tau_{XY} \times \cos 2\theta = 0$$

$$\Rightarrow \quad \tan 2\theta = 2\,\tau_{XY}\,/\,(\sigma_X - \sigma_Y)$$

It also corresponds to the plane where, $\tau = [(\sigma_X - \sigma_Y)/2] \times \sin 2\theta - \tau_{XY} \times \cos 2\theta = 0$

Corresponding to this angle,

$$\sigma_1 = (\sigma_X + \sigma_Y)/2 + \sqrt{[(\sigma_X - \sigma_Y)/2]^2 + \tau^2}$$

$$\sigma_2 = (\sigma_X + \sigma_Y)/2 - \sqrt{[(\sigma_X - \sigma_Y)/2]^2 + \tau^2} \qquad .....(7.1)$$

Max shear stress occurs at 45^0 from the direction of principal axes and is given by

$$\tau_{max} = (\sigma_1 - \sigma_2)/2 = \sqrt{[(\sigma_X - \sigma_Y)/2]^2 + \tau^2}$$

From uni-axial tensile test, where load is applied along Y-direction only,

$$\sigma_X = 0 \quad \text{and} \quad \tau = 0$$

Then, we get $\sigma_1 = \sigma_{max} = \sigma_Y$; $\quad \sigma_2 = 0 \quad$ and $\quad \tau_{max} = \sigma_1/2 = 0.5\,\sigma_{max}$

Example 7.1

A bolt of 20 mm diameter is subjected to a tensile force of 15 kN and shear force of 10 kN. Calculate normal and shear stresses on a plane inclined at 60^0 to the bolt axis. Also calculate max normal stress and max shear stress.

Solution

Normal stress, $\quad \sigma = P_T/A = 15{,}000 / (\pi \times 20^2/4) = 47.75$ N/mm^2

Shear stress, $\quad \tau = P_S/A = 10{,}000 / (\pi \times 20^2/4) = 31.83$ N/mm^2

On inclined plane, $\quad \sigma_n = \sigma \times (1 + \cos 2\theta)/2 + \tau \times \sin 2\theta = 39.5$ N/mm^2

and $\quad \tau_n = (\sigma \times \sin 2\theta)/2 - \tau \times \cos 2\theta = 36.6$ N/mm^2

Max normal stress, $\quad \sigma_{max} = \sigma_1 = (\sigma/2) + \sqrt{(\sigma/2)^2 + \tau^2} = 63.66$ N/mm^2

Max shear stress, $\quad \tau_{max} = \sqrt{(\sigma/2)^2 + \tau^2} = 39.789$ N/mm^2

Example 7.2

Stress state at a point in a body consists of normal stresses +200 N/mm^2 and −50 N/mm^2 on two perpendicular axes and a shear stress of 80 N/mm^2. Calculate principal stresses, orientation of principal planes and max shear stress.

Solution

Max normal stress, $\sigma_{max} = \sigma_1 = (\sigma_X + \sigma_Y)/2 + \sqrt{[(\sigma_X - \sigma_Y)/2]^2 + \tau^2}$

$$= 75 + 134.63 = 209.63 \text{ N/mm}^2$$

Min normal stress, $\sigma_{min} = \sigma_2 = (\sigma_X + \sigma_Y)/2 - \sqrt{[(\sigma_X - \sigma_Y)/2]^2 + \tau^2}$

$$= 75 - 134.63 = -59.63 \text{ N/mm}^2$$

Max shear stress, $\tau_{max} = \sqrt{[(\sigma_X - \sigma_Y)/2]^2 + \tau^2} = 134.63 \text{ N/mm}^2$

Orientation of principal planes, $\theta = (1/2)\tan^{-1}[2\tau_{XY}/(\sigma_X - \sigma_Y)] = 16.3^0 \text{ or } 106.3^0$

7.10   MOHR'S CIRCLE FOR REPRESENTATION OF 2-D STRESSES

The principal stresses can also be obtained by graphical method using Mohr's circle. Here, normal stresses are represented on X-axis and shear stresses on Y-axis. Principal or maximum normal stress (represented by line CD) is inclined to the given stress state by an angle θ, while maximum shear stress is the maximum ordinate of the circle and exists on a plane inclined at $(45 - \theta)^0$ from the given stress state.

Procedure

Plot OA and OB to represent normal stresses σ_X and σ_Y along X-axis. Plot AC and BD to represent shear stress τ_{XY}, parallel to Y-axis. With CD as diameter construct a circle (Ref.Fig.7.4). Then, OF and OG, extreme points of the circle on X-axis, indicate maximum and minimum principal stresses, inclined at an angle θ with the plane CD representing the given stress plane. Length EH, radius of the circle, indicates max shear stress.

It can be observed that sum of normal stresses on any plane at a point is constant and is called *1st stress invariant*, I_1

i.e., $I_1 = \sigma_X + \sigma_Y = \sigma_1 + \sigma_2$ for 2-D stress case

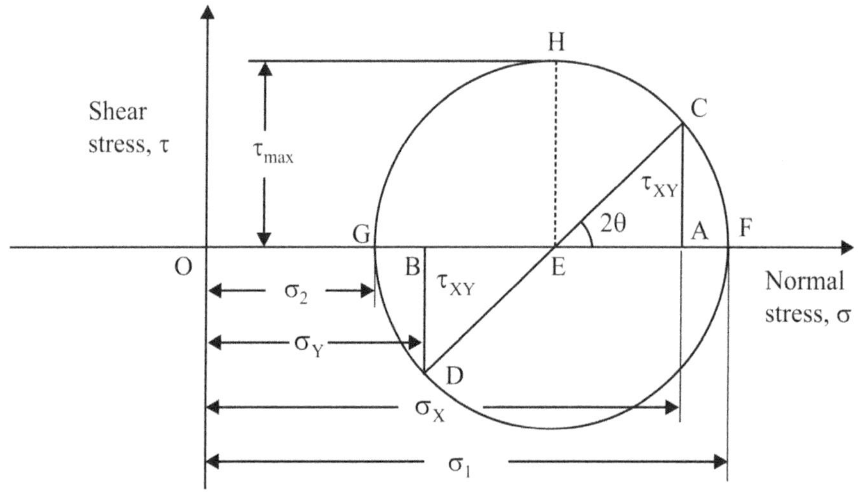

FIGURE 7.4 Mohr's circle for 2-D stress representation

Two special cases of Mohr's circle are of special importance

(a) For uniaxial tensile test, from which material properties are usually evaluated, the load is applied along one axis (usually Y-axis) and other components of stress are all zero. Then, $\sigma_1 = \sigma_Y$; $\sigma_2 = 0$ and $\tau_{max} = \sigma_Y / 2$ as can be seen from Fig.7.5

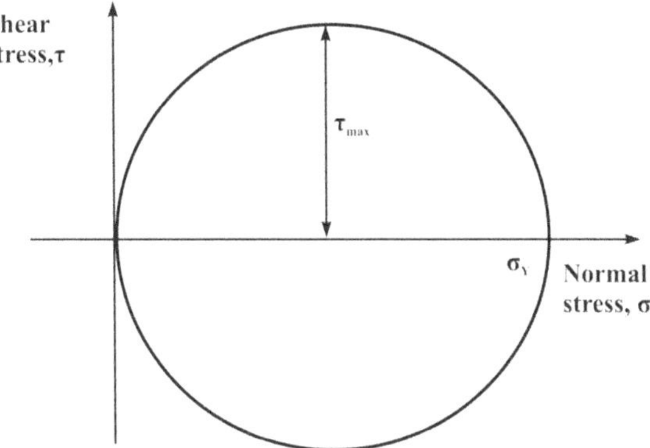

FIGURE 7.5 Mohr's circle for uni-axial tensile test

(b) In the earlier cases, it is seen that normal stress corresponding to maximum shear stress is not zero. If the normal stress, associated with maximum shear stress, is zero, then it is called ***pure shear state***. It is possible when a component is subjected to torsion only. In this case of pure shear, on a plane inclined at 45^0 to the plane of maximum shear stress, radius of the circle is

$$\sigma_1 = -\sigma_2 = \tau_{max}$$ as can be seen from Fig.7.6

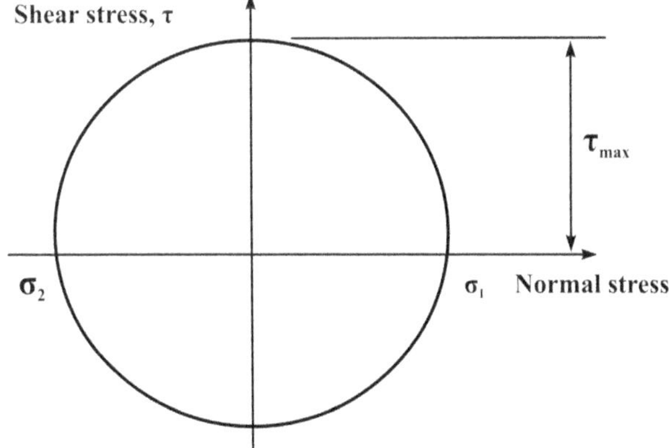

FIGURE 7.6 Mohr's circle for pure shear

This can also be understood by considering stresses acting at a point, identified by a small cube around it, when the component is subjected to torsion. Torsion is represented by a couple formed by two equal and opposite shear forces acting on opposite faces of the cube, as shown in Fig.7.7 (a). An equal and opposite couple is automatically formed, if the component is in static equilibrium, as shown in Fig. 7.7 (b). The shear forces resulting from this couple are called *complementary shear forces (stresses)*. If two different free bodies of half this cube about its diagonals are considered, resultant of the two shear forces on adjacent surfaces will result in tensile stress on one diagonal (BD) and a compressive stress on the other diagonal (AC), as shown in Fig. 7.7 (c).

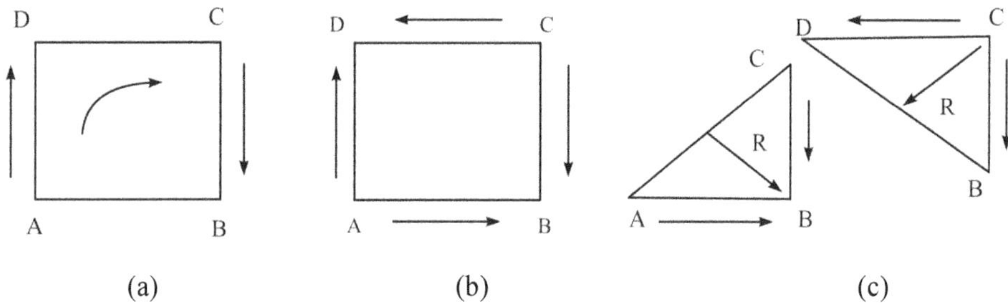

FIGURE 7.7 Representation of shear, complementary shear and principal normal stresses

If 'q' is the shear stress on the face of the cube and 's' is the side of the cube, then

Resultant force on the diagonal,

$$R = 2 \times q \times s^2 \times \cos \theta.$$

Stress on this diagonal,

$$\sigma = R/A = 2 \times q \times s^2 \times \cos \theta / 2 \times s^2 \times \cos \theta = q$$

Thus, both these stresses (maximum or principal normal stresses) are equal in magnitude to the shear stress on the surfaces and are inclined at 45^0 to the shear stresses, as already seen in the corresponding Mohr's circle.

Maximum shear stress theory, the most conservative and commonly used theory of failure, suggests that a component fails when the maximum shear stress at any point in a component exceeds the allowable maximum shear stress value of the material.

7.11   PRINCIPAL STRAINS

Similar to stresses on an inclined plane, strains on a plane inclined at angle θ with X-axis can be calculated from strains ε_X, ε_Y and γ_{XY} using

$$\varepsilon_n = (\varepsilon_X + \varepsilon_Y)/2 + [(\varepsilon_X - \varepsilon_Y)/2] \times \cos 2\theta + (\gamma_{XY}/2) \times \cos 2\theta$$

$$\text{and} \quad \gamma_n/2 = [(\varepsilon_X - \varepsilon_Y)/2] \times \sin 2\theta - (\gamma_{XY}/2) \times \cos 2\theta$$

Maximum normal strains or Principal strains can be obtained for the condition

$$\gamma_n = 0 \quad \text{or} \quad \tan 2\theta = \gamma_{XY} / (\varepsilon_X - \varepsilon_Y)$$

$$\text{as} \quad \varepsilon_{1,2} = (\varepsilon_X + \varepsilon_Y)/2 \pm (1/2)\sqrt{(\varepsilon_X - \varepsilon_Y)^2 + \gamma_{XY}^2}$$

Maximum shear strain can be obtained from

$$(\gamma/2)_{max} = (1/2)\sqrt{(\varepsilon_X - \varepsilon_Y)^2 + \gamma_{XY}^2} = (\varepsilon_1 - \varepsilon_2)/2$$

These can also be obtained from Mohr's circle of strains.

Measurement of three normal strains on the surface of a component, using **strain gauge rosette**, can also be used to calculate surface stresses. If ε_a, ε_b and ε_c are three normal strains measured along the directions as shown in the figure and ε_1 and ε_2 are the principal strains (along X and Y directions with $\gamma_{XY} = 0$,

$$\varepsilon_a = (\varepsilon_1 + \varepsilon_2)/2 + [(\varepsilon_1 - \varepsilon_2)/2] \times \cos 2\theta$$

$$\varepsilon_b = (\varepsilon_1 + \varepsilon_2)/2 + [(\varepsilon_1 - \varepsilon_2)/2] \times \cos 2(\theta + \alpha)$$

$$\varepsilon_c = (\varepsilon_1 + \varepsilon_2)/2 + [(\varepsilon_1 - \varepsilon_2)/2] \times \cos 2(\theta + \alpha + \beta)$$

Using the measured values of strains ε_a, ε_b and ε_c and knowing the angles α and β between the strain gauges, principal strains ε_1 and ε_2 and the angle θ can be calculated from the three equations.

If $\alpha = \beta = 45^0$ (**rectangular strain gauge rosette**),

$$\varepsilon_{1,2} = (\varepsilon_a + \varepsilon_c)/2 \pm (1/2)\sqrt{(\varepsilon_a - \varepsilon_b)^2 + (\varepsilon_c - \varepsilon_b)^2}$$

$$\text{and} \quad \tan 2\theta = (2\varepsilon_b - \varepsilon_a - \varepsilon_c) / (\varepsilon_a - \varepsilon_c)$$

Using the stress-strain relations for 2-D plane stress condition, surface principal stresses can be obtained from

$$\sigma_1 = [E/(1 - v^2)] \times (\varepsilon_1 + v\,\varepsilon_2) \quad \text{and} \quad \sigma_2 = [E/(1 - v^2)] \times (\varepsilon_2 + v\,\varepsilon_1)$$

Example 7.3

Stress state at a point in a body consists of normal stresses $+ 60$ N/mm^2 and $- 40$ N/mm^2 on two perpendicular axes and a shear stress of 50 N/mm^2. If young's modulus is 200GPa and Poisson's ratio is 0.3, calculate strains in that plane as well as principal strains and orientation of principal planes.

Solution

Normal strains, $\varepsilon_X = (\sigma_X - v\,\sigma_Y)/E = 36 \times 10^{-5}$; $\varepsilon_Y = (\sigma_Y - v\,\sigma_X)/E = -29 \times 10^{-5}$

Shear strain, $\gamma_{XY} = \tau_{XY}/G = \tau_{XY} / [E / \{2(1+v)\}] = 65 \times 10^{-5}$

Principal strains, $\varepsilon_{1,2} = (\varepsilon_X + \varepsilon_Y)/2 \pm (1/2)\sqrt{(\varepsilon_X - \varepsilon_Y)^2 + \gamma_{XY}^2}$

$$= 49.5 \times 10^{-5},\ -42.5 \times 10^{-5}$$

Orientation of principal planes is given by

$$\theta = (1/2) \times \tan^{-1}[\gamma_{XY}/(\varepsilon_X - \varepsilon_Y)] = 22.5^0 \ \text{ or }\ 112.5^0$$

Example 7.4

A rectangular strain gauge rosette attached to the surface of an aluminium bar, subjected to axial and shear load, gave readings $\varepsilon_a = 10 \times 10^{-4}$; $\varepsilon_b = -2 \times 10^{-4}$ and $\varepsilon_c = -3 \times 10^{-4}$, where a and c are along and perpendicular to the axis of the bar. Calculate axial and shear stresses in the bar. Also calculate axial and transverse loads applied on the bar, if its cross section is rectangular 15mm × 5mm

Solution

Assume E = 70 GPa for Aluminium

Principal strains, $\varepsilon_{1,2} = (\varepsilon_a + \varepsilon_c)/2 \pm (1/2)\sqrt{(\varepsilon_a - \varepsilon_b)^2 + (\varepsilon_c - \varepsilon_b)^2}$

$$= 10^{-4} \times [(10-3)/2 \pm (1/2)\sqrt{(10+2)^2 + (-3+2)^2}]$$

$$= 10^{-4} \times [3.5 \pm 8.51] = 12.01 \times 10^{-4},\ -5.01 \times 10^{-4}$$

Principal stresses, $\sigma_1 = [E/(1-v^2)] \times (\varepsilon_1 + v\,\varepsilon_2) = 80.85$ N/mm^2

and $\sigma_2 = [E/(1-v^2)] \times (\varepsilon_2 + v\,\varepsilon_1) = -10.85$ N/mm^2

Since $\sigma_Y = 0$, principal stresses $\sigma_{1,2} = (\sigma_X/2) \pm \sqrt{(\sigma_X/2)^2 + \tau_{XY}^2}$

$$\Rightarrow \quad \sigma_X = \sigma_1 + \sigma_2 = 70 \text{ N/mm}^2$$

and $\tau_{XY} = 29.62$ N/mm^2

Axial load applied on the bar, $P_X = \sigma_X \times A = 70 \times (15 \times 5) = 3250$ N

Transverse load applied, $P_Y = \tau_{XY} \times A = 29.62 \times (15 \times 5) = 2221$ N

7.12   VonMISES STRESS

It is the stress related to the three principal stresses at any point. It is used in Maximum distortion energy theory, which states that a component fails only due to distortion in shape and is independent of volumetric expansion or contraction. VonMises stress or Equivalent stress is given by

$$\sigma_{eq} = \sqrt{\frac{[(\sigma_1 - \sigma_2)^2 + (\sigma_2 - \sigma_3)^2 + (\sigma_3 - \sigma_1)^2]}{2}} \leq \sigma_Y \qquad .....(7.2)$$

It is also represented in terms of 1st and 2nd stress invariants (I_1 and I_2), as

$$\sigma_{eq} = \sqrt{I_1^2 - 3\,I_2}$$

where, $I_1 = \sigma_X + \sigma_Y + \sigma_Z = \sigma_1 + \sigma_2 + \sigma_3$

and $I_2 = \sigma_X \times \sigma_Y + \sigma_Y \times \sigma_Z + \sigma_Z \times \sigma_X - \tau_{XY}^2 - \tau_{YZ}^2 - \tau_{ZX}^2$

$\qquad\qquad = \sigma_1 \times \sigma_2 + \sigma_2 \times \sigma_3 + \sigma_3 \times \sigma_1$

In 2-D *plane stress case*, $\sigma_Z = 0$ and $\sigma_3 = 0$

$\qquad I_1 = \sigma_X + \sigma_Y = \sigma_1 + \sigma_2$ and $I_2 = \sigma_X \times \sigma_Y - \tau_{XY}^2 = \sigma_1 \times \sigma_2$

In 2-D *plane strain case*, $\sigma_z = \nu\,(\sigma_X + \sigma_Y)$

$\qquad I_1 = \sigma_X + \sigma_Y + \sigma_Z = \sigma_1 + \sigma_2 + \sigma_3$

and $I_2 = \sigma_X \times \sigma_Y + \sigma_Y \times \sigma_Z + \sigma_Z \times \sigma_X - \tau_{XY}^2 = \sigma_1 \times \sigma_2 + \sigma_2 \times \sigma_3 + \sigma_3 \times \sigma_1$

Comparing distortion energy of specimen in uni-axial tensile test,

$\qquad$ where $\sigma_1 = \sigma_Y$; $\qquad \sigma_2 = \sigma_3 = 0,$

with pure shear state, where $\qquad \tau_{max} = \sigma_1 = \sigma_2$,

$\qquad$ we get, $\tau_{max} \leq \sigma_Y / \sqrt{3}$ or $0.577\,\sigma_Y$

This stress value is used in the ***distortion energy theory of failure*** and is very popular.

7.13   THEORY OF ELASTICITY

The definitions of stress, strain and the relationships between displacement, strain and stress are explained in Strength of materials with special reference to 1-D structures. Before analysing continuum structures, a more general understanding of these concepts is essential. In the following, a brief discussion of these concepts is presented. Reader is advised to go through any book on theory of elasticity for a more detailed presentation of these concepts.

7.13.1   RIGIDITY MODULUS, BULK MODULUS

Shear modulus or Rigidity modulus, G = shear stress / shear strain

and Bulk modulus, K = Normal stress / Volumetric strain

where, volumetric strain $\varepsilon_V = \delta v/v \simeq \varepsilon_X + \varepsilon_Y + \varepsilon_Z$

neglecting higher order small terms

$$= \frac{(\sigma_X - v\,\sigma_Y - v\,\sigma_Z)}{E} + \frac{(\sigma_Y - v\,\sigma_Z - v\,\sigma_X)}{E} + \frac{(\sigma_Z - v\,\sigma_X - v\,\sigma_Y)}{E}$$

$$= (\sigma_X + \sigma_Y + \sigma_Z) \times \frac{(1 - 2v)}{E} \qquad\qquad .....(7.3)$$

$= 3\,\sigma \times (1 - 2v) / E$ when same load is acting along X, Y and Z directions

$$= \sigma_X \frac{(1 - 2v)}{E} \qquad \text{when load is acting along X direction only}$$

Bulk Modulus, $K = \sigma / \varepsilon_V = E / [3(1 - 2v)]$

Rigidity modulus and bulk modulus also have the same units as stress (N/m^2 or Pa)

Bulk modulus has very limited applications in structural analysis. For any given material, modulus of elasticity and modulus of rigidity reduce at higher temperatures.

The three material constants (modulii) are mutually related by the following expression

$$E = 2G \times (1 + v) = 3K \times (1 - 2v) \quad \text{or} \quad E \times (G + 3K) = 9G \times K \qquad .....(7.4)$$

Example 7.5

A 2m long M.S. rod of 15mm diameter is pulled by an axial force of 50 kN. Determine change in length, change in diameter and change in volume

Solution

Assuming E for mild steel $= 2 \times 10^5$ N/mm^2,

Change in length, $\Delta L = P \times L / (A \times E) = 2.83$ mm

Change in diameter, $\Delta D = D \times \varepsilon_D = D \times (v \times \varepsilon_L) = D \times (v \times \Delta L/L) = -0.00679$ mm

Change in volume, $\Delta V = V \times \varepsilon_V = V \times (1 - 2v) \times \varepsilon_L = 180$ mm^3

7.13.2 STRAIN - DISPLACEMENT RELATIONS

Strains can also be expressed as functions of displacement components at a point in the three Cartesian coordinate directions. If u, v and w (all functions of location of point, represented by its x, y and z coordinates) represent the displacement components along X, Y and Z directions, then

$$\varepsilon_X = \frac{\partial u}{\partial x} \qquad\qquad \varepsilon_Y = \frac{\partial v}{\partial y} \qquad\qquad \varepsilon_Z = \frac{\partial w}{\partial z}$$

$$\gamma_{XY} = \frac{\partial u}{\partial y} + \frac{\partial v}{\partial x} \qquad \gamma_{YZ} = \frac{\partial v}{\partial z} + \frac{\partial w}{\partial y} \qquad \gamma_{ZX} = \frac{\partial w}{\partial x} + \frac{\partial u}{\partial z} \quad .....(7.5)$$

7.13.3 THERMAL STRESS

Thermal strains do not induce any stresses unless thermal expansion is constrained. Also, temperature change produces same effect in all the directions of a component and Poisson's effect is not applicable (Ref. Fig.7.8 case-a)

For example stress in a uniform bar, subjected to temperature rise by ΔT, is dependent on the end condition as shown below. Let α be the coefficient of linear thermal expansion. Total elongation of a bar of length L due to increase in its temperature by ΔT is $L \times \alpha \times \Delta T$. Then, stress in the bar depends on the constraint (boundary condition) for its expansion, as shown in Fig.7.8. In each case, stress $\sigma = E \times \varepsilon = (E/L) \times$ Restrained part of expansion

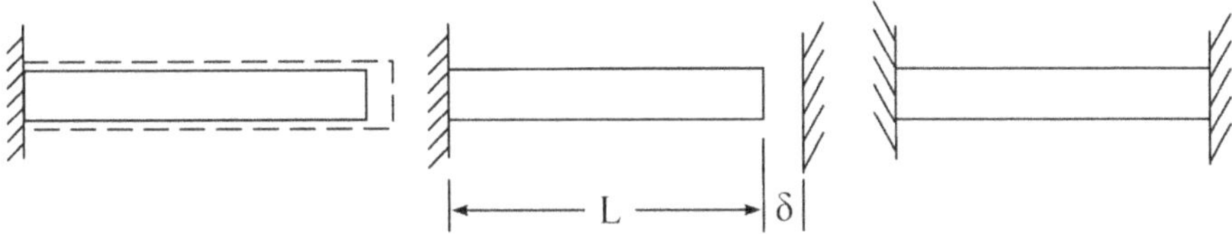

Case-a: Unconstrained   Case-b: Partially constrained   Case-c: Fully constrained

$$(\delta < L \times \alpha \times \Delta T)$$

Stress, $\sigma = 0$ $\qquad\qquad \sigma = E \times (L \times \alpha \times \Delta T - \delta) / L \qquad\qquad \sigma = E \times \alpha \times \Delta T$

FIGURE 7.8 Stress due to thermal expansion

7.13.4 STRESS-STRAIN RELATIONS

These are also called Constitutive equations (from generalised Hooke's law)

- For linearly elastic and isotropic material

$$\varepsilon_X = \frac{(\sigma_X - \nu\sigma_Y - \nu\sigma_Z)}{E}; \qquad \gamma_{XY} = \frac{\tau_{XY}}{G}$$

$$\varepsilon_Y = \frac{(-\nu\sigma_X + \sigma_Y - \nu\sigma_Z)}{E}; \qquad \gamma_{XY} = \frac{\tau_{XY}}{G} \qquad .....(7.6)$$

$$\varepsilon_Z = \frac{\left(-v\sigma_x - v\sigma_y + \sigma_z\right)}{E}; \qquad \gamma_{ZX} = \frac{\tau_{ZX}}{G}$$

where, $\quad G = \dfrac{E}{2(1+v)}$ is the shear modulus or rigidity modulus

Sum of the three equations gives, $\varepsilon_X + \varepsilon_Y + \varepsilon_Z = \left(1 - 2v\right)\left(\dfrac{\sigma_X + \sigma_Y + \sigma_Z}{E}\right)$

These equations can also be written in terms of stresses as functions of strains.

$$\sigma_X = \frac{E\left[\left(1-v\right)\varepsilon_X + v\varepsilon_Y + v\varepsilon_Z\right]}{(1+v)(1-2v)} \; ; \qquad \tau_{XY} = \frac{E\,\gamma_{XY}}{2(1+v)}$$

$$\sigma_Y = \frac{E\left[v\varepsilon_X + \left(1-v\right)\varepsilon_Y + v\varepsilon_Z\right]}{(1+v)(1-2v)} \; ; \qquad \tau_{YZ} = \frac{E\,\gamma_{YZ}}{2(1+v)} \qquad \dots\dots(7.7)$$

$$\sigma_Z = \frac{E[v\varepsilon_X + v\varepsilon_Y + \left(1-v\right)\varepsilon_Z]}{(1+v)(1-2v)} \; ; \qquad \tau_{ZX} = \frac{E\,\gamma_{ZX}}{2(1+v)}$$

The six equations of eq 7.7 can also be expressed as

$$\sigma_X = (\lambda + 2\mu) \times \varepsilon_X + \lambda \times \varepsilon_Y + \lambda \times \varepsilon_Z$$

$$\sigma_Y = (\lambda + 2\mu) \times \varepsilon_Y + \lambda \times \varepsilon_Z + \lambda \times \varepsilon_X$$

$$\sigma_Z = (\lambda + 2\mu) \times \varepsilon_Z + \lambda \times \varepsilon_X + \lambda \times \varepsilon_Y$$

$$\tau_{XY} = \mu \times \gamma_{XY} \; ; \qquad \tau_{YZ} = \mu \times \gamma_{YZ} \; ; \qquad \tau_{ZX} = \mu \times \gamma_{ZX}$$

where, $\quad \lambda = v \times E / [(1 + v)(1 - 2v)] \quad$ and $\quad \mu = E / [2(1 + v)] = G$

are called ***Lame's constants***

These equations are expressed more conveniently in matrix notation as

$$\{\sigma\} = [\,D\,]\{\varepsilon\} \qquad \dots\dots(7.8)$$

In general, $\{\sigma\} = [D]\,(\{\varepsilon\} - \{\varepsilon_0\}) \qquad \dots\dots(7.9)$

where, $\quad \{\varepsilon_0\} = [\,\alpha \times \Delta T, \alpha \times \Delta T, \alpha \times \Delta T, 0, 0, 0]^{\mathrm{T}}$

is the initial or stress-free strain vector.

and $\quad \Delta T$ is the change in temperature of the component

since thermal expansion produces only normal strain (with no shear strain and no Poisson's effect) and thermal strains do not induce any stresses unless thermal expansion is constrained

The 3-D stress-strain relations, simplified for 1-D and 2-D cases are:

1-D case:

P ←————————————————→ P ——→ X

FIGURE 7.9 1-Dimensional bar element

$$\sigma = E \times \varepsilon \qquad\qquad .....(7.10)$$

2-D cases:

(i) **Plane stress case,** represented by a thin plate in X-Y plane subjected to in-plane loads along X- and/or Y-directions, and no load (and, hence, stress) along the normal to the plane (in Z-direction), as shown in Fig.7.10

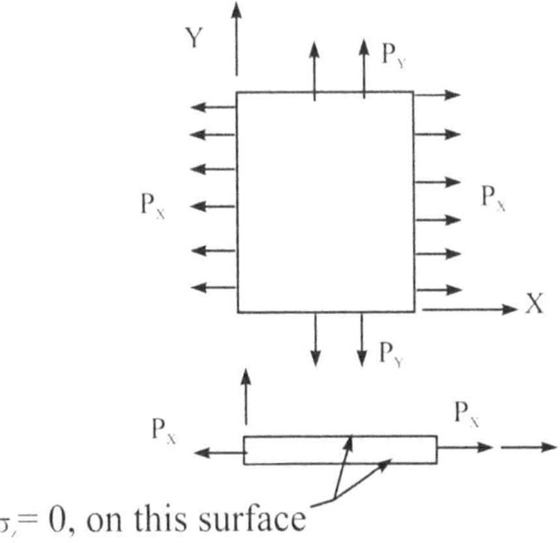

FIGURE 7.10 Plane stress case

$$\text{i.e.,}\qquad (\sigma_Z = 0,\ \varepsilon_Z \neq 0) \qquad\qquad .....(7.11)$$

$$[D] = \frac{E}{(1-v^2)}\begin{bmatrix} 1 & v & 0 \\ v & 1 & 0 \\ 0 & 0 & \dfrac{(1-v)}{2} \end{bmatrix} \qquad\qquad .....(7.12)$$

$$\{\varepsilon_0\} = [\ \alpha \times \Delta T,\ \alpha \times \Delta T,\ 0]^{T} \qquad\qquad .....(7.13)$$

(ii) **Plane strain case,** represented by a thin plate in X-Y plane, which is constrained along the normal to the plane in Z-direction (i.e., no strain along the normal). It is an approximation of a 3-dimensional solid of vary large dimension along Z compared to its dimensions along X and Y and loaded in X-Y plane. (Ref. Fig 7.11)

Example: A hydro dam between two hills, which can be considered as a set of slices or plates in the flow direction (X). Each slice in X-Y plane is modeled by plane strain elements.

Here, $(\sigma_Z \neq 0, \varepsilon_Z = 0)$ .....(7.14)

$$[D] = \frac{E}{(1+v)(1-2v)} \begin{bmatrix} 1-v & v & 0 \\ v & 1-v & 0 \\ 0 & 0 & \dfrac{(1-2v)}{2} \end{bmatrix}$$ .....(7.15)

and $\{\varepsilon_0\} = (1+v) \times \begin{bmatrix} \alpha \times \Delta T & \alpha \times \Delta T & 0 \end{bmatrix}^T$ .....(7.16)

This is obtained by using $\varepsilon_Z = (v \times \sigma_X - v \times \sigma_Y + \sigma_Z)/E = 0$ to represent σ_Z in terms of σ_X and σ_Y as $\sigma_Z = v \times (\sigma_X - \sigma_Y)$ and substituting for σ_Z in the relations for ε_X and ε_Y.

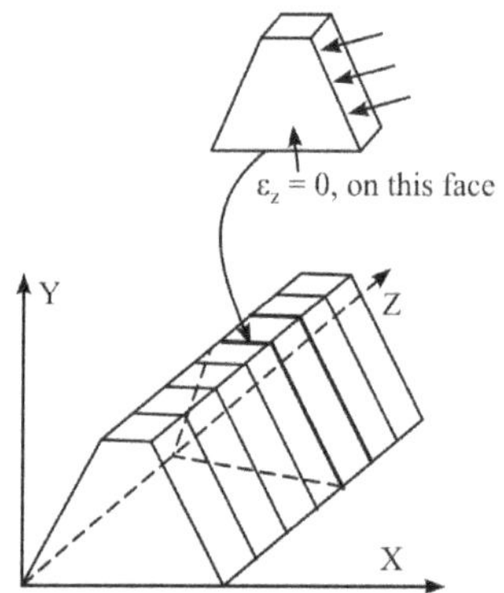

FIGURE 7.11 Plane strain case

Another example of plane strain is a segment of an axi-symmetric solid which is self constrained in the circumferential or hoop direction.

- **Orthotropic materials**

(a) 3-D case

Eq. (7.6) are modified to account for variation of E and v along material axes (1-2-3) as

$$
\begin{Bmatrix} \varepsilon_1 \\ \varepsilon_2 \\ \varepsilon_3 \\ \gamma_{12} \\ \gamma_{23} \\ \gamma_{31} \end{Bmatrix}
=
\begin{bmatrix}
\dfrac{1}{E_1} & \dfrac{-v_{21}}{E_2} & \dfrac{-v_{31}}{E_3} & 0 & 0 & 0 \\
\dfrac{-v_{12}}{E_1} & \dfrac{1}{E_2} & \dfrac{-v_{32}}{E_3} & 0 & 0 & 0 \\
\dfrac{-v_{13}}{E_1} & \dfrac{-v_{23}}{E_2} & \dfrac{1}{E_3} & 0 & 0 & 0 \\
0 & 0 & 0 & \dfrac{1}{G_{12}} & 0 & 0 \\
0 & 0 & 0 & 0 & \dfrac{1}{G_{23}} & 0 \\
0 & 0 & 0 & 0 & 0 & \dfrac{1}{G_{31}}
\end{bmatrix}
\begin{Bmatrix} \sigma_1 \\ \sigma_2 \\ \sigma_3 \\ \tau_{12} \\ \tau_{23} \\ \tau_{31} \end{Bmatrix}
\qquad .....(7.17)
$$

where, $E_1 \times v_{21} = E_2 \times v_{12}$; $\quad E_2 \times v_{32} = E_3 \times v_{23}$ and $\quad E_3 \times v_{13} = E_1 \times v_{31}$

(b) 2-D Plane stress case

Eq. (7.12) is modified to account for variation of E and v along material axes (1-2) as

$$
[D] = \frac{1}{\left(1 - v_{12}v_{21}\right)}
\begin{bmatrix}
E_1 & v_{21}E_1 & 0 \\
v_{12}E_2 & E_2 & 0 \\
0 & 0 & G\left(1 - v_{12}v_{21}\right)
\end{bmatrix}
\qquad .....(7.18)
$$

where, $v_{12} \times E_2 = v_{21} \times E_1$

When an orthotropic plate is loaded parallel to its material axes, it results only in normal strains. If the material axes (1, 2) are oriented at an angle θ w.r.t. global (X,Y) axes, then

$$
[D]^* = [T]^T [D]\, [T] \qquad .....(7.19)
$$

$$
\text{where, } \quad [T] =
\begin{bmatrix}
\cos^2\theta & \sin^2\theta & 2\sin\theta\cos\theta \\
\sin^2\theta & \cos^2\theta & -2\sin\theta\cos\theta \\
2\sin\theta\cos\theta & -2\sin\theta\cos\theta & \cos^2\theta - \sin^2\theta
\end{bmatrix}
\qquad .....(7.20)
$$

7.13.5 COMPATIBILITY RELATIONS

Compatibility is defined as the continuity of the structure as a whole and each individual member, while satisfying applied constraints.

While in 1-D elements, only one stress component and one strain component are used in the strain energy calculation, stresses and strains in 2-D and 3-D elements have more components due to the effect of Poisson's ratio even for a simple loading. While six strain components can be obtained from the three displacement components by partial differentiation, in a general 3-D case. the reverse requirement of ***calculating three displacement components is possible only when the six strain components are inter-related*** through the following six additional conditions, called compatibility equations.

$$\partial^2\varepsilon_X/\partial y^2 + \partial^2\varepsilon_Y/\partial x^2 = (\partial^2/\partial y^2)\,(\partial u/\partial x) + (\partial^2/\partial x^2)\,(\partial v/\partial y)$$

$$= (\partial^2/\partial y\,\partial x)\,(\partial u/\partial y) + (\partial^2/\partial x\,\partial y)\,(\partial v/\partial x)$$

$$= (\partial^2/\partial y\,\partial x)\,[(\partial u/\partial y) + (\partial v/\partial x)] = (\partial^2\gamma_{XY}/\partial y\,\partial x)$$

i.e.,
$$\frac{\partial^2\varepsilon_X}{\partial y^2} + \frac{\partial^2\varepsilon_Y}{\partial x^2} = \frac{\partial^2\gamma_{XY}}{\partial x\partial y}$$

Similarly,
$$\frac{\partial^2\varepsilon_Y}{\partial z^2} + \frac{\partial^2\varepsilon_Z}{\partial y^2} = \frac{\partial^2\gamma_{YZ}}{\partial y\partial z} \qquad \qquad .....(7.21)$$

$$\frac{\partial^2\varepsilon_Z}{\partial x^2} + \frac{\partial^2\varepsilon_X}{\partial z^2} = \frac{\partial^2\gamma_{ZX}}{\partial z\partial x}$$

$$(\partial/\partial x)\,[\,-(\partial\gamma_{YZ}/\partial x) + (\partial\gamma_{ZX}/\partial y) + (\partial\gamma_{XY}/\partial z)\,]$$

$$= (\partial/\partial x)\,[\,-\{(\partial^2 v/\partial z\,\partial x) + (\partial^2 w/\partial y\,\partial x)\} +$$

$$\{(\partial^2 w/\partial x\,\partial y) + (\partial^2 u/\partial z\,\partial y)\} + \{(\partial^2 u/\partial y\,\partial z) + (\partial^2 v/\partial x\,\partial z)\}\,]$$

$$= (\partial/\partial x)\,[\,(\partial^2 u/\partial z\,\partial y) + (\partial^2 u/\partial y\,\partial z)\,]$$

$$= (\partial^2/\partial y\,\partial z)\,[\,(\partial u/\partial x) + (\partial u/\partial x)\,] = (\partial^2/\partial y\,\partial z)\,(\varepsilon_X + \varepsilon_X)$$

$$= 2\,\partial^2\varepsilon_X/\partial y\,\partial z$$

i.e.,
$$(\partial/\partial x)\,[\,-(\partial\gamma_{YZ}/\partial x) + (\partial\gamma_{ZX}/\partial y) + (\partial\gamma_{XY}/\partial z)\,] = 2\,\partial^2\varepsilon_X/\partial y\,\partial z$$

Similarly,
$$(\partial/\partial y)\,[\,-(\partial\gamma_{ZX}/\partial y) + (\partial\gamma_{XY}/\partial z) + (\partial\gamma_{YZ}/\partial x)\,] = 2\,\partial^2\varepsilon_Y/\partial z\,\partial x$$

$$(\partial/\partial z)\,[\,-(\partial\gamma_{XY}/\partial z) + (\partial\gamma_{YZ}/\partial x) + (\partial\gamma_{ZX}/\partial y)\,] = 2\,\partial^2\varepsilon_Z/\partial x\,\partial y$$

7.13.6 EQUILIBRIUM EQUATIONS

Every member or particle in the universe is in static or dynamic equilibrium. Consequently, the algebraic sum of all the forces and of all the moments acting on a member must equal zero. The six equations of equilibrium are summarized as $\sum F_x = 0$; $\sum F_y = 0$; $\sum F_z = 0$ and $\sum M_x = 0$; $\sum M_y = 0$; $\sum M_z = 0$ in rectangular or Cartesian coordinate system.

Stress at a point in a component is described by the stress tensor – one normal stress component and two shear stress components on each of the six faces of a cube around that point. For equilibrium of this cube, these eighteen stress components should satisfy the following equilibrium conditions, where F_x, F_y and F_z are the forces acting on the cube.

$$\frac{\partial \sigma_X}{\partial x} + \frac{\partial \tau_{XY}}{\partial y} + \frac{\partial \tau_{XZ}}{\partial z} = F_X$$

$$\frac{\partial \sigma_X}{\partial x} + \frac{\partial \tau_{XY}}{\partial y} + \frac{\partial \tau_{XZ}}{\partial z} = F_X \qquad \qquad \text{.....(7.22)}$$

$$\frac{\partial \tau_{ZX}}{\partial x} + \frac{\partial \tau_{ZY}}{\partial y} + \frac{\partial \sigma_Z}{\partial z} = F_Z$$

and $\tau_{XY} = \tau_{YX}$; $\quad \tau_{YZ} = \tau_{ZY}$; $\quad \tau_{ZX} = \tau_{XZ}$

Stress analysis has to satisfy both compatibility equations and equilibrium equations. This can be done by two ways –

- Force method, where redundant reactions or member forces are treated as unknowns and are obtained by satisfying compatibility of structural nodes
- Displacement method, where nodal displacements are regarded as unknowns and are obtained by satisfying equilibrium conditions

These two methods are illustrated through the following example

Example 7.6

An axial load of P is applied on two long rods of areas A_1 and A_2, which are fixed between two rigid supports as shown. Calculate load shared by each rod, assuming both the rods are of same material.

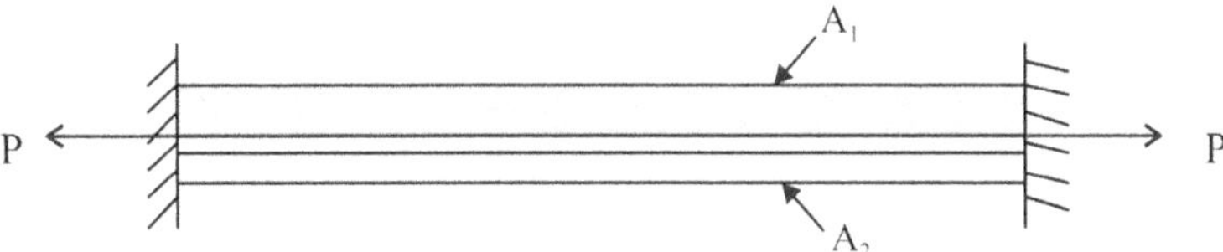

Solution

The structure is redundant to the first degree as one of the rods can safely be removed. Let E be the modulus of elasticity of the material of the rods. Let P_1 and P_2 be the loads shared by the two rods so that $P_1 + P_2 = P$ and δ_1 and δ_2 be the displacements of the rods with areas A_1 and A_2

   (a) Force method

$$\delta_1 = P_1 \times L / (A_1 \times E) \qquad \text{and} \qquad \delta_2 = P_2 \times L / (A_2 \times E)$$

As the two rods are located between two rigid supports,

$$\delta_1 = \delta_2 \quad \text{(Compatibility condition)}$$

or $P_1 \times L / (A_1 \times E) = P_2 \times L / (A_2 \times E) = (P - P_1) \times L / (A_2 \times E)$

$\Rightarrow$ $P_1 = P \times A_1 / (A_1 + A_2)$ and $P_2 = P \times A_2 / (A_1 + A_2)$

(b) Displacement method

Let δ be the elongation of the rods

Then, $P_1 = \delta \times A_1 \times E / L$ and $P_2 = \delta \times A_2 \times E / L$

But $P = P_1 + P_2$ (Equilibrium condition)

$\qquad = (\delta \times A_1 \times E / L) + (\delta \times A_2 \times E / L) = \delta \times (A_1 + A_2) \times E / L$

Therefore, $\delta = P \times L / [(A_1 + A_2) \times E]$

$\Rightarrow$ $P_1 = P \times A_1 / (A_1 + A_2)$ and $P_2 = P \times A_3 / (A_1 + A_2)$

It can be seen that same results are obtained by both the methods

7.13.7 AIRY STRESS FUNCTION

It is possible to define stress field by a function φ, in the form of a polynomial function or Fourier series, such that

$$\sigma_X = \partial^2 \varphi / \partial y^2; \quad \sigma_Y = \partial^2 \varphi / \partial x^2 \text{ and } \tau_{XY} = - \partial^2 \varphi / \partial x\, \partial y$$

If body forces F_X and F_Y exist, which can be defined by a potential function V such that $F_X = - \partial V / \partial y$ and $F_Y = - \partial V / \partial x$, then the stresses are defined by

$$\sigma_X = \partial^2 \varphi / \partial y^2 + V ; \quad \sigma_Y = \partial^2 \varphi / \partial x^2 + V \text{ and } \tau_{XY} = - \partial^2 \varphi / \partial x\, \partial y$$

This function φ (called Airy stress function) always satisfies the equilibrium conditions. The function should be selected to represent the given loading condition and satisfy the given boundary conditions. Airy stress function φ is usually selected in the form of a polynomial of the form

$$\sum \varphi_n \text{ for } n = 1,N \quad \text{or} \quad \sum (\sum A_{in} \times x^{n-1} \times y^i \text{ for } i = 0,n) \text{ for } n = 1,N$$

I Stress function φ with n=2 for in-plane loads

(a) $\varphi = A\, y^2$ to represent stress field along X direction in a thin plate, such that 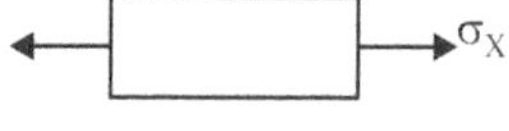

$$\sigma_X = \partial^2 \varphi / \partial y^2 = 2A$$

(where $A = A_{22}$ and all other $A_{ij} = 0$)

(b) $\varphi = A \times x^2 + B \times y^2$ to represent bi-axial normal stress field over a thin plate, such that 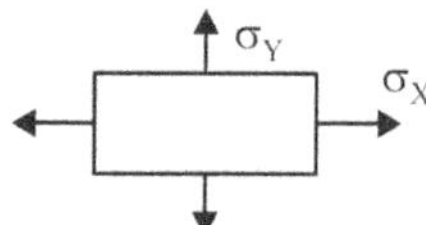

$$\sigma_X = \partial^2 \varphi / \partial y^2 = 2B \text{ and } \sigma_Y = \partial^2 \varphi / \partial x^2 = 2A$$

(where $A = A_{02}$; $B = A_{22}$ and all other $A_{ij} = 0$)

(c) $\varphi = A \times x \times y$ to represent shear stress field over a two dimensional thin plate, such that $\tau_{XY} = - \partial^2 \varphi / \partial x\, \partial y = - A$ 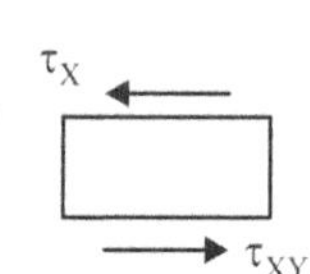
(where $A = A_{12}$ and all other $A_{ij} = 0$)

(d) $\varphi = A \times x^2 + B \times x \times y + C \times y^2$ to represent stress field over a two dimensional thin plate, such that

$\sigma_X = 2C$; $\sigma_Y = 2A$ and $\tau_{XY} = -B$

(where $A = A_{02}$; $B = A_{12}$; $C = A_{22}$ and all other $A_{ij} = 0$)

II Stress function φ with n = 4 for bending loads

(e) $\varphi = A \times y^3 + B \times y^3 \times x + C \times y \times x$ to represent stress field over a cantilever with a concentrated normal load at its free end, such that

$$\sigma_X = P \times (L - x) \times y / I ; \quad \sigma_Y = 0$$

and $$\tau_{XY} = P \times (d^2 - 4y^2) / 8I$$

with $$A = A_{33} = P \times L / (t \times d^3)$$

$$B = A_{34} = -2P / (t \times d^3)$$

$$C = A_{12} = 3P / (2\,t \times d) \quad \text{and all other } A_{ij} = 0)$$

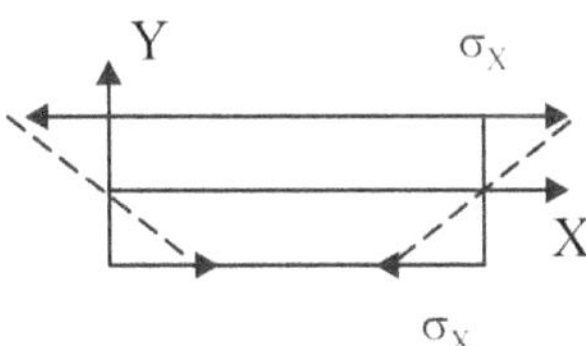

(f) $\varphi = A \times y^3 / 6$ to represent stress field in X-direction of a thin plate, varying from a +ve maximum on the top edge to a –ve maximum on the bottom edge, such that $\sigma_X = A \times y$, equivalent to $\sigma_X = (M/I) \times y$

(where $A = A_{33}$ and all other $A_{ij} = 0$)

For plane stress problem, the compatibility condition

$$\frac{\partial^2 \varepsilon_X}{\partial y^2} + \frac{\partial^2 \varepsilon_Y}{\partial x^2} = \frac{\partial^2 \gamma_{XY}}{\partial x \partial y}$$

can be expressed in terms of stresses, using Hooke's law equations

$$\varepsilon_X = (\sigma_X - v \times \sigma_Y)/E ; \quad \varepsilon_Y = (\sigma_Y - v \times \sigma_X)/E \quad \text{and} \quad \gamma_{XY} = \tau_{XY}/G, \text{ as}$$

$$[\partial^2(\sigma_X - v \times \sigma_Y)/\partial y^2 + \partial^2(\sigma_Y - v \times \sigma_X)/\partial x^2]/E = \partial^2[2(1-v) \times \tau_{XY}/E]/\partial x\,\partial y$$

Substituting for stresses in terms of stress function φ, we get

$$[\partial^4\varphi/\partial y^4 - v \times \partial^4\varphi/\partial x^2\partial y^2] + [\partial^4\varphi/\partial x^4 - v \times \partial^4\varphi/\partial x^2\partial y^2]$$

$$= -2(1-v) \times \partial^4\varphi/\partial x^2\partial y^2$$

$$\Rightarrow [\partial^4\varphi/\partial x^4 + 2\,\partial^4\varphi/\partial x^2\partial y^2 + \partial^4\varphi/\partial x^4] = 0$$

By expressing stresses in terms of stress function, the compatibility condition reduces to $\nabla^4\phi = 0$ where, $\nabla^2 = (\partial^2/\partial x^2 + \partial^2/\partial y^2)$ is the 2-D Laplace operator.

If body forces are present, $\nabla^4\phi = -(1-v) \times \nabla^2 V$

7.14   THEORIES OF FAILURE

In every component and at every point, three normal stresses and three shear stresses (some of them not significant in specific cases) can be evaluated by various methods due to any set of external loads. To decide whether the design is safe or not, based on these multiple stress values is not easy. Different researchers have formulated different design criteria which, if satisfied, ensure safety of a component. Some of them are better suited for ductile materials and some others for brittle materials. Three popular stress-based theories are explained here. These theories are based on yield stress in tension (S_{YT}) or ultimate tensile stress (S_{UT}) from uni-axial tensile test on a standard test specimen, reduced by a factor of safety (f_s) to account for material non-homogeneity at the micro level, scatter in experimental data, nature of loading, manufacturing process used to produce that component, etc.

7.14.1   MAXIMUM NORMAL STRESS THEORY

In this theory, proposed by **Rankine**, failure is expected to occur when maximum principal stress at any point in the component exceeds allowable yield stress in tension or allowably ultimate stress in tension. i.e.,

$$\sigma_1 \leq S_T/f_s \qquad\qquad .....(7.23)$$

where σ_1, σ_2, and σ_3 are the principal stresses in tri-axial stress state   and $\sigma_1 > \sigma_2 > \sigma_3$

It is graphically represented by the plot shown. Here, S_T represents ultimate stress (S_{UT}) for brittle materials and yield stress (S_{YT}) for operational utility of a component made from a ductile material.

7.14.2   MAXIMUM SHEAR STRESS THEORY

In this theory, proposed by **Coulomb, Tresca and Guest**, failure is expected to occur when maximum shear stress in the component at any point exceeds allowable shear stress for the material. i.e., $\tau_{max} \leq S_{SY}/f_s$

where, $S_{SY} = S_T / 2$ (S_T is the yield stress in tension for ductile material)

and τ_{max} = Max [$(\sigma_1 - \sigma_2)/2$, $(\sigma_2 - \sigma_3)/2$, $(\sigma_3 - \sigma_1)/2$]   .....(7.24)

It is graphically represented by the plot and differs from maximum principal stress theory only in 2nd and 4th quadrants (since the two principal stresses are of opposite sign).

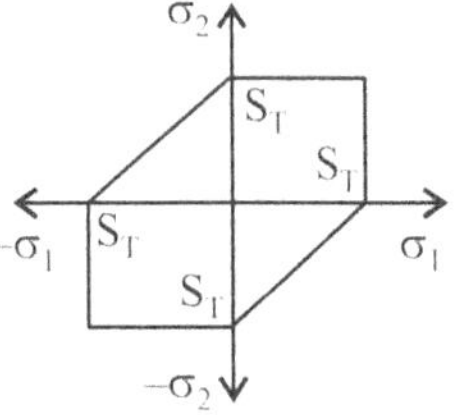

This check is popularly used with ductile materials. It is a conservative approach and is used by safety codes of many countries for pressure vessels.

7.14.3 DISTORTION ENERGY THEORY

In this theory, proposed by *Huber, Von Mises and Hencky*, failure is expected to occur when maximum distortion energy in the component at any point due to applied load exceeds allowable distortion energy of the material. It assumes that a component fails only because of its distortion and the volumetric stress and strain have no effect on the failure of a component. This is widely accepted by many engineers as a better method for prediction of failure of ductile components.

Distortion energy is a part of strain energy, specifically due to distortion of the component i.e. due to distortion part of strain.

Thus, $U_D \leq (1 + v) \times (S_{YT}/ f_S)^2 / (3E)$ .....(7.25)

where,   U = U_V (strain energy due to change in volume)

$+ U_D$ (Strain energy due to distortion)

Each of the three normal stresses can be considered as having two components - one component due to change in volume and another due to change of shape (distortion) without change in volume

i.e., $\sigma_1 = \sigma_{1d} + \sigma_V$; $\sigma_2 = \sigma_{2d} + \sigma_V$; $\sigma_3 = \sigma_{3d} + \sigma_V$

and $U_V = 3(\sigma_V \times \varepsilon_V /2)$

Since distortion of components involves no change in volume, $\varepsilon_{1d} + \varepsilon_{2d} + \varepsilon_{3d} = 0$

Expressing strains in terms of stresses,

$[\sigma_{1d} - v \times (\sigma_{2d} + \sigma_{3d})]/E + [\sigma_{2d} - v \times (\sigma_{3d} + \sigma_{1d})]/E + [\sigma_{3d} - v \times (\sigma_{1d} + \sigma_{2d})]/E = 0$

or $(1 - 2v) \times [\sigma_{1d} + \sigma_{2d} + \sigma_{3d}] / E = 0$

Since $(1 - 2v) / E \neq 0$, $\sigma_{1d} + \sigma_{2d} + \sigma_{3d} = 0$

$(\sigma_1 + \sigma_2 + \sigma_3) = [\sigma_{1d} + \sigma_{2d} + \sigma_{3d}] + 3 \sigma_V = 3 \sigma_V$

$\therefore$ $\sigma_V = (\sigma_1 + \sigma_2 + \sigma_3)/3$

$$\varepsilon_V = [\sigma_{1V} - v \times (\sigma_{2V} + \sigma_{3V})] / E = [\sigma_V - v \times (\sigma_V + \sigma_V)] / E = (1 - 2v) \times \sigma_V / E$$

$$U_D = U - U_V$$

$$= [(\sigma_1^2 + \sigma_2^2 + \sigma_3^2) - 2v(\sigma_1 \times \sigma_2 + \sigma_2 \times \sigma_3 + \sigma_3 \times \sigma_1)] / (2E)$$

$$- [(\sigma_1 + \sigma_2 + \sigma_3)/3] \times [(1 - 2v)(\sigma_1 + \sigma_2 + \sigma_3)/3E)]/2$$

$$= (1 + v) \times [(\sigma_1 - \sigma_2)^2 + (\sigma_2 - \sigma_3)^2 + (\sigma_3 - \sigma_1)^2] / (6E) \quad .....(7.26)$$

From (7.25) and (7.26),

$$\sqrt{\left[(\sigma_1 - \sigma_2)^2 + (\sigma_2 - \sigma_3)^2 + (\sigma_3 - \sigma_1)^2\right]/2} \leq S_{YT}/f_S \quad .....(7.27)$$

In simple tensile test, $\sigma_1 = S_{YT}$; $\sigma_2 = \sigma_3 = 0 \Rightarrow U_d = (1+v) \times S_{YT}^2 / E$

In bi-axial stress state, $\sigma_3 = 0$ So, $\sqrt{\left[\sigma_1^2 - \sigma_1\sigma_2 + \sigma_2^2\right]} \leq S_{YT} / f_S$

In a component subjected to pure shear, $\sigma_1 = -\sigma_2 = \tau_{XY}$ and $\sigma_3 = 0$

Therefore, $\sqrt{3}\tau_{XY} \leq S_{YT} / f_S$ or $S_{SY} = S_{YT} / \sqrt{3}$

The three stress based theories of failure are represented graphically for a bi-axial stress state ($\sigma_3 = 0$) with the principal stresses σ_1 and σ_2 represented along X-axis and Y-axis in Fig.7.12.

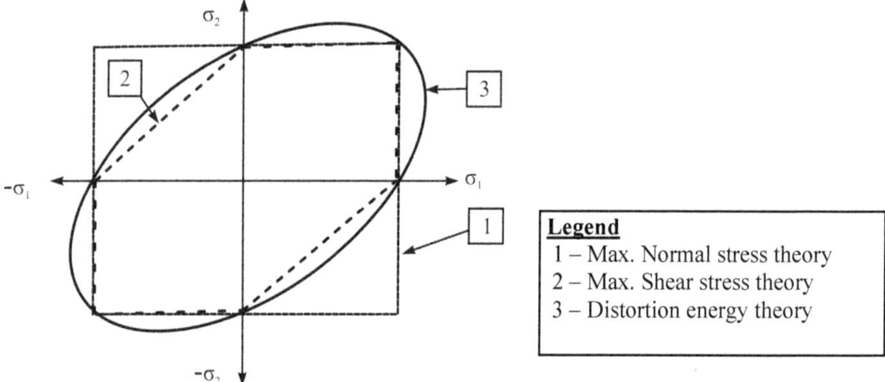

FIGURE 7.12 Graphical representation of theories of failure

7.15 CYLINDRICAL AND SPHERICAL SHELLS

Cylindrical shells are commonly used in pressure vessels, heat exchangers, turbines, condensers etc. while spherical shells are used for storing fluids at high pressure. These shells are classified, based on their ratio of diameter to thickness, as thin and thick. A ratio of 20 is used as a general thumb rule. However, the basic difference is in the assumption of stresses across the thickness. In thin shells, radial and hoop stresses are assumed uniform while in thick shells, they are assumed to vary as a function of radius. Axial stresses in both thin and thick cylindrical shells are assumed uniform across the thickness. The stresses can be evaluated from basic equations of force equilibrium.

7.15.1 THIN CYLINDRICAL SHELLS

Consider a long thin cylindrical shell of mean diameter 'd' and thickness 't' with its ends closed by a flat plate, hemispherical or tory-spherical dish, subjected to internal pressure 'P' (Ref Fig 7.13).

(a) Axial stress in the shell can be obtained from force equilibrium on a section plane, normal to the axis. Whether the shell is closed by a flat plate, hemispherical or tory-spherical dish, the axial load on the end depends only on the projected area, normal to the axis.

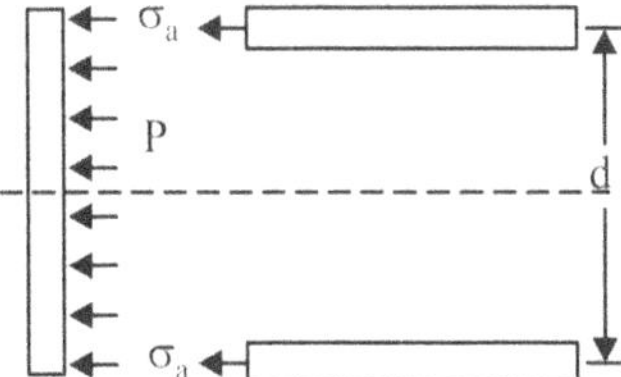

FIGURE 7.13 Axial stress in a thin cylindrical shell

For force equilibrium in the axial direction,

$$\sigma_a \times (\pi \times d \times t) = P \times (\pi \times d^2/4) \ \text{ or } \ \sigma_a = P \times d/(4 \times t) \ \text{ Tensile} \quad \text{.....(7.28)}$$

(b) Circumferential or hoop stress can be obtained from force equilibrium on a plane passing through diameter and length of the cylindrical shell, as shown in Fig 7.14. If length of the shell is 'L',

$$\sigma_h \times 2(t \times L) = P \times (d \times L) \ \text{ or } \ \sigma_h = P \times d / (2 \times t) \ \text{ (Tensile)} \quad \text{.....(7.29)}$$

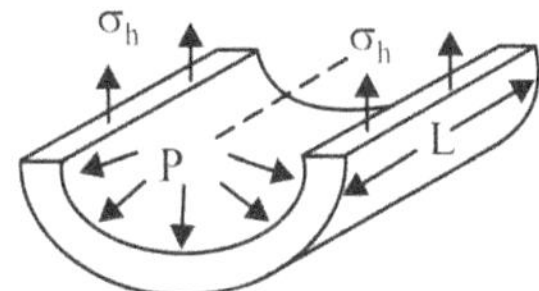

FIGURE 7.14 Hoop stress in a thin cylindrical shell

(c) For equilibrium of forces in the radial direction,

$$\sigma_r = P \ \text{(Compressive)} \quad \text{.....(7.30)}$$

7.15.2 THIN SPHERICAL SHELLS

Consider a thin spherical shell of mean diameter 'd' and thickness 't', subjected to internal pressure 'P'(Ref Fig 7.15). It will have only radial stress and hoop stress.

(a) Circumferential or hoop stress can be obtained from force equilibrium on any plane passing through diameter, as

$$\sigma_h \times (\pi \times d \times t) = P \times (\pi \times d^2/4) \quad \text{or} \quad \sigma_h = P \times d/(4 \times t) \text{ Tensile} \quad \ldots..(7.31)$$

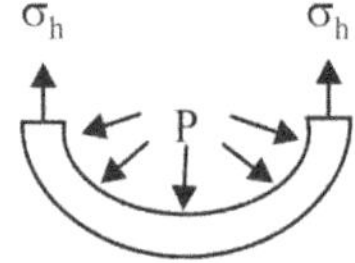

FIGURE 7.15 Hoop stress in a thin spherical shell

(b) For equilibrium of forces in the radial direction, $\sigma_r = P$ (Compressive)

It can be seen that the hoop stress in a spherical shell is half of that in a cylindrical shell. For the same diameter of shell, same material and same internal pressure, spherical shell requires lesser thickness and is, therefore, less heavy and less costly.

7.15.3 THICK CYLINDRICAL SHELLS

Consider a long thick cylindrical shell of internal radius 'R_i', outer radius 'R_o' with its ends closed by a flat plate, hemispherical or tory-spherical dish, subjected to internal pressure 'P_i' and external pressure 'P_o' (Ref Fig 7.15)..

(a) Assuming uniform axial stress as in a thin cylindrical shell,

$$\sigma_a \times \pi \times (R_o^2 - R_i^2) = P \times (\pi \times R_i^2)$$

$$\Rightarrow \ \sigma_a = P_i \times R_i^2 / (R_o^2 - R_i^2) = P_i \times R_i^2 / [(R_o - R_i) \times (R_o + R_i)]$$

$$= P_i \times R_i^2 / [t \times (R_o + R_i)]$$

$$\approx P_i \times d_m / (2 \times t) \quad \text{(Tensile)} \quad\quad \ldots..(7.32)$$

(b) Considering force equilibrium on a section plane along diameter and length of shell, at a radius 'r' and thickness 'dr' (Ref Fig 7.16),

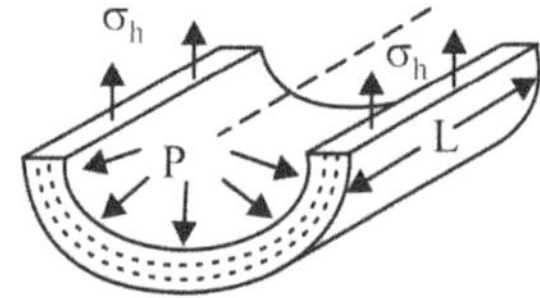

FIGURE 7.16 Hoop stress in a thick cylindrical shell

$$P \times (2 \times r \times L) - (P + \delta P) \times [2(r + \delta r) \times L] = \sigma_h \times (2 \times \delta r \times L)$$

$$\Rightarrow \quad \sigma_h \times \delta r \approx r \times \delta P - P \times \delta r$$

$$\sigma_h \approx - (P + r \times dP/dr) = - d(P \times r)/dr \qquad \ldots\ldots(7.33)$$

Also, $\varepsilon_a = (\sigma_a - v \times \sigma_h - v \times \sigma_r)/E = [\sigma_a - (\sigma_h - P) \times v]/E$

Since axial (longitudinal) stress and axial strain are assumed uniform across thickness, $\qquad \sigma_h - P = 2a,$ a constant $\qquad \ldots\ldots(7.34)$

From eq (7.33) and (7.34), $\quad - (2P + r \times dP/dr) = 2a$

$$\Rightarrow \quad dP/(P + a) = - 2 \ (dr/r)$$

Integrating both sides, we get

$$\log (P + a) = - \log r^2 + C \quad \text{where C is integration constant}$$

$$\Rightarrow \quad P = b/r^2 - a \qquad \ldots\ldots(7.35)$$

Using this value in eq (7.34), $\quad \sigma_h = b/r^2 + a \qquad \ldots\ldots(7.36)$

where   b is a constant

Eq (7.35) and (7.36) are called ***Lame's equations***

Substituting boundary conditions, $P = P_o$ at $r = r_o$ and $P = P_i$ at $r = r_i,$

$$b = r_i^2 \times r_o^2 \times (P_i - P_o) / (r_o^2 - r_i^2)$$

and $\qquad a = (P_i \, r_i^2 - P_o \, r_o^2) / (r_o^2 - r_i^2)$

Then, if $P_o = 0,$ as in most of the cases, at any radius 'r',

$$\sigma_r = P_r = - \frac{P_i \times r_i^2}{(r_0^2 - r_i^2)} + \frac{P_i \times r_i^2 \times r_o^2}{r^2 \times (r_o^2 - r_i^2)} \text{Compressive}$$

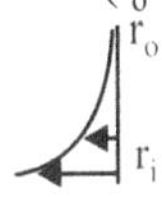

$$\sigma_h = \frac{P_i \times r_i^2}{(r_0^2 - r_i^2)} + \frac{P_i \times r_i^2 \times r_o^2}{r^2 \times (r_o^2 - r_i^2)} \qquad \text{Tensile}$$

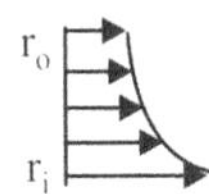

and   if $P_i = 0$,

$$\sigma_r = P_r = \frac{P_o \times r_o^2}{(r_0^2 - r_i^2)} - \frac{P_o \times r_i^2 \times r_o^2}{r^2 \times (r_o^2 - r_i^2)}$$ Compressive

$$\sigma_h = \frac{P_o \times r_o^2}{(r_0^2 - r_i^2)} + \frac{P_o \times r_i^2 \times r_o^2}{r^2 \times (r_o^2 - r_i^2)}$$ Compressive

7.15.4 THICK SPHERICAL SHELLS

Radial and hoop stresses in a thick spherical shell subjected to internal pressure 'P_i' can be obtained in a similar way as

$$\sigma_r = P_r = -\frac{P_i \times r_i^3}{(r_0^3 - r_i^3)} + \frac{P_i \times r_i^3 \times r_o^3}{r^3 \times (r_o^3 - r_i^3)}$$ Tensile

$$\sigma_h = \frac{P_i \times r_i^3}{(r_0^3 - r_i^3)} + \frac{P_i \times r_i^3 \times r_o^3}{r^3 \times (r_o^3 - r_i^3)}$$ Tensile

7.15.5 COMPOUND CYLINDERS

If a cylinder of some material is shrunk on a cylinder of same or different material, then tensile hoop stresses are induced on the outer cylinder to increase its inner diameter and compressive hoop stresses are induced on the inner cylinder to reduce its outer diameter, even without internal or external pressure. Compressive radial stress is introduced on the common surface (radius 'r_c') which varies on either side to zero value on the inner radius (r_i) and outer radius (r_o), as shown in Fig 7.17.

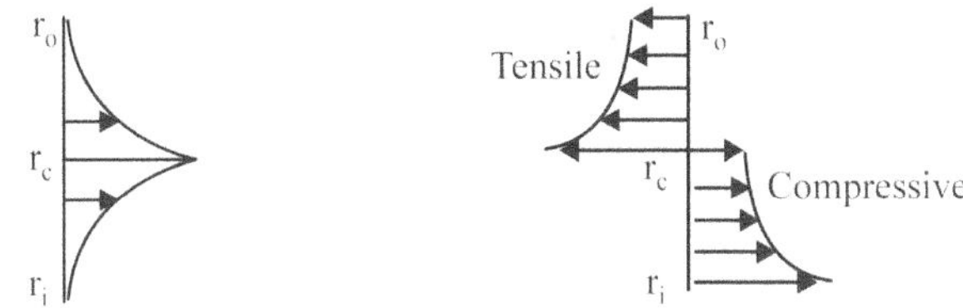

(a) Variation of radial stress   (b) Variation of hoop stress

FIGURE 7.17 Radial and hoop stresses in a compound shell

Radial and hoop stresses due to internal or external pressure applied on the compound cylinder are algebraically added to the stresses due to shrink fitting.

Part - II

Design and Analysis of Aeroplane Structures

CHAPTER 8

DESIGN ASPECTS OF AEROPLANE STRUCTURE

An aeroplane is a heavier-than-air, powered machine, capable of sustained and controlled flight through air, carrying payload (human beings, cargo, ammunition etc). Hot air balloons, gliders etc don't come under this category. Engineering aspects of aeroplane are also applicable to other flight vehicles such as helicopters, missiles, satellite launch vehicles etc with additional specific features.

Aerodynamic forces, essential for lifting an aeroplane into air, are generated primarily by the main wing and partly by the tail plane. In level (cruise) flight, lift component of aerodynamic forces balances weight of the aeroplane and drag component is balanced by the thrust of the engine. Changes in lift forces as per the operational needs are achieved by the operation of flaps, ailerons on main wing or tail plane and rudder. These changes in aerodynamic forces affect various structural members.

A conflicting phenomenon, as can be observed from theory of flight, is the altitude of flight for a particular velocity of aeroplane – at low altitudes, more lift is generated due to larger air pressure but more engine power is required to balance the inevitable larger drag; at high altitudes, less engine power is required to balance reduced drag but lesser lift is generated to balance weight. It is, therefore, desirable to operate

- low altitude flights at lower speeds for shorter distances (such as, domestic flights), with increased engine power and larger fuel consumption

- high altitude flights at higher speeds for longer distances (such as, international flights), with lesser engine power and reduced fuel consumption

Producing large lift with least weight of aeroplane (without payload) gives the best operational efficiency for any aeroplane and, hence, optimum design. This goal is difficult to achieve because of another conflicting phenomenon - larger lift necessitates larger area of aerodynamic surfaces; larger area necessitates bigger structure and increases weight of aeroplane. Emphasis is placed on designing more accurately, in addition to using lighter alloys in the structure, so as to minimise weight of the structure.

Aircraft structures, such as the typical wing box beam, are redundant structures. The order of redundancies is of the order of thousands and solving them by conventional methods is extremely tedious. Computer methods, such as the Finite Element Method and the method of successive approximations are the only reasonable methods to use in these cases.

Cost-effective and rapid preliminary sizing work is carried out by analytical methods, with many assumptions made to simplify the structure and arrive at an approximate solution. Computer analysis is a time consuming and costly process. In addition, it is a black-box operation, where the engineer has no way of justifying the results, in contrast to preliminary sizing which gives the engineer a feeling for the real world. Preliminary sizing, based on previous experience, can also be used in parallel to double check computer output and to assure that the result is reasonable. During preliminary design, it is not always possible to meet all the regulations of Aviation safety board.

A good working knowledge of stress analysis is essential for any engineer associated with aircraft engineering. The types of analysis used in airframe work are very different in many respects from those used in civil and mechanical structures, since structural weight saving (so that a high percentage of pay load can be carried) and fail-safe design with built-in redundancies are most important. The structure must have a serviceable life of 20 years with minimum maintenance.

Conceptual views of an aeroplane viewed by aerodynamics engineer and structural engineer are shown in Fig 8.1

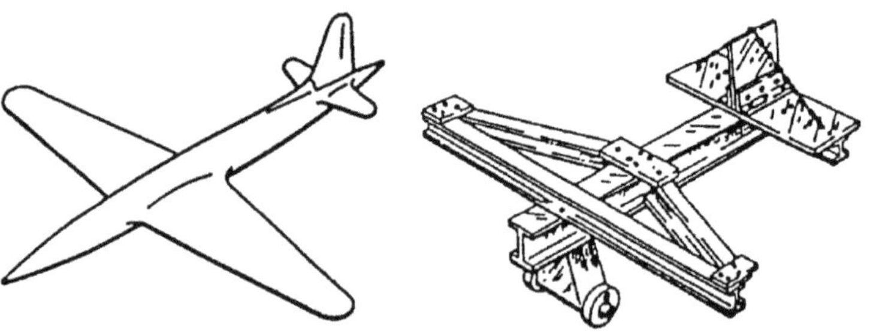

(a) View by Aerodynamics engineer   (b) View by Structural engineer

FIGURE 8.1 A conceptual aeroplane

8.1   MAIN PARTS OF AN AEROPLANE

In the conventional aeroplane, there are four main structural units (Ref Fig 8.2).

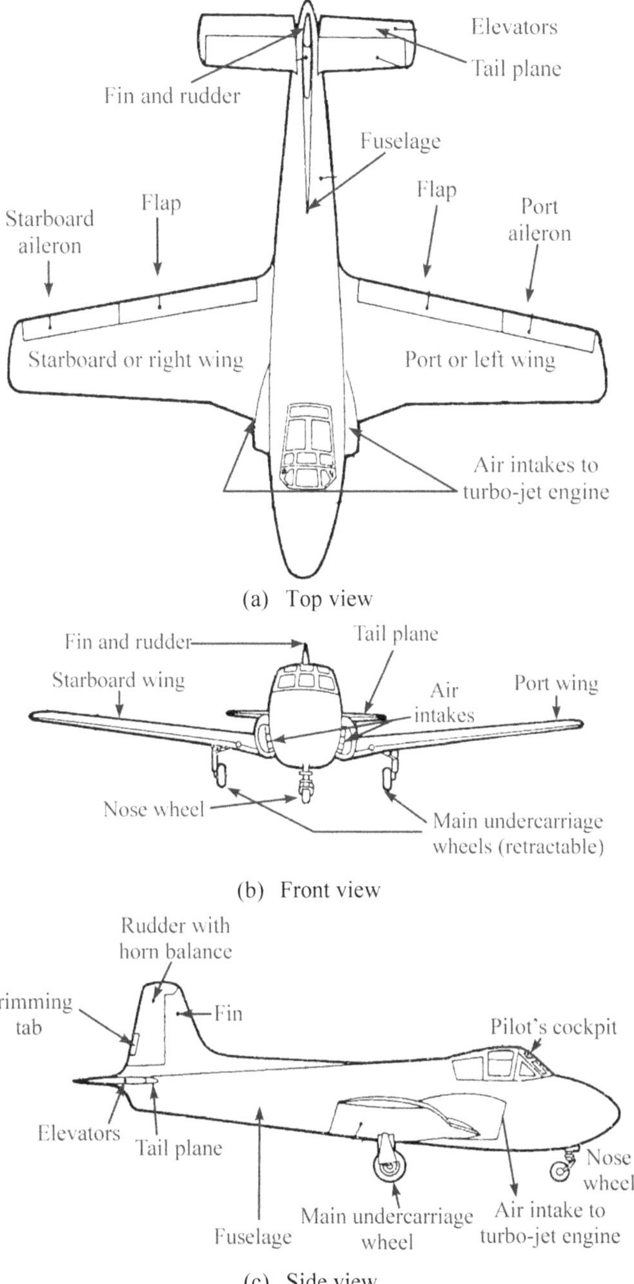

FIGURE 8.2 Parts of a typical aeroplane

(i)   The *wings* or *main planes* generate sufficient aerodynamic force (lift force to support weight) during flight. The lift force generated is a function of shape / size of wing and speed of travel. The main plane structure usually consists of one or more spars, a series of ribs, leading and trailing edges, and wing tip - all of which are covered by sheet metal skin (Ref Fig 8.3).

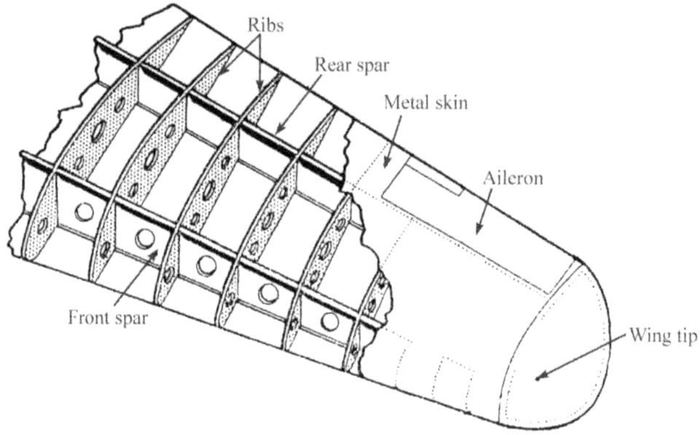

FIGURE 8.3 Main Plane Construction

Selection of the type of wing section (or aerofoil), depends on the particular nature of operation. **Wing loading** (weight divided by wing area) influences the landing speed significantly.

In general, the *front spar* is located at about 15% chord, the *rear spar* at about 55% to 60% chord. Wing *ribs* may be placed parallel to the flight path to ensure a smooth aerodynamic shape between the spars of a two-spar wing or perpendicular to the spars (Ref Fig 8.4).

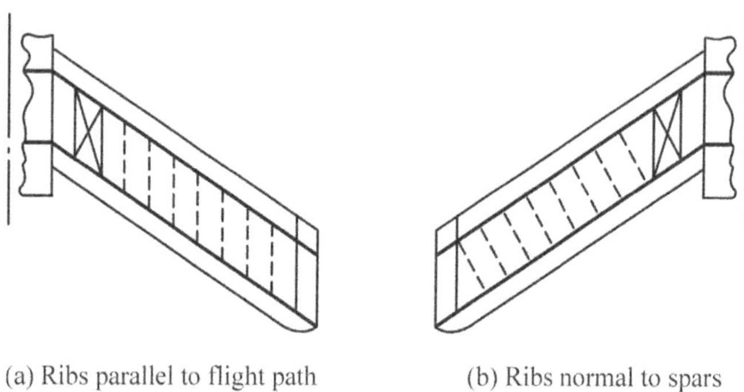

(a) Ribs parallel to flight path   (b) Ribs normal to spars

FIGURE 8.4 Arrangement of ribs in a wing

Aileron is normally of 30% chord length. Ribs are to be located at each aileron and flap hinge. The rib spacing is determined from panel size considerations, to prevent buckling. Reinforced ribs are used for engine-mount attachments, landing-gear attachments and fuel tank supports.

Span-wise *stringers* may be placed parallel to each other with run out at intersections with leading edge spar or at constant percentages of the wing chord (Ref Fig.8.5).

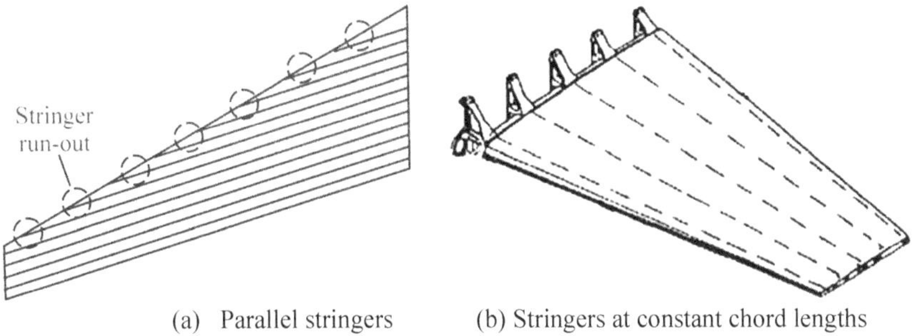

(a) Parallel stringers   (b) Stringers at constant chord lengths

FIGURE 8.5 Arrangement of stringers in a wing

Wing leading edge with *slats* and trailing edge with *flaps* are used to increase lift, whenever required, by changing one or more of chord length, camber and angle of incidence. Some of them, with sliding and/or rotary motions, are shown in Fig 8.6. Construction of slats is more complex due to the slat shape and anti-icing requirements. The trailing edge consists of more configurations than the leading edges. Deflection of a flap causes large nose down moments which must be balanced by the horizontal tail.

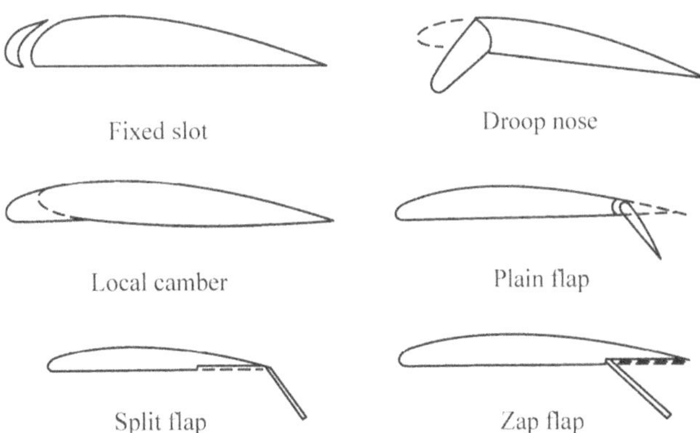

FIGURE 8.6 Some arrangements of leading edge and trailing edge

Some designs may have multiple segments of flaps which are independently deployed, as shown in Fig 8.7.

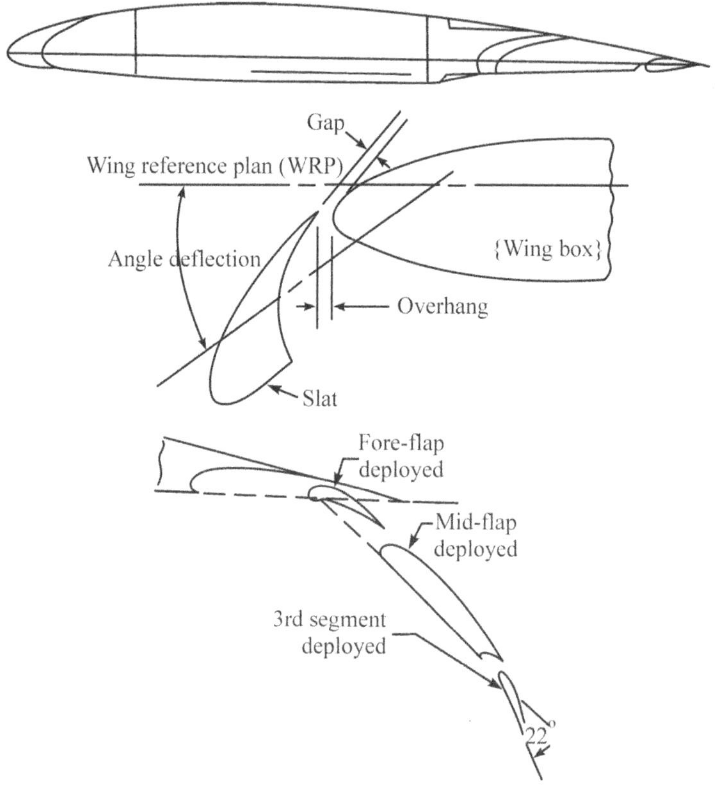

FIGURE 8.7 A wing design with multiple segments of flaps

Ailerons (Ref Fig 8.8) on the two wings deflect in opposite directions to provide rolling motion to the aeroplane. Outboard ailerons operate at low speeds, to provide rolling moment with small additional lift and large moment arm. Inboard ailerons operate at high speeds, to provide rolling moment with large additional lift and small moment arm without creating an adverse twisting moment on the wing

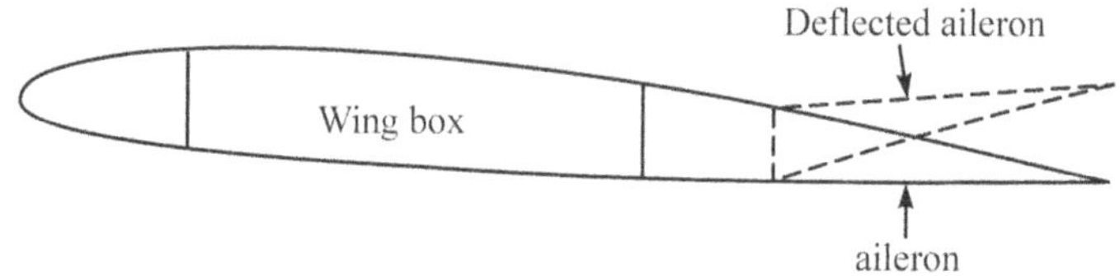

FIGURE 8.8 Operation of aileron

Engine mounts are attached to the box frame of the wings and parallel to the wing chord line, to support the engines (nacelle).

(ii)   The ***body*** or ***fuselage*** houses passengers, crew, cargo or other payload and instruments for safe flight. The modern fuselage consists of longitudinal members (**longerons** and stringers), transverse members (**frames** and bulkheads) and external **skin** (Ref Fig 8.9).

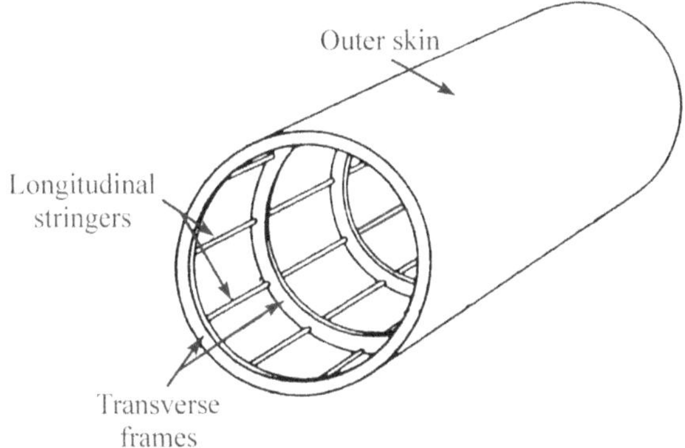

FIGURE 8.9 Fuselage construction

The longerons carry axial forces resulting from bending moment; Skin carries shear from applied external transverse and torsional forces as well as cabin pressure; Frames primarily serve to maintain shape of the fuselage and reduce column length of the stringers and skin panels to prevent instability (buckling) of the structure.

It is attached to the cockpit for operating the flight, wings for providing adequate aerodynamic lift and tail plane assembly for stabilising the aeroplane during flight

In a transport aeroplane flying at subsonic speeds, fuselage is pressurised for passenger comfort. Most efficient pressure-carrying structure has a cylindrical cross section with spherical end caps. This shape is compromised, in a fuselage to satisfy aerodynamics; to meet operational requirements (crew visibility,..); to accommodate doors, windows, cargo,.. and to facilitate nose cone, wings, landing gear, tail assembly,..etc. to arrive at a shape, as shown in Fig 8.10

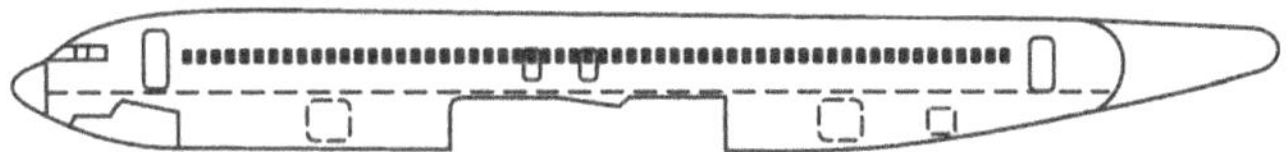

FIGURE 8.10 A typical Fuselage construction

Most common cross section consists of an upper lobe for accommodating passengers and a lower lobe for cargo (Ref Fig. 8.11)

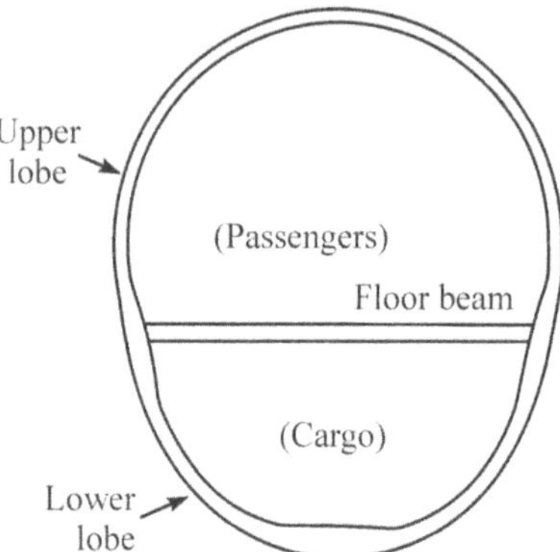

FIGURE 8.11 Cross section of a passenger aeroplane fuselage

(iii) The ***tail unit*** includes tail plane and rudder, essential for maneuvering the flight. It usually consists of a fixed tail plane and movable elevators in the horizontal plane and a fixed fin and movable rudder in the vertical plane (Ref Fig 8.12).

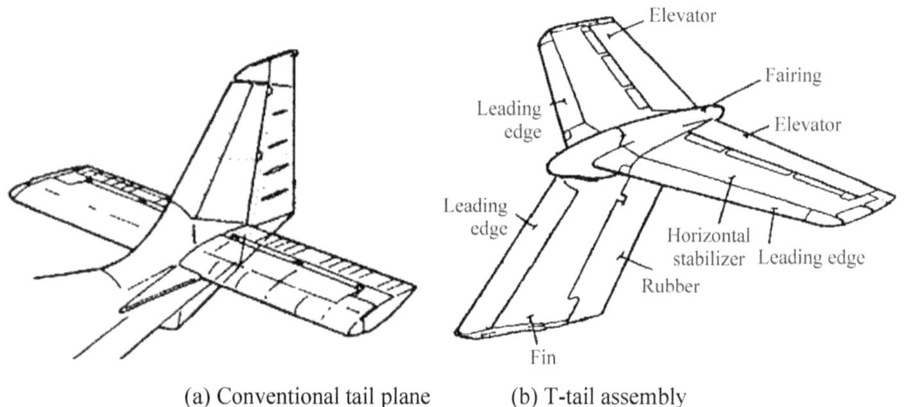

(a) Conventional tail plane   (b) T-tail assembly

FIGURE 8.12 Tail unit construction

Its construction is similar to that of main wing with lesser span. Aspect ratio of vertical and horizontal surfaces is smaller. There is no need for a leading edge device (slat or flap) and a single spar construction with auxiliary rear spar is generally used.

T-tail assembly (Fig 8.12 b) places the horizontal stabiliser in a favourable flow field during low-speed, high angle-of-attack operations. Span of the T-tail fin is approximately one-third shorter than conventional tails. It is designed for minimum horizontal size and minimum fin stiffness to avoid flutter. Due to the requirement of larger stiffness at the vertical fin root, weight increases.

(iv)   The ***undercarriage*** ensures safe take-off or landing of the aeroplane. It consists of the main wheels, a nose or tail wheel, axles and compression struts to take the vertical impact loads and brake loads while landing (Ref Fig 8.13).

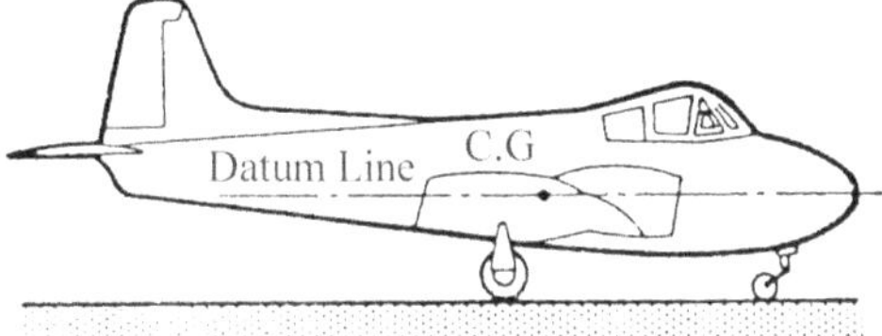

(a) Main wing and nose wheels

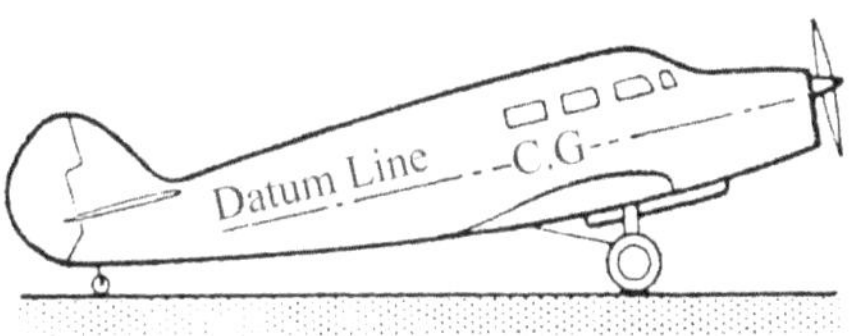

(b) Main wing and tail wheels

FIGURE 8.13 Nose and tail wheel undercarriages

In all modern planes, landing gear is retractable when the aeroplane is airborne (Ref Fig 8.14) and, so, heavier owing to a retraction gear, locking mechanism, valves,.. Decrease in drag (more at higher speeds) offsets increase in weight.

Design of landing gear depends on the type and specific function of aeroplane and soft or normal landings (with vertical speed at landing or sink speed of 0.3 - 1 m/sec) or hard landings (with sink speeds of 1 - 3 m/sec).

It should include towing on the runway, steering, locking of retract mechanism inadvertently, backup power supply for opening the gears while landing and provision for jacking for changing wheel, brake and shock absorbers. The number, configuration and inflation pressure of the tyres of the landing gear should be in accordance with the bearing capacity of the airfields.

During landing, every aeroplane experiences dynamic loads. Ratio of dynamic loads to static load is defined as the ***landing gear load factor.*** It has a value of about 1.5 for large transport aeroplanes; 3.0 for small utility aeroplanes and 5.0 for fighters and military trainers

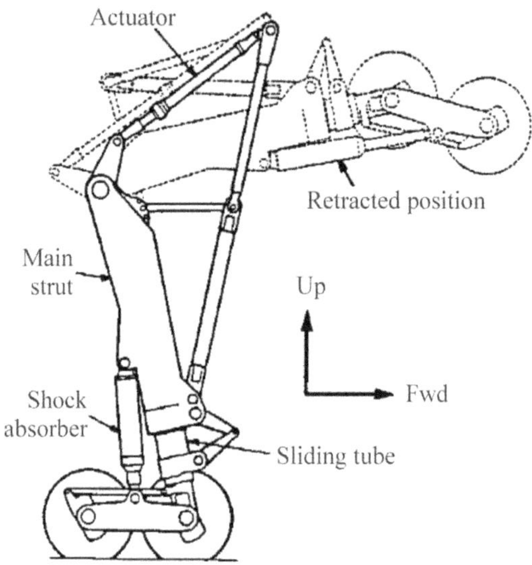

FIGURE 8.14 Retracting landing gear

In the older aeroplanes, the skeleton of the structure is covered by a fabric of Irish linen treated by various varnishes and dopes to make it air-tight and proof against weather and sunlight. It was required to take the pressure of air due to the wind flowing over it; but did not provide any real contribution towards strength of the skeleton structure. Then, three-ply wood or thin sheet metal was used as a covering, so that the skin also could contribute towards strength of the structure, eliminating the need for internal bracing. The metal 'stressed-skin' form of construction has the additional advantage of providing a smooth surface, which helped in increased speeds of aeroplanes.

The material used for skin over the wings and fuselage of an aeroplane should have adequate strength to take some structural load over extreme temperature conditions. With increasing altitude, there is less air above and therefore pressure reduces. Correspondingly, air expands and reduces temperature. Air temperature reduces with altitude upto 10 km altitude, approx at 6.5^0C / km (or 3.6^0F per 1000 ft). Thus, ambient air temperature changes from a maximum of about 50^0C at ground level in summer to a minimum of about -70^0C at cruise height of about 10km in winter. The minimum and maximum temperatures can be different in the case of ballistic missiles, fighter aeroplanes or satellite launch vehicles while skin friction drag may increase temperature of skin (particularly in the supersonic flight regime).

8.2 DESIGN OF AN AEROPLANE

Reduction of weight assumes primary importance, only in the design of an aeroplane. Every unnecessary addition to the weight of an aeroplane not only tends to spoil its performance, but adds to further increase in weight by making the parts stronger and hence heavier.

In order to obtain the best possible performance, an aeroplane must be shaped so that it offers as little resistance as possible to its motion through air, by streamlining the external shape. Often, what is the best shape from the structural point of view is bad from the aerodynamic point of view and the designer is forced to compromise.

In supersonic aeroplanes, the design should take care of economic cruise at Mach 2-3 as well as slow flight during take-off and landing. In order to achieve versatile performance, some fighter planes are designed with variable swept wings (Ref Fig 8.15).

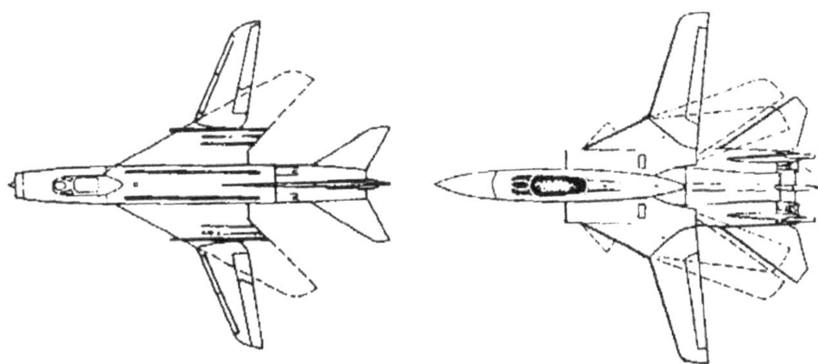

FIGURE 8.15 Fighter plane with variable swept wing

In aeroplane construction, cost is comparatively less important. Safety takes precedence. Safety can only be ensured by careful design, use of best materials and a high degree of workmanship, all of which increase cost of aeroplane. Similarly, best performance is desired even at increased cost. For military purposes, better performance means superiority over the enemy. For commercial or civil aeroplanes, better performance means better fuel economy as well as longer range.

Civil airplanes ***ensure passenger comfort*** through pressurising cabin and controlling temperature inside the cabin as well as low angle of climb / descent upto $\pm\, 15^0$ max. Military airplanes ***ensure safety from enemy*** through faster acceleration & lesser weight, steep angle of climb / descent upto $\pm\, 90^0$, for survival from enemy as well as smaller size and smaller radar image for avoiding detection by enemy.

8.2.1   STAGES IN THE DESIGN OF AN AEROPLANE

Each aeroplane is designed for a particular purpose or type of operation - based on its flying altitude, speed, fuel consumption, purpose (civilian, military transport, fighter) and will not be the appropriate aeroplane for a different purpose.

For example,

- low-altitude and low speed flight for transportation of relief material, spraying pesticides etc; medium-altitude high speed (subsonic) flight for short distance passenger aeroplanes and military transport planes and high-altitude and transonic or supersonic flight for long range civil transport planes or fighter aeroplanes

- Lower maneuverability with higher fuel efficiency for civil aeroplanes; high maneuverability, even with poor fuel economy, for fighter aeroplanes

- Slow take-off and landing for civil aeroplanes and faster take-off and landing for military aeroplanes and aeroplanes flying from carrier (ship).

Important stages in the design of an aeroplane are –

1. Design of an aeroplane starts with an *'estimated weight'*. Every part of aeroplane will be influenced by it. If the finished aeroplane turns out to be heavier than the initial estimate, then each part of the aeroplane having been designed to carry a lighter aeroplane will be too weak and has to be redesigned with an increased 'estimated weight'. A very accurate estimate of weight can be obtained after detailed drawings are completed, even without actually manufacturing the same.

2. Selection of the type of *wing section* (or *aerofoil*), depending on the particular nature of operation. A deep camber wing will give good lift at low speeds but a poor maximum speed whereas a thin wing will give a higher maximum speed but a high landing speed, unless an excessively large wing area is used. To get best of both the options, a more common design is to use a reasonably thin wing and flaps as a means of increasing the camber when required. *Wing loading* (i.e., weight divided by wing area) influences the landing speed significantly.

3. The *general layout* covering type of wing (span, chord, taper, angle of incidence, and dihedral angle), location of wing (high-, low- or mid-wing type) and location of engine is to be finalized, considering the operational requirements of the aeroplane (Ref Fig 8.16).

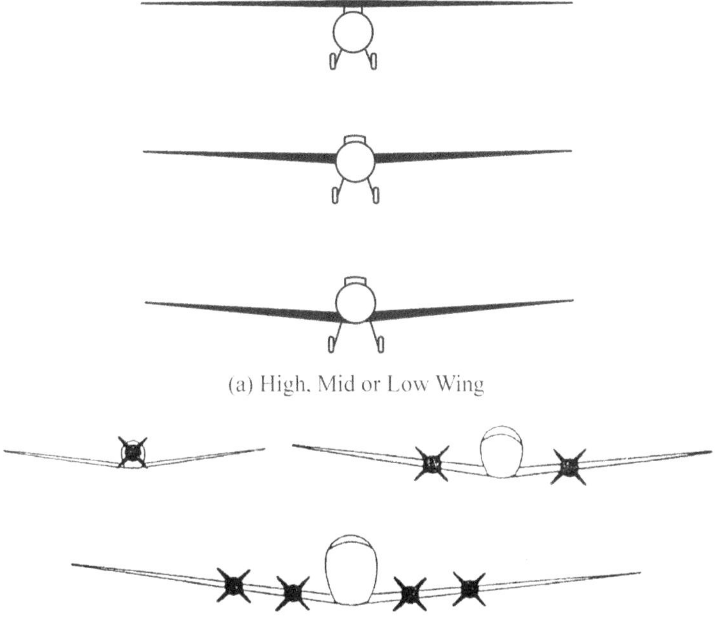

FIGURE 8.16 General arrangement of wing and engine

4. The design activity now deals with **stressing** or proper distribution of total weight of aeroplane and its component parts to the various points in the structure and drawing stress diagrams, shear and bending moment diagrams. It is possible to assess from these diagrams which members are in tension or compression and which members have a tendency to fail by bending, shearing or buckling. This operation needs to be carried out for several conditions – normal flight (cruising), take-off and for landing. The values obtained are multiplied by a **load factor**, which is a product of a factor to allow for extra-ordinary conditions of flight and a material safety factor to account for possible variation in the material properties from those assumed or used in the design calculations.

5. From the detailed drawings, actual weight of the aeroplane and its center of gravity (which is a key factor in effective control of aeroplane during various flight conditions) are estimated. If they do not meet operational requirements, all the above steps need to be repeated.

8.3   PERFECT DESIGN

Every structure is designed to resist or transfer applied loads to support points, without significant deformation which may adversely affect operational requirements of the structure. An optimum design is a function of configuration or layout of members, material properties and optimization criterion (such as safety, serviceability, fail-safe, minimum weight, minimum cost, length of life, reliability etc.). Any one particular design can not be optimal for all loading conditions.

Concurrent design is a design considering optimum combination of shape or layout, material and production simultaneously while ensuring that the component performs the desired function. Fig 8.17 shows the schematic diagram of concurrent design.

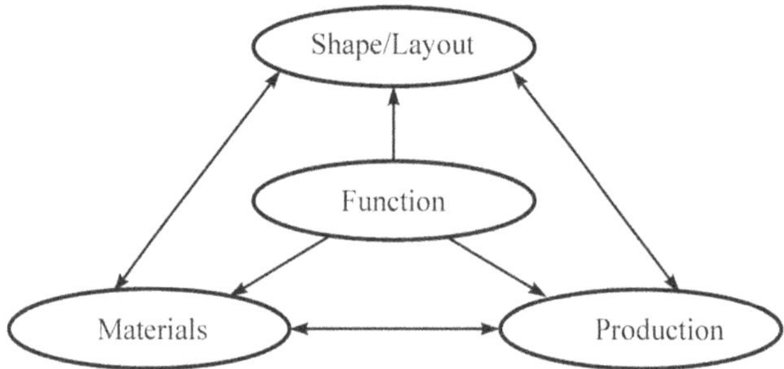

FIGURE 8.17 Concurrent design

Perfect design or ***ideal design*** is a fully stressed structure, which readjusts its configuration or material properties, depending on changes in load or environment. After a change in loading pattern is sensed as an error or deviation from the existing load structure by the feedback mechanism, an ideal structure readjusts itself to a new optimum layout, which resists the new load set effectively. Biological structures (like human skeleton) come very close to the ideal structure - orientation of hands or legs adjusts to suit the load; Stiffness of muscles changes with the load; etc..

Design of aeroplane wing structure is a step in this direction, which changes its layout with flaps and slats, depending on operational conditions. However, this change of shape is more for changing aerodynamic forces than for changing structural behaviour. Use of composite materials is also a step in arriving at an optimum design, since it can be tailor-made with specific properties to suit specific design requirement. However, it is still far from an ideal design, since it can not change its properties when the loads change.

Structures, whose members have gross displacements relative to each other and whose main function is transfer of motion without significant loss of power, are considered as ***mechanisms*** and are not included in this book.

Scope of this book is thus limited to design of static or quasi-static structures, whose members do not have gross relative displacements. A flying aeroplane has gross rigid body motion but does not have gross relative displacements between various members. It is therefore analysed as *a structure subjected to motion-dependent aerodynamic loads but satisfying equations of static equilibrium*

8.4 TYPES OF MEMBERS

Conventional terminology of different types of members of an aeroplane structure is -

Stiffener — a longitudinal or transverse reinforcing member

Flange — a longitudinal stiffener running along an edge of an open or closed shell, capable of resisting bending

Stringer — a longitudinal stiffener which is not a flange

Boom — a stringer or a flange in an idealised shell

Rib — a transverse stiffener in an open or closed shell

Bulkhead — a transverse member in a closed shell, such as rings in fuselage

The aeroplane structure consists of a skeleton framework – with a sheet metal covering to give it shape. The skeleton may consist of many types of components, each designed to resist a particular type of load – axial, bending, torsion, etc.. Analysis of a structure is aimed at evaluating stresses at various points of the structure to ensure that these values are within the permissible limits. Skeleton of many structures consists mainly of three different types of members – *Ties* which resist axial tensile loads; *struts* which resist axial compressive loads and *beams* which resist loads perpendicular to their length and cause bending. Sheet metal panels in semi-monocoque structures transfer distributed aerodynamic load from the surface of wing, fuselage and control surfaces to the structural members such as spars, shear web beams, etc.

It is not always possible to classify all the members exactly under these categories. A member may act as ties under certain conditions (for example, normal flight) and as struts under some other conditions (for example, resting on ground). Some members may serve two purposes (for example, tie and a beam or strut and a beam) at the same time. In addition to these three types, there may be members subject to torsion (moment about the axis of the member). A member may also be in compression due to one particular load and in tension due to another load, acting simultaneously on the component. Net result will be that they neutralize partially or fully each other and only the greater of the two is effective, to the extent of difference in load.

8.5 TYPES OF FRAMES

Ideal frames or trusses consist of only struts and ties, pin-jointed together and loaded at the joints. In practice, members may be connected at the joints by more than one rivet, which doesn't act as a perfect hinge; loads may be acting along the member resulting in some bending; etc. Yet, many of them may be approximated as ideal frames. Even an ideal frame is classified as ***perfect frame***, (Ref Fig. 8.18 (a)) which has just sufficient number of members (a triangular framework) to keep its shape whatever loads are applied; a ***deficient frame***, (Ref Fig. 8.18 (b)) which has less than sufficient members and so acts like a mechanism whenever loads are applied and a ***redundant frame***, (Ref Fig. 8.18 (c)) which has more than the minimum required number of members, to act as a frame.

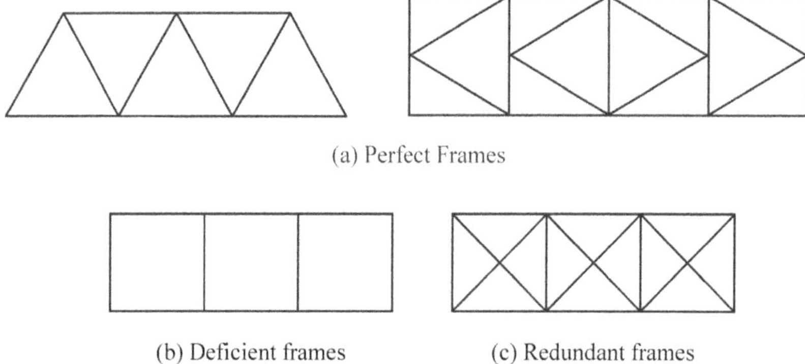

(a) Perfect Frames

(b) Deficient frames   (c) Redundant frames

FIGURE 8.18 Types of frames

In aeroplane structures, redundant frames are most often used. Its main advantage is that it forms a perfect frame, even when one member fails and can serve its purpose – a consequence of ***fail-safe principle*** followed in aeroplanes. Its main disadvantage is - assembling of the frame is very difficult unless very high dimensional accuracy in the lengths of members is maintained, which naturally increases cost of manufacturing.

In a normal redundant frame, by redistributing the total applied load among all the members, the weight of the frame may remain same or even reduce. Because of fail-safe principle in aeroplanes, since the frame is designed to operate as a perfect frame even when one member fails, the redundant frame members are not optimized for minimum weight. Instead, they are designed to withstand the loads as perfect frames (neglecting the presence of redundant members). Thus, total weight of the aeroplane structure increases.

Wire-braced frame used in earlier models of aircrafts (biplanes) is not actually a redundant frame, since the wire can not act as a strut. Such a frame

(Ref.Fig.8.19) may not behave like a perfect frame for certain loads, when the redundant brace is cut. It is not used in the present day aeroplanes. It has some of the features of a perfect frame and some of those of the redundant frame, with a good strength/weight ratio.

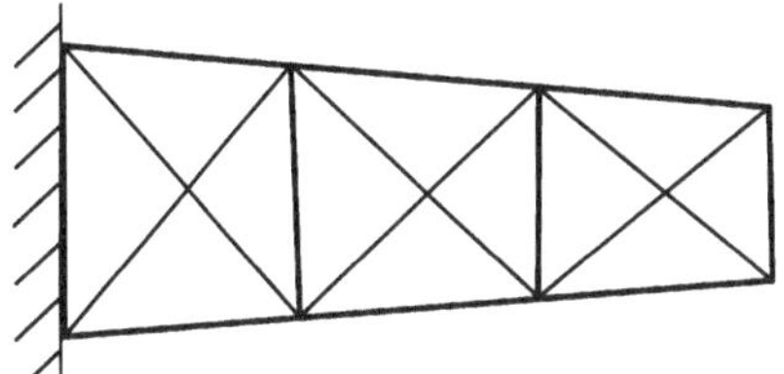 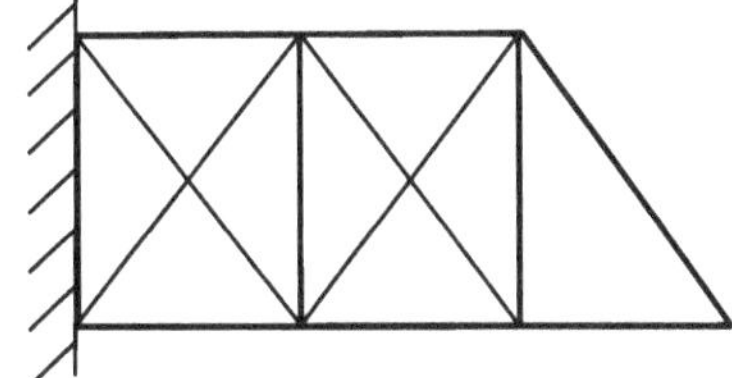

FIGURE 8.19 Wire-braced frames

8.6 TYPES OF CONSTRUCTION

While the fabric cover in the earlier aeroplanes did not serve as a structural member, sheet metal skin in the present day aircraft could resist some load and formed part of the structure. In *monocoque* construction, the skin is made so strong that it can carry all the loads in the structure without the help of internal members, leaving the interior completely free of obstruction. Even in these structures, some reinforcing and stiffening of the skin is advisable if local crippling (or buckling) is to be avoided. The usual practice in fuselage is to have a series of ribs in the form of hoops or ring frames at regular intervals inside the skin, connected together by longitudinal members (stringers), running along the length of the fuselage.

In main planes, depth is not usually sufficient for pure monocoque construction and *stressed-skin method* is employed, which is a compromise, in which the spars take main lift loads and the skin takes the smaller drag loads and prevents the wing from twisting.

One problem with any construction in which the skin carries part of the load is the need for *openings* (also called *cut-outs*) such as for doors, windows, etc. A large hole not only breaks the load path across the panel, but concentrates the load around the hole causing high stresses and greatly increasing the tendency to buckle, unless the skin is sufficiently reinforced around the hole.

8.7 DESIGNING A TIE

It is by far the easiest of the three types of structural members – tie, strut and beam. The design is based on limiting tensile stress in the member (tensile force divided by its area of cross section) to the allowable tensile stress of the material. If the member is required to take tension only, a wire may be used; but, if it may be required to take tension or compression, a rigid member such as a tube must be used. Care must be taken to see that the end fittings are at least as strong as the ties themselves.

8.8   DESIGNING A STRUT

It is difficult to design a strut, because it is liable to bend or buckle, even though only axial load is applied. Its strength depends not only on the kind of material and area of cross section but also on the length, shape of cross section (which changes moment of inertia) and type of connection (fixed, hinged or free) used at each end. Based on these parameters, struts are classified as *long struts*, which always fail by bending or buckling (Ref Fig. 8.20); *short struts*, which always fail by crushing or compression and *medium-length struts* which sometimes fail by bending, sometimes by crushing and sometimes by a combination of both. A numerical relation to divide a strut into long, medium or short strut can not be given because the liability to bend depends also on the shape of cross section and whether it is solid or hollow. Hollow sections, with larger moment of inertia for the same area of cross section (and, hence, weight) than solid sections, resist more compressive load and are preferred. A strut is equally liable to bend in any direction and, hence, *hollow tube is the best shape for a strut*. As a rough guide, hollow steel tube strut may be considered long if its length is greater than 30 times its outer diameter and short if its length is less than 5 times its outer diameter.

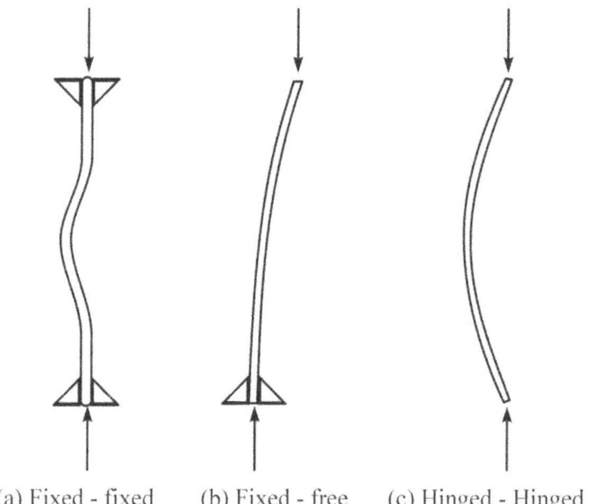

FIGURE 8.20 Effect of end fixity on buckling of a Strut

Load is equally distributed all over the area of cross section in a tie or short strut while stress is maximum on innermost or outermost layers of a strut in bending. Hence, less material is sufficient at the center. For the same area of cross section, as diameter increases, thickness of hollow struts can be reduced.

However, below a limiting thickness, depending on the material, thin walls may fail due to ***crinkling***. This is observed in hollow mild steel struts when the outer diameter is greater than 100 times its thickness (limited to 60 times, on a conservative design). Both these failures have to be considered while designing a strut.

Also ***tapered struts,*** with cross section increasing towards mid point from both ends (Ref Fig. 8.21 a) can take more buckling load than ***prismatic struts,*** with constant area of cross section (Ref Fig. 8.21 b). This tapering can be easily done in a wooden strut; but difficult to produce in a metal strut and, hence, less common.

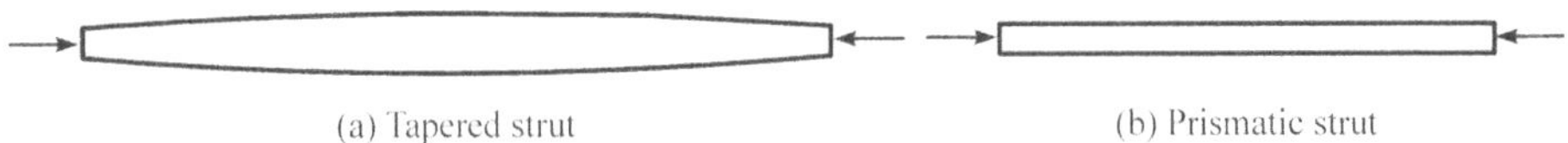

FIGURE 8.21 Types of struts

8.9 DESIGNING A BEAM

For loads acting normal to the axis, design of a beam is also based on type of section, end connections and material. But, it bends in the plane of loading and the section such as I, H,.. which gives sufficient moment of inertia to resist bending in that plane is preferred for minimising weight. Bending depends not just on the applied load; but on the moment it produces at any section – the larger the bending moment, the larger is the requirement of moment of inertia. By choosing a variable section, in proportion to the bending moment at that section, weight of the beam can be reduced.

Usually bending loads act along with axial tensile or compressive loads. Bending of a tie will increase its tensile stress and the member's cross section has to be proportionately increased; whereas bending of a strut will increase its buckling tendency. Bending stress is always associated with shearing stress, which can at times become critical.

8.10 DESIGNING A PANEL

The term 'shell' is used to denote an elongated structure consisting of longitudinal members, transverse members and connecting sheet. When one set of reinforcing members is absent, the sheet itself takes over the function of missing members and serves a dual purpose. When both sets of members are absent, the structure becomes a 'pure shell'.

The term 'panel' includes any plane sheet between stringers in a web or a curved sheet between stiffeners in a shell. Panels in aeroplane structures are assumed so thin, compared to other dimensions, that variation of stress across their thickness is neglected. Design of a panel includes calculation of bending and buckling stresses for loads acting normal to the plane and calculation of shear flow in the panel between two longitudinal spars. In order to improve resistance to buckling, stiffeners or stringers of different sections – angle, Z, hollow square etc - are attached to these panels. These stiffened panels are treated in more detail in the subsequent chapters.

8.11   STRUCTURAL INDEX

Structural index, a function of the intensity of loads and dimensions of the structure, offers the designer a guide to the optimum type of structure. It helps in determining what construction and/or material will fit a particular loading. It also helps in determining the relative efficiencies of different types of structures. Design proportions that are optimum for a particular structure are also optimum for structures of any size, provided they all have the same structural index and additional stress analysis is not necessary.

For an individual structural member, the structural index is based on its own function -

- For shear beams, on the shear load and depth of beam

- For columns, on the axial load and length of the column

- For circular shells or tubes, on the bending moment and diameter of the shell

For any given value of the structural index, it is possible to find the combination of material and cross-sectional shape (or its proportions) that will give the lightest structure. For example, for the simple case of pure tension, the optimum configuration is a straight line and the proportions of the cross section have no effect on the weight. In the case of pure compression, the optimum configuration is a straight line, but the size and shape of the cross section play an important part in the buckling strength of the member.

In choosing a material, the structural engineer is strongly influenced by many factors such as material cost, availability, suitability for the particular environment, formability, machinability etc., besides the weight/strength factor.

These indexes are shown in Fig 8.22 for a few common cases, and are tabulated for different categories of aeroplanes.

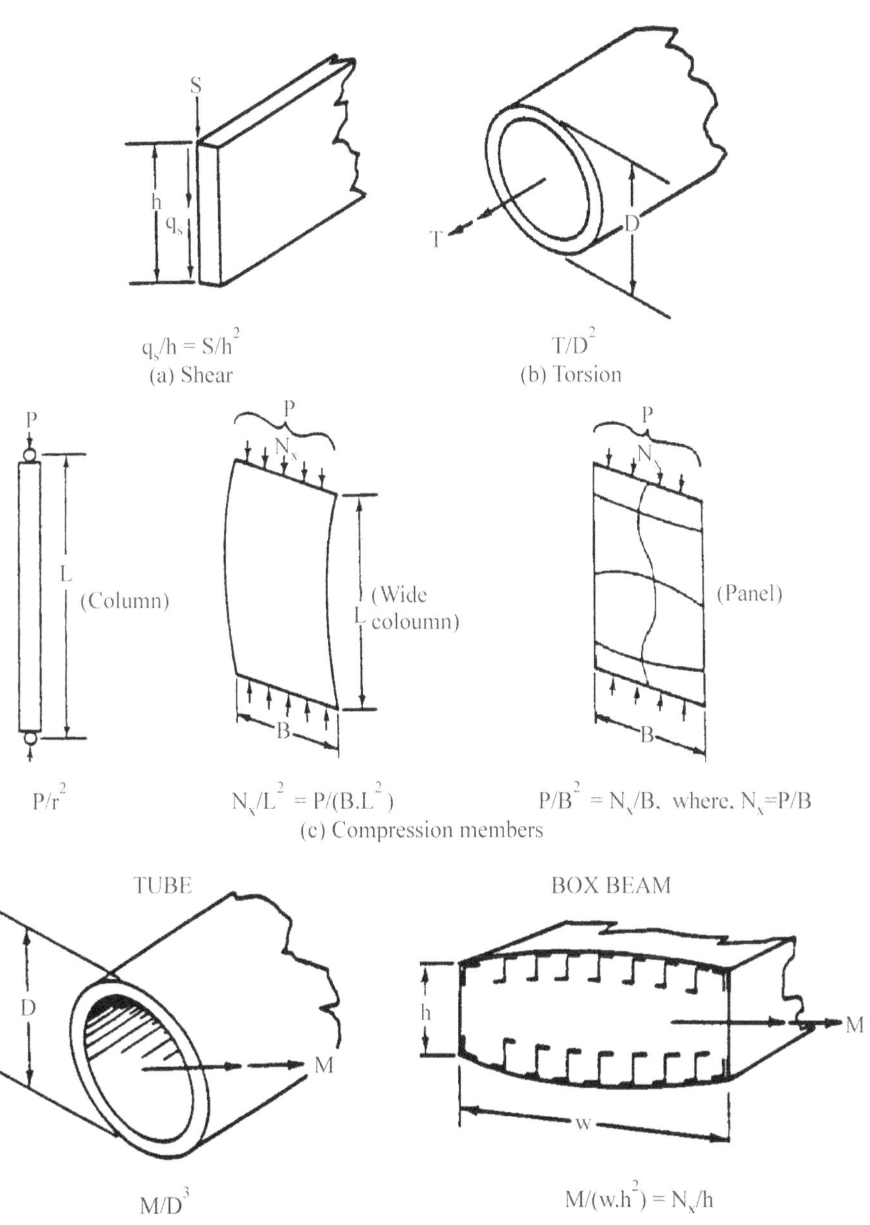

FIGURE 8.22 Structural Index

CHAPTER 9

BASICS OF
AERODYNAMICS

Aerodynamic forces (or surface forces), self weight and inertia forces (or body forces) are the main loads acting on an aeroplane structure. Of these, aerodynamic forces acting on a particular aeroplane vary significantly with different operating conditions such as take-off, level flight (cruise), landing and pitch / turning / yaw motions.

Aerodynamics is one branch of fluid mechanics which deals with the reactions between fluids and solids immersed in them. The force exerted by a fluid on a body does not depend on the absolute velocity of either fluid or body, but on the relative velocity between them. Thus, actual flight condition of an aeroplane moving in still air is identical to the flow of air at the same speed on a static wing structure in a wind tunnel. The relative velocity in both the cases produces the same type of aerodynamic effect.

The normal force component per unit area is called the pressure while the tangential force component (due to skin friction) per unit area is called the shearing stress. As a first approximation, air is treated as incompressible and perfect fluid, where 'perfect' means 'having zero viscosity'. For a perfect fluid, all shearing stresses vanish. Treating air as incompressible fluid is reasonable at subsonic speeds (unto about 600kmph) of aeroplane. Compressibility factor becomes significant due to shock waves when the aeroplane flies at supersonic speeds.

The aerodynamic forces are produced when air flows over the aeroplane. Main contribution comes from the wing in order to support weight of the aeroplane. Aerodynamic loads are the resultants of differential pressure distribution over the top and bottom surfaces of the wing. These resultants cause direct loads (bending, shear and torsion) in different parts of the structure, in addition to local normal pressure on the skin. A brief review of aerodynamics will help in better understanding of nature of loads on an aeroplane structure in different operating conditions.

9.1   PRESSURE DISTRIBUTION OVER AN AEROFOIL

The force due to flowing air on a submerged solid is due to variation of air velocity and, hence, pressure on the top and bottom surfaces. Increased pressure on the lower surface and negative pressure on the upper surface will produce a resultant force 'F' at the center of pressure (CP). An aerofoil shape is found to generate large resultant force and is commonly used for wings. Nomenclature of aerofoil section is given in Fig 9.1 (a) and pressure distribution on aerofoil is shown in Fig 9.1 (b).

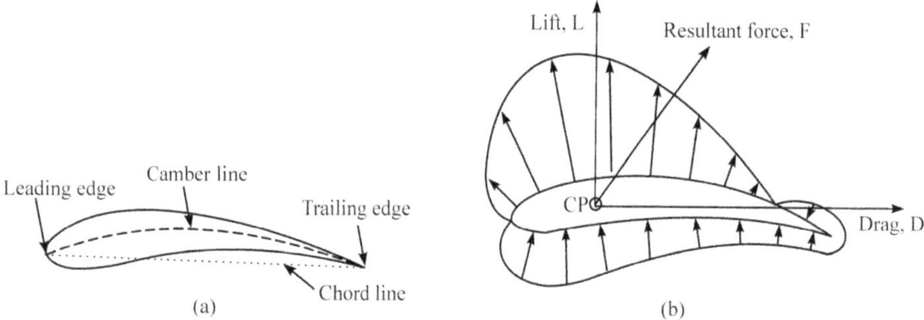

FIGURE 9.1 A typical aerofoil section and pressure distribution on it

9.2   RESULTANT FORCE - LIFT AND DRAG COMPONENTS

The resultant force 'F' of differential pressure over the two surfaces of wing section acts at the center of pressure (CP) and depends on the particular aerofoil section of wing. Its magnitude is proportional to the relative velocity of air 'V', density of air at that altitude 'ρ', effective wing area 'S' and angle of attack or incidence of air 'α' (defined w.r.t. the chord line of the wing). Thus,

$$F = C_F \left(\tfrac{1}{2} \times \rho \times V^2 \times S \right) \text{ where, } C_F \text{ is a non-dimensional coefficient}$$

The angle of attack α is not explicitly reflected in the above equation. Instead, the coefficient C_F is taken as a function of angle of attack. The force F is resolved into two components along and perpendicular to the direction of motion of aeroplane. Design engineer is primarily concerned with these two components of resultant force. They are –

> ***Lift, L*** - force component perpendicular to the direction of motion, and usually vertical

Drag, D - force component in the direction of motion, taken as positive when the force is in the downstream direction

The resultant lift force on the two parts of main plane are shown in Fig. 9.2 (a) while the lift force distribution along the length of the wing is shown in Fig 9.2 (b).

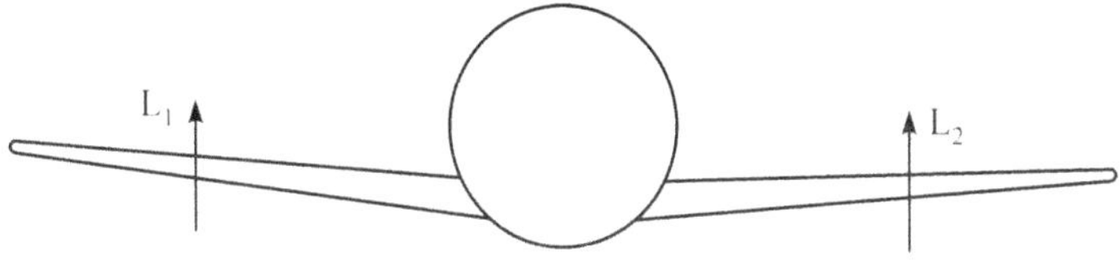

(a) Resultant lift force on main plane

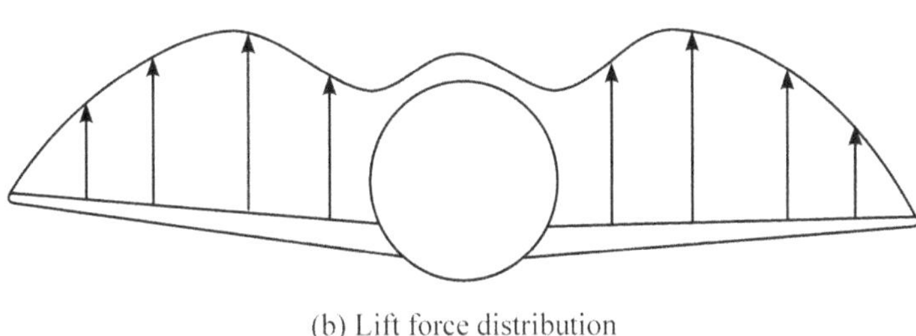

(b) Lift force distribution

FIGURE 9.2 Lift force and its distribution on a wing-fuselage

If 'θ' is the angle between the resultant force F and the lift component L, then

$$L = F \times \cos\theta \quad \text{and} \quad D = F \times \sin\theta$$

However, the lift and drag force components are usually defined, without referring to the resultant force F, by

$$L = C_L \left(\tfrac{1}{2} \times \rho \times V^2 \times S \right) \quad \text{and} \quad D = C_D \left(\tfrac{1}{2} \times \rho \times V^2 \times S \right)$$

where, the non-dimensional coefficients C_L and C_D are functions of angle of attack

Obviously, the coefficients C_L and C_D have different values compared to C_F. Angle θ is not separately indicated in these equations, as this angle is a function of angle of attack.

In the steady level (cruise) flight, lift force 'L' due to main planes, tail planes and fuselage balances weight 'W' of the aeroplane while engine thrust 'T' balances the drag force 'D', as shown in Fig. 9.3. Thus, in level flight,

Lift (L) = Weight (W)   and   Drag (D) = Thrust (T)

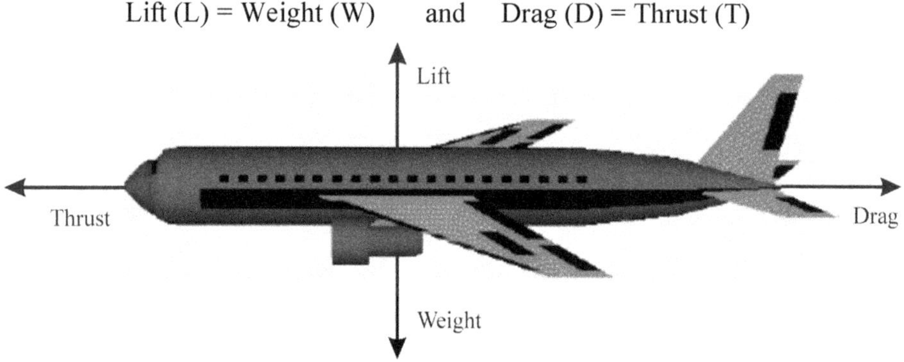

FIGURE 9.3 Forces in steady level flight

Flight condition is influenced by the magnitudes of aerodynamic forces (lift and drag force components) as shown

Aerodynamic forces	Effect on Aeroplane
Lift > Weight	Ascends to higher altitude
Lift < Weight	Descends to lower altitude
Drag > Thrust	Slows down for landing
Drag < Thrust	Accelerates for take-off

For a given aeroplane, lift and drag forces depend on density of surrounding air (ρ) as well as square of velocity of airplane (v^2). However, density of air is a function of temperature and, therefore, changes with altitude, altering lift and drag forces. **At high altitude** of about 10 km, air density is 3 times lower than at ground level. Hence, Lift and Drag are less. So, ***airplane has to fly faster*** at higher altitudes in order to balance weight and **needs less power** (to cover a distance) due to reduced drag.

Altering lift forces on the starboard wing, port wing and tail plane by operating ailerons, rudder etc are crucial for different maneuvers of the aeroplane in flight (Ref Fig 9.4).

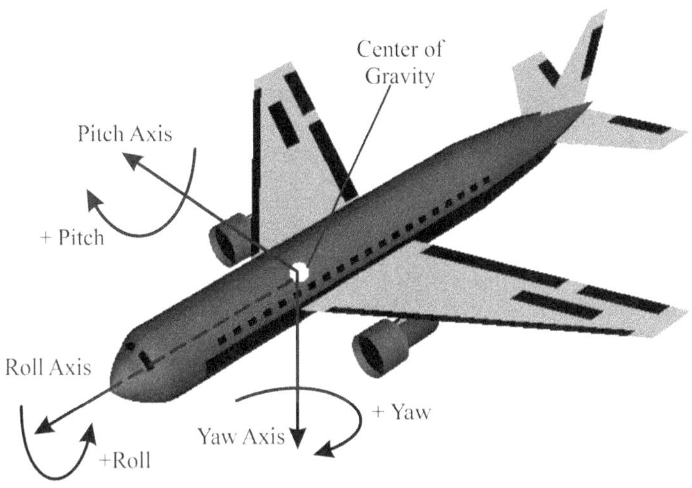

FIGURE 9.4 Rotation of an aeroplane in flight about its three axes

Roll is achieved by the operation of ailerons in opposite directions on main wing (on right & left sides of fuselage)

Pitch for take-off / landing is achieved by operation of flaps on wings and elevators on the tail planes

Yaw or rotation of plane in horizontal plane is achieved by deflection of rudder

9.3 PITCHING MOMENT

Resultant aerodynamic force F acts at a point on the aerofoil section, called the center of pressure (CP). The location of center of pressure depends on the velocity of air and its angle of attack. It will be difficult to consider changing location of resultant force components L and D in design calculations. It has been observed that there exists a point on aerofoil section, about which moment of lift and drag forces remains constant for all angles of attack. This point is called **_aerodynamic center_** (Ref. Fig 9.5). The lift and drag forces acting at different points in different conditions are, therefore, conveniently expressed by transferring them to the aerodynamic center and supplementing them with a moment.

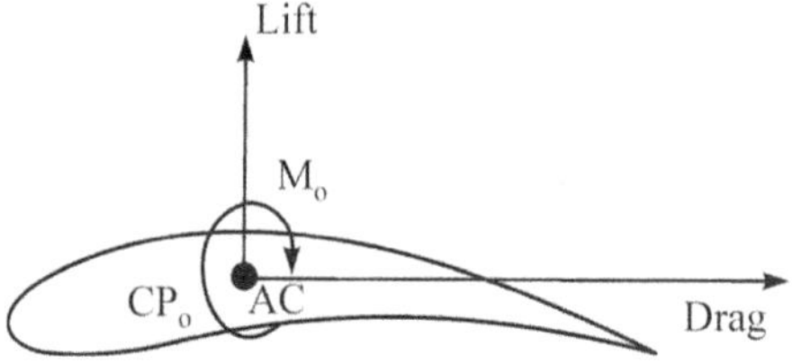

FIGURE 9.5 Center of pressure and Aerodynamic center of a wing section

The moment, which has a tendency to rotate the aeroplane about the wing axis, is called the ***Pitching moment 'M'***. Similar to the lift and drag components of resultant aerodynamic force, pitching moment is also defined by

$$M = C_M (\tfrac{1}{2} \times \rho \times V^2 \times S \times C)$$

where, C_M is a non-dimensional coefficient

and   C is the length of chord of wing section

9.4   EFFECT OF ANGLE OF ATTACK

In general, the coefficients C_L, C_D and C_M are functions of shape of section and angle of attack, α. Typical plot of C_L as a function of α, for a particular aerofoil section, is given in Fig 9.6. As the angle of attack increases, lift and drag increase. This option is used to increase lift for raising the altitude of flight (or take-off) or decrease lift for lowering the altitude of flight (or landing). However, there is a limit, beyond which increasing positive angle of attack (leading edge up) reduces lift. This is called ***positive stall angle***. Similarly, beyond a limit increasing negative angle of attack (leading edge down) increases lift. This is called ***negative stall angle***. Stall is defined as the aerodynamic loss of lift. Every aeroplane operates within these limits of positive high and negative low angles of attack, to avoid stall.

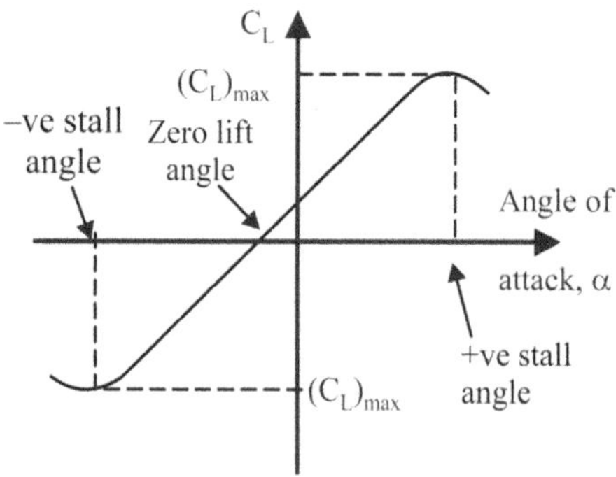

FIGURE 9.6 Plot of variation of C_L with angle of attack (α)

CHAPTER 10

LOADS ON AEROPLANE STRUCTURE

10.1 TYPES OF LOADS

An understanding of types of loads in different flight conditions is essential for safe design of an aeroplane. Operational requirements (and, hence, loads) vary with the configuration of the aeroplane (single/twin engine; propeller/jet; piston engine/gas turbine,..) as well as its nature of function (civil transport, cargo, spying, fighter, food supply during natural calamities, spraying pesticides,..) which decide speed, altitude, total payload and manoeuvrability.

The structure of an aeroplane is required to support two distinct classes of load – *ground loads* (during movement on ground such as taxiing and landing) and *aerodynamic forces* (or air loads) induced during flight. In addition, aeroplane designed for a particular role encounter some peculiar loads. Carrier (ship) borne aeroplane are, for example, subject to catapult take-off and arrested landing loads. These loads are further subdivided into *surface forces* or aerodynamic forces (which act on the surface of the structure) and *body forces*, produced by gravitational and inertia effects. Over and above these basic in-flight loads, fuselages of passenger aeroplanes are pressurized and hence support hoop stresses. Wings may carry engines, weapons and/or extra fuel tanks, which produce additional aerodynamic and body forces.

10.2 AERODYNAMIC FORCES

Basically, all air loads are the resultants of the pressure distribution over the surface of the skin. These resultants cause direct loads (bending, shear and torsion) in different parts of the structure, in addition to local normal pressure loads on the skin.

Conventionally, an aeroplane usually consists of fuselage, wings and tail plane. The fuselage contains crew and payload (passengers, cargo, weapons and fuel, depending on the type of aircraft and its function). Wings provide the necessary lift to ensure that the airplane floats. Tail plane is the main contributor to directional control. In addition, ailerons, elevators and rudder enable the pilot to manoeuvre the aircraft and maintain its stability in flight, while wing flaps help in increasing or decreasing lift in different operating conditions such as take-off, level flight (cruise) and landing (Ref Fig 10.1).

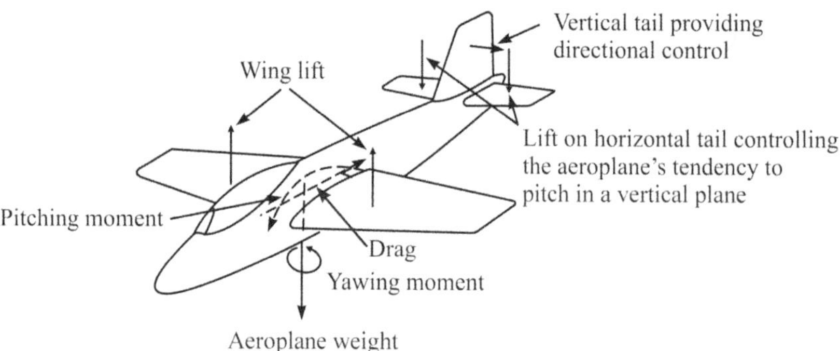

FIGURE 10.1 Aerodynamic forces during flight

The force on an aerodynamic surface (wing, vertical or horizontal tail) results from a differential pressure distribution caused by incidence, camber or a combination of both. Such a pressure distribution has vertical (**lift**) and horizontal (**drag**) resultants acting at a point, called centre of pressure (CP). The position of CP changes as the pressure distribution varies with speed or wing angle of attack (Ref Fig. 10.2). There is a point in the aerofoil section, called aerodynamic center (AC), about which the **moment** due to lift and drag forces remains constant (M_0). Thus, lift and drag acting at CP are replaced by lift and drag forces as well as a constant moment M_0 at AC.

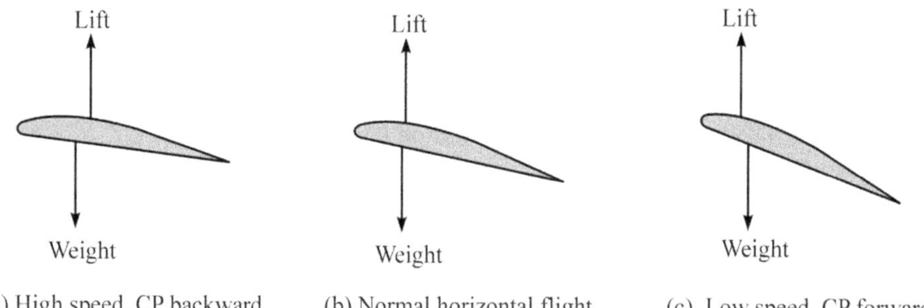

FIGURE 10.2 Pressure distribution on an aerofoil & its resultant Lift force

While the chord-wise pressure distribution fixes the position of the resultant aerodynamic load on the wing cross section, the span-wise distribution locates its position in relation to the wing root. A typical distribution for a wing-fuselage combination is shown in Fig. 10.3.

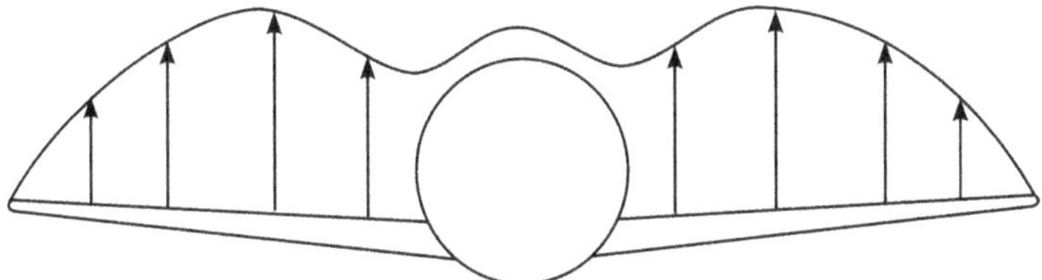

FIGURE 10.3 Typical lift distribution on wing-fuselage combination

The aerodynamic data required for the structural analysis are lift, drag and pitching moment distributions for the complete aeroplane through the range of angles of attack from the negative stalling angle to the positive stalling angle. While these data can be calculated accurately for a wing with a conventional aerofoil section, similar data for the combination of wing and fuselage or the wing, fuselage and nacelles are more difficult to calculate because of uncertain effects of aerodynamic interference of various components. It is therefore desirable to obtain wind tunnel data on a model of the complete aeroplane. Information from published data can be used to obtain approximate air loads for preliminary design purposes.

One of the four basic conditions, representing symmetrical flight manoeuvres – positive high angle of attack (PHAA), positive low angle of attack (PLAA), negative high angle of attack (NHAA), negative low angle of attack (NLAA) – produce the highest load in any part of the aeroplane for any flight condition. (Ref.Fig.10.4)

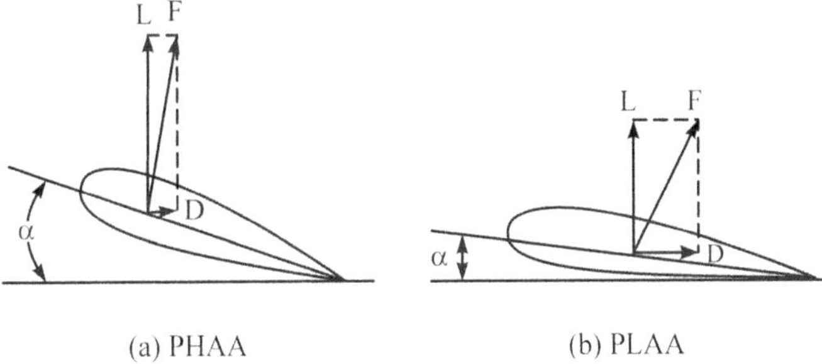

(a) PHAA   (b) PLAA

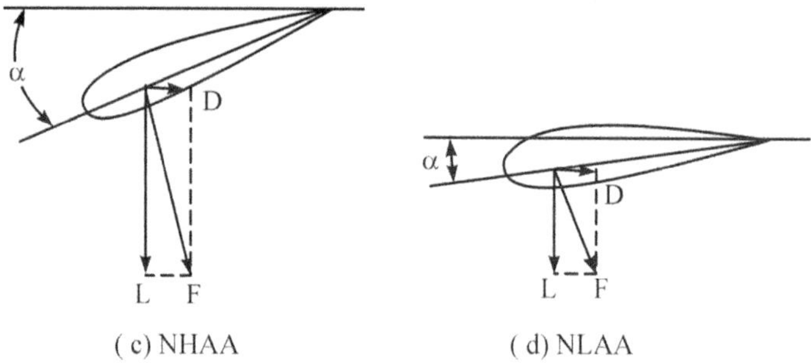

FIGURE 10.4 Change of lift with angle of attack (α)

Corresponding axial loads in stringers in different locations of wing cross section for different angles of attack (α) are shown in Fig. 10.5

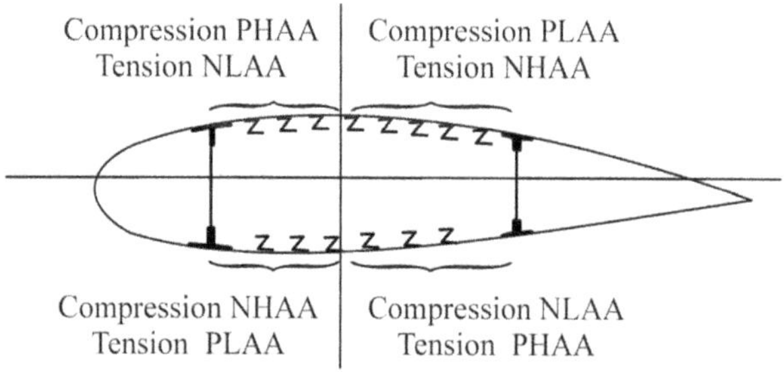

FIGURE 10.5 Axial loads on stringers for different angles of attack (α)

10.3   MAJOR AEROPLANE WEIGHT

Both aeroplane gross weight and its detailed distribution have a large influence on structural design loads. For example, wing down-bending on ground and wing up-bending during level flight will be a direct function of aeroplane gross weight and its distribution, as shown in Fig 10.6.

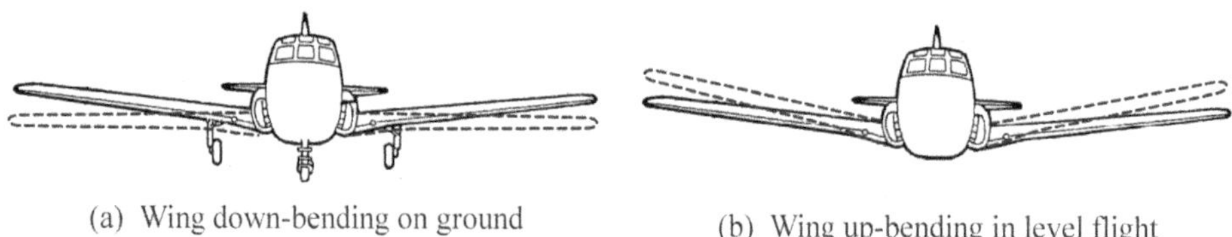

(a) Wing down-bending on ground

(b) Wing up-bending in level flight

FIGURE 10.6 Lifting forces Vs Gross weight of aeroplane

Gross weight can also be seen as the **vertical load** imposed on the landing gear when it comes in contact with the ground at a given sinking speed (based on the mission requirements of the aeroplane), **take-off weight** which includes maximum fuel weight and pay load, design **landing weight** which includes minimum fuel weight (using fuel dump provisions, if necessary, during emergency landing) and is critical for down-bending of wing and fuselage and **tow loads** obtained by multiplying take-off weight by various factors.

10.3.1 CENTER OF GRAVITY ENVELOPE

The combinations of design weight and aeroplane center of gravity (CG) are of considerable importance. An envelope enclosing all extremes of variation of CG with design weight is shown in the plot of Fig 10.7.

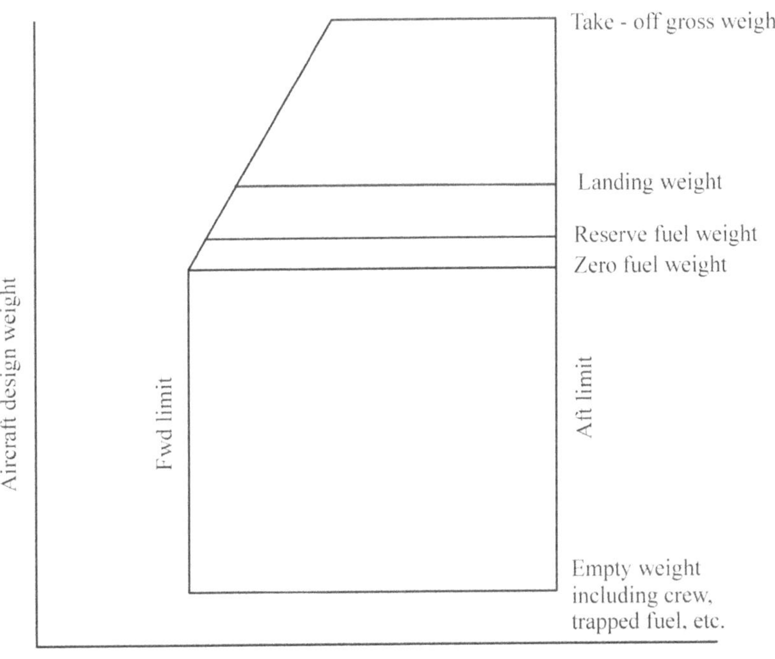

FIGURE 10.7   A typical center of gravity envelope

10.3.2 WEIGHT DISTRIBUTION

The dead weight of fuselage, wings, cargo etc., contributes to a large part of the weight distribution and influences greatly the magnitude of down-bending experienced by the fuselage fore-body or aft-body during a hard landing.

The amount and disposition of fuel weight in the wing is important to provide bending relief from aerodynamic forces, during flight. Placing fuel as far outboard as possible and using fuel from the most inboard tanks first provides the optimum arrangement for wing bending during flight (Ref Fig 10.8).

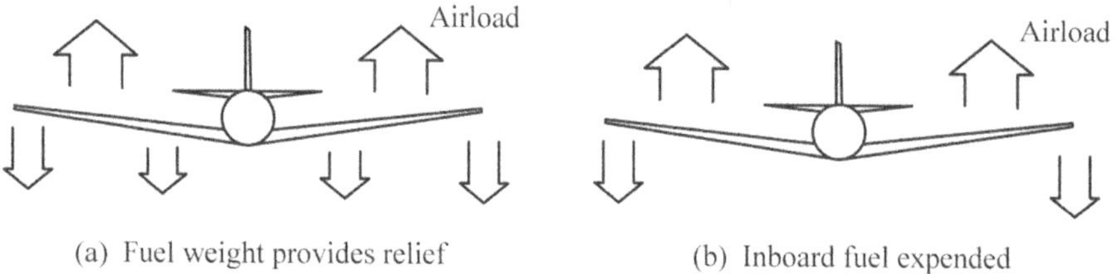

FIGURE 10.8   Wing up-bending relief due to fuel weight

10.4   GUST LOADS

Movements of air in turbulence are known as gusts and produce changes in wing incidence and consequent sudden changes in lift. These may be critical for large high speed aircraft. They do not introduce different loads but result only in changes of magnitude and position of the existing loads.

Gust analysis determines loads due to a single or 'discrete' gust of a given profile (distribution of vertical gust velocity over a given period of time) and the response to it by the aircraft. Initially, analysis was carried out for a ***'sharp-edged'*** or ***'step' gust***, sudden change in vertical air velocity and the aerodynamic forces are determined by the instantaneous incidence of the particular lifting surface, assuming that the aircraft's structure is rigid. It neglected gradual growth of circulation and, hence, lift to a steady state value (Wagner effect) and resulted in overestimation of gust loads.

It was realized that the gust velocity built up to a maximum over a period of time and over a specified gust gradient distance 'H' (of about 20 to 40m) and is defined as the ***'graded'*** or ***'ramp' gust***. By the time gust attains its maximum value, the aircraft attains a vertical component of velocity decreasing severity of the gust and experiences pitching, depending on its longitudinal stability characteristics, which affect aerodynamic forces. Usually graded gust is analysed by an equivalent sharp-edged gust (or a large number of small 'steps' and superimposing the responses, called convolution or Duhamel integration) producing approximately the same effect.

10.4.1 'SHARP-EDGED' GUST

Let us consider an aircraft flying at a speed V with wing incidence α_0 in still air. After entering the gust of upward velocity u, the incidence increases by an amount $\tan^{-1}(u/V)$ or u/V. This is accompanied by an increase in aircraft speed from V to $\sqrt{(V^2 + u^2)}$, which is negligible. The increase in wing lift is

$$\Delta L = (\rho \times V^2 \times S/2) \times (\partial C_L/\partial\alpha) \times (u / V) = (\rho \times V \times S/2) \times (\partial C_L/\partial\alpha) \times u$$

where, $\partial C_L / \partial\alpha$ is the wing lift-curve slope

Neglecting change of lift on the plane, since change in tail plane incidence is not equal to the change in wing incidence due to downwash effects at the tail, gust load factor Δn produced by this change of lift in an aircraft of weight W is

$$\Delta n = [(\rho \times V \times S / 2) \times (\partial C_L / \partial\alpha) \times u] / W$$

$$= [(\rho \times V / 2) \times (\partial C_L / \partial\alpha) \times u] / w$$

where,   w = W/S is the wing loading

The gust load factor is directly proportional to air speed but inversely proportional to wing loading. Hence, high speed aircraft with low or moderate wing loadings are most likely to be affected by gust loads.

If flight conditions are expressed in terms of equivalent sea-level conditions with suffix 's', $\Delta n = [(\rho_S \times V_S / 2) \times (\partial C_L / \partial\alpha) \times u_S] / w$

Therefore, as a result of the upward gust, total gust load factor $n = 1 + \Delta n$

Similarly, for a downward gust, total gust load factor $n = 1 - \Delta n$

10.4.2 'GRADED' GUST

The graded gust may be converted to an equivalent 'sharp-edged' gust by multiplying the maximum velocity in the gust by a ***gust alleviation factor***, F. It takes care of some of the dynamic properties of the aircraft, including unsteady lift and heaving motion (upward motion with zero rate of pitch). Thus, total gust load factor changes to

$$\Delta n = 1 + [(\rho_S \times V_S / 2) \times (\partial C_L / \partial\alpha) \times u_S] \times F / w$$

Horizontal gusts cause lateral loads on the vertical tail or fin and can be calculated in a similar manner.

10.5 GROUND LOADS

These are encountered in landing and taxiing and subject the aircraft to concentrated shock loads through the undercarriage system, usually located in the wings (close to the wing root) with a nose wheel or tail wheel in the vertical plane of symmetry.

Other loads may include engine thrust on the wings or fuselage which acts in the plane of symmetry but may, in the case of one engine failure, cause severe fuselage bending moments.

10.6 LOADS DURING MANOEUVRES

Every aeroplane is designed for many different planned and unplanned flight conditions and manoeuvres. These conditions need to be understood while designing aeroplane structure. Some of them are briefly mentioned here.

- *Nose-dive* condition - If an airplane is dived vertically towards the ground, its speed will increase until it reaches a steady maximum velocity (called *terminal velocity*) irrespective of whether engine is running or not. In practice, most modern aeroplanes will hit the ground before they reach their terminal velocity. In a vertical dive, the weight of the aeroplane must be supported by the drag forces and will therefore be much larger than in normal horizontal flight and the drag bracing has to be adequately strengthened.

- *Turning an aeroplane* (*Yawing* motion) - In the horizontal plane, rudder generates side loads by deflecting its fin. In most aeroplanes, the rudder is above the center line of fuselage. So, the side force causes bending and twisting of the fuselage, inducing severe stresses in the skin as well as fuselage structure.

- *Upside-down flight* – It reverses the loads of normal flight. In order to generate sufficient lift to balance weight of aeroplane, a large angle of attack on the wings is used, at considerable expense of engine power. The reversal of lift forces has little effect on the spars but considerable effect on the skin and ribs.

- *Landing forces* – These depend on the vertical velocity, smoothness and travel of the shock absorbing mechanism. The shock load, received first by the tyres, is transferred to the axle, undercarriage struts and then to the fuselage. Because of the tail-wheel or nose-wheel striking the ground, there will be loads upon the rear or front portion of the fuselage tending to bend it upwards, affecting most of the structural parts of the fuselage.

10.7 LANDING LOADS

Load during landing is related to the descent velocity, type of landing and shock absorbing characteristics of the main and nose landing gears (Ref Fig 10.9).

- During *level landing* conditions (two point and three point landing cases), vertical loads (V) and horizontal loads (D) are applied at the wheel axles, which are reacted by aeroplane inertia loads nW and T. In two point level landing, the two main gears come in contact with the ground first; while in three point level landing, nose gear and the two main gears come in contact with the ground simultaneously.

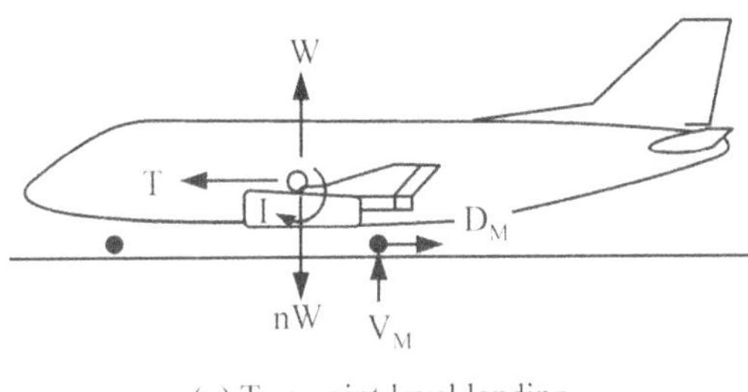

(a) Two point level landing

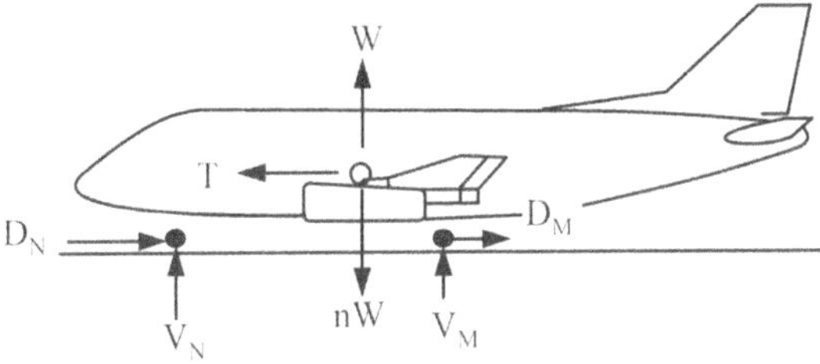

(b) Three point level landing

- *Tail-down landing* is made at an extreme angle of attack, limited by stall angle or clearance with the ground

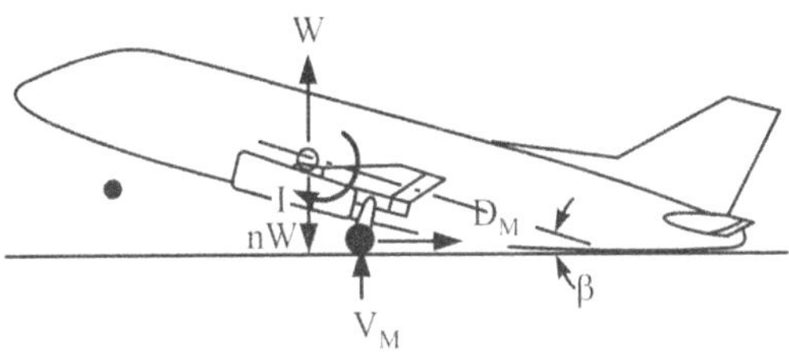

(c) Tail down landing

- *One wheel landing* is equal to landing on one side main gear during level landing. The unbalanced moment is reacted by aeroplane rolling inertia

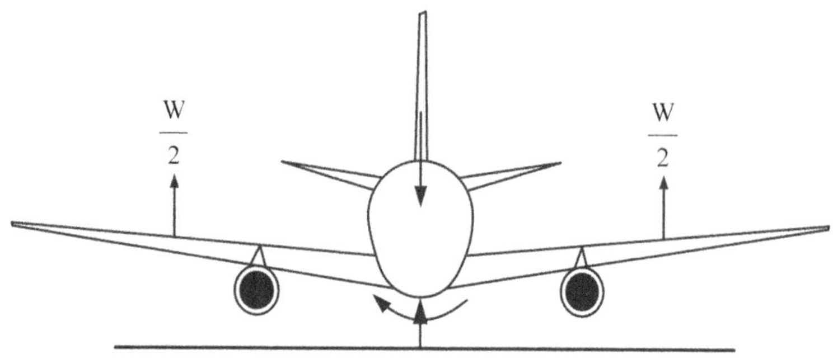

(d) One wheel landing

FIGURE 10.9 Landing loads

- The landing gear and its supporting structure must be checked for *rebound landing*, which occurs after a hard landing rebounds aeroplane into the air. With the landing gear fully extended and not in contact with the ground, a load factor of 20 must be applied to the unsprung weights of the gear in the direction of Oleo movement

10.8 LIMIT LOADS AND ULTIMATE LOADS

Limit loads are the maximum loads anticipated on the aeroplane during its service life. The aeroplane structure shall be capable of supporting the limit loads without suffering detrimental deformation. Gust loads are arbitrary, based on assumed gust velocity from past experience; It is possible that an aeroplane, in its life time, experiences higher gust loads. Hence, aeroplane is designed for **ultimate load**, which is equal to the limit load multiplied by a factor of safety. In general, the factor of safety is 1.5

10.9 WEATHER EFFECTS

Very bad *weather effects*, gusty winds, bumps and air pockets may cause severe stresses in flight and may be worse than those in recognized manoeuvres. All these gusts will have the effect of sudden change in the angle of attack and, hence, lift.

The speed at which sound travels in air varies with temperature, which is a function of altitude – about 340 m/sec or 1200 kmph at ground level to about 295 m/sec or 1050 kmph in stratosphere. As the aeroplane travels through the air, the pressure waves created by it are sent out in all directions at the speed of sound, causing pattern of air inflow ahead to change before the aeroplane arrives. *Flight at subsonic speeds* is based on the assumption that air behaves as an incompressible fluid (like water).

When the aeroplane travels at or near the speed of sound (*transonic speed* in the range of 220 to 355 m/sec or 800 to 1280 kmph), the pressure waves can not get ahead of the aeroplane and so the aeroplane comes against air with shock. It creates a shock wave, a sharp dividing line representing sudden drop in the speed of airflow and an increase in the pressure and density of the air, which now behaves like a compressible fluid. From the structural point of view, the airflow behind the shock wave becomes turbulent, lift reduces and drag increases. This phenomenon is called shock stall, which can be postponed by having thin wings, slim fuselages and, most effective of all, a large degree of sweepback.

In the *supersonic range of speeds*, there are no longer violent fluctuations of lift and drag and the airflow is steady, although altogether different from that at subsonic speeds. Streamlined shape is no longer applicable; bodies should be sharp-pointed; plan shape of the wings becomes more important than their cross section and aspect ratio (ratio of square of wing span to the area of wing or L^2/A) should be low. Loads on the aircraft are very large and all parts of the structure must be strong and rigid.

10.10 LOAD FACTOR AND FACTOR OF SAFETY

It is difficult to estimate *forces in all the maneuvers*, which an aeroplane is capable of. An instrument, called accelerometer, is able to detect the loads on various parts of the structure. It is seen that the loads during the maneuvers are greater than those of normal flight and for the same maneuver, loads depend on the skill of the pilot.

The amount of additional load depends on the severity of the maneuvers or turbulence. Its magnitude is measured in terms of aerodynamic load during normal straight and level flight (equal to weight of aeroplane), called *load factor.* The maximum maneuvering load factor to which an aeroplane is designed depends on its intended usage. It may be in the range of 6 to 12, depending on the role of the particular aeroplane.

There are many structures in an aeroplane, in which self weight of the structure is a major load. A heavier aeroplane needs larger wing surface to ensure adequate lift and a stronger and, hence, heavier structure. Decrease in weight of aircraft structure not only reduces material cost and labour cost but also improves performance. The structure must also withstand not only the quasi-static design loads but also 'reasonable' increase in design load due to change in operating conditions; fatigue damage due to repeated fluctuating loads; and impact nature of some design loads (such as those during landing). These factors as well as those related to deterioration of material properties with changes in operating temperature are all considered in the *factor of safety*.

The basic strength and flight performance limits for a particular aircraft are selected by Airworthiness authorities. The maximum load that an aircraft is expected to experience in normal operation (including gust load of a given severity) is called the *limit load*. The aircraft's structure must withstand *proof load* (1.0-1.25 times the limit load) without detrimental distortion and should not fail until the *ultimate load* (usually 1.5 times the limit load) has been achieved.

10.10.1 SAFE-LIFE DESIGN

Structure is designed to have a minimum life during which no catastrophic damage will occur. At the end of this life, the structure must be replaced even though there may be no detectable signs of fatigue. It is possible to have different safe life periods for different parts of structure, lesser life for undercarriage system so that two light-weight undercarriage systems of shorter life can be replaced a few times during the life of aircraft than carry a heavier undercarriage with same life as the aircraft.

10.10.2 FAIL-SAFE DESIGN

Any aeroplane should be capable of reaching the ground safely even after some parts of it have failed during flight in the air. It is impossible to guard completely against breakage of any part, for that would mean either complete duplication of all parts or an increase in structural dimensions; In either case, aeroplane would be too heavy for any useful purpose. Therefore, the designer should only duplicate or strengthen such parts which are liable for accidental breakage. It is known that statically indeterminate structures are inherently safer while carrying load after damage, provided that the remaining members are able to carry the load shed by the failed member until the failed member is identified and replaced.

A built-up spar has the advantage of providing alternate load paths for tension in case any single web beam should fail. '*Crack stoppers*' are used to arrest propagation of cracks in skin panels. A ***fail-safe strap*** or ***tear strap*** is used to arrest propagation of skin crack in a fuselage.

10.11 FLIGHT ENVELOPE OR V-n DIAGRAM

Various loading conditions for an aeroplane are represented on a graph of limit-load factor 'n' plotted against indicated airspeed 'V' and is called V-n diagram (Ref Fig. 10.10). Indicated airspeed is the speed of airplane at its flying altitude, with lower air density ρ' adjusted to air density at standard sea level ρ, so that $\rho V^2/2$ is same. V-n diagrams have different shapes at different altitudes. However, they are same for all altitudes, if compressibility effects (which are based on actual air speed) are neglected.

If the aeroplane has no angular acceleration, both the inertia and gravity forces will be distributed in the same manner as the weights of components of the aeroplane and will have resultants acting through the center of gravity of the aeroplane (G). It is convenient to combine the inertia and gravity forces as the product of a load factor n and the weight $W \times$ Z-component of the resultant gravity and initial force is (n × W) acting at G.

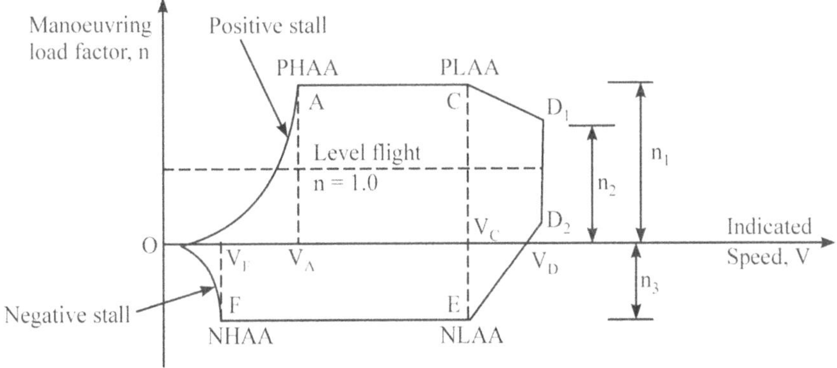

FIGURE 10.10 Flight envelope (V-n diagram)

The load factor n is obtained from a summation of forces along the z-axis, as

$$n \times W = C_{zo} \times q \times S \quad \Rightarrow \quad n = C_{zo} \times \rho \times S \times V^2 / (2W) \quad \text{since} \quad q = \rho \times V^2 / 2$$

For level flight at a unit-load factor, the value of V corresponding to $(C_{zo})_{max}$ would be the stalling speed of the aeroplane. In accelerated flight, the maximum coefficient might be obtained at higher speeds. For $(C_{zo})_{max}$ to be obtained at twice the stalling speed, a load factor n = 4 would be developed (as seen from the above equation). It is possible to manoeuver the aeroplane at speeds and load factors to the right of OA but it is impossible to manoeuver the airplane at speeds and load factors to the left of OA. The line OA represents a limiting condition. The line AC represents the limit on the maximum manoeuvring load factor for which the aeroplane is designed.

For speeds below V_A (max positive wing incidence) and V_F (max negative wing incidence), the max loads which can be applied to the aeroplane are governed by $(C_L)_{Max}$. AC and FE represent max operational load factors of n_1 and n_3 for the aircraft. Above the design cruising speed V_C, cut-off lines CD_1 and D_2E relieve the design cases to be covered.

Because of the effects of compressibility on wing lift and the effects of varying air density on aeroplane speed capability, $(C_L)_{Max}$ is generally reduced with an increase of altitude and the speed of sound decreases with altitude thereby reducing the critical Mach number. Hence, a particular flight envelope is applicable to one altitude only. The design diving speed V_D is usually specified as $1.2 - 1.5$ times the max indicated air speed in level flight. Lines OA and OF indicate that the limit-load factor specified for negative manoeuvers is considerably less than for positive manoeuvers.

Most severe structural load (ng) conditions are represented by the corners A, C, E and F of the diagram. Point A represents positive high angle of attack (PHAA) condition. Point F represents negative high angle of attack (NHAA) condition. Point C represents positive low angle of attack (PLAA) condition. Point E represents negative low angle of attack (NLAA) condition.

10.12 GUST ENVELOPE

Airworthiness requirements usually specify that gust loads shall be calculated at certain combinations of gust and flight speed. Since gust load factor n is proportional to aircraft speed for a given gust velocity, a gust envelope similar to flight envelope can be plotted (Ref Fig 10.11).

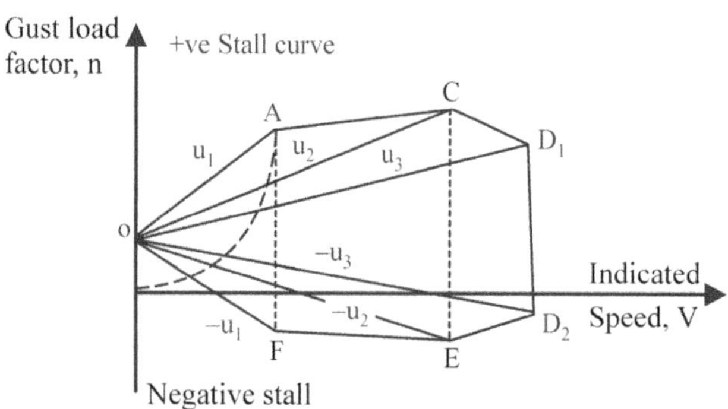

FIGURE 10.11 Gust envelope (V-n diagram)

The gust speeds $\pm u_1$, $\pm u_2$ and $\pm u_3$ are high, medium and low gust velocities respectively. Typical values of u_1, u_2 and u_3 are about 20 m/sec, 15 m/sec and 7.5 m/sec. Cut-offs occur at points where gust velocity lines meet specific aircraft speeds.

The lift coefficient – incidence curve is affected by compressibility and, therefore, altitude so that a series of gust envelopes should be drawn for different altitudes and different wing loadings. It can be seen from the gust envelope that the maximum gust load factor occurs at the cruising speed V_C. Although the same combination of V and n in the flight and gust envelopes will produce the same total lift on an aircraft, the individual wing load and tail plane load will be different. In the flight envelope case, the tail load is downwards whereas in the gust load case it is upwards.

CHAPTER 11

ANALYSIS OF DETERMINATE AND INDETERMINATE BEAMS

Trusses, made up of spars, are designed for loads which act at the joints and the members are hinged (or are assumed hinged) at the joints permitting relative rotation between different members at that joint. However, there are many situations in which load (such as wind load) acts on the structural members at any intermediate point or distributed along its length. Such members, which are acted upon by concentrated or distributed transverse loads along its length, are called *beams*. The difference between a truss and a *frame* lies in the manner in which the external loads are resisted rather than the geometrical configuration of the members.

A *simple beam* is a slender, homogeneous bar that bends without twisting when acted upon by loads applied perpendicular to its axis and in a single plane containing the axis. The end of a beam may be free or may have a simple (or hinged) support with restraint only on deflection or a fixed support with restraint on both deflection and slope. In frames, loads include moments and displacements include slopes or rotations.

11.1 CLASSIFICATION OF BEAMS

Based on the types of supports, beams are generally classified as follows:

(a) *Cantilever beam*: with one end fixed for deflection as well as rotation

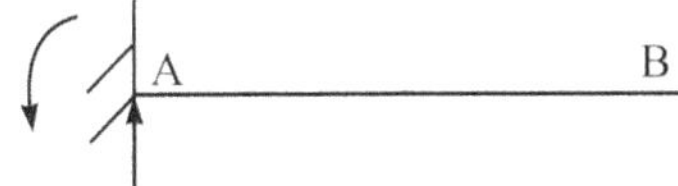

(b)   *Simply supported beam*: with both ends supported only for deflection, permitting free rotation

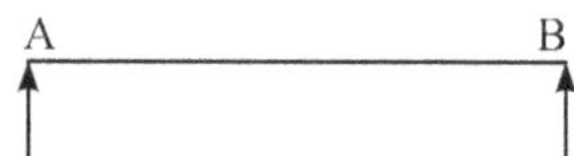

(c)   *Fixed beam*: with both ends fixed for deflection as well as rotation

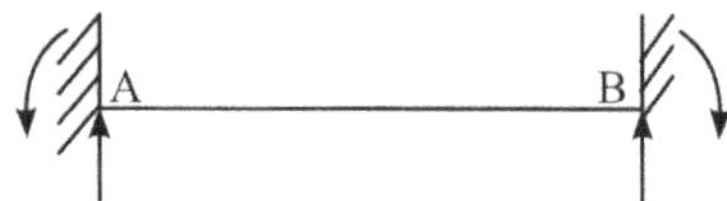

(d)   *Propped cantilever*: a cantilever with an additional support for deflection at or near the free end; the prop may be rigid, sinking or elastic

(e)   *Continuous beam*: simply supported beam with more than t wo supports for normal deflection

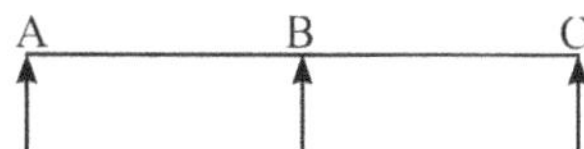

(f)   *Overhanging beam*: with the beam extending on one or both sides of a simply supported or continuous beam

The loads and reactions on a beam structure will always form a system of non-concurrent forces. Reactions are calculated by considering the entire frame and using the equations of equilibrium, applicable to the system of non-concurrent forces ($\Sigma F_X = 0$, $\Sigma F_Y = 0$ and $\Sigma M_Z = 0$ for a plane frame in X-Y plane and $\Sigma F_X = 0$, $\Sigma F_Y = 0$, $\Sigma F_Z = 0$, $\Sigma M_X = 0$, $\Sigma M_Y = 0$ and $\Sigma M_Z = 0$ for a space frame). Many of the beam elements may not be subjected to axial loads (in X-direction). In such a beam member in X-Y plane, there are only two equations of equilibrium ($\Sigma F_Y = 0$ and $\Sigma M_Z = 0$) assuming X-axis as the longitudinal axis of the beam member.

If the number of unknown reactions is equal to the number of equations of static equilibrium, such as in a simply supported beam or in a cantilever beam (Fig. 11.1 a), that structure is called a '***statically determinate beam***'. If the number of unknown reactions is more than the number of equations of static equilibrium, such as in a continuous beam or in a propped cantilever beam or a fixed beam (Fig. 11.1 b), that structure is called a *'statically indeterminate beam'*.

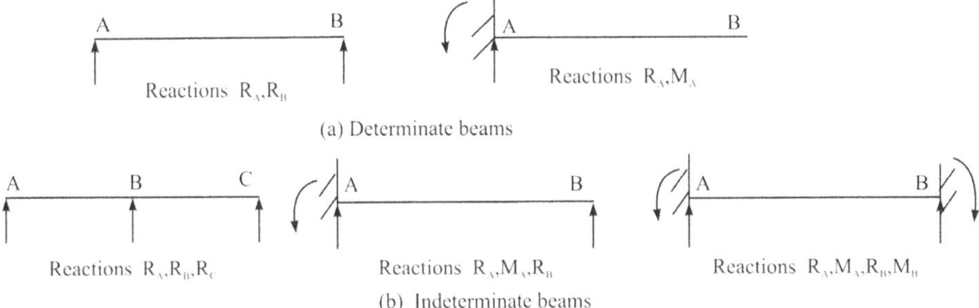

FIGURE 11.1 Types of beams

The difference between the number of unknown reactions and the number of relevant equilibrium equations is called **'degree of indeterminacy'**. The above figures are drawn assuming that there is no horizontal load acting on the beam and, hence, no horizontal reactions. Thus, the relevant equations of equilibrium for a plane frame are only 2, excluding the irrelevant equation $\Sigma F_X = 0$ along the axis of the beam. The degree of indeterminacy, which is the difference between the number of reactions (3 in the first two cases and 4 in the last case) and the number of equilibrium equations, is 1 for the first two cases and 2 for the last case. Adequate number (equal to degree of redundancy) of additional equations, necessary to evaluate support reactions, are generated from the displacement or slope conditions at the supports, such as $\delta_B = 0$, $\theta_A = 0$ and $\theta_A = \theta_B = 0$ in these three examples respectively.

11.2 SHEAR FORCE AND BENDING MOMENT

If we imagine a cut in the beam at some section, free body diagram shows a shear force (parallel to the surface) and a bending moment at the imaginary section, to ensure equilibrium of the free body. Plots of variation of shear force (S) and bending moment (M) along the length of a beam, as shown in Fig 11.2, are called shear force diagram (SFD) and bending moment diagram (BMD).

For equilibrium of the free body of the cantilever beam shown in Fig.11.2a,

$$\Sigma F_Y = S - P = 0 \quad \text{or} \quad S = P \quad \text{and} \quad \Sigma M_Z = M - P{\times}x = 0 \quad \text{or} \quad M = P{\times}x$$

For this particular case, therefore, shear force S is constant all along the beam while bending moment M varies linearly from M = 0 at the free end to $M = M_{MAX} = P.L$ at the fixed end.

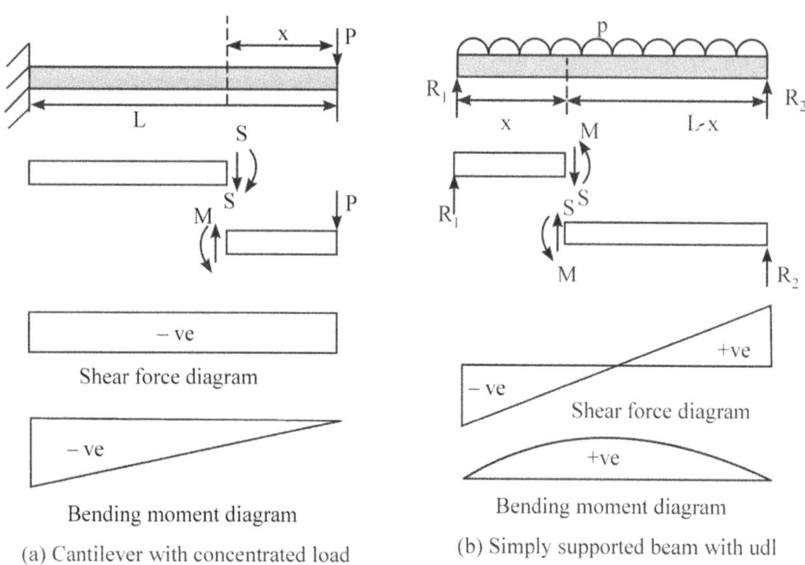

FIGURE 11.2 Shear force and bending moment diagrams

For equilibrium of the free body of the simply supported beam shown in Fig.11.2b, with uniformly distributed load (u.d.l.) of magnitude 'p' along its length

$$\sum F_X = S - R_1 + p \times x = 0 \quad \text{or} \quad S = R_1 - p.x \qquad \text{where} \quad R_1 = p \times L / 2$$

$$\text{and} \quad \sum M_Z = M - R_1 \times x + (p \times x) \times x/2 = 0$$

$$\text{or} \quad M = R_1 \times x - p \times x^2/2 = p \times x \times (L - x)/2$$

For this particular case, therefore, shear force S varies linearly along the beam while bending moment M varies parabolically from M = 0 at the fixed ends to $M = M_{MAX} = p \times L^2/8$ at x = L/2 or the section at mid-point along the axis of the beam.

Shear force and bending moment are represented with the following sign convention (Ref Fig.11.3).

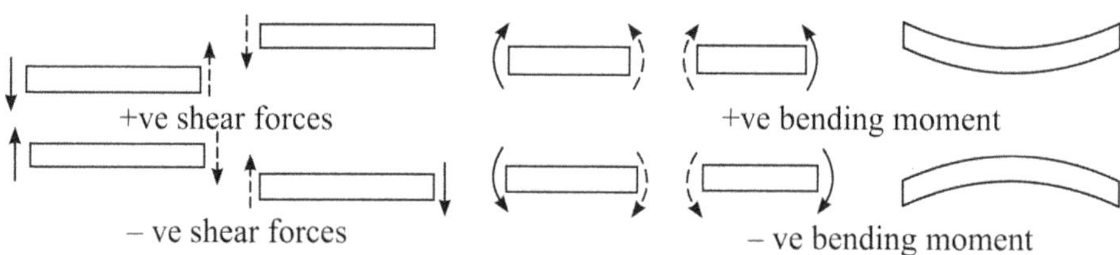

FIGURE 11.3 Sign conventions for shear force & bending moment

Shear force diagram and bending moment diagram for a few simple cases are shown in Fig.11.4. The values can be easily calculated, as explained earlier.

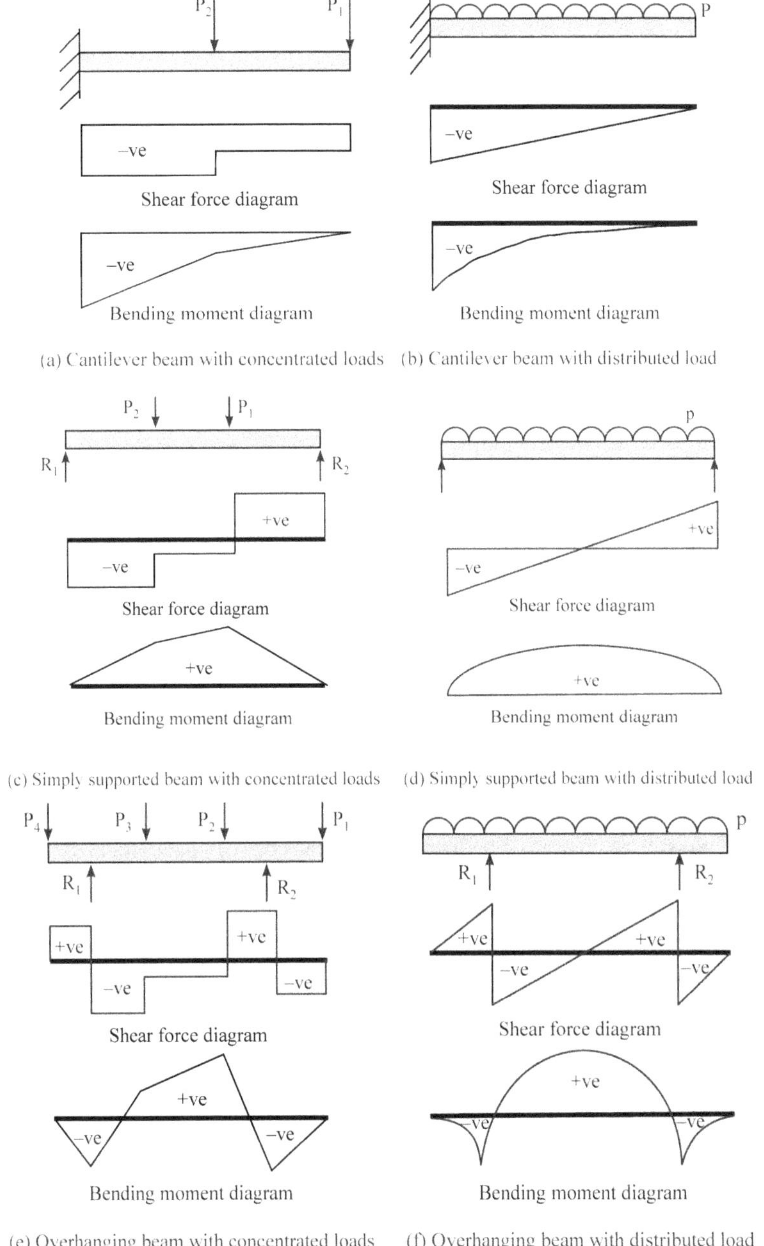

FIGURE 11.4 Shear force & bending moment diagrams for some common applications

11.3 VARIATION OF SHEAR FORCE AND BENDING MOMENT

Consider an infinitesimal length 'dx' of a beam, with shear force 'S' and moment 'M' on the right side, which increase to 'S+dS' and 'M+dM' on the left side due to applied normal load 'p(x)' along the length 'dx' of the element, as shown in Fig.11.5. Equilibrium equations for this segment are:

$$\sum F_Y = (S + dS) - p(x) \times dx - S = 0 \quad \Rightarrow \quad dS/dx = p(x)$$

$$\text{and} \quad \sum M_Z = M - p(x) \times dx \times \left(\frac{dx}{2}\right) + S \times dx - (M + dM) = 0$$

$$\Rightarrow \quad dM/dx = S, \quad \text{neglecting higher order terms}$$

Therefore, $p(x) = dS/dx = d^2M/dx^2$

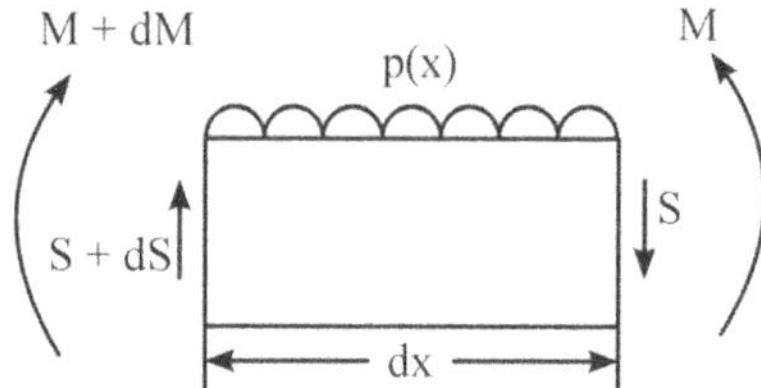

FIGURE 11.5 Variation of shear force and bending moment along a small element

We can conclude from this equation that

(a) when S is constant, M varies linearly, as can be seen from Fig.11.2a

and

(b) when S = 0, M is maximum or minimum as can be seen from Fig.11.2b

In case of ***multiple loads acting on a beam***, shear force diagram and bending moment diagram are drawn by linear superposition of the effects of individual loads. These diagrams are not drawn to any scale, but proportional with values at points of discontinuity or maximum values indicated on the diagrams. A ***moment applied*** on a beam is indicated in the bending moment diagram by a sudden jump and will not have any contribution in the shear force diagram.

The points in a beam (other than the extreme ends) at which the bending moment is zero, are called ***points of contra flexure*** or ***points of inflexion.***

11.4 THEORY OF SIMPLE BENDING

The deflection δ at any cross section of a beam is the displacement of the centroid of that cross section with respect to the given reference axes, resulting from deformation of the elements of the beam. When the beam is loaded, if no element of the beam is stressed beyond the elastic limit of the material, the deflection curve of the centroidal axis is often called the ***elastic curve*** and is a curve of radius 'R'.

The slope (θ) is the angle between a given reference axis and the tangent to the deflected centroidal axis at the section under consideration, taken as positive when the tangent slopes upward to the right and negative when the tangent slopes downward to the right. The curvature of a beam is the reciprocal, 1/R, of the radius of curvature of its centroidal axis. Usually, if M is the bending moment, E is the modulus of elasticity and I is the moment of inertia of the cross section of a originally straight beam stressed below the elastic limit,

$$1/R = M/EI = d\theta/dx$$

Since the slope and deflection are relative, both must be measured with respect to definite axes of reference. Unless otherwise specified, the x-axis corresponds to the position of the centroidal axis of the unloaded beam and y-axis is perpendicular to it, in the plane of loading. For 'small deflections', the angle θ in radians equals sin θ or tan θ, and cos θ is so near unity that 'ds' along the deflected axis is almost equal to its projection 'dx' on the X-axis.

Changes in slopes and deflections are purely geometrical consequences of bends in the beam axis. Whatever their cause, bending to a particular radius of curvature will produce the same changes in the slopes and deflections of identical size beams even though they may produce different stresses in different beams. Slope at any section for originally straight beam is the rate of change of deflection per unit length, $\theta = dy/dx$.

'Simple bending' distinguishes itself from 'pure bending' in which the shear is zero and the bending moment is constant, involving no axial load or twisting. Efficient design of a simple beam requires that its cross sectional area be concentrated in two 'chords' or 'flanges', spaced as far apart as possible from the beam axis, with a minimum area in the 'web' joining the two chords to make them act together (such as an I, ⌐ , ⊏ , sections). The bending moment is then resisted primarily by tensile and compressive stresses in the chords and the shears are resisted by the web.

In general, computation of normal stress and shear stress is based on Euler-Bernoulli beam theory. It assumes that *plane cross sections perpendicular to the axis of the beam before bending remain plane and perpendicular to the axis even after bending*. It is governed by the fourth order differential equation

$$\frac{d^2}{dx^2}\left[EI\frac{d^2 v}{dx^2} \right] = p(x)$$

where, p(x) is the load over a small segment of length 'dx' of the beam

and   v is the transverse deflection of the beam

This equation is based on the relations

$$S = dM/dx \; ; \quad dS/dx = d^2M/dx^2 = p(x) \quad \text{and} \quad M = EI\,(d^2v/dx^2)$$

where, S is the shear force at a section

It implies that all transverse shear strains are zero and warping of planar sections because of shear strains is ignored.

However, in view of the presence of transverse shear strains, if the above assumption is relaxed i.e., ***plane sections remain plane but not necessarily normal to the longitudinal axis after deformation***, the rotation of a transverse normal plane about Y-axis is not equal to –(dw/dx). Beam theory based on this relaxed assumption is called *shear deformation beam theory or Timoshenko beam theory*. In this theory, rotation about the Y-axis is denoted by an independent function $\psi(x)$, resulting in two second order differential equations.

$$\frac{d}{dx}\left[GAK_S \left(\psi + \frac{dv}{dx} \right) \right] + P(x) = 0$$

$$\text{and} \qquad \frac{d}{dx}\left(EI\frac{d\psi}{dx} \right) - GAK_S \left(\psi + \frac{dv}{dx} \right) = 0$$

Substituting the second equation in the first equation and $\psi = dv/dx$, we get equation of the Euler-Bernoulli beam theory. Lagrange interpolation is used for the transverse deflection and the rotation. Since the rotation function is like the derivative of the transverse deflection, the degree of the interpolation used for the rotation should be one less (usually first order) than that used for the transverse deflection (usually second order). Such selective interpolation of the variables is called *consistent interpolation*.

11.5 EULER-BERNOULLI BEAM THEORY

An initially straight beam is deflected to a circular arc when loaded in pure bending. Cross sections ABCD and FGHJ (Ref.Fig.11.6), vertical planes before bending, would become radial planes after bending. If the beam is considered as made up of longitudinal fibers and the stress is below the elastic limit, stress in each longitudinal fiber is proportional to strain and consequently proportional to the distance 'y' from the centroidal axis or elastic curve. The resultant horizontal component of stress on any cross section due to bending loads is zero, so that the equilibrium equation $\Sigma F_x = 0$ is always satisfied.

11.6 BEAM DEFLECTION EQUATION

Bending stress is derived from the assumption that stress is proportional to strain whereas bending strain depends on the deflection. If the deflections are assumed to be small in comparison to the original dimensions, then shearing deformation can be neglected. If shearing deformations are significant, they can be calculated separately and superimposed.

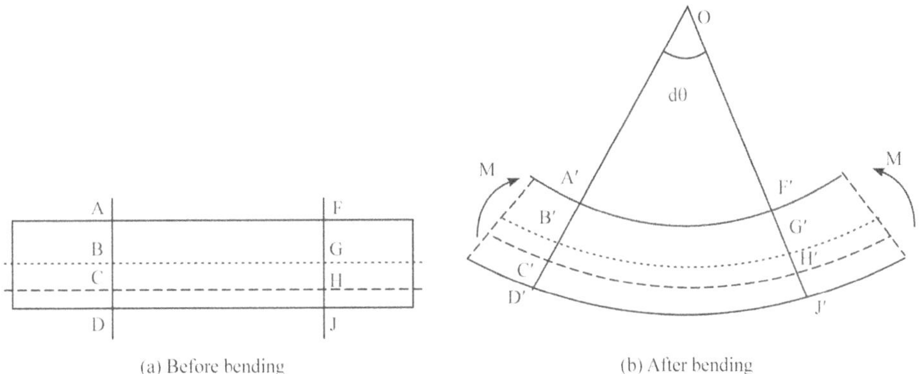

FIGURE 11.6 Beam flexure

In an initially straight beam with two parallel sections ABCD and FGHJ in the unstressed condition, the bending stress varies from compressive on top surface to tensile on bottom surface with +ve bending moment and vice versa for −ve bending moment. Obviously, there is a layer or plane which has no strain and, hence, no stress. This is called ***neutral plane.*** Intersection of neutral plane with the plane of bending is called ***neutral axis***. The neutral axis passes through the centre of gravity of cross section of the beam. Let OB=OG be the radius of curvature, R due to bending. Then sections ABCD and FGHJ which are parallel before bending (Ref Fig.11.6a) are rotated about O as A′B′C′D′ and E′F′G′H′ after bending (Fig.11.6b), producing an angle dθ at the center of radius, O.

For small angle dθ, strain in the segment CH at distance y from neutral axis (N.A.) BG or B′G′,

$$\varepsilon_X = du/dx \approx (C'H' - CH) / CH = [(R + y).d\theta - R.d\theta] / (R.d\theta) = y/R$$

and, within the elastic limit, stress $\sigma_X = E.\varepsilon_X = E.y/R$ .....(11.1)

Thus, bending stress is a linear function of distance y from N.A. and changes its sign from +ve when the particular layer CH is farther than neutral axis from center of curvature O to −ve when the particular layer is closer to the center of curvature O than neutral axis.

The total axial internal force in a beam across any cross section,

$$F_X = \int dF_X = \int \sigma_X \times dA = \int (E \times y/R) \times dA$$

Since E and R are constant across any section,

$$F_X = (E/R) \times \int y \times dA = 0$$

$$\Rightarrow \int y \times dA = 0 \quad \text{or} \quad BG, \text{ the neutral axis, coincides with centroidal axis}$$

The resultant of compressive stresses and tensile stresses on a cross section form two equal and parallel forces, thus producing a couple called **moment of resistance**. Bending moment due to external applied loads is independent of beam section details while moment of resistance, for moment equilibrium, is equal to the bending moment and depends on shape of cross section and material of beam. If bending moment is more than moment of resistance, the beam fails.

Moment of resistance at any section is given by,

$$M_R = \int y \times dF = \int y \times (\sigma_X \times dA) = \int y \times (E \times y/R) \times dA$$

$$= (E/R) \times \int y^2 \times dA$$

$$= (E/R) \times I \qquad\qquad .....(11.2)$$

Combining equations (11.1) and (11.2), we get $\mathbf{M_Z/I_Z = \sigma_X/y = E/R}$

This equation is called Engineer's *theory of simple bending*

If v is the vertical displacement of neutral axis at any section, dv/dx gives the slope and d^2v/dx^2 is called the curvature (1/R). If y is measured +ve upwards, positive bending moment produces a +ve curvature.

Thus, $M_Z = E \times I_Z \times (d^2v/dx^2)$, which can be used by double integration to calculate deflection v. Integration constants are evaluated based on the specified boundary conditions. The product $E \times I$ is called *flexural rigidity of beams* and used as a single property 'EI'.

11.7   SHEAR STRESSES IN A BEAM SECTION

It is seen that strain in bending along the axis of the beam is a linear function of distance from the neutral plane, tensile on one side and compressive on the other side. Within the elastic limit, since stress is proportional to strain, *bending(normal) stress* (σ_X) is also a linear function of distance from the neutral plane, tensile on one side and compressive on the other side. Varying normal stress along two parallel layers produces shear stress (τ_{YX}) between them, which can be seen to be zero on the outermost layers and maximum along the plane where bending stress changes from tensile to compressive. This shear stress produces complementary shear stress (τ_{XY}) on the perpendicular face, whose sum over the cross sectional area is called shear force at that section (Fig.11.7). Thus, the stresses τ_{XY} and σ_X are related to the shear force and bending moment at a section of the beam subjected to bending in X-Y plane (X being the axial direction), by

$$S_Y = \int \tau_{XY}\, dA \quad \text{and} \quad M_Z = \int \sigma_X \times y \times dA$$

Similarly, for bending in X-Z plane, $\quad S_Z = \int \tau_{XZ}\, dA \quad$ and $\quad M_Y = \int \sigma_X \times z \times dA$

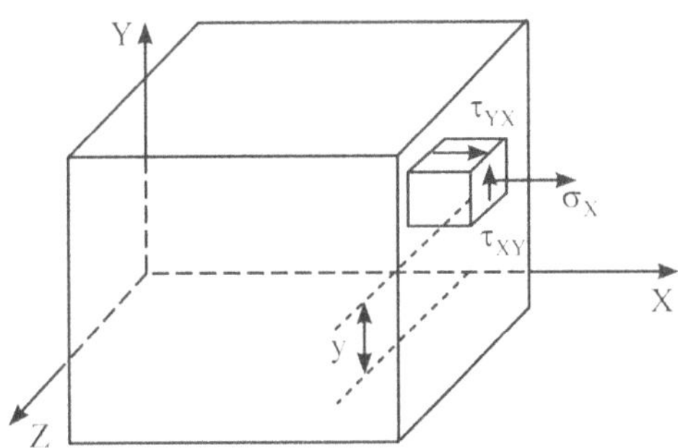

FIGURE 11.7 Shear stresses in a beam

Average shear stress across any section is obtained from S_Y/A. Actual (variable) shear stress on a layer is calculated by considering small section of beam of length 'dx' and an imaginary cut along that layer. Let σ be the bending stress on a layer of thickness 'dy', at a distance 'y' from the neutral axis. For the equilibrium of free body of the cut section of the beam, axial (or bending) stress on the free body is to be balanced by the shear force along the cutting plane (Ref Fig 11.8). Shear stress on that layer is then equal to the shear force divided by the area of the cutting plane. Let 'b' be the width of the beam in the Z-direction. M_Z and I_Z are henceforth written as M and I, for convenience.

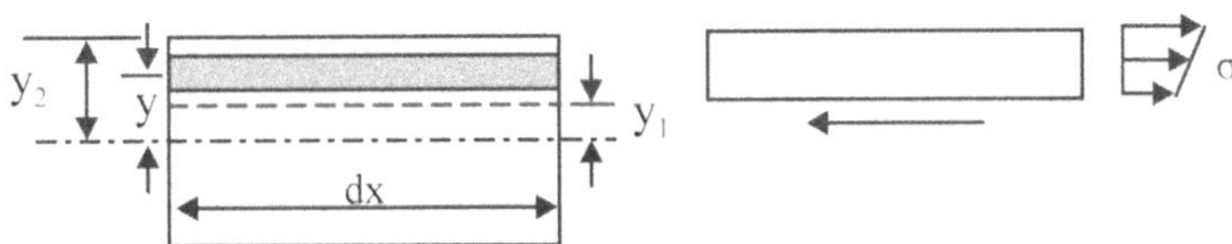

FIGURE 11.8 Shear stresses on a layer

$$\tau \times b \times dx = \int_{y_1}^{y_2} \sigma \times dA = \int_{y_1}^{y_2} (dM \times y\,/\,I) \times dA = (dM\,/\,I)\int_{y_1}^{y_2} y \times dA = (dM\,/\,I) \times A \times \overline{y}$$

where $\overline{y}$ is the distance of centroid of that small segment from the reference or neutral axis

$$\text{or} \quad \tau = \frac{dM\,/\,I}{b \times dx} \times A \times \overline{y} = \frac{dM}{dx} \times \frac{1}{I \times b} \times A \times \overline{y} = S_Y \times \frac{1}{I \times b} \times A \times \overline{y}$$

For a ***uniform rectangular section*** (width 'b' × depth 'd'), $\;y_2 = d/2$

$$\tau = S_Y \times \frac{1}{I \times b} \times A \times \bar{y}$$

$$= S_Y \times \frac{1}{I \times b} \times [(d/2) - y_1] \times b \times \frac{[(d/2) + y_1]}{2} = (S_Y / 8I) \times (d^2 - 4y_1^2)$$

Maximum shear stress occurs when $y_1 = 0$

$$\tau_{Max} = (S_Y / 2I) \times \left[(d/2)^2 - 0 \right] = S_Y \times d^2 / 8I$$

Minimum shear stress occurs when $y_1 = \pm\, d/2$ $\qquad \Rightarrow \qquad \tau_{Min} = 0$

Average shear stress, $\tau_{Ave} = S_Y / A = S_Y / (b \times d)$

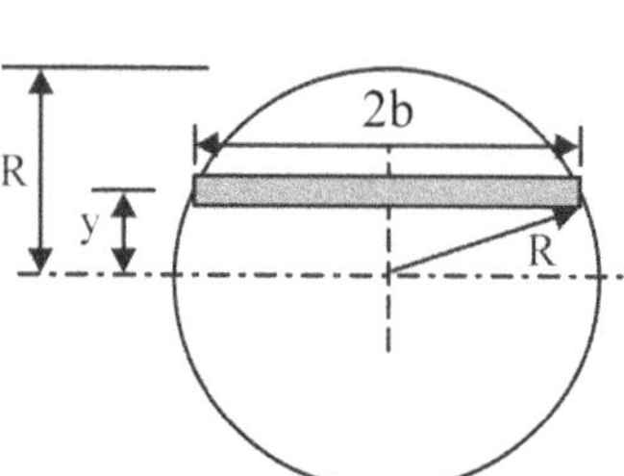

$$\frac{\tau_{Max}}{\tau_{Ave}} = \frac{S_Y \times d^2 / 8I}{S_Y / (b \times d)} = \frac{d^3}{8I \times b} = \frac{d^3}{8 \times (bd^3 / 12) \times b} = 3/2$$

For a ***solid circular section*** of radius 'R',

$$\tau \times 2b \times dx = \int_y^R \sigma_X \times dA = \int_y^R (dM \times y / I) \times dA = (dM / I)\int_y^R y \times dA$$

Here, $dA = 2b \times dy$, where, $b = \sqrt{R^2 - y^2}$

$$= (dM / I)\int_y^R y \times \left[2 \times \sqrt{(R^2 - y^2)} \right] \times dy$$

$$\tau = (dM / dx) \times [1 / (I \times 2b)] \times \left[-\left(R^2 - y^2 \right)^{3/2} / (3/2) \right]$$

$$= S_Y \times \frac{1}{I \times \left[2 \times \sqrt{R^2 - y^2} \right]} \times \frac{2}{3} \times (R^2 - y^2)^{3/2} = \frac{S_Y}{3I} \times (R^2 - y^2)$$

$$= \frac{S_Y}{3 \times (\pi R^4 / 4)} \times (R^2 - y^2)$$

Maximum shear stress occurs when $y = 0$

$$\Rightarrow \quad \tau_{Max} = \left[\frac{4S_Y}{3\pi R^4} \right] \times \left[R^2 - 0 \right] = (4/3) \times (S_Y / \pi R^2)$$

Minimum shear stress occurs when $y = R \quad \Rightarrow \quad \tau_{Min} = 0$

Average shear stress, $\tau_{Ave} = S_Y / A = S_Y / (\pi R^2)$

$$\frac{\tau_{Max}}{\tau_{Ave}} = \frac{(4/3) \times (S_Y / \pi R^2)}{S_Y / (\pi R^2)} = 4/3$$

Maximum compressive stress may not always be equal to the maximum tensile stress (as in unsymmetric T, L sections) except for beam sections which are symmetric w.r.t. the neutral plane (as in rectangular, circular, I sections). In an **unsymmetric section**, width of section on one side of neutral axis is more and smaller stress acting on the larger width will produce the same total force as larger stress on smaller width. Thus, stress is maximum on the outermost layer on the side of smaller width (or more depth) w.r.t. neutral axis, as can be seen from Fig.11.9.

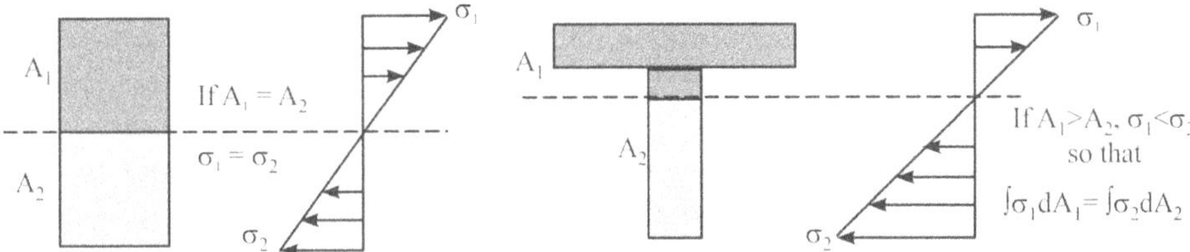

FIGURE **11.9** Variation of bending stress in symmetrical & unsymmetrical sections

Example 11.1

A wooden beam of rectangular section, 200mm × 300mm is simply supported over a span of 4m. If allowable shear stress for the material is 50 N/cm^2, find safe point load at its mid span

Solution

Let the point load at mid-span of a simply supported beam be 'P'. Reaction at each end will be P/2 and the shear force S_Y is constant all along the span and equal to P/2

Moment of inertia, $I_{ZZ} = b \times d^3/12 = 200 \times 300^3 / 12 = 450 \times 10^6$ mm^4

Max shear stress, $\tau_{Max} = S_Y \times d^2/(8 \times I)$

or $50 \times 10^{-2} = (P/2) \times 300^2/(8 \times 450 \times 10^6)$ $\Rightarrow$ P = 500 N

Example 11.2

A 6m long circular rod of 100mm diameter is used as a cantilever. It carries a concentrated load of 200kN at the free end and a uniformly distributed load of 50kN/m over its entire length. Determine max shear stress and shear stress at a distance of 20mm from neutral axis

Solution

Maximum shear force, $S_Y = 200 + 50 \times 6 = 500$ kN

Average shear stress, $\tau_{Ave} = S_Y/A = S_Y/\pi R^2 = 63.64$ N/mm^2

Maximum shear stress, $\tau_{Max} = (4/3) \times \tau_{Ave} = 84.84$ N/mm^2
Shear stress at a distance y of 20 mm from neutral axis,
$$\tau = (S_Y/3I) \times (R^2 - y^2) = [S_Y/(3\pi R^4/4)] \times (R^2 - y^2) = 71.3 \text{ N/mm}^2$$

11. 8   SHEAR STRESSES IN A SOLID (3-D) CONTINUUM

In a general condition of three-dimensional stress on an element, the shear stresses τ_{ij} on the faces of the element may be designated as shown in Fig.11.10 where the first subscript indicates the plane on which the shear stress acts and the last subscript indicates the direction of shear stress. The other three faces of the element are assumed to be under the action of shear stresses, which are equal and opposite to those shown. The normal stresses on the faces of the elements are not shown but are assumed to be equal on opposite faces so that they do not affect the equilibrium of the element. For the moments about an axis through the center of the element in the z-direction to be zero,

$$(\tau_{XY} \, dy \, dz) \, dx = (\tau_{YX} \, dx \, dz) \, dy \quad \text{or} \quad \tau_{XY} = \tau_{YX.}$$

Similarly, equating moments about axes through the center of the element in the X and Y directions, $\tau_{YZ} = \tau_{ZY}$ and $\tau_{XZ} = \tau_{ZX}$

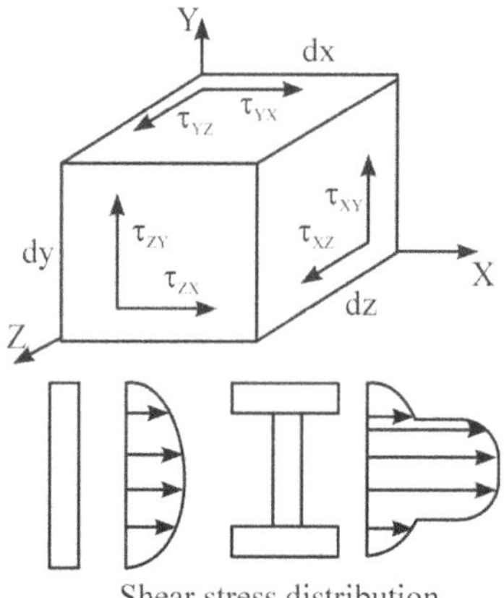

FIGURE 11.10 Shear stress distribution in symmetrical sections

Different normal stresses along adjacent layers, parallel to neutral plane, induce shear force (S_Y) between the layers. The shear stress (S_Y/A) produced across the cross section depends on thickness distribution of the cross section. The shear stress distribution for rectangular section and I-section are given, to indicate how thickness of each layer affects. This shear stress (τ_{YX}) induces, for equilibrium of an infinitesimal element, shear stresses τ_{YZ}, τ_{ZY}, τ_{ZX}, τ_{XZ} and τ_{XY} in other directions.

On a free unloaded surface of any structural member, there can be no shear stress. Consequently, the relations $\tau_{XY} = \tau_{YX} = 0$; $\tau_{YZ} = \tau_{ZY} = 0$ and $\tau_{XZ} = \tau_{ZX} = 0$ must apply at the surface. The remaining shear stress components must therefore be parallel to the free surface. Since, there can be no abrupt changes in shear distribution, the shear stress must be zero at any sharp corner and may be very high at a sharp re-entrant angle. At the juncture of the web and the flange, a radius is desirable in order to permit a good distribution of shear stress.

11.9 COMBINED BENDING AND AXIAL LOADS

Quite often, a beam may be subjected to a combination of axial (tensile or compressive) load and bending load. Since bending load produces tensile stress on one side of neutral axis and compressive stress on the other side, these stresses ($\sigma_b = M \times y/I$) can be superimposed on the axial (tensile or compressive) stress $\sigma_a = P/A$ (Fig.11.11). *The linear superposition is valid only when total stress is within the elastic limit.* Maximum axial tensile stress $(\sigma_t)_{max}$ and maximum axial compressive stress $(\sigma_c)_{max}$, given by the following relations, have to be limited to the allowable stress of the material.

$$(\sigma_t)_{max} = P/A + M \times y/I \qquad \text{(Ref Fig.11.11 a)}$$

$$\text{or} \quad (\sigma_c)_{max} = -P/A - M \times y/I \qquad \text{(Ref Fig.11.11 b)}$$

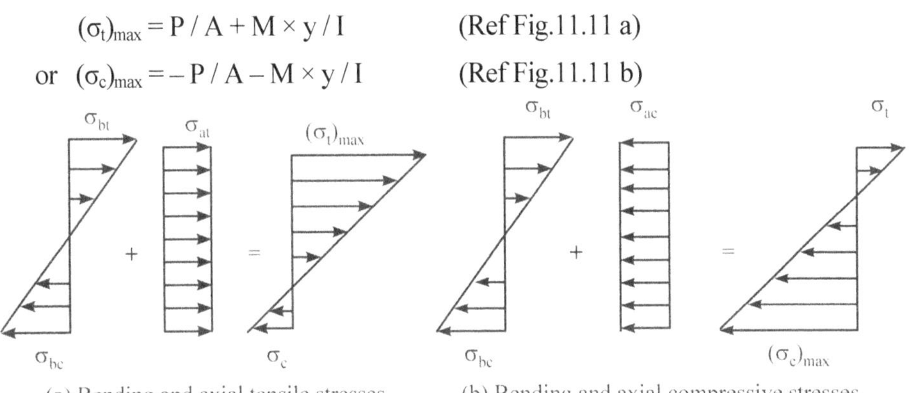

FIGURE 11.11 Superposition of bending and axial stresses

11.10 COMPOSITE OR FLITCH BEAM

Sometimes, a beam may be made up of different materials (Ref Fig.11.12). In such cases, the neutral axis is found from the condition that the total force across any section is zero.

The total axial internal force in a beam across any cross section,

$$\text{i.e.,} \quad F = \int dF = \int \sigma \times dA = \int (E \times y/R) \times dA$$

$$= \sum (1/R) \times \int E_i \times y \times (b \times dy_i) = \sum (1/R) \times \int E \times y \times (b_i \times dy_i) = 0$$

where, $b_i = (E_i/E) \times b$

(assuming radius of curvature, R and width of section, b to be constant)

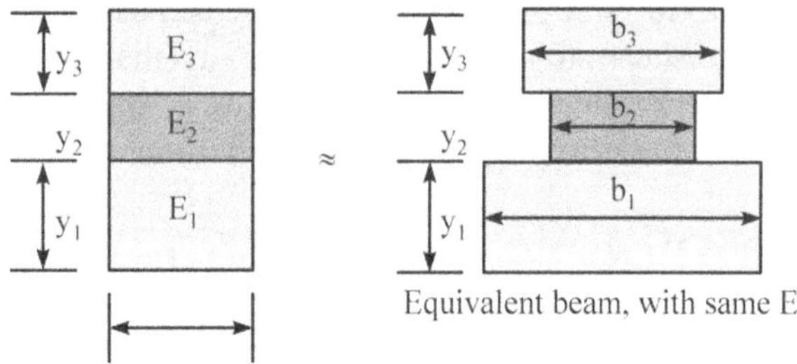

FIGURE 11.12 Composite beam with different materials (E values)

Then, $I_{eq} = \sum \int y_i^2 \times (b_i \times dy_i)$

$$\sigma_i = E_i \times \varepsilon_i = E_i \times (M \times y_i / E \times I_{eq})$$

11.11 ANALYSIS OF DETERMINATE BEAMS

11.11.1 DEFLECTION BY DOUBLE INTEGRATION METHOD

Displacement v, normal to the beam axis, is calculated at any section by calculating bending moment at that section and using the relation

$$EI \, (d^2v/dx^2) = M$$

By integrating the above equation, we get slope of the elastic curve (deflected shape of beam) at that section, $\theta = dv/dx = (1/EI) \times [\int M \times dx + C_1]$

Integrating again, we get displacement at the section as

$$v = (1/EI) \times [\int (\int M \times dx) \times dx + \int C_1 \times dx + C_2]$$

From the above equations, it can be seen that the values of slope and displacement, naturally, depend on the values of the constants C_1 and C_2. These constants are evaluated from the end conditions (displacement and/or slope).

Many books use 'y' for displacement of beam as well as for the distance of a layer from the neutral axis. To avoid this confusion, normal convention of indicating displacement in Y-direction (normal to the X-axis, representing length of beam) by 'v' is used in this text. Displacement v is always referred to the undeformed neutral axis.

Example 11.3

Calculate slope and deflection of a simply supported beam AB of length 'L', with a concentrated load P acting at C, at a distance of 'a' (a < L/2) from the left support, by double integration method. Deduce the values for the particular case of a = L/2

Solution

Taking moments about B, $\sum M = R_A \times L - P \times (L - a) = 0 \Rightarrow R_A = P \times (L - a)/L$

Then, $\sum F_Y = R_A + R_B - P = 0 \Rightarrow R_B = P \times a/L$

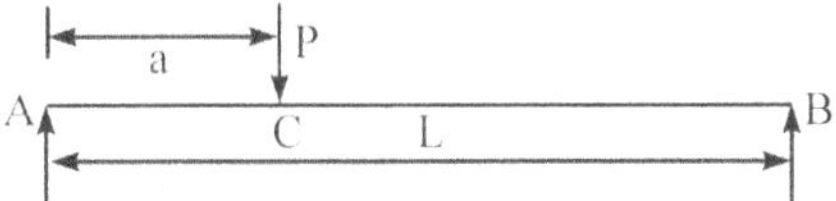

Bending moment in section AC, at a distance 'x' from A,

$$M_{AC} = R_A \times x \qquad \text{for} \quad 0 \leq x \leq a$$

Bending moment in section CB, at a distance 'x' from A,

$$M_{CB} = R_A \times x - P \times (x - a) \qquad \text{for} \quad a < x \leq L$$

Integrating the equation EI (d^2v/dx^2) = M for different parts of the beam separately, since the value of M changes in parts AC and CB of the beam, is tedious and it is also difficult to evaluate integration constants in each part. Macaulay has suggested a notation for bending moment M which can be used over the entire beam. In this particular example,

$$M_{AB} = R_A \times x - <P \times (x - a)>$$

According to Macaulay's notation, expression in $<\ >$ need to be considered only for positive values i.e., $x > a$

Thus, $M_{AB} = R_A \times x - <P \times (x - a)> = R_A \times x = M_{AC}$ for $x < a$

$M_{AB} = R_A \times x - <P \times (x - a)> = R_A \times x - P \times (x - a) = M_{CB}$ for $x > a$

Integrating beam deflection equation, EI $(d^2v/dx^2) = M_{AB}$ once, we get

$$EI \times \theta = EI \times (dv/dx) = \int M_{AB} \times dx + C_1$$

$$= \int [\, R_A \times x - <P \times (x - a)> \,] \times dx + C_1$$

$$= [\, R_A \times x^2/2 - <P \times (x - a)^2/2> \,] + C_1$$

Integrating the above expression once again, we get

$$EI \times v = \int [\, R_A \times x^2/2 - <P \times (x - a)^2/2> \,] \times dx + C_1 \times x + C_2$$

$$= [\, R_A \times x^3/6 - <P \times (x - a)^3/6> \,] + C_1 \times x + C_2$$

The end condition, $v = 0$ at $x = 0$, gives

$$0 = [R_A \times 0] + C_1 \times 0 + C_2 \qquad \Rightarrow C_2 = 0$$

Here, $x - a$ is –ve for $x = 0$ and, hence, the 2nd term is ignored

The end condition, $v = 0$ at $x = L$, gives

$$0 = [\, R_A \times L^3/6 - P \times (L - a)^3/6 \,] + C_1 \times L + 0$$

from which C_1 can be evaluated if P, L and a are given

Here, $x - a$ is +ve for $x = L$ and, hence, the 2nd term is included

Special case:

When $a = b = L/2$, $R_A = R_B = P/2$

$$\theta_A = \theta_B = P \times (L/2) \times [L^2 - (L/2)^2] / (6EIL) = PL^2/(16EI)$$

Corresponding to maximum deflection,

$$3(L/2)x^2 - 3L[x - (L/2)]^2 - (L/2) \times [L^2 - (L/2)^2] = 0 \qquad \Rightarrow \quad x = L/2$$

At this location,

$$\mathbf{v_{max}} = [R_B \times (x^3/6) - P \times (x - b)^3/6 + C_1 \times x + C_2] / (EI)$$

$$= [R_B \times (x^3/6) - P \times (x - b)^3/6 - x \times P \times a \times (L^2 - a^2)/6L] / (EI)$$

$$= [(P/2) \times \{(L/2)^3 / 6\} - P\{(L/2) - (L/2)\}^3 / 6$$

$$- (L/2) \times P \times (L/2) \times \{L^2 - (L/2)^2\}/6L] / (EI)$$

$$= - PL^3 / (48EI)$$

For $a = L/2$, $R_A = R_B = P/2$ and $C_1 = - P \times L^2/16$

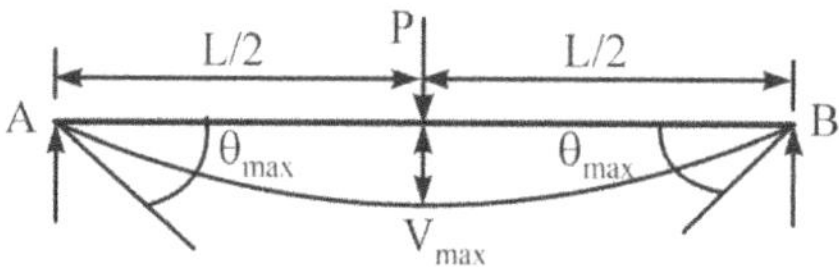

Then, displacement at any section,

$$v = [1/(EI)] \times [R_A \times x^3/6 - <P \times \{x - (L/2)\}^3/6 > - (P \times L^2/16) \times x]$$

and   slope at any section,

$$\theta = [1/(EI)] \times [\, R_A \times x^2/2 - <P \times \{x - (L/2)\}^2/2 > - P \times L^2/16]$$

Maximum displacement, $v_{max} = -P \times L^3 / (48EI)$ downwards at $x = L/2$

and maximum slope, $\theta_{max} = -P \times L^2 / (16EI)$ at $x = 0$

Also, $\theta_{max} = P \times L^2 / (16EI)$ at $x = L$

At the point of maximum displacement, slope changes from –ve to +ve

and, hence, $\theta = 0$ at $x = L/2$

11.11.2 DEFLECTION BY MACAULAY'S METHOD

This method is similar to the double integration method. Basic changes are
- Origin is to be taken at the extreme end of the beam
- Section at which bending moment is found out is selected between the last load and the other extreme end
- While integrating terms of bending moment, brackets are not to be expanded
- If, for any value of x, some terms have negative value they are omitted
- If a beam carries distributed load over a portion of its length, the load is extended unto the extreme end of the beam and superimposed with a distributed load, equal and opposite to that which has been added. It can be analysed only if the load starts from one end of the beam

This method is explained through the following examples.

Example 11.4

Calculate max deflection and max slope in a simply supported beam AB of length 'L' with a concentrated load 'P' at C, at a distance 'a' from end A, by Macaulay's method.

Solution

Taking moments about A, $M = R_B \times L - P \times a = 0 \implies R_B = P \times a / L$

Taking moments about B, $M = P \times b - R_A \times L = 0 \implies R_A = P \times b / L$

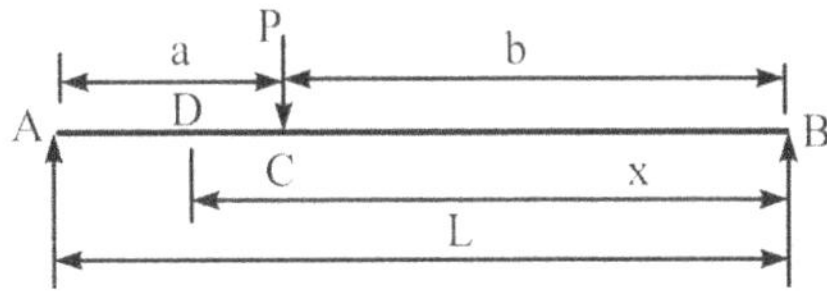

Taking B as the origin and measuring x towards A, Equation of elastic curve gives, $EI\, d^2v/dx^2 = M = R_B \times x - P \times (x - b)$

Integrating both sides, $EI\,(dv/dx) = R_B \times (x^2/2) - P \times (x - b)^2/2 + C_1$

Integrating both sides again,

$$EI \times v = R_B \times (x^3/6) - P \times (x - b)^3/6 + C_1 \times x + C_2$$

The constants of integration C_1 and C_2 can be evaluated from the known end conditions.

$v = 0$ at $x = 0$ $\Rightarrow$ $C_2 = 0$ neglecting 2nd term whose value is negative

$$v = 0 \text{ at } x = L \Rightarrow C_1 = -[R_B \times (L^3/6) - P \times (L - b)^3/6] / L$$

$$= -[(Pa/L) \times (L^3/6) - P \times a^3/6] / L$$

$$= -P \times a \times (L^2 - a^2)/6L$$

Substituting these values in the above equations,

$$\theta_B = (dv/dx)_{x=0} = -P \times a \times (L^2 - a^2) / (6EIL)$$

$$\theta_A = (dv/dx)_{x=L} = [R_B \times (L^2/2) - P \times (L - b)^2/2 - P \times a \times (L^2 - a^2)/6L] / (EI)$$

$$= [(P \times a/L) \times (L^2/2) - P \times a^2/2 - P \times a \times (L^2 - a^2)/6L] / (EI)$$

$$= P [3aL^2 - 3a^2L - a \times (L^2 - a^2)] / (6EIL)$$

$$= P [2aL^2 - 3a^2L + a^3] / (6EIL)$$

$$= P [2aL(L - a) - a^2(L - a)]/(6EIL) = P [ba(2L - a)] / (6EIL)$$

$$= P \times b \times (L - b) \times (L + L - a) / (6EIL)$$

$$= P \times b \times (L^2 - b^2) / (6EIL)$$

Corresponding to maximum deflection, slope is zero.

i.e., $(dv/dx) = [R_B \times (x^2/2) - P \times (x - b)^2/2 + C_1] / (EI) = 0$

$$\Rightarrow [(P \times a/L) \times (x^2/2) - P \times (x - b)^2/2 - P \times a \times (L^2 - a^2)/6L] / (EI) = 0$$

$$\Rightarrow 3a \times x^2 - 3L \times (x - b)^2 - a \times (L^2 - a^2) = 0$$

which gives the value of x for specific values of a, b and L

Using this value of x in the equation for deflection, maximum deflection can be evaluated

Special case:

When $a = b = L/2$, $R_A = R_B = P/2$

$$\theta_A = \theta_B = P \times (L/2) \times [L^2 - (L/2)^2]/(6EIL) = PL^2/(16EI)$$

Corresponding to maximum deflection,

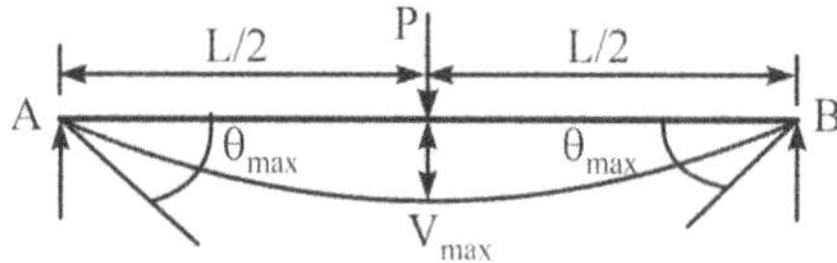

$$3(L/2)x^2 - 3L[x - (L/2)]^2 - (L/2) \times [L^2 - (L/2)^2] = 0 \quad \Rightarrow \quad x = L/2$$

$$v_{max} = [R_B \times (x^3/6) - P \times (x - b)^3/6 + C_1 \times x + C_2] / (EI)$$

$$= [R_B \times (x^3/6) - P \times (x - b)^3/6 - x \times P \times a \times (L^2 - a^2)/6L] / (EI)$$

$$= [(P/2) \times \{(L/2)^3/6\} - P\{(L/2) - (L/2)\}^3/6$$

$$- (L/2) \times P \times (L/2) \times \{L^2 - (L/2)^2\}/6L] / (EI)$$

$$= - PL^3 / (48EI)$$

Example 11.5

Calculate max deflection and its location for a simply supported beam AB of length 6 m when concentrated loads of 10 kN and 2 kN act at 2 m and 4 m respectively from end A. Take $I = 20000$ cm^4 and $E = 200$ GPa

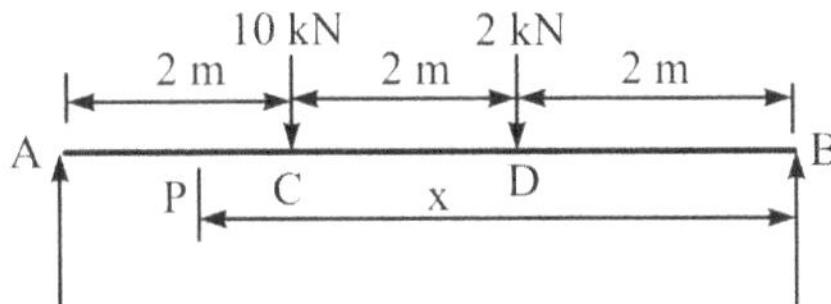

Solution

Taking moments about A,

$$R_B = (10 \times 2 + 2 \times 4)/6 = 14/3 \text{ kN}$$

From $\sum F_Y = 0$, $R_A = (10 + 2) - (14/3) = 22/3$ kN

Taking B as the origin and measuring x towards A, equation of elastic curve gives,

$$EI \, (d^2v/dx^2) = M = R_B \times x - 2 \times (x{-}2) - 10 \times (x - 4)$$

Integrating both sides,

$$EI \, (dv/dx) = R_B \times x^2/2 - 2 \times (x - 2)^2/2 - 10 \times (x - 4)^2/2 + C_1$$

Integrating again,

$$EI \times v = R_B \times x^3/6 - 2 \times (x{-}2)^3/6 - 10 \times (x - 4)^3/6 + C_1 \times x + C_2$$

Flexural rigidity, mmEI $= 2 \times 10^7$ (N/cm^2) $\times 20000$ (cm^4) $= 40 \times 10^6$ N m^2

The constants of integration C_1 and C_2 can be evaluated from the known end conditions.

$v = 0$ at $x = 0$ $\quad \Rightarrow$ $C_2 = 0$ neglecting 2nd term whose value is negative

$v = 0$ at $x = 6$m $\Rightarrow$ $C_1 = - [R_B \times L^3/6 - 2 \times (L - 2)^3/6 - 10 \times (L - 4)^3/6] / L$

$$= - [(14/3) \times 6^3/6 - 2 \times 4^3/6 - 10 \times 2^3/6] / 6$$

$$= - 22.22$$

Deflection at C, with x = 4m,

$v_C = [R_B \times x^3/6 - 2 \times (x-2)^3/6 - 10 \times (x-4)^3/6 + C_1 \times x] / (EI)$

$\quad = [(14/3) \times 4^3/6 - 2 \times (4-2)^3/6 - 10 \times (4-4)^3/6 - 22.22 \times 4] / (40 \times 10^6) \, m$

$\quad = -0.1044 \, cm$

Deflection at D, with x = 2m,

$v_D = [R_B \times x^3/6 - 2 \times (x-2)^3/6 - 10 \times (x-4)^3/6 + C_1 \times x] / (EI)$

$\quad = [(14/3) \times 2^3/6 - 2 \times (2-2)^3/6 - 10 \times (2-4)^3/6 - 22.22 \times 2] / (40 \times 10^6) \, m$

$\quad = [(14/3) \times 2^3/6 - 22.22 \times 2] / (40 \times 10^6) \, m = -0.09556 \, cm$

At the point of maximum deflection,

$dv/dx = [R_B \times x^2/2 - 2 \times (x-2)^2/2 - 10 \times (x-4)^2/2 + C_1] / (EI) = 0$

$\Rightarrow \quad (14/3) \times x^2/2 - 2 \times (x-2)^2/2 - 10 \times (x-4)^2/2 - 22.22 = 0$

$\Rightarrow \quad x = 8.65 \, m \quad or \quad 3.35 \, m$

Since x = 8.65 is not a valid solution, maximum deflection occurs at x = 3.35m

At x = 3.35m, $v_{max} = [R_B \times x^3/6 - 2 \times (x-2)^3/6 - 10 \times (x-4)^3/6 + C_1 \times x] / (EI)$

$\quad = [(14/3) \times 3.35^3/6 - 2 \times (3.35-2)^3/6$

$\quad\quad\quad\quad - 10 \times (3.35-4)^3/6 - 22.22 \times 3.35] / (40 \times 10^6) \, m$

$\quad = -0.115 \, cm$

Example 11.6

Calculate max deflection and its location in a simply supported beam AB of length 'L' with a uniformly distributed load of 'w' per unit length over a length 'a' from end B. Deduce the values for the particular case of a = L

Solution

Taking moments about A, $\quad R_B = (w \times a) \times (L - a/2)/L$

Taking moments about B, $\quad R_A = (w \times a) \times (a/2)/L$

Taking A as the origin and measuring x towards B, equation of elastic curve gives,

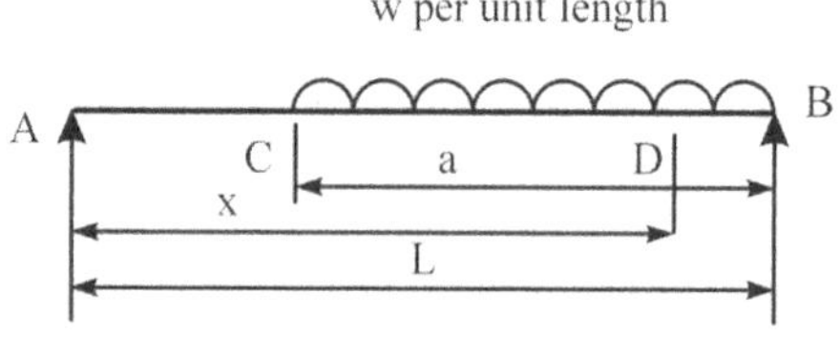

$EI \times d^2v/dx^2 = M = R_A \times x - w \times \{x - (L-a)\}^2/2$

Integrating both sides, $EI \times (dv/dx) = R_A \times (x^2/2) - w \times \{x - (L-a)\}^3/6 + C_1$

Integrating both sides again,

$$EI \times v = R_A \times (x^3/6) - w \times \{x - (L - a)\}^4/24 + C_1 \times x + C_2$$

The constants of integration C_1 and C_2 can be evaluated from the known end conditions.

$v = 0$ at $x = 0 \Rightarrow C_2 = 0$ neglecting 2nd term whose value is negative

$$v = 0 \text{ at } x = L \Rightarrow C_1 = - [R_A \times (x^3/6) - w \times \{x - (L - a)\}^4/24] / L$$
$$= - [(w \times a^2/2L) \times (L^3/6) - w \times \{L - (L - a)\}^4/24] / L$$
$$= - w \times a^2 \times L/12 + w \times a^4 / (24L)$$

At the point of maximum deflection,

$$dv/dx = [R_A \times (x^2/2) - w \times \{x - (L - a)\}^3/6$$
$$- w \times a^2 \times L/12 + w \times a^4/24L] / (EI) = 0$$
$$\Rightarrow (w \times a^2/2L) \times (x^2/2) - w \times \{x - (L - a)\}^3/6 - w \times a^2 \times L/12 + w \times a^4 / (24L) = 0$$

which gives the value of x for specific a and L

Using this value of x in the equation for deflection, maximum deflection can be evaluated

Special case:

When $a = L$, $R_A = R_B = w \times L/2$

$$\theta_A = - \theta_B = C_1 / (EI) = w \times [-L^3/12 + L^3/24] / (EI) = - w \times L^3 / (24EI)$$

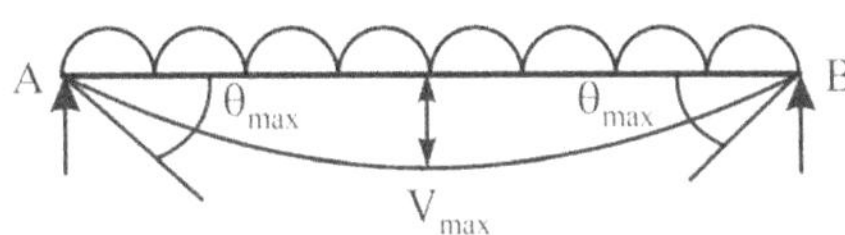

Corresponding to maximum deflection,

$$(E \times I) \times \theta = (w \times L/2) \times (x^2/2) - w \times \{x - (L - L)\}^3/6$$
$$- w \times L^2 \times L/12 + w \times L^4/ (24L) = 0$$
$$\Rightarrow w \times L \times x^2/4 - w \times x^3/6 - w \times L^3/12 + w \times L^3/24 = 0$$
$$\Rightarrow 6L \times x^2 - 4x^3 - L^3 = 0 \qquad \Rightarrow x = L/2$$
$$V_{max} = [R_A \times (x^3/6) - w \times \{x - (L - L)\}^4/24 + C_1 \times x + C_2] / (EI)$$
$$= [(w \times L/2) \times (L^3/48) - w \times (L/2)^4/24 + \{-w \times L^2 \times L/12$$
$$+ w \times L^4/(24L)\} \times (L/2) + 0] / (EI)$$
$$= [(w \times L^4/96) - (w \times L^4/384) - (w \times L^4/48)] / (EI) = - 5wL^4 / (384\ EI)$$

Example 11.7

Calculate max deflection and its location in a simply supported beam AB of length 'L' with a uniformly distributed load of 'w' per unit length over a length of 'L/4' starting from 'L/2' from end A.

Solution

Taking moments about A,

$$R_B = (w \times L/4) \times (L/2 + L/8)/L = 5wL/32$$

From $\sum F_Y = 0$, $R_A = w \times L/4 - 5wL/32 = 3wL/32$

In order to analyse this beam by Macaulay's method, the load should be taken from one of the ends. So, the given loading in the left figure is taken as a combination of two distributed loads in opposite directions from end B as shown in the right figure.

Taking A as the origin and measuring x towards B, equation of elastic curve gives,

$$EI\, d^2v/dx^2 = M = R_A \times x - w \times \{x - (L/2)\}^2/2 + w \times \{x - (3L/4)\}^2/2$$

Integrating both sides,

$$EI\,(dv/dx) = R_A \times (x^2/2) - w \times \{x - (L/2)\}^3/6 + w \times \{x - (3L/4)\}^3/6 + C_1$$

Integrating again,

$$EI \times v = R_A \times (x^3/6) - w \times \{x - (L/2)\}^4/24$$
$$+ w \times \{x - (3L/4)\}^4/24 + C_1 \times x + C_2$$

The constants of integration C_1 and C_2 can be evaluated from the known end conditions.

$v = 0$ at $x = 0 \Rightarrow C_2 = 0$ neglecting 2nd term whose value is negative

$v = 0$ at $x = L \Rightarrow$

$$C_1 = -[R_A \times (L^3/6) - w \times \{L - (L/2)\}^4/24 + w \times \{L - (3L/4)\}^4/24]\,/\,L$$
$$= -[(3wL/32) \times (L^3/6) - w \times \{L^4/(16 \times 24) - L^4/(256 \times 24)\}]\,/\,L$$
$$= -27w \times L^3/2048$$

At D, $x = 3L/4$ and

$$v_D = [R_A \times \{(3L/4)^3/6\} - w \times \{(3L/4) - (L/2)\}^4\,/\,24$$
$$- (27w \times L^3/2048) \times (3L/4)]\,/\,(EI)$$

At the point of maximum deflection,

$$dv/dx = [R_A \times (x^2/2) - w \times \{x - (L/2)\}^3/6 + w \times \{x - (3L/4)\}^3/3$$

$$- 27w \times L^3/2048] / EI = 0$$

$$\Rightarrow \quad (3wL/32) \times (x^2/2) - w \times \{x - (L/2)\}^3/6 + w \times \{x - (3L/4)\}^3/3$$

$$- 27w \times L^3/2048 = 0$$

which gives the value of x

Using this value of x in the equation for deflection, maximum deflection can be evaluated.

11.11.3 DEFLECTION BY MOMENT-AREA METHOD

Change of slope and change of displacement between any two sections, at distance 'dx' apart, along the length of beam axis are calculated graphically in this method (Ref Fig 11.13), from a plot of bending moment diagram, using the elastic curve equation

$$d^2v/dx^2 = M / EI$$

1st theorem of Moment area – Integrating elastic curve equation w.r.t. x, between sections at a and b, $\quad \int (d^2v/dx^2) \times dx = \int [M/(EI)] \times dx$

which reduces to $\quad (dv/dx)_a - (dv/dx)_b = \theta_1 - \theta_2 = \int [M/(EI)] \times dx$

This equation gives change in slope between two sections, represented by the area under the curve (M/EI) between the two sections. Note that the area can be positive or negative, depending on the direction of bending moment. If the slope at one section, such as at the fixed support ($\theta = 0$), slope at other section can be obtained from the above relation.

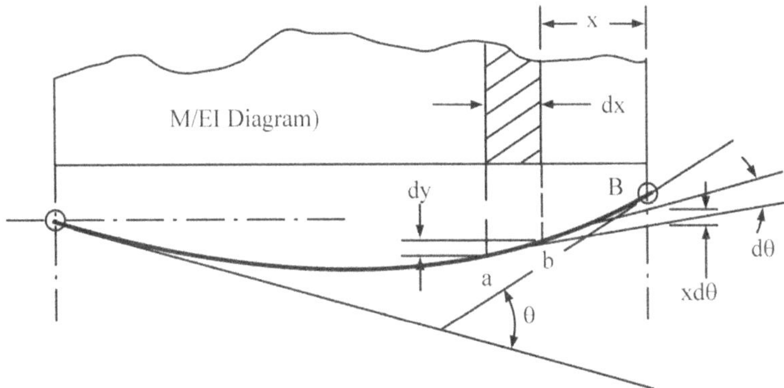

FIGURE 11.13 Beam slope and deflection

2nd *theorem of Moment area* – Integrating elastic curve equation, after multiplying both sides by x, between sections at a and b, we get

$$\int x \times [M_x/(EI)] \times dx = \int x \times (d^2v/dx^2) \times dx$$

Integrating by parts and substituting limits at a and b,

$$\int x \times [M_x/(EI)] \times dx = (dv/dx) \times x - \int (dv/dx) \times dx$$

$$= [\theta_a \times (x + dx) - \theta_b \times x] - (v_a - v_b)$$

Therefore, $(v_a - v_b) = [\theta_a \times (x + dx) - \theta_b \times x] - \int x \times [M/(EI)] \times dx$

$\int x \times [M/(EI)] \times dx$ can be written as $\sum [M/(EI)]_k \times x_k$

where x_k is the distance of centroid of each individual area $(M/EI)_k$, along x

Example 11.8

Calculate deflection at the free end of a stepped cantilever ABC of length L which has a moment of inertia I for a length of L/2 from fixed end and a moment of inertia I/2 for the remaining length of L/2, when a concentrated load P acts at its free end.

Solution

Area of bending moment diagram between A and C, identified by 3 segments (1, 2 and 3), is

$$\int [M/(EI)] \times dx = (1/2) \times [- PL/(AE)] \times (L/2) + [- PL/(2AE)] \times (L/2)$$

$$+ (1/2) \times [- PL/(2AE)] \times (L/2)$$

$$= - PL^2/(4EI) - PL^2/(4EI) - PL^2/(8EI) = - 5PL^2/(8EI)$$

At the fixed end of the cantilever, $\theta_A = 0$ and $v_A = 0$

Therefore, using 1st theorem of moment area between sections A and C,

$$\theta_C - \theta_A = \theta_C = -5PL^2/(8EI)$$

and using 2nd theorem of moment area,

$$v_C - v_A = v_C = [\theta_C \times x_C - \theta_A \times x_A] - \sum [M/(EI)]_k \times x_k$$
$$= [\{-5PL^2/(8EI)\} \times L - 0] - [\{(-PL^2/4EI) \times (L/2 + L/6)\}$$
$$+ \{(-PL^2/4EI) \times (L/4)\} + \{(-PL^2/8EI) \times (L/6)\}]$$
$$= -3PL^3/(8EI)$$

11.11.4 DEFLECTION BY ENERGY METHOD

It is possible to calculate displacement and slope at a section of beam from strain energy stored in the beam by using Castigliano's theorem.

Thus, $v = \partial U/\partial P$ and $\theta = \partial U/\partial M$

If these values are to be calculated at sections where load P or moment M is not applied, then a fictitious load has to be applied while expressing strain energy in the beam, which can be equated to zero later.

Strain energy due to bending can be calculated from

$$U = \int (1/2) \times \sigma \times \varepsilon \times dv = \int (1/2) \times (\sigma^2/E) \times dv = \int [1/(2E)] \times (M \times y/I)^2 \times dv$$
$$= \int\int [1/(2E)] \times (M \times y/I)^2 \times dv$$
$$= [1/(2EI^2)] \int M^2 \times [\int y^2 \times dA] \times dx \ = [1/(2EI^2)] \int M^2 \, dx \text{ since } \int y^2 \times dA = I$$

Therefore, $v = \partial U/\partial P = \int [M/(EI)] \times (\partial M/\partial P) \times dx$

Example 11.9

Calculate slope of a simply supported beam AB of length 'L', when a concentrated load P acts at C, at a distance of 'a' ($a < L/2$) from the left support. Deduce the values for the case of $a = L/2$

Solution

Taking moments about B,

$$\sum M = R_A \times L - P \times (L - a) = 0 \implies R_A = P \times (L - a)/L \text{ and,}$$
$$\sum F_y = R_A + R_B - P = 0 \qquad \implies R_B = P - R_A = P \times a/L$$

For the particular case of $a = L/2$, $R_A = R_B = P/2$

Bending moment in section AC, at a distance 'x' from A,

$$M_{AC} = R_A \times x = P \times x/2 \qquad\qquad \text{for} \quad 0 \le x \le a$$

Bending moment in section CB, at a distance 'x' from A,

$$M_{CB} = R_A \times x - P \times (x - a) = P \times x/2 - P \times (x - a) \qquad \text{for} \quad a < x \le L$$

Then, displacement at C,

$$v_C = \partial U / \partial P = -\left\{ \int (M_{AC} / EI) \times (\partial M_{AC} / \partial P) \times dx + \int (M_{CB} / EI) \times (\partial M_{CB} / \partial P) \times dx \right\}$$

$$= -\int_0^a \left\{ (P/2) \times x / EI \right\} \times (x/2) \times dx$$

$$\qquad -\int_a^L \left\{ \left[(P/2) \times x - P \times (x - L/2) \right] / EI \right\} \times \left[(x/2 - (x - L/2) \times dx) \right]$$

$$= -(P/4EI) \times \int_0^a x^2 dx - (P/4EI) \times \int_a^L (L - x)^2 dx$$

$$\qquad -\int_a^L \left\{ \left[\left(\frac{P}{2} \right) \times x - P \times \left(x - \frac{L}{2} \right) \right] / EI \right\} \times \left[\left(\frac{x}{2} - \left\{ x - \frac{L}{2} \right\} \times dx \right) \right]$$

$$= -(P/4EI) \times \int_0^a x^2 dx - (P/4EI) \times \int_a^L (L - x)^2 dx$$

$$= -(P/4EI) \times \left[x^3 / 3 \right]_0^a - (P/4EI) \times \left\{ L^2 \times [x]_a^L - 2L \times \left[x^2 / 2 \right]_a^L + \left[x^3 / 3 \right]_a^L \right\}$$

$$= -(P/4EI) \times \left[(a^3 / 3) + \left\{ L^2 \times (L - a) - L \times (L^2 - a^2) + (L^3 - a^3) / 3 \right\} \right]$$

For the particular case of $a = L/2$,

$$v_C = -[PL^3 / (4EI)] \times [1 + (12 - 18 + 7)] / 24 = -PL^3 / (48EI)$$

To calculate slope at A, a fictitious clockwise moment M' is applied at A, in addition to the actually applied load P at L/2 from A, while calculating strain energy U. The reactions at A and B change, to ensure moment equilibrium of the beam, as

$$R_A = P/2 - M'/L \quad \text{and} \quad R_B = P/2 + M'/L$$

Then, $M = -R_A \times x - M' + <P \times (x - a)>$

and $\theta = \partial U/\partial M' = \int (M / EI) \times (\partial M / \partial M') \times dx$

$$= -\left\{ \int (M_{AC} / EI) \times (\partial M_{AC} / \partial M') \times dx + \int (M_{CB} / EI) \times (\partial M_{CB} / \partial M') \times dx \right\}$$

$$= \int_0^{L/2} \left[\left\{ -(P/2) \times x - (M'/L) \times x \right\} / EI \right] \times (-x/L + 1) \times dx$$

$$+ \int_{L/2}^{L} \left\{ \left[-(P/2) \times x - (M'/L) \times x - M' + P \times (x - L/2) \right] / EI \right\} \times (-x/L + 1) \times dx$$

After substituting $M' = 0$

$$\theta = \left(\frac{P}{EI}\right) \times \int_0^a \left[\frac{x^2}{2L} - \frac{x}{2} \right] \times dx + \left(\frac{P}{EI}\right) \times \int_a^L \left[\left(\frac{x^2}{2L} - \frac{x^2}{L} + \frac{x}{2} \right) + \left(-\frac{x}{2} + x - \frac{L}{2} \right) \right] dx$$

$$= (P/EI) \left[\left\{ (a^3 - 0)/6L - (a^2 - 0)/4 \right\} + \left\{ (L^3 - a^3)/6L - (L^3 - a^3)/3L + (L^2 - a^2)/4 \right\} \right.$$

$$\left. + \left\{ -(L^2 - a^2)/4 + (L^2 - a^2)/2 - L(L-a)/2 \right\} \right]$$

For the particular case of $a = L/2$,

$$\theta = [PL^2/(EI)] \times [(1/48 - 1/16) + (7/48 - 7/24 + 3/16) + (-3/16 + 3/8 - 1/4)]$$

$$= [PL^2/(EI)] \times [1 - 3 + 7 - 14 + 9 - 9 + 18 - 12]/48 = (PL^2/EI) \times [35 - 38]/48$$

$$= -PL^2/(16EI)$$

11.11.5 DEFLECTION BY CONJUGATE BEAM METHOD

This method is applicable to simply supported beam or cantilever beam carrying point loads and is based on Mohr's theorems. A ***conjugate beam*** is an imaginary beam of the same length as the original beam but load at each point equal to the bending moment at that point of the original beam divided by flexural rigidity EI and boundary conditions reversed. Slope and deflection at any point along the length of beam are obtained from Mohr's theorems.

Mohr's theorem-1 states that slope at any point of the actual beam is equal to the shear force at that point of the conjugate beam.

Mohr's theorem-2 states that deflection at any point of the actual beam is equal to the bending moment at that point of the conjugate beam.

Application of this method is illustrated through the following two examples.

Example 11.10

Calculate slope and deflection of a simply supported beam of length 'L' with point load 'P' at mid–point of the beam.

Solution

For the original beam, due to symmetry in loading, $R_A = R_B = P/2$ and maximum bending moment,

$$M = (R_A) \times (L/2) = (R_B) \times (L/2) = P \times L/4, \text{ at the center of the beam.}$$

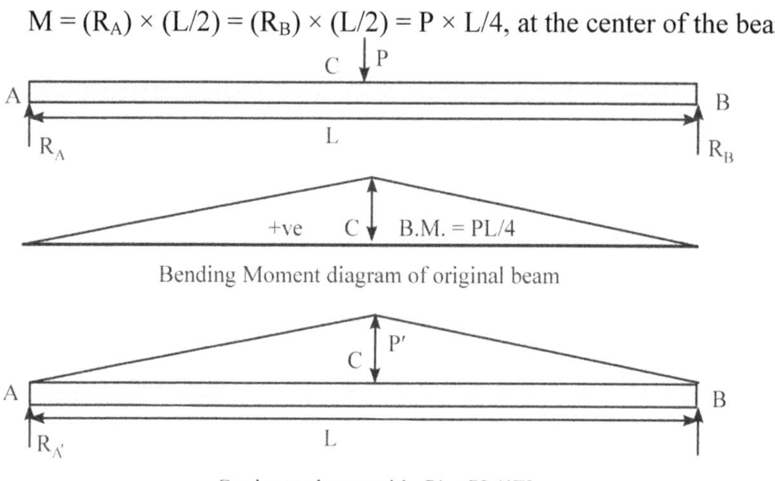

Bending Moment diagram of original beam

Conjugate beam with $P' = PL/4EI$

Conjugate beam is now defined as the beam of same length and flexural rigidity but with triangular load distribution equal to the moment distribution on the original beam, divided by flexural rigidity EI, as shown in the figure. The reactions of the conjugate beam are now obtained, due to symmetry of loading, as half the area of the load distribution.

Thus, $R_A' = R_B' = (1/2) \times (L/2) \times [PL/(4EI)] = PL^2/(16EI)$

According to Mohr's theorem-1,

slope at end A, θ_A = shear force at A on conjugate beam = $R_A' = PL^2/(16EI)$ and slope at mid-point C, θ_C = shear force at C on conjugate beam = 0

According to Mohr's theorem-2,

deflection at mid-point C, δ_C = bending moment at C on conjugate beam

$$= -R_A' \times (L/2) + (1/3) \times (L/2) \times (\text{Total load between A and C})$$

$$= -[PL^2/(16EI)] \times (L/2) + (1/3) \times (L/2) \times [PL^2/(16EI)]$$

$$= -PL^3/(48EI)$$

Slope at A and deflection at C are same as those obtained by the other methods.

Example 11.11

Calculate slope and deflection of a simply supported beam AB of length 'L' with point load 'P' at mid point C of the beam, if moment of inertia of one half of the beam is 3 times that of the other half and E is same for the entire beam.

Solution

For the original beam, due to symmetry in loading, $R_A = R_B = P/2$ and maximum bending moment,

$$M = (R_A) \times (L/2) = (R_B) \times (L/2) = P \times L/4 \quad \text{acts at the center of the beam.}$$

Let $(I)_{AC} = 3\,(I)_{CB} = 3I$

Conjugate beam is now defined as the beam of same length and flexural rigidity but with triangular load distribution equal to the moment distribution on the original beam, divided by flexural rigidity EI, as shown in the figure. Thus, load on AC varies from zero at A to PL/(12EI) at C while load on BC varies from zero at B to PL/(4EI), with a step at C. Calculation of reactions of conjugate beam and calculation of slope and deflection of the original beam is carried out as explained in the previous example.

The reactions of the conjugate beam are not same at the two ends, since the load is not symmetric.

Taking moments about A, with the resultant loads on the two parts A – C and C – B of the beam acting at the centroid of the triangle (shown by dotted lines),

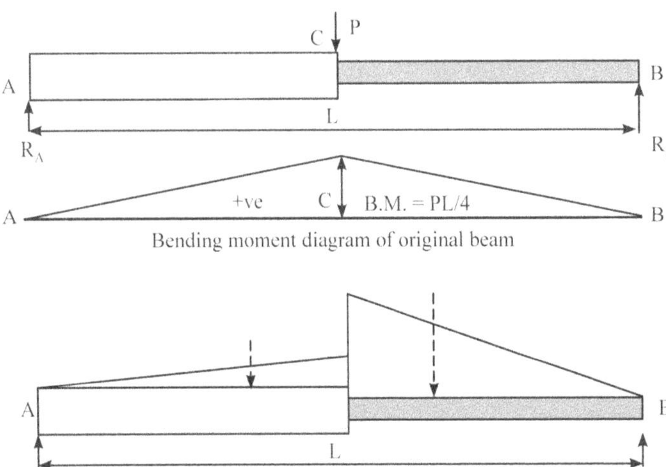

Bending moment diagram of original beam

Conjugate beam with $I_{AC} = 3\,I_{CB}$

$$\sum M_A = R_B' \times L - [(1/2) \times (L/2) \times \{PL/(4EI)\}] \times [(L/2) + (1/3) \times (L/2)]$$

$$- [(1/2) \times (L/2) \times \{PL/(12EI)\}] \times [(2/3) \times (L/2)] = 0$$

$$\Rightarrow \quad R_B' = [PL^2/(24EI)] + [PL^2/(144EI)] = 7PL^2/(144EI)$$

Then, from $\sum F_Y = 0$,

$$R_A' = \text{Total load} - R_B'$$

$$= [(1/2) \times (L/2) \times \{PL^2/(144EI)\}]$$

$$+ [(1/2) \times (L/2) \times \{PL^2/(12EI)\}] - [PL^2/(144EI)]$$

$$= 5PL^2/(144EI)$$

According to Mohr's theorem-1,

slope at end A, θ_A = shear force at A on conjugate beam = $R_A' = 5PL^2/(144EI)$

slope at end B, θ_B = shear force at B on conjugate beam = $R_B' = 7PL^2/(144EI)$

and slope at mid-point C,

$$\theta_C = \text{shear force at C on conjugate beam}$$

$$= R_A' - (\text{Total load on AC})$$

$$= 5PL^2/(144EI) - [(1/2) \times (L/2) \times \{PL/(12EI)\}] = PL^2/(72EI)$$

According to Mohr's theorem-2,

deflection at mid-point C, δ_C = bending moment at C on conjugate beam

$$= - R_A' \times (L/2) + (1/3) \times (L/2) \times (\text{Total load between A and C})$$

$$= - \{5PL^2/(144EI)\} \times (L/2)$$

$$+ (1/3) \times (L/2) \times [(1/2) \times (L/2) \times \{PL/(12EI)\}]$$

$$= PL^3/(72EI)$$

Note that slope and deflection at a point in the beam are functions of flexural rigidity (material and cross section) while shear force and bending moment are independent. This is reflected by the changes in slope and deflection values at the center of the beam in example-2, compared to those of example-1

Example 11.12

Calculate slope and deflection of a cantilever beam of length L and flexural rigidity EI, when a concentrated load acts at its free end by conjugate beam method.

Solution

Bending moment diagram of the cantilever A–B is a triangle with zero moment at the free end B and maximum bending moment of (P×L) at the fixed end A.

The conjugate beam will have a linearly distributed load with zero value at the fixed B and maximum value of $P \times L/(EI)$ at the free end A. Note that the free and fixed ends of the original cantilever beam are reversed for conjugate beam. In a simply supported beam, this change is not distinctly noticed.

Reaction R_B' at fixed end B of the conjugate beam is equal to the total applied load.

Thus, $R_B' = (1/2) \times L \times \{PL/(EI)\} = PL^2/(2EI)$

According to Mohr's theorem-1,

slope at end B, θ_B = shear force at B on conjugate beam = $R_B' = PL^2/(2EI)$

According to Mohr's theorem-2,

deflection at end B, δ_B = bending moment at B on conjugate beam

$$= -(2L/3) \times (\text{Total load between A and B})$$

$$= -(2L/3) \times \{PL^2/(2EI)\} = -PL^3/(3EI)$$

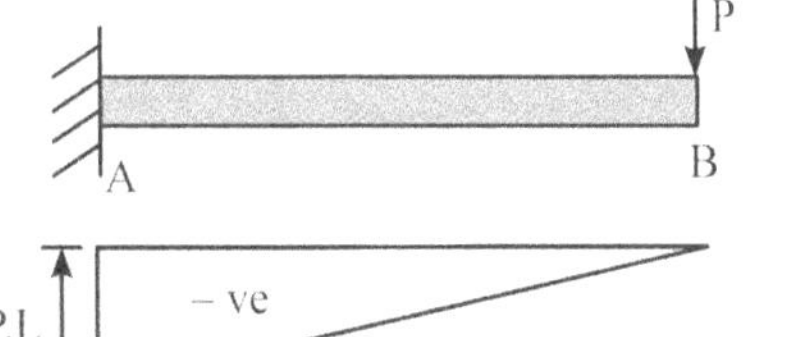

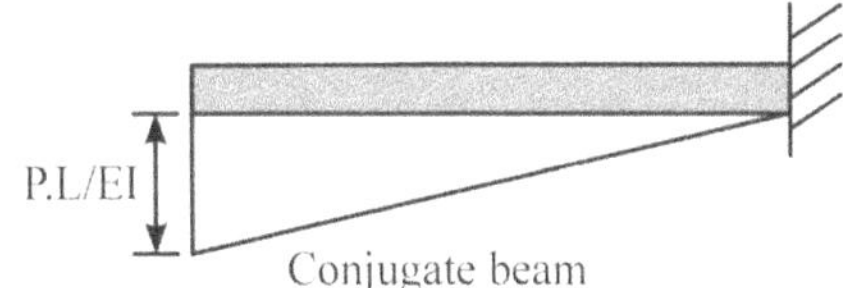

Slope and deflection at the free end B are same as those obtained by other methods.

11.11.6 SOLUTIONS FOR SOME DETERMINATE BEAMS

Support reactions, max. shear force, max. bending moment, max. slope and max. deflection obtained for a few common cases are given below. It is easy to obtain the same by any of the methods explained above.

 P A ⟵a⟶⟵b⟶ B C L = a+b	$R_A = P \times b/L$ $R_B = P \times a/L$ $M_{Max} = M_C = P \times a \times b/L$	$\theta_A = Pb(L^2 - b^2)/(6EIL)$ $\theta_B = -Pa(L^2 - a^2)/(6EIL)$

Beam	Reactions & Moments	Deflections & Slopes
Simply supported beam with central point load P at $L/2$; supports A and B, span L	$R_A = R_B = P/2$ $M_{Max} = P \times L/4$	$\delta_{Max} = PL^3/48EI$ at $x = L/2$ $\theta_{Max} = \theta_A = -\theta_B = PL^2/(16EI)$
Simply supported beam with uniformly distributed load w; supports A and B, span L	$R_A = R_B = wL/2$ $M_{Max} = wL^2/8$ at $x = L/2$	$\delta_{Max} = 5wL^4/(384EI)$ at $x = L/2$ $\theta_{Max} = \theta_A = -\theta_B = wL^3/(24EI)$
Simply supported beam with triangular distributed load w; supports A and B, span L	$R_A = wL/3$ $R_B = wL/6$ $M_{Max} = wL^2/9\sqrt{3}$ at $x = L/\sqrt{3}$ from B	$\delta_{Max} = 0.00652wL^4/(EI)$ at $x = 0.519L$ from B $\theta_{Max} = \theta_A = wL^3/(45EI)$ $\theta_B = 7wL^3/(360EI)$
Simply supported beam with applied moment M at center C; $AC = CB = L/2$	$-R_A = R_B = M/L$ $M_{Max} = M$	$\theta_{Max} = \theta_A = -\theta_B = ML/(24EI)$ $\delta_C = 0$ $\delta_{Max} = +ML^2/(128EI)$ at $x = L/4$ $= -ML^2/(128EI)$ at $x = 3L/4$
Beam with point load P at C; a from A, b to B, $L = a + b$	$R_A = P$ $M_A = P \times a$	$\delta_{Max} = \delta_B$ $= (Pa^2/EI) \times (a/3 + b/2)$ $\theta_{Max} = \theta_B = \theta_C = Pa^2/(2EI)$
Cantilever fixed at A with point load P at free end B, length L	$R_A = P$ $M_A = P \times L$	$\delta_{Max} = \delta_B = 5PL^3/(48EI)$ $\theta_{Max} = \theta_B = PL^2/(2EI)$
Cantilever fixed at A with point load P at C; a, b, $L = a + b$	$R_A = wL$ $M_A = wL^2/2$	$\delta_{Max} = \delta_B = wL^4/(8EI)$ $\theta_{Max} = \theta_B = wL^3/(6EI)$
Cantilever fixed at A with partial uniformly distributed load w over a; C, b, $L = a + b$	$R_A = wa$ $M_A = wa^2/2$	$\delta_C = wa^4/(8EI)$ $\theta_{Max} = \theta_B = \theta_C = wa^3/(6EI)$ $\delta_{Max} = \delta_B = wa^4/(8EI)$ $+ wa^3(L - a)/(6EI)$
Cantilever fixed at A with triangular distributed load w; free end B, length L	$R_A = wL/2$ $M_A = wL^3/6$	$\theta_{Max} = \theta_B = wL^3/(12EI)$ $\delta_{Max} = \delta_B = wL^4/(15EI)$

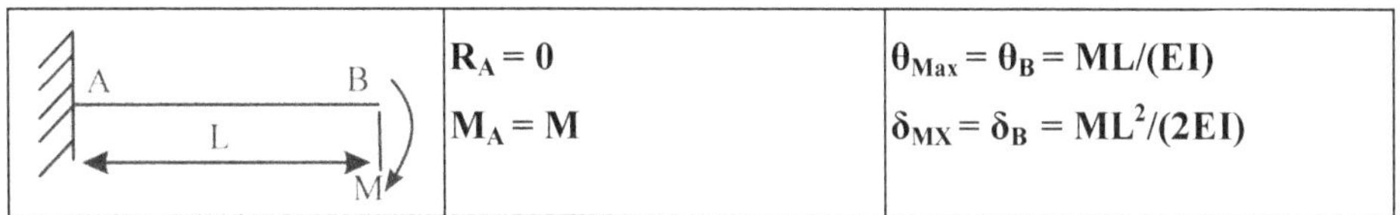

Example 11.13

A steel wide-flange beam of uniform thickness 20 mm, shown in figure, is subjected to a shear force (S) of 80 kN. Plot the shear stress distribution acting over the beam cross sectional area. Also determine the shear force resisted by the web

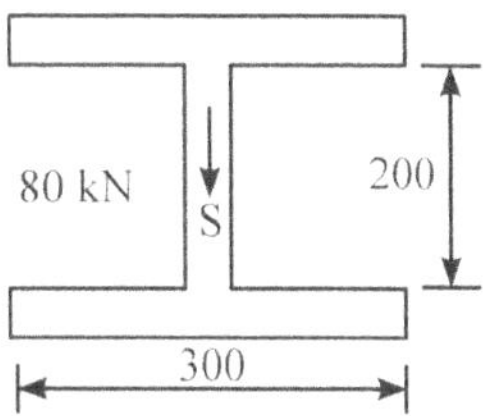

Solution

Due to symmetry we can conclude that max shear stress occurs in the web at the middle of section (along the neutral axis) and is

$$\tau_{max} = (S/8I) \times [B \times (D^2 - d^2)/t + d^2]$$

with   B = 300 mm, d = 200 mm, D = 240 mm, t = 20 mm, F = 80000 N

and $I = [300 \times 240^3 - 280 \times 200^3] / 12 = 158.9 \times 10^6$ mm^4

Therefore, $\tau_{max} = (80000/8 \times 158.9 \times 10^6) \times [300 \times (240^2 - 200^2) / 20 + 200^2]$

$$= 19.13 \text{ N/mm}^2$$

Alternatively, from basics, $\tau_{max} = S \times A \times y_G / (I \times B)$

where, A is the area above N.A;

   y_G is the distance of centroid of area A from N.A.

and   B is the width of the section at N.A.

$\tau_{max} = 80000 \times \{ 300 \times 20 \times 110 + 20 \times 100 \times 50 \} / (158.9 \times 10^6 \times 20)$

$$= 19.13 \text{ N/mm}^2$$

Example 11.14

A cantilever beam of rectangular section, shown in figure, is subjected to a load (F) of 1000 N, inclined at an angle of 30^0 to the vertical, at its free end. What is the stress due to bending at point D. Also calculate maximum deflection. All dimensions are in mm.

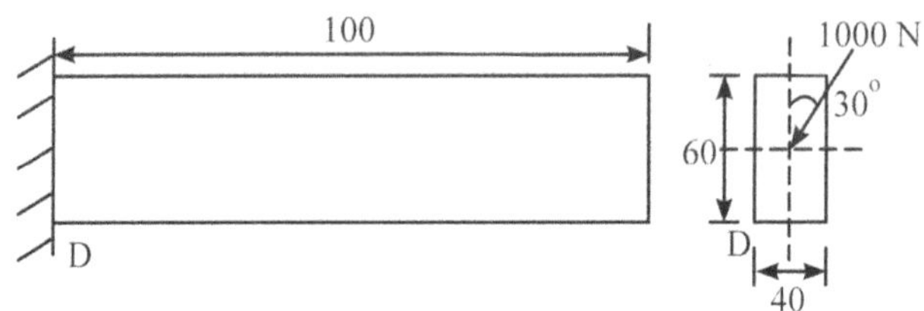

Solution

Resolving force F into two components $F_Y = F \cos 30^0$ and $F_Z = F \sin 30^0$

$$I_{ZZ} = 40 \times 60^3/12 \text{ mm}^4 \;; \quad I_{YY} = 60 \times 40^3 / 12 \text{ mm}^4$$

Bending stress at D due to F_Y,

$$(\sigma_D)_Y = (F_Y \times L) \times y / I_{ZZ} = 1000 \cos 30^0 \times 1000 \times (-30) / (40 \times 60^3/12)$$
$$= -36.084 \text{ N/mm}^2$$

Bending stress at D due to F_Z,

$$(\sigma_D)_Z = (F_Z \times L) \times z / I_{YY} = 1000 \sin 30^0 \times 1000 \times 20 / (60 \times 40^3/12)$$
$$= 31.25 \text{ N/mm}^2$$

Total stress at D, $\sigma_D = (\sigma_D)_Y + (\sigma_D)_Z = -4.834 \text{ N/mm}^2$ (Compressive)

Example 11.15

An aluminium cantilever of length 1m is loaded as shown in figure. It has C-section with uniform thickness of 12mm. Determine the maximum bending stress in strut. All dimensions are in mm.

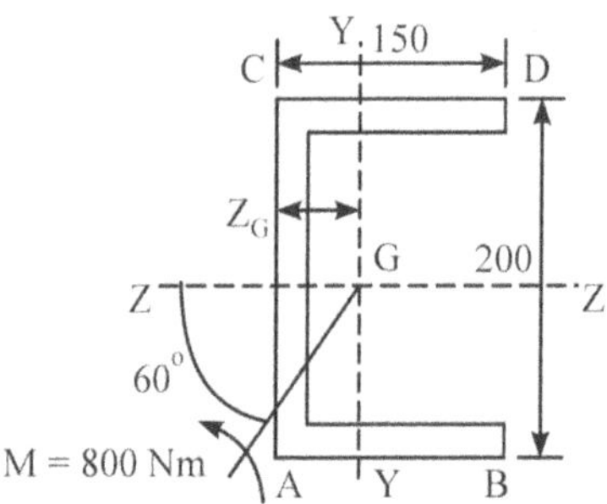

Solution

The applied moment M can be resolved into two components

$$M_Z = M \cos 60^0 \quad \text{and} \quad M_Y = M \sin 60^0$$

$$I_{ZZ} = (150 \times 200^3 - 138 \times 176^3)/12 = 37.3 \times 10^6 \text{ mm}^4$$

$$Z_G = \sum (A_i \times z_i) / \sum A_i$$

$$= [2(12 \times 150 \times 75) + 176 \times 12 \times 6] / [2(12 \times 150) + 176 \times 12]$$

$$= 49.5 \text{ mm}$$

$$I_{yy} = 2 \left[(12 \times 150^3 / 12) + (12 \times 150) \times (75 - Z_G)^2 \right]$$

$$+ \left[(176 \times 12^3 / 12) + (176 \times 12) \times (Z_G - 6)^2 \right]$$

$$= 3.537 \times 10^6 \text{ mm}^4$$

Due to M_y bending stress at A and C will be opposite to that at B and D

Due to M_z bending stress at A and B will be opposite to that at C and D

Thus, maximum bending stress is $M_Y \times z_A / I_{YY} + M_Z \times y_A / I_{ZZ}$ at A

or $- M_Y \times z_D / I_{YY} - M_Z \times y_D / I_{ZZ}$ at D

Therefore,

$$\sigma_A = (M \sin 60^0) \times Z_G / (3.537 \times 10^6) + (M \cos 60^0) \times 100 / (37.3 \times 10^6)$$

$$= + 10.768 \text{ N/mm}^2$$

$$\sigma_D = - (M \sin 60^0) \times (150 - Z_G) / (3.537 \times 10^6) - (M \cos 60^0) \times 100 / (37.3 \times 10^6)$$

$$= - 20.758 \text{ N/mm}^2$$

Maximum bending stress occurs at D

Example 11.16

A cantilever of length L has a circular cross section which varies from D at the free end to 2D at the fixed end. Determine (i) the section at which maximum stress is developed, (ii) deflection at the free end and (iii) slope at the free end, when a concentrated vertical load P is applied at the free end.

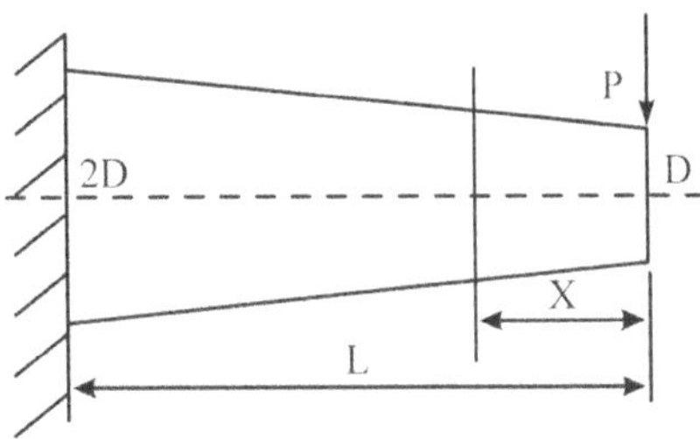

Solution

On a section at distance 'x' from the free end,

Diameter, $d = D + x \times [(2D - D)/L] = D + x \times D / L = D \times (1 + x/L)$

Moment of inertia at this section, $I = \pi \times d^4/64 = \pi \times D^4 \times (1 + x/L)^4 / 64$

$$= I_{tip} \times (1 + x/L)^4$$

Neutral axis passes through the center of the section (along horizontal or X-axis). Bending moment at this section, $M = P \times x$

Maximum bending stress at this section occurs at its outermost point, at distance $(d/2)$ from neutral axis and is given by,

$$\sigma = M \times (d/2)/I = (P \times x) \times [D \times (1 + x/L)/2] / [\pi \times D^4 \times (1 + x/L)^4 / 64]$$

$$= 32P / [\pi \times D \times (1 + x/L)^3]$$

(i)   At the section, which has maximum bending stress,

$$d\sigma/dx = [32P / \pi \times D^3] \times \{(1 + x/L)^{-3} + x \times (-3) \times (1 + x/L)^{-4} \times (1/2)\}$$

$$= 0$$

$\Rightarrow$ $1 - 3x \times (1 + x/L)^{-1} \times (1/2) = 0$ or $2(1 + x/L) - 3x = 0$

$\Rightarrow$ $x = 2L / (3L - 2)$

(ii)   Deflection and slope of the beam at this section are obtained from

$$1/R = -M/(EI) \quad \text{or} \quad d^2y/dx^2 = -(P \times x) / [E \times I_{tip} \times (1 + x/L)^4]$$

$$= -\{P \times L^4/(E\ I_{tip})\} \times [\{(L + x) - L\}/(L + x)^4]$$

$$= -\{P \times L^4/(E\ I_{tip})\} \times [(L + x)^{-3} - L \times (L + x)^{-4}]$$

Integrating once,

$$dy/dx = -\{P \times L^4/(E\ I_{tip})\} \times [\{(L + x)^{-2}/(-2)\} - \{L(L + x)^{-3}/(-3)\}] + C_1$$

The constant C_1 can be evaluated from the end condition, $dy/dx = 0$ at $x = L$

$\Rightarrow$ $C_1 = P \times L^2 / (12\ E\ I_{tip})$

Therefore,

$$dy/dx = -\{P \times L^4/(E\ I_{tip})\} \times [\{L(L + x)^{-3}/3\} - \{(L + x)^{-2}/2\} + (1/12L^2)]$$

Slope at the tip, for $x = 0$, is $(dy/dx)_{tip} = P \times L^2 / (12E\ I_{tip})$

Integrating the differential equation again,

$$y = -\{P \times L^4 / (E\ I_{tip})\} \times [\{L(L + x)^{-2} / (-6)\}$$

$$- \{(L + x)^{-1} / (-2)\} + (x/12L^2)] + C_2$$

The constant C_2 can be evaluated from the end condition, $y = 0$ at $x = L$

$$\Rightarrow C_2 = 7P \times L^3 / (24\, E\, I_{tip})$$

Therefore, $y = - \{P \times L^4/(E\, I_{tip})\} \times [\{(L+x)^{-1}/2\} - \{L(L+x)^{-2}/6\}$

$$+ \{x/(12L^2)\} - \{7/24L)\}]$$

Deflection at the free end, for $x = 0$, $y_{tip} = - P \times L^3 / (24\, E\, I_{tip})$

Example 11.17

For the same bending stress, compare the moments of resistance of a beam of square section when placed (i) with two sides of the square horizontal and (ii) with one of its diagonal horizontal.

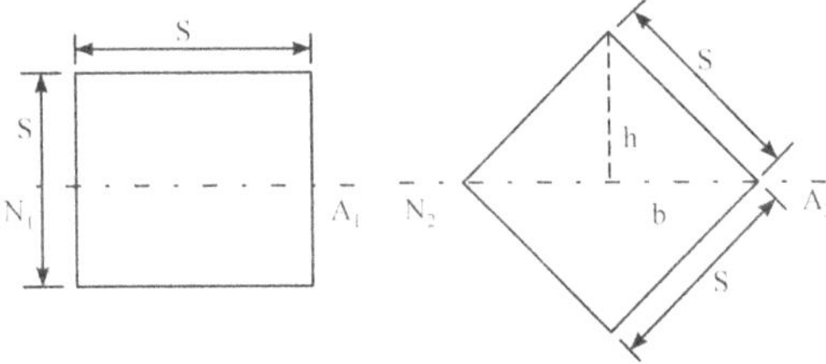

Solution

For the first case, with sides of square parallel to neutral axis,

Moment of inertia, $\quad I_1 = (1/12) \times b \times d^3$

$$= (1/12) \times S \times S^3 \qquad \text{with } b = S \text{ and } d = S$$

$$= S^4/12$$

Max Bending stress, $\sigma_1 = M_1 \times y_{max}/I_1$

$$= M_1 \times (S/2) \times (S^4/12) \quad \text{with } y_{max} = d/2 = S/2$$

$$= M_1 \times S^5/24$$

For the second case, with diagonal of square parallel to neutral axis,

Moment of inertia, $\quad I_2 = 2\,[(1/12) \times b \times h^3] \quad$ with $\ b = S\sqrt{2}$ and $h = S/\sqrt{2}$

$$= 2\,[(1/12) \times (S\sqrt{2}) \times (S/\sqrt{2})^3]$$

$$= S^4/12$$

Max Bending stress, $\sigma_2 = M_2 \times y_{max}/I_2$

$$= M_2 \times (S/\sqrt{2}) \times (S^4/12) \quad \text{with } y_{max} = h = S/\sqrt{2}$$

$$= M_2 \times S^5 / (12\sqrt{2})$$

Moment of resistance of the two sections, for the same bending stress, is given by $\sigma_1 = \sigma_2$

$$\text{or} \qquad M_1 \times (S^5/24) = M_2 \times S^5 / (12\sqrt{2})$$
$$\Rightarrow \qquad M_1/M_2 = \sqrt{2} = 1.414$$

11.12 ANALYSIS OF INDETERMINATE BEAMS

A structure may be ***externally redundant*** (for example, continuous beam, propped cantilever and beam with its both ends rigidly fixed), if it has more number of support reactions than the number of relevant equations of static equilibrium. Number of reactions in excess of the number of static equilibrium equations is called the degree of indeterminacy or order of redundancy. Such an indeterminate beam remains in static equilibrium even if all such redundant reactions are dispensed with. Externally redundant structures are analysed by the following methods.

11.12.1 SUPERPOSITION METHOD

The principle of superposition states that the effect of a combination of loads acting on a structure is equal to the algebraic sum of the effects of individual loads on the structure, provided the load-displacement relationship is linear for the entire load range. This principle is useful for analysing a statically indeterminate structure as a combination of two or more statically determinate structures. Its application to the analysis of a structure is better explained through the following examples.

Example 11.18

Consider the case of a ***propped cantilever*** beam, with a concentrated load at the middle of the beam. The beam has three unknown reactions – vertical reaction and a moment at the fixed support and a vertical reaction at the propped end. The two equations of static equilibrium $\sum F_Y = 0$ and $M_Z = 0$, assuming there is no horizontal load (and, hence no horizontal reaction), are not adequate to determine the three unknown support reactions. Thus, the propped cantilever is an indeterminate beam with degree of indeterminacy equal to 1.

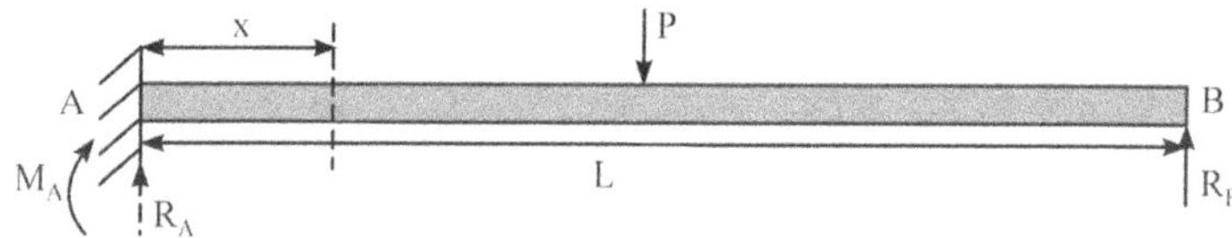

(i) ***Rigid prop*** – The cantilever is first analysed without support at end B and deflection δ_B at end B is calculated. Then, the reaction R_B required at end B to produce same δ_B upwards is calculated. Superposition of these two cases results in zero deflection at end B, which is the real situation. Reaction and moment at end A for both the cases are also added together with corresponding signs (representing their directions). This method is applicable as long as the net bending stress is in the linear stress-strain regime of the particular beam material.

In case-1, deflection at end B,

δ_B = Deflection at load point + deflection due to constant slope upto B

$$= - \{ [P \times (L/2)^3/(3EI)] + [P \times (L/2)^2/(2EI)] \times (L/2) \}$$

$$= - \{ P \times L^3/(24EI) + P \times L^3/(16EI)\}$$

$$= - \ 5P \times L^3/(48EI)$$

In case-2, Load to be applied at end B to get $\delta_B = 5P \times L^3/(48EI)$ is given by

$$R_B = 3EI \ \delta_B/L^3 = 3EI \times [5P \times L^3 / (48EI)] / L^3 = (5/16) \times P$$

Then, $\sum F_Y = 0$ gives, $R_A = P - R_B = P - (5/16) \times P = (11/16) \times P$

and $\sum M_Z = 0$ at A gives, $M_A = R_B \times L - P \times (L/2) = (5/16) \times P \times L - P \times L/2$

$$= - (3/16) \times P \times L$$

Alternatively, reaction R_A can also be calculated as the algebraic sum of the reactions of the two cases, as

$$R_A = (R_A)_1 + (R_A)_2 = P + (- R_B) = P - (5/16) \times P = (11/16) \times P$$

Shear force and bending moment diagrams are drawn, as shown, with

$$(S.F.)_{Max} = R_B = 5P/16 \qquad \text{on +ve side}$$

and R_A or $(R_B - P) = -11P/16$ on –ve side

$$M_{Max} = R_B \times (L/2) = 5PL/32 \qquad \text{on +ve side}$$

and $M_A = R_B \times L - P \times (L/2) = -3PL/16$ on –ve side

Deflection at the load point is obtained by superposition of values of case-1 and case-2 as

$$\delta = - [P \times (L/2)^3 / (3EI)] + [5R_B \times L^3 / (48EI)] = - 7P \times L^3 / (768 \ EI)$$

Point of contra flexure D is at a distance 'x' from B towards A such that,

$$M_X = R_B \times x_D - P \times (x_D - L/2) = 0 \ \Rightarrow \ 5x_D - 16x_D + 8L = 0 \text{ or } x_D = 8L/11$$

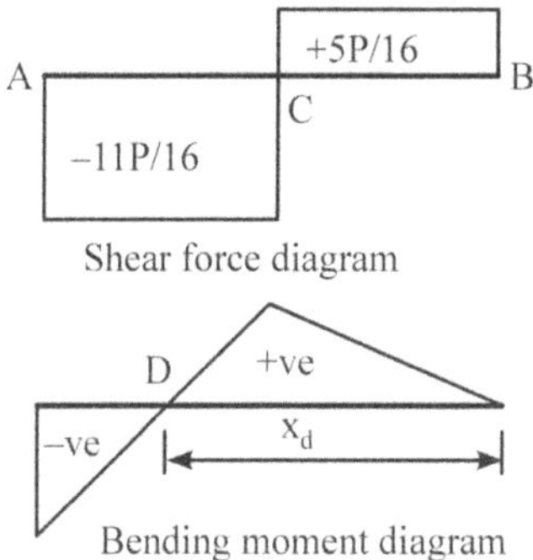

Shear force diagram

Bending moment diagram

Alternative method

It is also possible to solve this problem choosing M_A as the redundant unknown, instead of reaction R_B chosen earlier. Here, the magnitude of moment at A is calculated to compensate the downward slope at A due to P and ensure slope at A is zero. Corresponding reactions at the two ends are obtained by equating those of the two cases, as explained here.

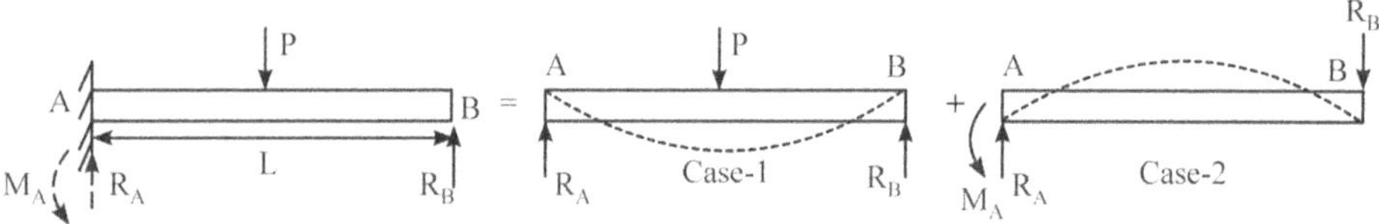

In case-1, reactions at the two ends by symmetry are $(R_A)_1 = (R_B)_1 = P/2$

and   slope at A, $\theta_A = -[P \times L^2/(16EI)]$

In case-2, reactions at the two ends, for an applied moment M_A, are

$$(R_A)_2 = -(R_B)_2 = M_A / L$$

Moment to be applied at end A to get $\theta_A = P \times L^2/(16EI)$ is given by

$$M_A = 3EI\, \theta_A/L = 3EI \times [P \times L^2/(16EI)] / L = (3/16)\, P \times L$$

By superposition of the two cases, we get

$$R_A = (R_A)_1 + (R_A)_2 = P/2 + \{(3/16)\, P \times L\}/L = (11/16)\, P$$

and $R_B = (R_B)_1 - (R_B)_2 = P/2 - \{(3/16)\, P \times L\}/L = (5/16)\, P$

It can be seen that these values are same as those obtained earlier.

Example 11.19

Consider the case of a ***propped cantilever*** beam, with a uniformly distributed load over its entire length. With M_A as the redundant reaction, solution for the propped cantilever of degree of indeterminacy equal to 1 can be obtained by the superposition of the following two cases.

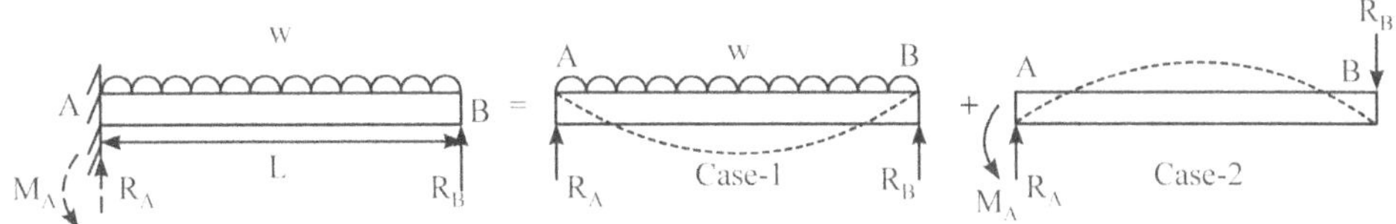

(i)   *Rigid support* – In case-1, reactions at the two ends by symmetry are

$$(R_A)_1 = (R_B)_1 = wL/2 \quad \text{and} \quad \text{slope at A,} \quad \theta_A = -[w \times L^3/24EI]$$

In case-2, reactions at the two ends, for an applied moment M_A, are

$$(R_A)_2 = -(R_B)_2 = M_A / L$$

Moment to be applied at end A to get $\theta_A = w \times L^3 / (24EI)$ is given by

$$M_A = 3EI \times \theta_A/L = 3EI \times [w \times L^3 / (24EI)] / L = w \times L^2 / 8$$

By superposition of the two cases, we get

$$R_A = (R_A)_1 + (R_A)_2 = w \times L/2 + (w \times L^2/8) / L = (5/8) \times w \times L$$

and $R_B = (R_B)_1 - (R_B)_2 = w \times L/2 - (w \times L^2/8) / L = (3/8) \times w \times L$

(ii)   *Sinking prop* – A support resting on a weak foundation may sink due to the applied load, resulting in a change in the level of the supports. This can be analysed from the relation,

sink = Downward deflection due to applied load
    – Upward deflection due to support reaction

In the above example, if the prop is rigid (zero sink), reaction at prop is calculated from sink, $\delta = -[5P \times L^3 / (48EI)] + [R_B \times L^3 / (3EI)] = 0$

$\Rightarrow \ R_B = 5P/16$ which is independent of values of E and I

However, in the same example, if the prop sinks by 5mm, reaction at prop is calculated from sink, $5 = -[5P \times L^3 / (48EI)] + [R_B \times L^3 / (3EI)]$

In this case, R_B can be calculated only if E and I are given

(iii) *Elastic prop* – It is also possible that a prop sinks by some value, depending on the value of applied load. In this case, if k_P is the stiffness of the prop,

Sink, $\delta = -5P \times L^3 / 48EI + R_B \times L^3 / 3EI = R_B \times k_P$

R_B can then be calculated if prop stiffness k_P, E and I are specified

Alternatively, k_P can be calculated if R_B, E and I are specified

Example 11.20

Consider the case of a **continuous beam with 3 simple supports**. It can be treated as a simply supported (s.s.) beam, with a prop in between the supports.

As an example, let us take a simply supported beam of length L, with an additional support at L/2 from either support, subjected to uniformly distributed load 'w', as shown.

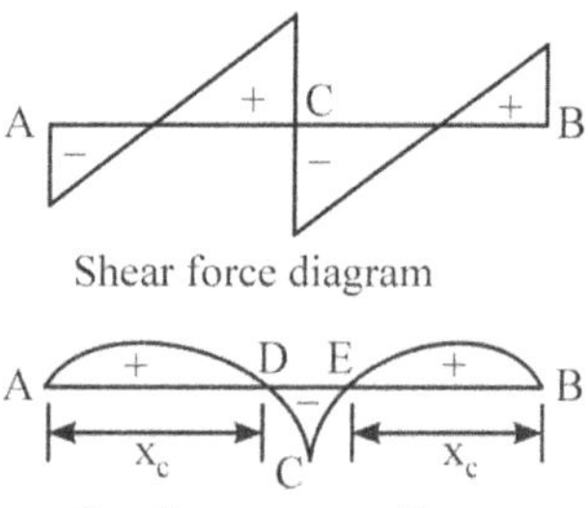

Deflection at C, δ_C = –Downward deflection of s.s.beam A – B at C due to u.d.l.

  + Upward deflection of s.s.beam A – B at C due to reaction R_C

$$= -[5wL^4 / (384EI)] + [R_C \times L^3 / (48EI)] = 0$$

$\Rightarrow$ $R_C = 5wL/8$

Due to symmetry, $R_A = R_B = (w \times L - R_C)/2 = 3wL/16$

Bending moment at the middle support C is,

$$M_C = R_B \times (L/2) - w \times (L/2) \times (L/4) = -wL^2/32$$

Point of contra flexure is at a distance 'x_E' from B towards C such that,

$$M_X = R_B \times x_E - w \times x_E \times (x_E /2) = 0$$

$\Rightarrow$ $3L - 8 x_E = 0$ or $x_E = 3L/8$

Due to symmetry, it can be concluded that another point of contra flexure also exists at a distance of $x_D = 3L/8$ from A (or 5L/8 from B), which can be obtained by equating bending moment for a section in A – C part of beam to zero.

Shear force diagram

Bending moment diagram

Example 11.21

Consider the case of a **beam with both ends fixed** with a uniformly distributed load of 'w' over its entire span 'L'. It has four unknown reactions – force and moment at each end, against two equations of equilibrium. It is thus a problem of degree of indeterminacy equal to 2. Solution is therefore obtained by superposition of three simple cases, as shown here.

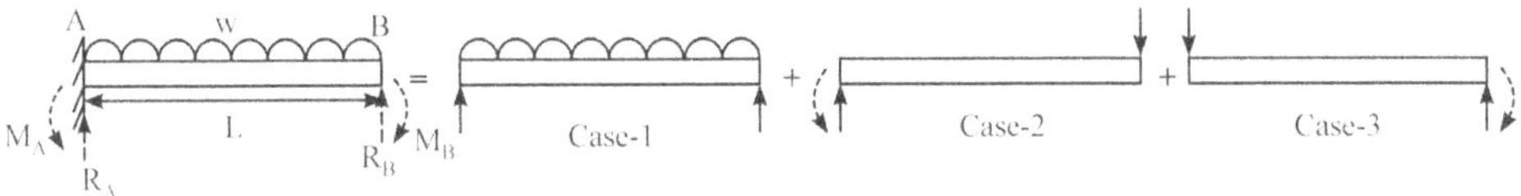

The end reactions M_A and M_B are calculated from the compatibility conditions

$$\theta_A = (\theta_A)_1 - (\theta_A)_2 - (\theta_A)_3 = 0 \quad \text{and} \quad \theta_B = (\theta_B)_1 - (\theta_B)_2 - (\theta_B)_3 = 0$$

$$\Rightarrow \quad \theta_A = wL^3/(24EI) - M_A L/(3EI) - M_B L/(6EI) = 0$$

$$\text{and} \quad \theta_B = wL^3/(24EI) - M_A L/(6EI) - M_A L/(3EI) = 0$$

Solving these two equations, we get $\quad M_A = M_B = wL^2/12$

Corresponding reactions are obtained as

$$R_A = (R_A)_1 + (R_A)_2 - (R_A)_3 = wL/2 + wL/12 - wL/12 = wL/2$$

$$\text{and} \quad R_B = (R_B)_1 - (R_B)_2 + (R_B)_3 = wL/2 - wL/12 + wL/12 = wL/2$$

ADDITIONAL PROBLEMS FOR PRACTICE

1. Propped cantilever with udl 'w' over its entire length 'L'

    Ans: $R_A = (5/8)wL$; $R_B = (3/8)wL$; $M_A = wL^2/8$

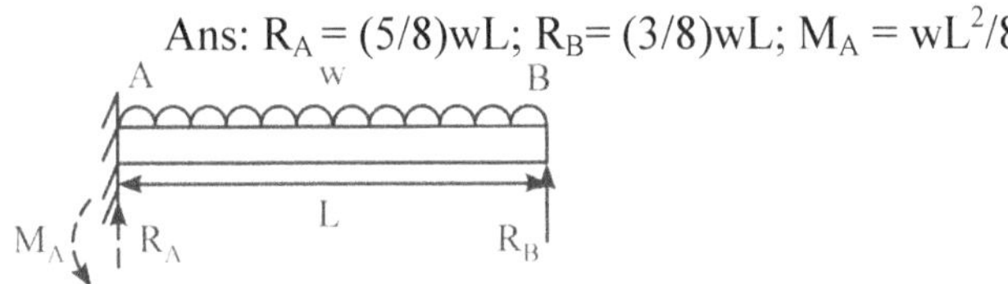

2. Fixed end beam with a concentrated load 'P' at a distance 'a' from end A or at distance 'b' from end 'B'.

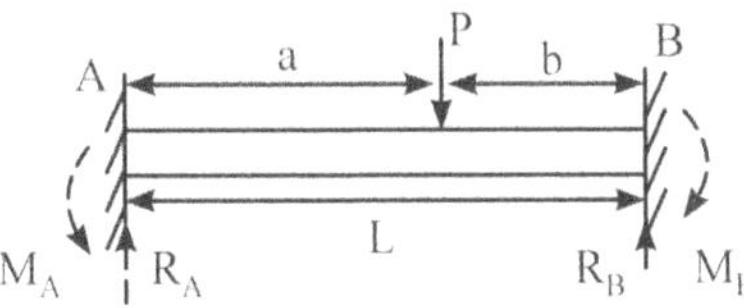

3. Propped cantilever of length 5m with overhang. Calculate reactions for udl of 10kN/m over the entire length, if the prop is located at 3m from the fixed support. Ans : $R_A = 8.59$kN; $R_B = 41.41$kN

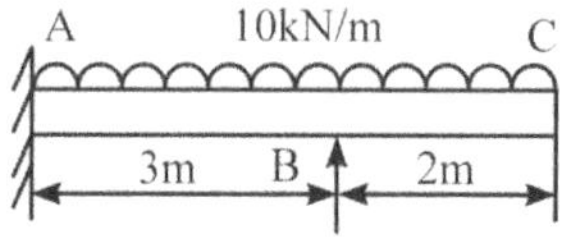

4. Propped cantilever of length 5m. Calculate reactions for udl of 10kN/m over a length of 3m from the fixed end, if the prop is located at the free end of cantilever.   Ans : R_A = 27.559kN; R_C = 2.441kN

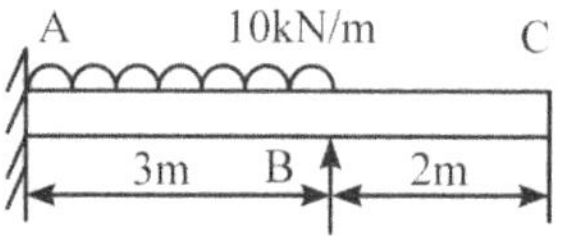

5. Propped cantilever of length 'L' with overhang of length 'L/4'. Calculate reactions for a concentrated load of 'P' applied at its free end.
                          Ans : R_A = – 0.5P; R_B = 1.5P; M_A = – PL/8

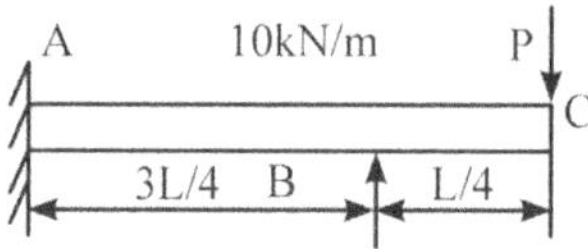

11.12.2 DOUBLE INTEGRATION METHOD

In this method, the three kinematic conditions (viz., vertical displacement at both ends and rotation at the fixed end are zero for a propped cantilever; vertical displacement at the three fixed supports are zero for a 3-support continuous beam) along with the two equations of static equilibrium can be used to evaluate the two integration constants as well as the three support reactions. However, superposition method is more convenient and is more popular. Application of this method is explained through the example of a propped cantilever with udl over its complete length. At any section at distance 'x' from fixed end,

$$EI \times (d^2v/dx^2) = - M = - w\,(L - x)^2/2 + R_B\,(L - x)$$

Integrating once, we get

$$EI \times (dv/dx) = w\,(L - x)^3/6 - R_B\,(L - x)^2/2 + C_1$$

Integrating again, we get

$$EI \times v = w\,(L - x)^4/24 - R_B\,(L - x)^3/6 + C_1 \times x + C_2$$

The three unknowns R_B, C_1 and C_2 in the above two equations can be obtained from the following three kinematic conditions

$$dv/dx = 0 \ \text{ at } \ x = 0 \ \Rightarrow \ w\,L^3/6 - R_B\,L^2/2 + C_1 = 0 \qquad \qquad .....(i)$$

$v = 0$ at $x = 0$ $\Rightarrow$ $w\,L^4/24 - R_B\,L^3/6 + C_2 = 0$ .....(ii)

$v = 0$ at $x = L$ $\Rightarrow$ $C_1\,L + C_2 = 0$ .....(iii)

Solving the above three equations, we get

$$R_B = 3wL/8 \,;\; C_1 = 5wL^3/24 \text{ and } C_2 = -\,wL^4/48$$

From equations of equilibrium,

$$R_A = wL - R_B = 5wL/8$$

and $M_A = wL^2/2 - R_B \times L = \; wL^2/2 - (3wL/8) \times L = wL^2/8$

11.12.3 SLOPE-DEFLECTION METHOD

It is a widely used displacement method for analyzing all types of statically indeterminate beams and frames and can be easily programmed. All joints are considered rigid such that all angles between members at a joint are assumed to remain same when loads are applied to the structure. For a member bounded by two end joints, the end moments can be expressed in terms of the end rotations. For static equilibrium, sum of end moments of all the members meeting at a joint must be equal to zero. The equations of static equilibrium provide the necessary conditions to handle the unknown joint reactions. When the joint rotations are found, the end moments can be computed from the slope-deflection equations.

The end deformations and moments are obtained by using the following equations.

Case-1 : Derivation by rotation

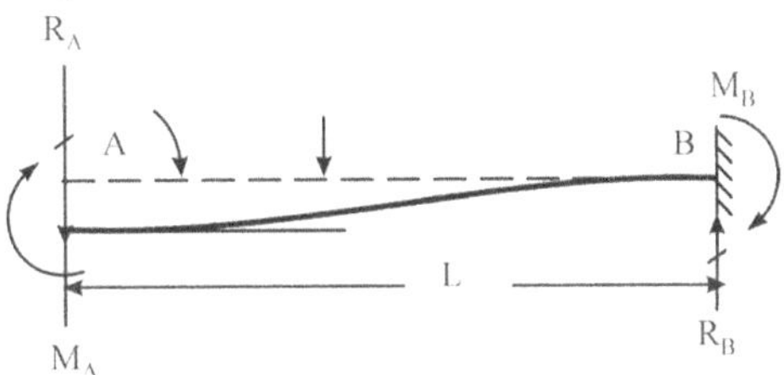

Using 1st moment-area theorem, $(M_A + M_B) \times L/2EI = \theta_A$

Using 2nd moment-area theorem and taking moments about A,

$$[M_A L/(2EI)] \times (L/3) + [M_B L/(2EI)] \times (2L/3) = 0$$

Solving these two equations, we get $M_A = 4EI \times \theta_A/L$ and $M_B = 2EI \times \theta_A/L$

$$\Rightarrow \quad R_A = -\,R_B = 6EI \times \theta_A/L^2$$

Case-2 : Derivation by deflection

Using 1st moment-area theorem, $(M_A + M_B) \times [L/(2EI)] = \theta_A = 0$

Using 2nd moment-area theorem and taking moments about A,

$$[M_A L / (2EI)] \times (L/3) + [M_B L / (2EI)] \times (2L/3) = \delta$$

Solving these two equations, we get $M_A = 6E \times I \times \delta/L^2$ and $M_B = -2EI \times \delta/L^2$

$$\Rightarrow \quad R_A = - R_B = 12EI \times \delta/L^3$$

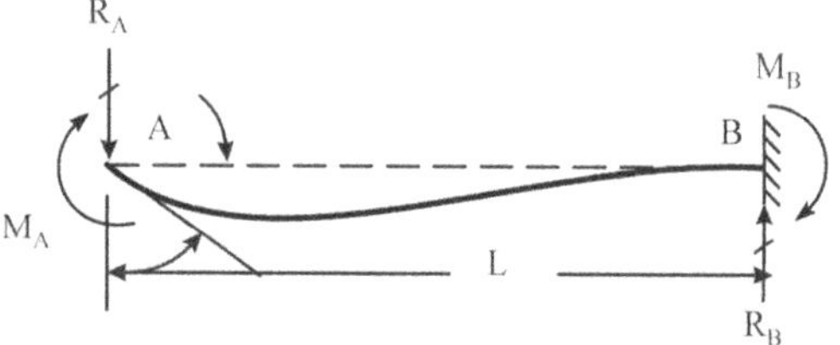

Using the above relationship, moments M_{AB} and M_{BA} for the beam AB can be expressed as follows

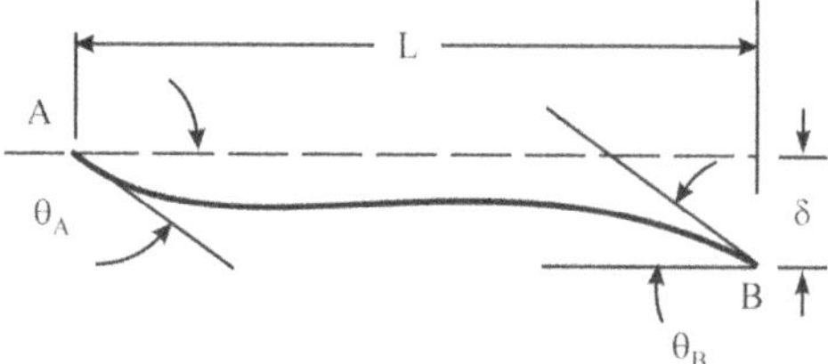

$$M_{AB} = 4EI \times \theta_A/L + 2E \times I \times \theta_B/L - 6EI \times \delta/L^2 \; = 2k \times (2\theta_A + \theta_B - 3\varphi)$$

$$M_{BA} = 2EI \times \theta_A/L + 4E \times I \times \theta_B/L - 6EI \times \delta/L^2 \; = 2k \times (\theta_A + 2\theta_B - 3\varphi)$$

$$\text{where,} \quad k = E I / L \quad \text{and} \quad \varphi = \delta / L$$

Fixed end moments M_{AB}' and M_{BA}' of beam AB are calculated when the beam is held fixed at beam ends, under the applied loading.

Example 11.22

Determine moments at supports of the 4m long continuous beam ABC with a uniformly distributed load of 3kN/m over the part AB of length 2m and a concentrated load of 8kN at the mid-point of part BC. Assume $(EI)_{AB} = 10Nm^2$ and $(EI)_{BC} = 5Nm^2$

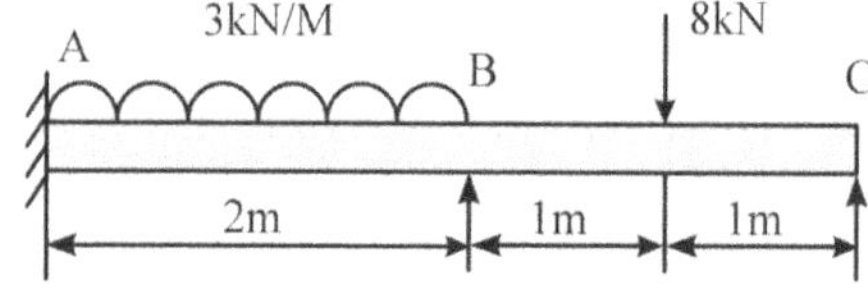

Solution

Fixed end moments in the two parts AB and BC of the continuous beam are :

$$M_{AB}' = - w \times L_{AB}^2 / 12 = - 3 \times 2^2 / 12 = - 1 \text{ kN m} ; \qquad M_{BA}' = 1 \text{ kN m}$$

$$M_{BC}' = - P \times L_{BC} / 8 = - 8 \times 2 / 8 = - 2 \text{ kN m} ; \qquad M_{CB}' = 2 \text{ kN m}$$

$$k_{AB} = (EI)_{AB}/L_{AB} = 10/2 = 5 \text{ kN m}; \ k_{BC} = (EI)_{BC}/L_{BC} = 5/2 = 2.5 \text{ kN m}$$

and $\varphi = \delta / L = 0$ for all supports, since there is no deflection

Slope-deflection equations :

$$M_{AB} = M_{AB}' + 2k_{AB} \times (2\theta_A + \theta_B - 3\varphi) = - 1 + 2 \times 5 \times \theta_B$$

$$M_{BA} = M_{BA}' + 2k_{AB} \times (\theta_A + 2\theta_B - 3\varphi) = \ 1 + 2 \times 5 \times 2\theta_B$$

$$M_{BC} = M_{BC}' + 2k_{BC} \times (2\theta_B + \theta_C - 3\varphi) = - 2 + 2 \times 2.5 \times (2\theta_B + \theta_C)$$

$$M_{CB} = M_{CB}' + 2k_{BC} \times (\theta_B + 2\theta_C - 3\varphi) = \ 2 + 2 \times 2.5 \times (\theta_B + 2\theta_C)$$

For equilibrium, $\sum M_B = M_{BA} + M_{BC} = 0$ and $\sum M_C = M_{CB} = 0$

$$\Rightarrow \quad (1 + 20 \ \theta_B) + (- 2 + 10 \ \theta_B + 5 \ \theta_C) = 0$$

$$\text{or} \qquad 30 \ \theta_B + 5 \ \theta_C \ = 1 \qquad\qquad .....(i)$$

$$\text{and} \qquad 5 \ \theta_B + 10 \ \theta_C \ = - 2 \qquad\qquad .....(ii)$$

Solving eq (i) and (ii), we get $\theta_B = 4/55$ and $\theta_C = - 13/55$

Then, $M_{AB} = - 1 + 10 \ \theta_B = - 1 + 10 \times (4/55) = - 3/11 \text{ kN m}$

and $M_{BA} = \ 1 + 20 \ \theta_B = \ 1 + 20 \times (4/55) = \ 27/11 \text{ kN m}$

Check : $M_{BC} = - 2 + 10 \ \theta_B + 5 \ \theta_C = - 2 + 10 \times (4/55) + 5 \times (-13/55)$

$$= - 2 + (8/11) + (-13/11) = - 27/11 = - M_{BA}$$

11.12.4 MOMENT DISTRIBUTION METHOD

It is a simple and convenient displacement approach, mathematically called a relaxation technique, for analysing redundant structures such as continuous beams, stiff-jointed frames etc. and give an approximate solution. It was widely used, with reasonable accuracy, prior to the advent of computer calculation.

In this method, a moment is assumed at the end of a member which is considered as the moment the member exerts on the joint or support. In order to balance any joint, it is only necessary to determine the magnitude and sign of the unbalanced internal moment that must be distributed to the various members intersecting at that joint to make $\sum M$ for the internal moments equal to zero.

This unbalanced moment, with sign reversed, is distributed to the members intersecting at that joint in accordance with the distribution factors.

This carry-over operation will usually upset the previous balance of one or more joints and these must be rebalanced. The carry-over will continue through all the joints and the corrections become smaller and smaller until they are no longer significant.

It uses the following displacement results in its derivations

(a)   The far end is fixed :

$$M_A = 4EI \times \theta/L; \quad R_A = 4EI \times \theta/L^2 \text{ and } M_B = 4EI \times \theta/L; \quad R_B = -4EI \times \theta/L^2$$

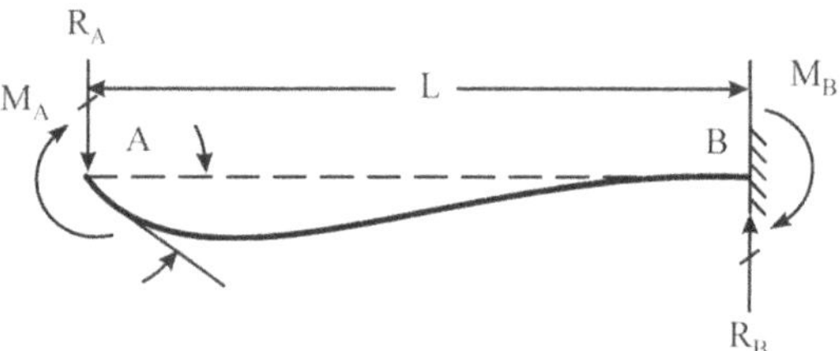

(b)   The far end is pinned :

$$M_A = 3EI \times \theta/L \; ; \quad R_A = 3EI \times \theta/L^2 \; \text{ and } M_B = 0; \quad R_B = -3EI \times \theta/L^2$$

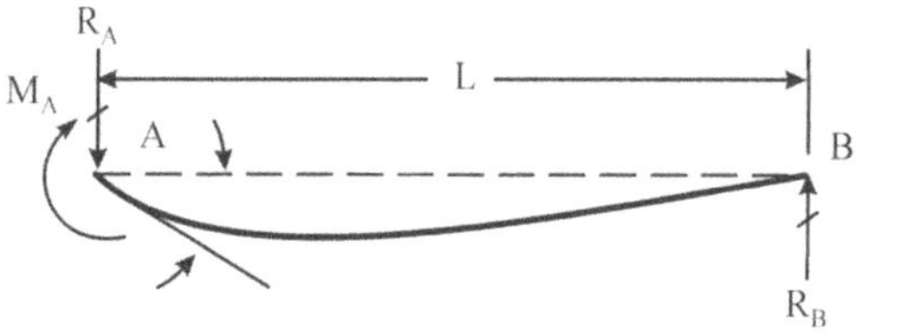

The applied moment M carried by each member is proportional to its stiffness K and, thus, the **distribution factor** (f) for a member at a joint is the ratio of the member stiffness to joint stiffness. For example, if three members AB, AC and AD are meeting at a joint A, distribution factors of the three members are

$$f_{AB} = K_{AB} / K_A \; ; \; f_{AC} = K_{AC} / K_A \; ; \; f_{AD} = K_{AD} / K_A$$

and the moments carried by them are

$$M_{AB} = M \times f_{AB} \; ; \quad M_{AC} = M \times f_{AC} \; ; \quad M_{AD} = M \times f_{AD}$$

Carry-over factors are ½ for fixed end and 0 for pinned end. Thus, $M_{BA} = M_{AB} / 2$, if end B is fixed   and $M_{CA} = 0$, if end C is pinned.

The method is explained through the following example

Example 11.23

Determine moments and reactions at the supports of a 4.5m long continuous beam ABCD with a uniformly distributed load of 600N/m over the part AB of length 2m, concentrated loads of 300N at 0.6m and 600N at 0.9m from the end B of the part BC and a uniformly distributed load of 300N/m over the 1m long part CD. End A is a fixed support while B, C and D are simple supports. Assume $(EI)_{AB} = 3Nm^2$, $(EI)_{BC} = 2Nm^2$ and $(EI)_{CD} = 1Nm^2$

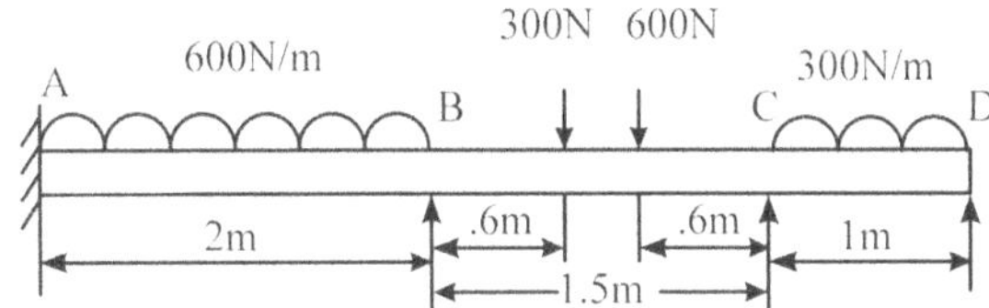

Solution

(a)   Distribution factors (f) :

$$K_{BA} = 4\,(EI)_{AB} / L_{AB} = 4 \times 3 / 2 = 6 \text{ Nm}$$

$$K_{BC} = 4\,(EI)_{BC} / L_{BC} = 4 \times 2 / 1.5 = 16/3 \text{ Nm}$$

$$K_B = K_{BA} + K_{BC} = 6 + (16/3) = 34/3 \text{ Nm}$$

$$f_{BA} = K_{BA} / K_B = 6 / (34/3) = 0.5294$$

$$f_{BC} = K_{Bc} / K_B = (16/3) / (34/3) = 0.4706$$

$$K_{CB} = 4\,(EI)_{CB} / L_{CB} = 4 \times 2 / 1.5 = 16/3 \text{ Nm}$$

$$K_{CD} = 4\,(EI)_{CD} / L_{CD} = 4 \times 1 / 1 = 4 \text{ Nm}$$

$$K_C = K_{CB} + K_{CD} = (16/3) + 4 = 28/3 \text{ Nm}$$

$$f_{CB} = K_{CB} / K_C = (16/3) / (28/3) = 0.5714$$

$$f_{CD} = K_{CD} / K_C = 4 / (28/3) = 0.4286$$

(b)   Fixed end moments :

$$M_{AB} = M_{BA} = w \times L_{AB}^2 / 12 = 600 \times 2^2 / 12 = 200 \text{ Nm}$$

$$M_{BC} = \sum P_i \times a_i \times b_i^2 / L^2$$

$$= 300 \times 0.6 \times 0.9^2 / 1.5^2 + 600 \times 0.9 \times 0.6^2 / 1.5^2$$

$$= 64.8 + 86.4 = 151.2 \text{ Nm}$$

$$M_{CB} = \sum P_i \times a_i \times b_i^2 / L^2$$

$$= 300 \times 0.9 \times 0.6^2 / 1.5^2 + 600 \times 0.6 \times 0.9^2 / 1.5^2$$

$$= 43.2 + 129.6 = 172.8 \text{ Nm}$$

$$M_{CD} = M_{DC} = w \times L_{CD}^2 / 12 = 300 \times 1^2 / 12 = 25 \text{ Nm}$$

The calculation can be done in a tabular form as follows, three columns representing beam segments AB, BC and CD, until desired accuracy is obtained

Distribution factors (f)	1.0	0.5294	0.4706	0.5714	0.4286	0
Fixed end moments (M')	−200	200	−151.2	172.8	−25	25
Balance at D carried over to C					−12.5	
Starting moment	−200	200	−151.2	172.8	−37.5	0
Balance and distribute carry over	−12.9	−25.8	−23 −43.3	−86.6 −11.5	−48.7	
Sum	**−212.9**	**174.2**	**−217.5**	**74.7**	**−86.2**	**0**
Balance and distribute carry over	11.5	23	20.3 3.7	7.4 10.2	4.1	
Sum	**−201.4**	**197.2**	**−193.5**	**92.3**	**−82.1**	**0**
Balance and distribute carry over	−1.0	−2.0	−1.7 −3.3	−6.5 −0.9	−3.7	
Sum	**−202.4**	**195.2**	**−198.5**	**84.9**	**−85.8**	**0**
Balance and distribute carry over	0.8	1.7	1.6 0.3	0.6 0.8	0.3	
Sum	**−201.6**	**196.9**	**−196.6**	**86.3**	**−85.5**	**0**
Balance and distribute carry over		−0.2	−0.1	−0.5	−0.3	
Sum	**−201.6**	**196.7**	**−196.7**	**85.8**	**−85.8**	**0**

Support reactions are obtained from free body diagrams, taking moments M_B and M_C values as averages of values from the table, as follows

For the beam segment AB,

$$R_A = (w \times L^2/2 + M_A − M_B)/L$$

$$= (600 \times 2^2/2 + 201.6 − 196.7) / 2$$

$$= 602.45 \text{ N}$$

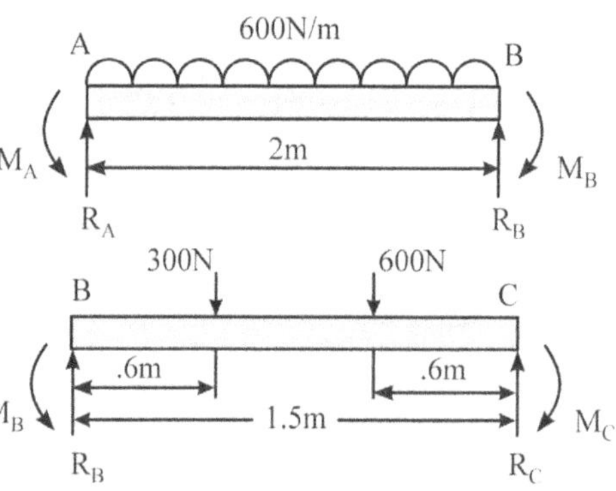

$R_{BA} = (w \times L^2/2 - M_A + M_B)/L = (600 \times 2^2/2 - 201.6 + 196.7) / 2$

$\qquad = 597.55 \text{ N}$

Alternatively, $R_{BA} = w \times L - R_A = 597.55 \text{ N}$

For the beam segment BC,

$R_{BC} = (\sum P \times b + M_B - M_C)/L$

$\qquad = (300 \times 0.9 + 600 \times 0.6 + 196.7 - 85.8) / 1.5 = 493.9 \text{ N}$

$R_{CB} = (\sum P \times b - M_B + M_C)/L$

$\qquad = (300 \times 0.6 + 600 \times 0.9 - 196.7 + 85.8) / 1.5 = 406.1 \text{ N}$

Alternatively, $R_{CB} = \sum P - R_{BC} = 406.1 \text{ N}$

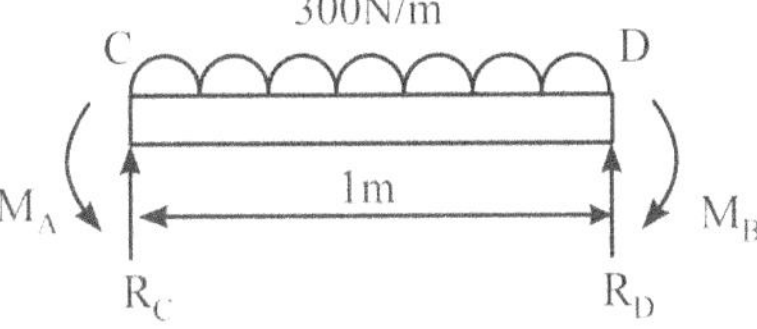

For the beam segment CD,

$R_{CD} = (w \times L^2/2 + M_C) / L$

$\qquad = (300 \times 1^2/2 + 85.8) / 1.0 = 235.8 \text{ N}$

$R_D = (w \times L^2/2 + M_C) / L = (300 \times 1^2/2 - 85.8) / 1.0 = 64.2 \text{ N}$

Alternatively, $R_D = w \times L - R_{CD} = 64.2 \text{ N}$

Net beam reactions at supports are

$R_A = 602.45 \text{ N} ; \qquad R_D = 64.2 \text{ N}$

$R_B = R_{BA} + R_{BC} = 597.55 \text{ N} + 493.9 \text{ N} = 1091.45 \text{ N}$

$R_C = R_{CB} + R_{CD} = 406.1 \text{N} + 235.8 \text{ N} = 641.9 \text{ N}$

Check :

$\sum F_Y = 600 \times 2 + 300 + 600 + 300 \times 1 - 602.45 - 1091.45 - 641.9 - 64.2 = 0$

11.13 BEAM OF UNIFORM STRESS

Dimensions of any uniform section beam are designed to withstand maximum bending stress, given by $\sigma = My/I = M/Z$, where $Z = I/y$ is called section modulus. The bending moment is a function of the nature of load and its location and is, in most applications, not constant over the length of the beam. So, considerable wastage of material can be observed in a long beam of uniform section. If the cross section of the beam is adjusted so as to produce uniform bending stress σ all along the beam length, it is called beam of uniform stress (also called ***inappropriately***, beam of uniform strength). The additional cost of making a beam of varying section may not always justify its use. However, in aeroplanes, where any saving in structure weight offsets additional manufacturing cost, beam of uniform stress is advantageous for long beams.

11.14　CURVED BEAM

Beam analysis by all the earlier methods are based on the assumption that axis of the beam before bending is straight or has negligible curvature. However, when the radius of curvature is of the same order of magnitude as the depth of beam section, the stress distribution differs considerably from that of the straight beam. The stresses on the concave side of the beam are higher than those for a similar straight beam and the stresses on the convex side are lower. When the maximum stress exceeds the elastic limit, local yielding occurs and causes redistribution of stress. Thus the beam curvature has the effect of reducing the yield strength but not of appreciably changing the ultimate bending strength.

In a beam of initial radius of curvature 'R' of the centroidal line (Ref Fig. 11.14), a plane cross section p-p remains plane after bending and its relative position after bending is indicated by n-n. A longitudinal fiber of initial length 'L' is extended by a distance 'δ'. Since δ is measured between straight lines p-p and n-n, it varies linearly with the distance 'y' from the centroidal axis. Then,
$$L = k \times (R + y) \quad \text{and} \quad \delta = k_1 + k_2 \times y$$

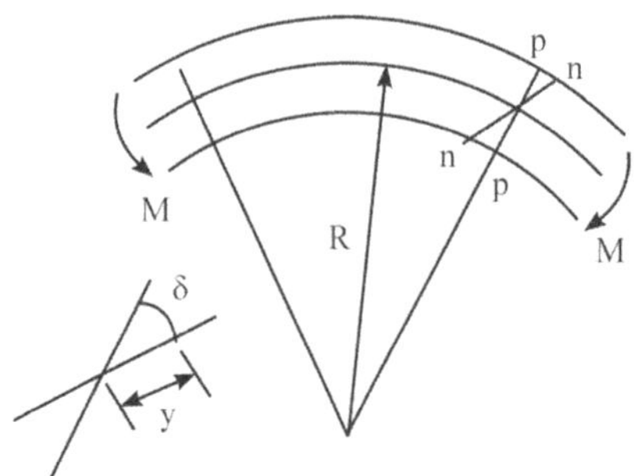

FIGURE 11.14 Bending of curved beam

Bending stress is then obtained from,
$$\sigma = E \times \varepsilon = E \times (\delta/L) = E \times (k_1 + k_2\, y) / [k \times (R + y)]$$
$$= a + b / (R + y) \qquad\qquad \text{.....(11.3)}$$
where a and b are unknown constants

Resultant force acting at the centroid of the area,

$$P = \int \sigma \, dA = a \int dA + b \int dA/(R + y) \qquad\qquad .....(11.4)$$

The external bending moment M about the centroidal axis must be equal to the moment of internal forces.

$$M = \int y \, dP = \int y \, (\sigma \, dA) = a \int y \, dA + b \int y \, dA/(R + y)$$

$$\int y \, dA = 0 \quad \text{when } y \text{ is measured from the centroidal axis}$$

$$\text{Therefore, } M = b \int y \, dA/(R + y) = b \, [\int \{ 1 - R/(R + y) \} \, dA \,]$$

$$= b \times A - b \times R \int dA/(R + y) \qquad\qquad ..... (11.5)$$

from which the unknown constant 'b' can be evaluated, using the known data of geometry and loading. The other unknown constant 'a' can be evaluated from eq (11.4). The effect of axial load P is to change the stress σ by an amount P/A. The same stress distribution can be obtained by superimposing the bending stresses for P = 0 and the stresses P/A resulting from the axial load P at the centroid of the area.

11.15 **DESIGNING A BEAM**

Analysis of any beam structure involves calculation of deflection, bending stress and shear stress at all points of each member for the applied shear force and bending moment. All these values are primarily dependent on the distribution of bending moment over the beam, flexural rigidity, shape of cross section and the end conditions.

Note: Every mathematical calculation has its limitations in terms of assumptions made in the relevant theory. While designing a beam based on beam theory explained above, care should be taken that the *assumptions – such as long beam, small deflection and no buckling in the region of compressive stress* – are applicable in the particular case.

Exercises for Practice

1. A simply supported beam of length 4 m carries inclined loads of 100 N, 200 N and 300 N inclined at 30^0, 45^0 and 60^0 to the right of vertical respectively. The loads act at 1 m, 2 m and 3 m from the left support respectively. Draw the shear force diagram and bending moment diagram for the beam, mentioning their values at salient points

2. A beam ABCD of length 8 m is simply supported at B and C. If the overhang portions AB and CD are 1.5 m each and the beam is loaded with udl of 900 N/m over a length of 3 m from A and udl of 300 N/m over the remaining length of the beam, draw the shear force diagram and

bending moment diagram for the beam, mentioning their values at salient points

3. A 50 mm diameter bar of length 2 m extends by 0.2 mm under a pull of 10 kN. Find the maximum deflection when it is used as a cantilever with point load of 1 kN at a distance of 0.5 m from the free end.

4. A beam ABC of 10 m length is simply supported at A and B and has overhanging portion BC such that AB = 8 m and BC = 2 m. The beam carries udl of 2 kN/m over a length of 4 m from A, two concentrated loads of 4 kN and 2 kN at a distance of 6 m from A and at the free end C respectively.

   (a) Find reactions at the supports A and B

   (b) Draw shear force diagram and bending moment diagram for the beam,   mentioning their values at salient points

   (c) Find the location of point of contra flexure

5. A simply supported beam ABCDE of length 4 m, with equal segments AB, BC, CD and overhang portion DE of 1 m each. Loads of 1 kN each are applied at B and at the free end E while a counter-clockwise moment of 3 kNm is applied at C. Draw the shear force diagram and bending moment diagram for the beam, mentioning their values at salient points

6. A timber beam of rectangular section is to be simply supported at the ends and to carry a load of 15 kN at the middle of 4.8 m spar. If the max.stress is not to exceed 1.2 kN/cm^2 and depth is to be twice the breadth, determine suitable dimensions of the beam section

7. A rectangular beam 100 mm width and 200 mm depth is simply supported over a span of 3.5 m and carries a udl of 1.5 kN/m over a length of 2.5 m from the left end support. Find (a) bending stress developed at a section 1m from the right end support and (b) position and magnitude of max.stress developed in the beam

8. A concentrated load of 'w' is placed at the middle of a simply supported beam of 6 m length. If the beam has I-section of 300 mm depth and $I = 15000$ cm^4, what is the max.value of 'w' so that deflection is limited to 5 mm and stress is limited to 5 kN/cm^2. Take E = 200 GPa

9. A beam 10 m long is simply supported at points 2 m from each end. It carries two point loads of 20 kN at each end. If E = 200 GPa and $I = 2500$ cm^4, calculate deflection at the mid-span. What load is to be applied at the mid-point of the beam, if the deflection is to be reduced by half?

10. A 3m long cantilever deflects at the free end by 0.5 mm under a load of 2 kN at the free end. What is the max.deflection produced and its location, if the cantilever is used as a simply supported beam with a load of 100 kN applied at a distance of 1.2 m from the left support

11. A cantilever 5 m long is propped at its free end and is loaded with udl of 10 kN/m over a length of 3 m from the fixed end. Determine (a) support reactions and (b) max.deflection and its location

12. Compare the bending strength of a rectangular section with depth 1.5 times the breadth and a circular section, both of the same area

13. What is the maximum bending stress induced in a steel flat of 1 m long, 100mm width and 10mm thick, if it is bent into a circular arc of 10m radius. Take $E = 2 \times 10^5 \, \text{N/cm}^2$

14. A simple beam of 10m span carries point loads of 100 kN and 60 kN at distances of 2 m and 5m from the left end. Determine the deflection under each load as well as max. deflection using Macaulay's approach. Take $E = 200$ GPa and $I = 18 \times 10^8 \, \text{mm}^4$.

Chapter **12**

ANALYSIS OF SHAFTS

12.1 SHAFT AS A CANTILEVER

Shaft is a long rotating cylinder which transmits motion and power, by the application of torque or torsional moment (couple) about its axis. ***Ideally, shafts are of circular cross section.*** When a shaft is acted upon by two equal and opposite couples at its two ends, the shaft will either be at rest or rotates with uniform r.p.m., depending on the end conditions. This rotation will tend to move one plane perpendicular to the axis w.r.t. the adjacent plane, thus introducing shear stresses between them. Shaft rotating at uniform r.p.m. can be treated, for the purpose of calculating stresses and twist, as a cantilever fixed at one end and free at the other end with torque applied at the free end. The design of a shaft for torque load closely resembles design of a beam for bending load.

Analysis of shaft involves shear stress (τ) and angle of twist (θ) similar to the bending stress (σ) and deflection (v) in the analysis of beams. The shaft design is based on modulus of rigidity 'G' and polar moment of inertia 'J' just as beam design is based on modulus of elasticity 'E' and moment of inertia 'I'.

12.2 SHEAR STRESS DISTRIBUTION

Similar to the beam equation ($M/I = \sigma/y = E/R$), we have torsion equation for the shaft relating geometric and material parameters to the resulting stress and twist as

$$T/J = \tau/r = G \times \theta/L \qquad\qquad\qquad ......(12.1)$$

where, $R_i \leq r \leq R_o$ is the particular radius of the shaft, with inner radius R_i and outer radius R_o, at which stress is to be calculated

$R_i = 0$ and $R_o = R$ for a solid shaft

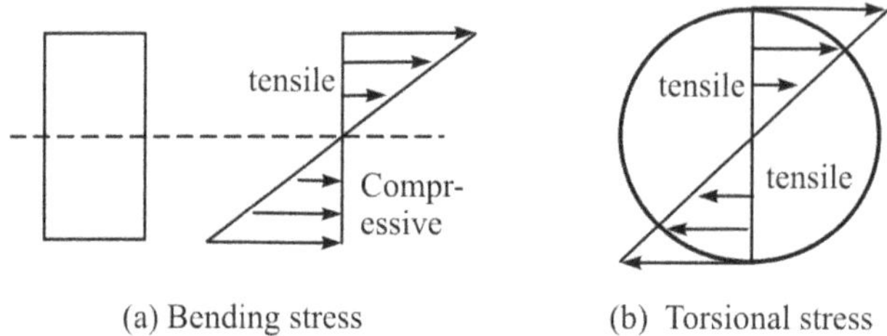

(a) Bending stress   (b) Torsional stress

FIGURE 12.1 Stress distribution in bending and torsion

Just as bending stress on each layer of beam parallel to neutral axis varies with the distance of the layer, with zero stress on the neutral axis, shear stress on each cross section of a solid shaft varies linearly with the radius of the layer, with zero stress on the axis (Ref. Fig. 12.1).

Similar to the beam equation, torque equation (12.1) is based on the following assumptions

- A plane cross section of the shaft remains plane, even after applying torque

- The maximum shear stress induced in the shaft is within the elastic limit of the material of the shaft

- The shaft is reasonably long, compared to its outer radius ($L > 40R_o$)

- The axial distance between any two cross sections of the shaft does not change due to application of torque. Non-circular sections of shaft do not satisfy this assumption. The resulting distortion of a cross sectional plane, called warping, is a function of maximum shear stress and may be neglected for shafts of small cross section subjected to low torque (such as square section shafts in clocks,...)

The major difference in the distribution of stresses is that ***bending stress is a normal stress (perpendicular to the cross section) and is tensile on one side of neutral axis and compressive on the other side*** while ***torsional stress is in the plane of the cross section and is constant at each point on a ring of same radius***. Hence, there is no stress reversal and all the rings tend to rotate in the same direction.

The torque equation (Ref Fig 12.2) is obtained from

$$\text{shear strain } \gamma = CC'/L = r \times \theta/L = \tau/G \qquad \text{and} \qquad CC'/R = BB'/r$$

or $\quad G \times \theta/L = \tau/r = \tau_{max}/R \qquad\qquad\qquad\qquad\qquad .....(12.2)$

and $\quad$ torque, $T = \int [\tau \times (2\pi \times r) \times dr] \times r = \int [(\tau_{max} \times r/R) \times (2\pi \times r) \times dr] \times r$

$$= \tau_{max} \times J/R \qquad \text{since} \quad J = \int A \times r^2 = \int [(2\pi \times r) \times dr] \times r^2$$

or $\quad T/J = \tau_{max}/R \qquad\qquad\qquad\qquad\qquad\qquad\qquad .....(12.3)$

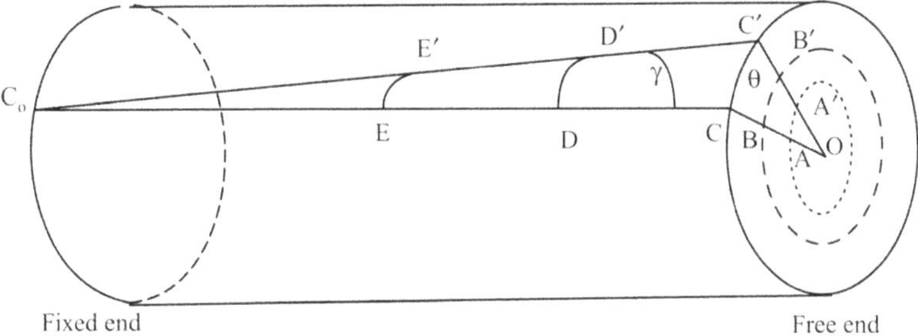

FIGURE 12.2 Torsional deformation of a shaft

It can be seen from the torsion equation that the rotational displacement or twist (E − E′, D − D′, C − C′, ..) increases linearly with length, L, from the fixed end to the free end, in the direction of applied torque. At a particular section, even though actual twist (A–A′, B–B′, C–C′,..) increases linearly with radius, from the axis of the shaft, the angle of twist ($\angle$ AOA′ = $\angle$ BOB′ = $\angle$ COC′ =.. θ) is constant from the axis to the outermost radius at any section. Shear strain, γ, is defined as the rotary displacement per unit length given by the ratio C C′/L or R × θ/L. Thus, large diameter and long shaft will have larger twist associated with a given torque and, hence, higher stress. Any one of the two ends of the shaft can be considered as the fixed end of equivalent cantilever. Depending on the fixed end, direction of torque, twist and shear stress may change but the magnitudes remain the same. Hence, ***design of shaft for torque load is independent of the fixed end selected***.

Polar moment of inertia for a solid shaft of diameter 'd', $J = \pi \times d^4/32$

For a hollow shaft of outer diameter 'D' and inner diameter 'd',

$$J = \pi \times (D^4 - d^4) / 32$$

Similar to modulus of section in bending ($Z = I/R_{max}$) which relates maximum bending stress in a round bar of radius R to the applied bending moment ($\sigma = M/Z$), polar modulus ($Z_P = J/R_{max}$) relates maximum shear stress at a section in a shaft to the applied torque T, by $\tau = T/Z_P$. Shear stress distribution in a solid shaft and a hollow shaft are shown in Fig 12.3.

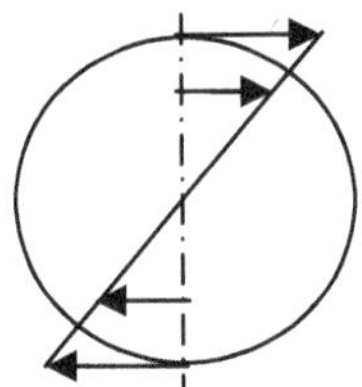

(a) Shear stress in a solid shaft

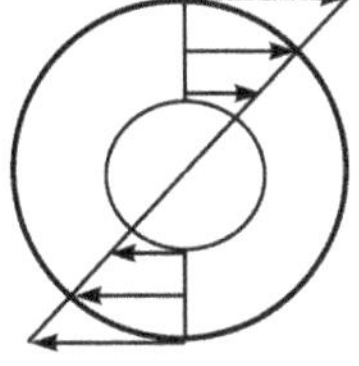

(b) Shear stress in a hollow shaft

FIGURE 12.3 Shear stress distribution due to torque

For a solid shaft, $Z_P = (\pi \times d^4/32) / (d/2) = \pi\, d^3/16$

For a hollow shaft, $Z_P = [\pi \times (D^4 - d^4)/32]/(D/2) = \pi \times (D^4 - d^4) / (16D)$

$$.....(12.4)$$

12.3 POWER TRANSMITTED BY A ROTATING SHAFT

Torque load on a static cylinder is directly specified. But, torque on a rotating shaft may not be directly known but related to the power transmitted by the shaft through

Power = Torque $\times$ angular velocity

$$= T \times \omega = 2\pi \times n \times T/60 \ \ \text{Watts,} \quad \text{if torque, T is in Nm}$$

where, $\omega = 2\pi \times n/60$ is the angular velocity in rad/sec

and   n = speed of shaft in revolutions per minute (rpm)

Power rating of any equipment is usually specified w.r.t. the average torque, whereas the maximum torque (such as while starting a motor) is usually higher. Design of the shaft should be based on the maximum torque.

Example 12.1

The maximum allowable shear stress of a 100 mm diameter and 2 m long solid shaft is 50 N/mm^2. Determine the maximum torque that the shaft can transmit without failure and the corresponding maximum twist. Also calculate power transmitted at 300 rpm, if the average torque is 46% of maximum torque. Design an equivalent hollow shaft with D =120 mm and compare their weights and torsional rigidities. Assume G = 80 GPa

Solution

For the solid shaft of diameter D_S = 100mm or radius R_S = 50mm,

Maximum Torque transmitted, $T_{max} = Z_P \times \tau = (\pi \times D_S^3/16) \times \tau = 9.82$ kNm

Average torque, $\qquad T_{ave} = 0.46\, T_{max} = 4.5172$ kNm

Power transmitted, $\qquad P = 2\pi \times n \times T_{ave}/60 = 141.97$ kW

Maximum twist, $\qquad \varphi = R_S \times \theta = R_S \times [\, T_{max} \times L / (G \times J)\,]$

$$= (T_{max} \times L \times R_S/G) / (\pi \times D_S^4/32) = 1.25 \text{ mm}$$

For the hollow shaft of outer dia D_O =120mm or R_O = 60mm and inner dia 'd',

Maximum Torque transmitted, $T_{max} = Z_P \times \tau = [\, \pi \times (D_O^4 - d^4) / (16D_O)\,] \times \tau$

$$= 9.82 \text{ kNm} \ \ (\text{max torque of solid shaft})$$

$\Rightarrow \qquad d = 96.7$ mm

Maximum twist, $\qquad \varphi = R_O \times \theta = R_O \times [\, T_{max} \times L / (G \times J)\,]$

$$= R_O \times T_{max} \times L / [G \times \pi \times (D_O^4 - d^4)/32\,]$$

$$= 1.04 \text{ mm}$$

Wt of solid shaft / wt of hollow shaft $= (\pi \times D^2 \times L \times \rho/4) / [\pi \times (D^2 - d^2) \times L \times \rho/4]$

$$= D_S^2 / (D^2 - d^2) = 100^2 / (120^2 - 96.7^2) \approx 1.98$$

Torsional rigidity of solid shaft / torsional rigidity of hollow shaft

$$= G \times J_S / (G \times J_H) = (G \times \pi \times D_S^4/32) / [G \times \pi \times (D^4 - d^4)/32]$$

$$= D_S^4/(D^4 - d^4) = 0.834$$

Example 12.2

A stepped shaft ABC having portion AB of 4 m long and 50 mm diameter and portion BC of 2 m long and 20 mm diameter is subjected to a torque of 200 Nm at the free end C. Determine maximum angle of twist, if $G = 8 \times 10^6$ N/cm^2. Compare the results if torques of 100 Nm are applied at B and C; and torques of +200 Nm at C and −100 Nm at B are applied.

Solution

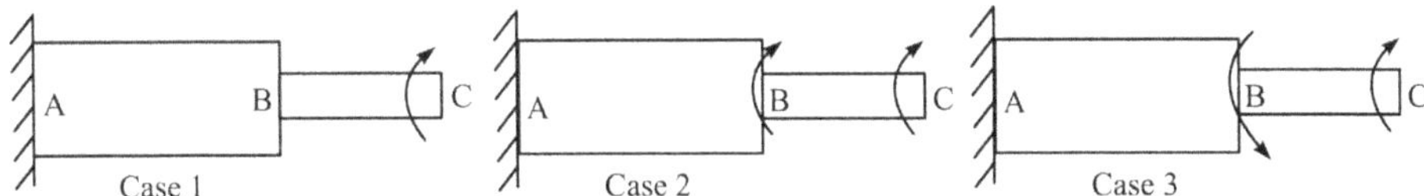

Case-1

Even though torque remains same throughout the stepped shaft, twist in the two portions AB and BC is different due to change in polar Moment of Inertia J

$$
\begin{aligned}
\text{Angle of twist at } C &= \text{Angle of twist in AB} + \text{Angle of twist in BC} \\
&= T \times L_{AB} / (G \times J_{AB}) + T \times L_{BC} / (G \times J_{BC}) \\
&= (T/G) \times [L_{AB}/J_{AB} + L_{BC}/J_{BC}] \\
&= [200/(8 \times 10^{10})] \times [4/(\pi \times 0.05^4/32) + 2/(\pi \times 0.02^4/32)] \\
&= 0.3345 \text{ radians} \quad \text{or} \quad 19.16^0
\end{aligned}
$$

Case-2

Torque in BC, $T_{BC} = 100$ Nm   while the torque in AB, $T_{AB} = 100 + 100$ Nm

$$
\begin{aligned}
\text{Angle of twist at } C &= \text{Angle of twist in AB} + \text{Angle of twist in BC} \\
&= T_{AB} \times L_{AB} / (G \times J_{AB}) + T_{BC} \times L_{BC} / (G \times J_{BC}) \\
&= [T_{AB} \times L_{AB} / J_{AB} + T_{BC} \times L_{BC} / J_{BC}] / G \\
&= [4 \times 200 / (\pi \times 0.05^4/32) \\
&\qquad\qquad + 2 \times 100 / (\pi \times 0.02^4/32)] / (8 \times 10^{10}) \\
&= 0.1754 \text{ radians} \quad \text{or} \quad 10.045^0
\end{aligned}
$$

Case-3

Torque in BC, $T_{BC} = 200$ Nm   while the torque in AB, $T_{AB} = +200 - 100$ Nm

$$
\begin{aligned}
\text{Angle of twist at} \quad C &= \text{Angle of twist in AB} + \text{Angle of twist in BC} \\
&= T_{AB} \times L_{AB} / (G \times J_{AB}) + T_{BC} \times L_{BC} / (G \times J_{BC}) \\
&= [T_{AB} \times L_{AB} / J_{AB} + T_{BC} \times L_{BC} / J_{BC}] / G \\
&= [-4 \times 100 / (\pi \times 0.05^4/32) \\
&\qquad\qquad + 2 \times 200 / (\pi \times 0.02^4/32)] / (8 \times 10^{10}) \\
&= 0.310 \text{ radians} \quad \text{or} \quad 17.756^0
\end{aligned}
$$

12.4 SHAFTS OF COMPOSITE SECTIONS

In some cases hollow shaft of some material may enclose a solid shaft of a different material for some specific advantage like wear resistance, protection from corrosive or chemically reactive environment etc (Fig. 12.4). Assuming there is no slip between the two shafts, the angle of twist remains same in the two shafts while the torque transmitted by each shaft varies depending on rigidity modulus 'G' and polar moment of inertia 'J'. The total torque transmitted by the composite shaft is the sum of torques transmitted by the two shafts

$$
\begin{aligned}
\text{Thus,} \quad T_{total} &= T_{solid} + T_{hollow} \\
&= (G \times J \times \theta / L)_{solid} + (G \times J \times \theta / L)_{hollow} \\
&= (\theta/L) \times [(G \times J)_{solid} + (G \times J)_{solid}]
\end{aligned}
$$

since θ and L are same for both the shafts

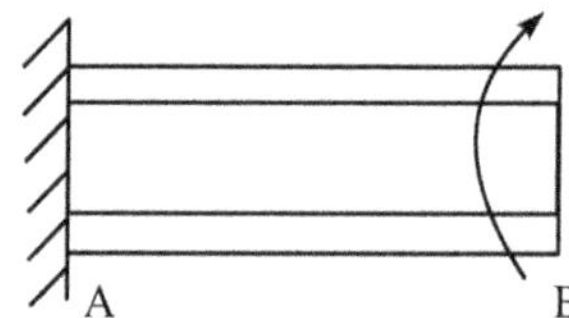

FIGURE 12.4 Composite shaft

12.5 SHAFTS IN SERIES AND IN PARALLEL

When shafts are connected such that torque applied at the free end remains same throughout the shaft while the maximum angle of twist at the free end equals sum of angles of twist in different parts from the fixed end to the free end

(Fig. 12.5a), they are said to be in *series*. Different parts of the shaft may be made of different materials or different sizes

Thus, for shafts in series $\theta_{Max} = \sum\theta_i$ and $T = T_i$ ....(12.5)

In the example of Fig 12.5 (a), θ_C or $\theta_{Max} = \theta_{AB} + \theta_{BC}$ and $T = T_{AB} = T_{BC}$

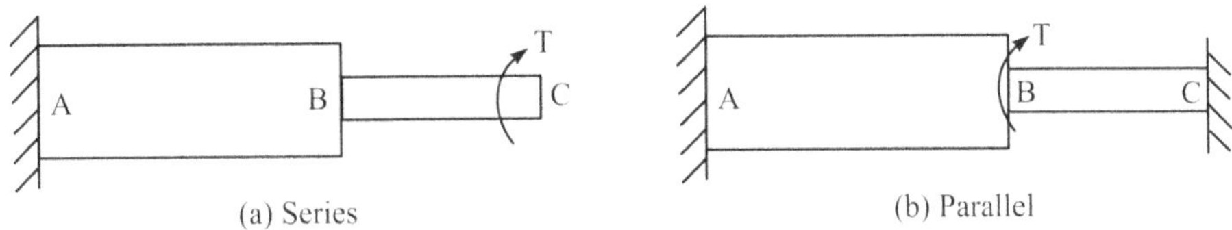

FIGURE 12.5 Shafts in series and in parallel

When shafts are connected such that their angular displacements are same throughout the shaft while the total torque applied equals sum of the torques shared by different parts (Ref Fig. 12.5b), they are said to be in *parallel*. Different parts of the shaft may be made of different materials or different sizes.

Thus, for shafts in parallel, $\theta_{Max} = \theta_i$ and $T = \sum T_i$ .....(12.6)

In the example of Fig 12.5 (b), $\theta_B = \theta_{AB} = \theta_{BC}$ and $T = T_{AB} + T_{BC}$

12.6 TORSIONAL STRAIN ENERGY

For any gradually applied load, Strain energy is defined as

$$U = (1/2) \int (\text{Stress} \times \text{strain}) \times dV$$

For a solid shaft, strain energy due to torque load

$$U = (1/2) \times \int (\tau \times \gamma)\, dV = [1/(2 \times G)] \times \int \tau^2\, dV$$
$$= [1/(2 \times G)] \times \int (\tau_{max} \times r / R)^2 \times (2\pi \times r \times L\, dr) \qquad \text{for } 0 \leq r \leq R$$
$$= [\pi \times L \times \tau_{max}^2 / (G \times R^2)] \times \int r^3\, dr$$
$$= [\pi \times L \times \tau_{max}^2 / (G \times R^2)] \times (R^4 / 4)$$
$$= \tau_{max}^2 \times V / (4G) \qquad \text{since, volume of shaft } V = \pi \times R^2 \times L$$

For a hollow circular shaft, strain energy due to torque load

$$U = (1/2) \times \int (\tau \times \gamma)\, dV = [1/(2 \times G)] \times \int \tau^2\, dV$$
$$= [1/(2 \times G)] \times \int (\tau_{max} \times r / R)^2 \times (2\pi \times r \times L\, dr) \qquad \text{for } R_i \leq r \leq R_o$$
$$= [\pi \times L \times \tau_{max}^2 / (G \times R^2)] \times \int r^3\, dr$$
$$= [\pi \times L \times \tau_{max}^2 / (G \times R^2)] \times (R_o^4 - R_i^4) / 4$$
$$= [\tau_{max}^2 \times V / (4G)] \times (R_o^2 + R_i^2) / R_o^2$$
$$\text{since, volume of shaft } V = \pi \times (R_o^2 - R_i^2) \times L$$

12.7 COMBINED BENDING AND TWISTING

Most shafts are also subjected to bending due to local load (belt drive, gear wheel, etc..) or distributed load (armature of a motor, blade rows of a turbine, self weight of shaft, etc..) on the shaft. Even though bending stress is usually less, bending deflection demands more clearance between the rotating shaft and the casing, reducing functional utility of the component. For large torque applications, hollow shaft is preferred so as to increase moment of inertia without corresponding increase in area of cross section and self weight, thereby reducing maximum shear stress and maximum bending deflection in the shaft.

Design of a shaft is based on bending behaviour (bending stress and deflection) as well as torsional behaviour (shear stress and twist). Thus, the design is dependent on flexural rigidity 'EI' as well as torsional rigidity 'GJ'.

Bending results in stress normal to the cross section while torsion results in shear stress in the plane of cross section. In order to design a shaft for the combined effect of bending and torsion, these two stresses need to be combined as per Mohr's circle to get maximum principal stress or maximum shear stress for use with appropriate theory of failure. For a solid shaft,

Max. principal stress,

$$\sigma_{max} = \sigma/2 + \sqrt{(\sigma/2)^2 + \tau^2}$$

$$= (1/2) \times [\,32M / (\pi \times d^3)\,] + \sqrt{(1/4) \times \left[32M/(\pi \times d^3)\right]^2 + \left[16T/(\pi \times d^3)\right]^2}$$

$$= \left[16T/(\pi \times d^3)\right] \times \left[M + \sqrt{M^2 + T^2}\right]$$

Max. shear stress, $\tau_{max} = \sqrt{(\sigma/2)^2 + \tau^2} = [\,16/(\pi \times d^3)\,] \times \sqrt{M^2 + T^2}$

Example 12.3

A hollow shaft having outer diameter twice the inner diameter is subjected to a torque of 4000 Nm and a bending moment of 3000 Nm. Determine the outer and inner diameters of the shaft if the maximum permissible shear stress is 8.2 kN/cm^2

Solution

For a hollow shaft with $D = 2d$,   Max. shear stress,

$$\tau_{max} = \sqrt{(\sigma/2)^2 + \tau^2}$$

$$= \sqrt{(1/4)\times\left[32M\times D / \left\{\pi\times(D^4 - d^4)\right\}\right]^2 + \left[16T\times D / \left\{\pi\times(D^4 - d^4)\right\}\right]^2}$$

$$= \sqrt{\left[16M\times(2d) / \left\{\pi(16d^4 - d^4)\right\}\right]^2 + \left[16T\times(2d) / \left\{\pi\times(16d^4 - d^4)\right\}\right]^2}$$

$$= [32 / (15\,\pi\,d^3)] \times \sqrt{M^2 + T^2}$$

or $d^3 = [32 / (15\,\pi\,\tau_{max})] \times \sqrt{M^2 + T^2} = 41.39$ cm^3

Hence, $d = 3.46$ cm   and $D = 6.92$ cm

12.8 WARPING OF NON-CIRCULAR SECTIONS DUE TO PURE TORSION

A bar of circular cross section, subjected to pure torsion, is analysed using the assumption that plane section before applying torsion remains plane after applying the load. i.e, *In a bar of circular section, there is no warping* of the section or all points on the cross section will have either zero or equal displacements along the axis of the bar. The same is *not* true with a bar of non-circular cross section. Displacements at a point $P(y,z)$ of a bar of non-circular cross section on a section at distance x from one end, subjected to torsion, are defined by

$$u = -\theta \times z \times y; \quad v = \theta \times x \times z \quad \text{and} \quad w = \theta \times \psi(x,y)$$

where $\psi(x,y)$ is the movement of cross section in the axial direction (Z-axis) per unit twist, called *warping function*

and θ is the angle of twist per unit length

For pure torsion, $\varepsilon_X = \varepsilon_Y = \varepsilon_Z = \gamma_{XY} = 0$

$$\gamma_{ZX} = \partial w/\partial x + \partial u/\partial z = \partial w/\partial x - \theta \times y$$

$$\gamma_{YZ} = \partial w/\partial y + \partial v/\partial z = \partial w/\partial y + \theta \times x$$

From Hooke's law, corresponding stress components

$$\sigma_X = \sigma_Y = \sigma_Z = \tau_{XY} = 0$$

$$\tau_{ZX} = G \times \gamma_{ZX} = G \times [\, \partial w/\partial x - \theta \times y \,]$$

and $\tau_{YZ} = G \times \gamma_{YZ} = G \times [\, \partial w/\partial y + \theta \times x \,]$

The general equation of motion for a static body without any body force is

$$\partial \tau_{ZX}/\partial x + \partial \tau_{YZ}/\partial y = 0$$

Substituting for the shear stresses in the equation of motion yields Laplace's equation $\partial^2 w/\partial x^2 + \partial^2 w/\partial y^2 = 0$

Using $\psi(x,y)$, we get $\quad \partial^2 \psi / \partial x^2 + \partial^2 \psi / \partial y^2 = \nabla^2 \psi = 0$

The stress-free condition of the periphery of the cross section requires that the tangential stress be zero on the lateral surface of the prism.

Alternatively, a **stress function** $\varphi(y,z)$ can be specified over each element (satisfying equilibrium conditions) and warping function $\psi(y,z)$ specified along inter element boundary (satisfying compatibility conditions). This model gives more accurate stress solutions, particularly in the vicinity of external boundary. The stress components, in terms of the stress function, are given by

$$\tau_{ZX} = \partial \varphi/\partial y \; ; \qquad \tau_{XY} = \partial \varphi/\partial z$$

Complete restraint of warping of a member, of non-circular cross section and subjected to torque, reduces rotation at the end but introduces axial normal stresses much larger than the torsional shear stress.

A few examples are given here for better understanding of the concepts.

Example 12.4

Show that $\varphi = k \times (r^2 - a^2)$, where k is an unknown constant, is a valid stress function for a uniform solid bar of radius 'a', subjected to torsion T.

Solution

Initially the stress function, φ, must be expressed in terms of Cartesian coordinates. Thus, from the equation of a circle of radius, a, and having the origin of its axes at its centre $\quad \varphi = k \times (z^2 + y^2 - a^2)$

Compatibility conditions, in terms of shape functions, are given by

$$\frac{\partial}{\partial y}\left(-\frac{\partial \gamma_{ZX}}{\partial y} + \frac{\partial \gamma_{XY}}{\partial z}\right) = -\frac{\partial}{\partial y}\left(\frac{\partial^2 \phi}{\partial y^2} + \frac{\partial^2 \phi}{\partial z^2}\right) = 0 \implies \frac{\partial^2 \phi}{\partial y^2} + \frac{\partial^2 \phi}{\partial z^2} = \text{constant}$$

Compatibility condition is satisfied with $\varphi = k \times (r^2 - a^2)$, since $\dfrac{\partial^2 \phi}{\partial y^2} + \dfrac{\partial^2 \phi}{\partial z^2} = 4k$

$$\frac{\partial^2 \phi}{\partial y^2} + \frac{\partial^2 \phi}{\partial z^2} = \frac{\partial}{\partial y}\left(\frac{\partial \phi}{\partial y}\right) - \frac{\partial}{\partial z}\left(-\frac{\partial \phi}{\partial z}\right) = \frac{\partial \tau_{XZ}}{\partial y} - \frac{\partial \tau_{XY}}{\partial z}$$

$$= G\frac{\partial\gamma_{xz}}{\partial y} - G\frac{\partial\gamma_{xy}}{\partial z} = G\left[\frac{\partial}{\partial y}\left(\frac{\partial u}{\partial z} + \frac{\partial w}{\partial x}\right) - \frac{\partial}{\partial z}\left(\frac{\partial u}{\partial y} + \frac{\partial v}{\partial x}\right)\right]$$

$$= G\left[\frac{\partial^2 w}{\partial x \partial y} - \frac{\partial^2 v}{\partial x \partial z}\right] = G(-\theta - \theta) \quad \text{since} \quad v = \theta \times x \times z \,;\quad w = -\theta \times x \times y$$

$$\Rightarrow \quad 4\,k = -\,2\,G \times \theta \qquad\qquad \text{or} \qquad k = -\,G \times \theta\,/\,2$$

The torsional moment is related to the stress function by

$$T = 2\iint \phi(dydz) = -G\theta\left[\iint z^2 dydz + \iint y^2 dydz - a^2 \iint dydz\right]$$

The first and second integrals in this equation are the second moments of area $I_{yy} = \pi \times a^4/4$ and $I_{zz} = \pi \times a^4/4$, while the third integral is the area of the cross-section $A = \pi \times a^2$. Replacing the integrals by these values gives

$$T = G \times \theta \times (\pi \times a^4 / 2)$$

$$\text{or} \quad \theta = T/ (G \times \pi \times a^4/2) = T/(G \times J)$$

$$\text{where, } J = \pi \times a^4/2 \text{ is the polar moment of inertia}$$

$$\tau_{xy} = -\,(\partial\varphi/\partial z) = -\,2k \times z = G \times \theta \times z = (T/J) \times z$$

$$\text{and} \quad \tau_{xz} = \partial\varphi/\partial y = 2k \times y = -\,G \times \theta \times y = -\,(T/J) \times y$$

$$\text{Then, } \tau_{xs} = \tau_{xy} \times l - \tau_{xz} \times m = (T/J) \times (z \times l + y \times m)$$

$$\text{where } l \text{ and } m \text{ are direction cosines}$$

The direction cosines are also related as $\quad l = dy/ds \quad$ and $\quad m = -\,(dz/ds)$

If the radius makes an angle α with the Y-axis,

$$\text{then} \quad l = \cos\alpha \quad \text{and} \quad m = \sin\alpha$$

Also, at any radius r, $\quad y = r \times \sin\alpha \quad$ and $\quad z = r \times \cos\alpha$

Substituting these values for x, l, y and m, we get

$$\tau \text{ or } \tau_{xs} = (T/J) \times (r \times \cos^2\alpha + r \times \sin^2\alpha) = (T/J) \times r$$

Substituting for τ_{zx}, τ_{xy} and $d\theta/dx$ in Eqs (3.10), we get

$$\partial u/\partial z = \tau_{xz}/G + (d\theta/dx) \times y = -\,[T \times y \,/\,(G \times J)] + [T\,/\,(G \times J)] \times y = 0$$

$$\text{and} \quad \partial u/\partial y = \tau_{xy}/G - (d\theta/dx) \times x = [T \times x \,/\,(G \times J)] - [T\,/\,(G \times J)] \times x = 0$$

The possible solutions of these two equations are u = 0 and u = constant. The latter solution implies a displacement of the whole bar along the X-axis which, for a torsional moment, cannot occur. Therefore, the first solution applies, i.e. ***the warping is zero at all points in the cross-section***.

Example 12.5

A uniform bar has the elliptical cross-section, with semi-major axis and semi-minor axis of a and b respectively, and is subjected to equal and opposite torques T at each of its free ends. Derive expressions for the rate of twist in the bar, the shear stress distribution and the warping displacement of its cross-section.

Solution

The semi-major and semi-minor axes are a and b, respectively, so that the equation of its boundary is $z^2/a^2 + y^2/b^2 = 1$

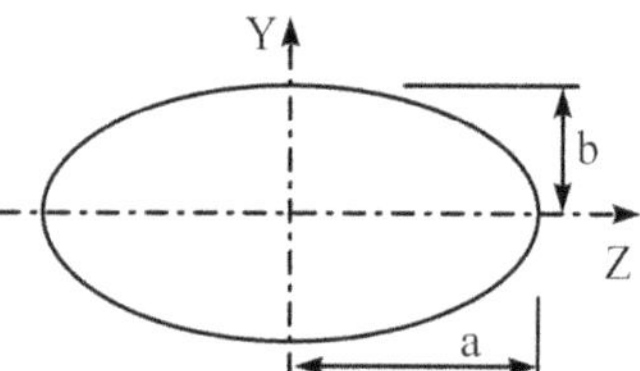

If we choose a stress function of the form $\phi = C\left(\dfrac{z^2}{a^2} + \dfrac{y^2}{b^2} - 1\right)$ then the

boundary condition $\varphi = 0$ is satisfied at every point on the boundary and the constant C may be chosen to fulfill the remaining requirement of compatibility. Thus,

$$\frac{\partial^2 \phi}{\partial y^2} + \frac{\partial^2 \phi}{\partial z^2} = 2C\left(\frac{1}{a^2} + \frac{1}{b^2}\right) = -2G\frac{d\theta}{dx}$$

or

$$C = -G\frac{d\theta}{dx}\frac{a^2 b^2}{(a^2 + b^2)}$$

giving

$$\phi = -G\frac{d\theta}{dx}\frac{a^2 b^2}{(a^2 + b^2)}\left(\frac{z^2}{a^2} + \frac{y^2}{b^2} - 1\right)$$

Substituting this expression for φ in the relationship between torque T and the rate of twist, we get

$$T = \iint 2\phi(dydz) = -2G\frac{d\theta}{dx}\frac{a^2 b^2}{(a^2 + b^2)}\left(\frac{1}{a^2}\iint z^2 dydz + \frac{1}{b^2}\iint y^2 dydz - \iint dydz\right)$$

The first and second integrals in this equation are the second moments of area $I_{yy} = \pi a^3 b/4$ and $I_{zz} = \pi a b^3/4$, while the third integral is the area of the cross-section $A = \pi ab$. Replacing the integrals by these values gives

$$T = G\frac{d\theta}{dx}\frac{\pi a^3 b^3}{(a^2 + b^2)}$$

Since, $T = GJ \dfrac{d\theta}{dx}$, we get $J = \dfrac{\pi \times a^3 \times b^3}{a^2 + b^2}$

The shear stress distribution is obtained in terms of the torque by substituting for the product $G\ (d\theta/dx)$ in the equation for φ and then differentiating φ, as

$$\tau_{ZX} = \frac{\partial \phi}{\partial y} = \frac{2T}{\pi ab^3}\, y \quad \text{and} \quad \tau_{XY} = -\frac{\partial \phi}{\partial z} = -\frac{2T}{\pi a^3 b}\, z$$

Warping distribution u over the cross-section is obtained by substituting for τ_{zx}, τ_{xy} and $d\theta/dz$, as

$$\frac{\partial u}{\partial z} = -\frac{2T}{\pi ab^3 G}\, y + \frac{T(a^2 + b^2)}{\pi a^3 b^3 G}\, y = -\frac{T(a^2 - b^2)}{\pi a^3 b^3 G}\, y$$

$$\frac{\partial u}{\partial y} = \frac{2T}{\pi a^3 bG}\, z - \frac{T(a^2 + b^2)}{\pi a^3 b^3 G}\, z = -\frac{T(a^2 - b^2)}{\pi a^3 b^3 G}\, z$$

Integrating these two equations, we get

$$u = -\frac{T(a^2 - b^2)}{\pi a^3 b^3 G}\, yz + f_1(y) \quad \text{and} \quad u = -\frac{T(a^2 - b^2)}{\pi a^3 b^3 G}\, yz + f_2(z)$$

The warping displacement given by each of these equations must have the same value at identical points (z, y). It follows that $f_1(y) = f_2(z) = 0$. Hence

$$u = -\frac{T(a^2 - b^2)}{\pi a^3 b^3 G}\, yz$$

Lines of constant u therefore describe hyperbolas with the major and minor axes of the elliptical cross-section as asymptotes. Further, for a positive (anticlockwise) torque the warping is negative in the first and third quadrants ($a > b$) and positive in the second and fourth.

Example 12.6

Show that warping function $\psi = k \times y \times z$, where k is an unknown constant, can be used to solve torsion of a uniform bar of elliptical section.

Solution

With $\psi = k \times y \times z$, the equilibrium equation $\dfrac{\partial^2 \psi}{\partial y^2} + \dfrac{\partial^2 \psi}{\partial z^2} = 0$ is satisfied.

Substituting for ψ in the summation of forces in X-direction, we get

$$\tau_{xy} \times l - \tau_{xz} \times m = [(\partial\psi/\partial z) - y] \times l + [(\partial\psi/\partial z) + z] \times m$$

$$= (k{\times}y - y) \times (dz/ds) + (k \times z + z) \times (dy/ds)$$

$$= -z(k+1)\frac{dz}{ds} + y(k-1)\frac{dy}{ds} = 0$$

or $$\frac{d}{ds}\left[-\frac{z^2}{2}(k+1) + \frac{y^2}{2}(k-1)\right] = 0$$

so that $$-\frac{z^2}{2}(k+1) + \frac{y^2}{2}(k-1) = \text{constant}, \quad \text{on the boundary of the bar}$$

Rearranging, $$z^2 + \frac{1-k}{1+k}y^2 = \text{constant}$$

Also, equation of the elliptical boundary of the bar is $$\frac{z^2}{a^2} + \frac{y^2}{b^2} = 1$$

or $$z^2 + \frac{a^2}{b^2}y^2 = a^2$$

Comparing the two equations, $$\frac{a^2}{b^2} = \frac{1-k}{1+k}$$

from which, we get $$k = \frac{b^2 - a^2}{b^2 + a^2} \quad \text{and} \quad \psi = \frac{b^2 - a^2}{b^2 + a^2}yz$$

The torsion constant, J, can then be obtained as

$$J = \iint\left[\left(\frac{b^2 - a^2}{b^2 + a^2} + 1\right)z^2 - \left(\frac{b^2 - a^2}{b^2 + a^2} - 1\right)y^2\right]dydz$$

$$= \iint\left[\left(\frac{2b^2}{b^2 + a^3}\right)z^2 + \left(\frac{2a^2}{b^2 + a^3}\right)y^2\right]dy\,dz$$

For an elliptical cross-section,

$$\iint z^2 dydz = I_{YY} = \frac{\pi a^3 b}{4} \quad \text{and} \quad \iint y^2 dydz = I_{ZZ} = \frac{\pi ab^3}{4}$$

Therefore, torsion constant simplifies to $$J = \frac{\pi \times a^3 \times b^3}{a^2 + b^2}$$

The rate of twist is given by $$\theta = \frac{T(a^2 + b^2)}{G\pi a^3 b^3}$$

The shear stresses are obtained as

$$\tau_{ZX} = \frac{GT(a^2 + b^2)}{G\pi a^3 b^3}\left[\frac{b^2 - a^2}{b^2 + a^2}y - y\right] = -\frac{2Ty}{\pi ab^3}$$

$$\text{and}\quad \tau_{XY} = \frac{GT(a^2 + b^2)}{G\pi a^3 b^3}\left[\frac{b^2 - a^2}{b^2 + a^2}z + z\right] = \frac{2Tz}{\pi a^3 b}$$

$$u = \theta \times \psi(y, z) = \frac{T(a^2 + b^2)}{G\pi a^3 b^3}\left[\frac{b^2 - a^2}{b^2 + a^2}\right]yz = \frac{T(b^2 - a^2)}{G\pi a^3 b^3}yz$$

Thus, warping varies at different points throughout the section, based on the coordinates (y,z) of the points.

A few problems are given here for practice

Problem-1

Determine the torque T to produce an angle of twist of 0.12 radians over a span of 2 m, for the cross section shown in figure. Assume uniform thickness of 5mm. Also determine average shear stress in the member and strain energy stored. All dimensions, shown in figure, are in mm.

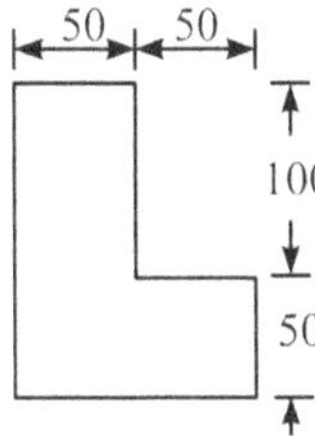

Problem-2

A thin tube is made from 5 mm thick steel plate such that it has cross section of equilateral triangle of side 90 mm. Determine maximum torque T to which it can be subjected, if the allowable shear stress is 90 MPa and the tube is restricted to twist no more than 2×10^{-3} radians. Length of the tube is 3 m. Take $G = 75 \times 10^9$ N/m^2.

Problem-3

A structural aluminium tubing of 60 × 100 mm rectangular cross section with uniform thickness of 4mm was fabricated by extrusion. Determine the shearing stress in each of the four walls of such tubing when it is subjected to a torque of 2700 Nm.

12.9   WARPING OF CLOSED SECTION THIN-WALLED (HOLLOW) BEAMS DUE TO SHEAR AND TORSION

A small element $\delta s \times \delta x \times t$ of the beam wall is maintained in equilibrium by a system of direct and shear stresses as shown in Fig. 12.6. Let the parameter 's' in the analysis be the distance measured around the cross section from some convenient origin and 'r' be the radius of curvature for the small element. Thickness 't' is assumed constant over a small length 'ds'. Direct stress σ_X is produced by the bending action of shear loads while the shear stresses τ_{XS} and τ_{SX} are due to shear and/or torsion. For moment equilibrium of the element, $\tau_{XS} = \tau_{SX} = \tau$. However, we use shear flow 'q' (shear force per unit length or $q = \tau \times t$) in our analysis with positive sign in the direction of 'S'.

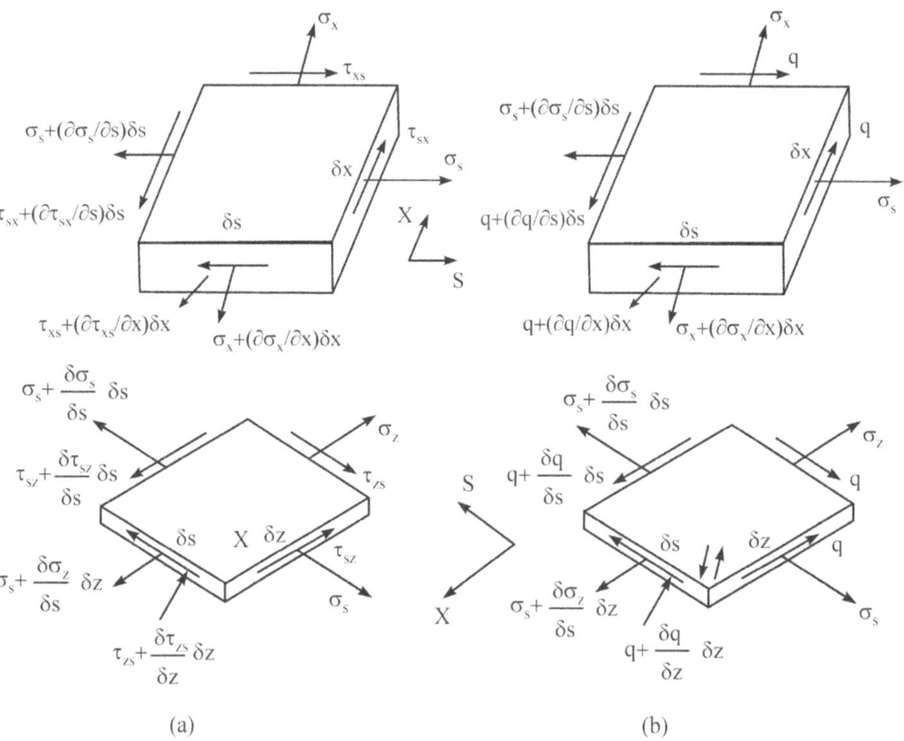

FIGURE 12.6 Shear stress and shear flow in a thin walled beam section

For equilibrium of the element in 'X' and 'S' directions and neglecting body forces,

$$\partial q/\partial s + t \times \partial \sigma_X/\partial x = 0 \quad \text{and} \quad \partial q/\partial x + t \times \partial \sigma_S/\partial s = 0$$

With displacements u and v along X and S directions on the element and w along the normal to the surface direction, corresponding strains are

$$\varepsilon_X = \partial u/\partial x \; ; \quad \varepsilon_S = \partial v/\partial s + u/r \quad \text{and} \quad \gamma = \partial u/\partial s + \partial v/\partial x$$

The shear flow q can now be expressed as

$$q = \tau \times t = G \times \gamma \times t = G \times t \times [\; \partial u/\partial s + \partial v/\partial x\;]$$

Warping distribution around the cross section from a reference value u_0 at $s = 0$ is

$$u_s - u_0 = \int (\partial u/\partial s)\,ds = \int (q/Gt)\,ds - \int (\partial v/\partial x)\,ds$$
$$= \int (q/Gt)\,ds - (A_O/A)\oint (q/Gt)ds$$

where A_O is the area between an arbitrary point 'o' in the section and segment of the surface from 'o' to 's' and A is the total area of cross section

Here, the second integral is performed along the contour of the surface.

In this equation, variation of G and t along the surface can be considered in the integration. However, if G and t are constant along the surface, equation for the warping distribution reduces to

$$u_s - u_0 = [1/(G \times t)] \times \left[\int q\,ds - (A_0/A)\oint q\,ds\right]$$

The unknown warping displacement at the reference point on the surface, 'u_0', can be found from the condition that the resultant of any direct stress system must be zero .i.e., $\oint (u_s - u_0)\,t\,ds = 0$

or $u_0 = \oint u_s t\,d_s \big/ \oint t\,ds$

$$= \oint u_s\,ds\,/S \;\; \text{if, t is constant and S is the length of surface of section}$$

For a solid thin rectangular section, warping distribution is shown in Fig 12.7.

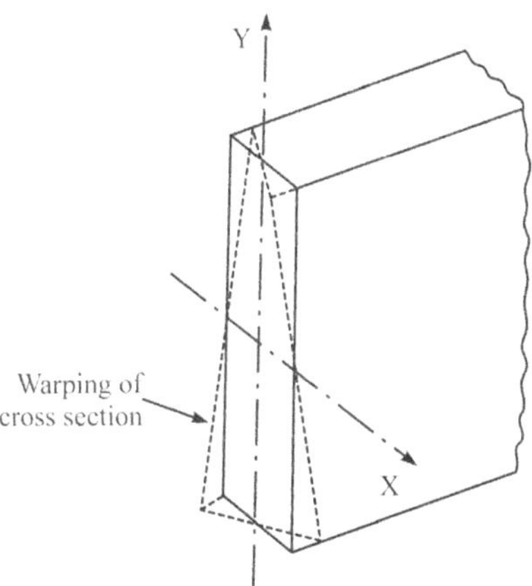

FIGURE 12.7 Warping distribution in a solid thin rectangular section

Special case : Hollow rectangular section, subjected to torsion

Consider a hollow rectangular section beam of sides a and b and thicknesses t_1 and t_2 for the horizontal and vertical segments, subjected to torsion 'T', as shown in fig 12.8. For the chosen origin 0, $u_0 = 0$ and

$$u_s - u_0 = (q/G) \left[\int (1/t)ds - (A_0 / A) \oint (1/t)ds \right]$$

By substituting $q = T/(2A)$,

$$u_s - u_0 = (T/2AG) \left[\int (1/t)ds - (A_0 / A) \oint (1/t)ds \right]$$

$$u_1 - u_0 = [T/(2A \times G)] \times [\delta_0 - (A_0 / A) \times \delta]$$

$$= [T \times \delta / (2A \times G)] \times [\delta_0/\delta - A_0/A]$$

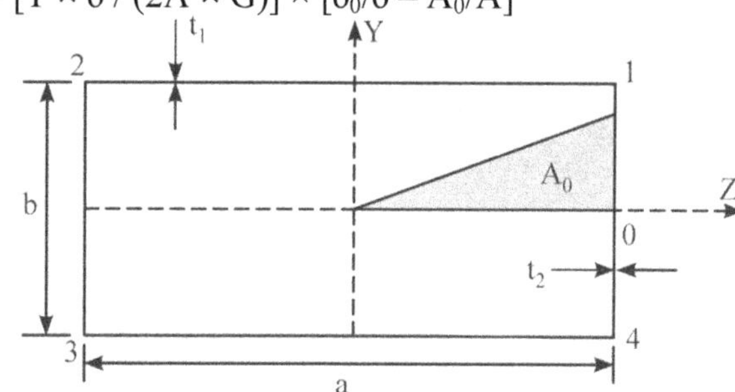

FIGURE 12.8 Hollow rectangular section of a thin walled beam

where $\delta_0 = \int ds/t$ in the limits $(0,s) = s_1/t_2$ for $0 \leq s_1 \leq b/2$ (from 0 to 1)

$$\delta = \oint (1/t)ds = 2[b/t_2 + a/t_1]$$

$$A_0 = (1/2) \times (a/2) \times s_1 = a \times s_1/4 \text{ and } A = a \times b$$

$$\Rightarrow u_1 = [T/(8a \times b \times G)] \times [b/t_2 - a/t_1]$$

and because of symmetry, $u_2 = -u_1 = -u_3 = u_4$

If the origin is chosen at 1, then,

$$u_0 = [1 / \{2(a \times t_1 + b \times t_2)\}] \times [2 \int u_{12} \times t_1 ds + 2 \int u_{23} \times t_2 ds]$$

$$= -[T/(8a \times b \times G)] \times [b/t_2 - a/t_1]$$

and $u_1 = -u_2 = [T/(8a \times b \times G)] \times [b/t_2 - a/t_1]$

If $\delta_0/\delta = A_0/A$ (i.e., $b/t_2 = a/t_1$), warping is zero at all points in the cross section. This is the ***condition for zero warping.***

Distribution of warping in a hollow rectangular section is shown in fig 12.9

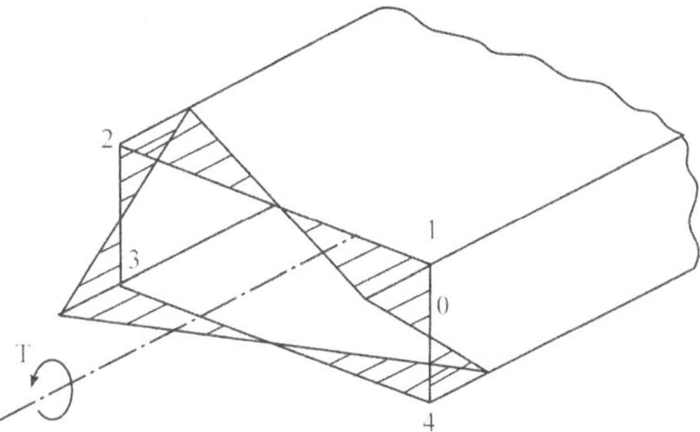

FIGURE 12.9 Warping distribution with S measured from 0

If the origin for S is chosen arbitrarily at 1, the distribution of warping changes as shown in fig 12.10

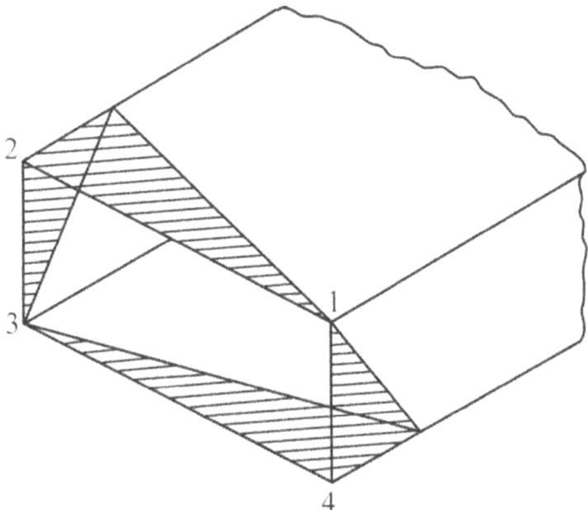

FIGURE 12.10 Warping distribution with S measured from 1

BUCKLING OF COLUMNS

A large proportion of aeroplane structure comprises thin webs stiffened by slender *longerons* or stringers. Buckling is the most critical mode of failure for both types of members. Axial loaded slender members are called columns (conventionally, vertically placed) or struts (inclined or horizontal). It is common experience that if an increasing axial compressive load is applied to a slender column, there is a load at which the column will suddenly bow or *buckle* in some direction, which can not be predetermined.

13.1 BEAM DEFLECTION EQUATION

Aircraft structural members do not fail due to direct compressive stress but only due to the combination of compressive and bending stresses. Bending stresses depend on the bending deflections, which are derived from the assumptions that stress is proportional to strain and deflections are small in comparison to the original dimensions. If shearing deformations are significant, they can be calculated separately and superimposed.

In an initially straight beam with two parallel sections in the unstressed condition (Ref Fig. 13.1).

The stress at a distance 'c' below the neutral axis, $\sigma = M \times c / I = E \times c / R$

Since curvature, $1/R = d\theta/dx$, from the above equation, we get

$$d\theta = \sigma \, dx / (E \times c)$$

If θ is small enough, $\theta = \sin \theta = \tan \theta = dy/dx$

and $d\theta/dx = d^2y/dx^2 = \sigma / (E \times c) = M / (EI)$ .....(13.1)

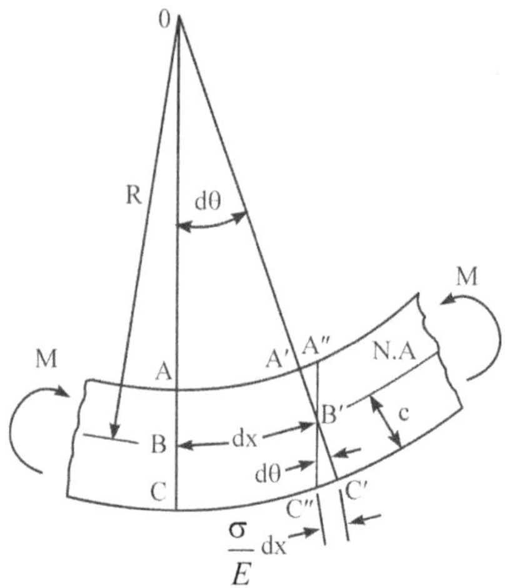

FIGURE 13.1 Simple bending

If y is measured +ve upwards, positive bending moment produces a +ve curvature d^2y/dx^2.

Here, $BO = B'O = R$; $BC = B'C' = c$; ABC is parallel to $A''B'C''$; $BB' = dx$

13.2 Critical or Euler Load

Buckling is the failure of compression members due to lateral bow induced by compression load. Columns, which are long in comparison to other dimensions, buckle even though compressive stresses are below the elastic limit. This phenomenon is called *elastic buckling*. Columns are classified as short columns or long columns based on linear or non-linear stress-strain relationship, represented by slenderness ratio L/k. *In a long column, failure occurs mainly due to buckling*. For a slender column AB of length 'L', subjected to axial load 'P' (Ref Fig.13.2), governing equation is

$$M = -P \times y \quad \text{or} \quad d^2y/dx^2 + P \times y / (EI) = 0$$

The general solution is $y = C_1 \times \sin ax + C_2 \times \cos ax$ where $a = \sqrt{P/EI}$

Constants C_1 and C_2 are evaluated from the end conditions.

$$y = 0 \text{ at } x = 0 \quad \Rightarrow \quad C_2 = 0$$
$$\text{while} \quad y = 0 \text{ at } x = L \quad \Rightarrow \quad C_1 \times \sin ax = 0$$

$$C_1 = 0 \quad \text{gives a trivial solution}$$

Therefore, $\sin(a \times L) = 0$ or $a \times L = L \times \sqrt{P/EI} = n\pi$ where, 'n' is any integer

For n = 1, $\quad P_{cr} = \pi^2 \times EI / L^2$ is called critical or *Euler load* and is associated with a state of *neutral equilibrium*.

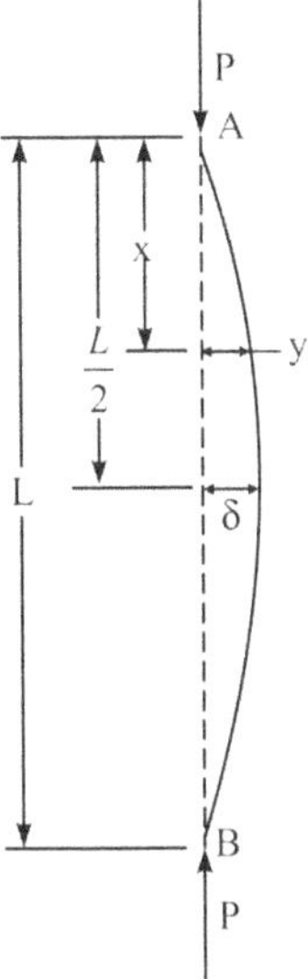

FIGURE 13.2 Buckling of a long column due to axial compression

This is an eigen value problem and the function $y = C_1 \sin ax$ is called an eigen function. Discrete values of buckling load $P = n^2 \times \pi^2 \times EI / L^2$ are called *eigen values* and corresponding displacement shapes are called *eigen modes*. Some of them are shown in Fig. 13.3.

Figure 13.3 Buckling modes

Corresponding buckling stress,

$$\sigma_{cr} = P_{cr} / A = \pi^2 \times EI / (L^2 \times A) = \pi^2 \times E / (L/k)^2 \qquad .....(13.2)$$

where, $\qquad k = \sqrt{(I/A)}$ is the radius of gyration

The value of C_1 can not be obtained at the critical load. C_1 is equal to the maximum deflection δ at the center of the column, which is indeterminate for the assumed conditions. For loads smaller than P_{cr}, the deflection C_1 or δ must be zero or else the column remains straight.

At the critical load, any deflection δ for which maximum stress is below the elastic limit will satisfy conditions of equilibrium. If the elastic limit is not exceeded, the column returns to its initial shape when the load is removed.

For a given material of modulus of elasticity 'E', the critical stress σ_{cr} increases as the slenderness ratio 'k' decreases i.e., as the column becomes shorter and thicker (Ref Fig. 13.4). Thus, a point is reached for low values of 'L/k' when $\sigma_{cr} > \sigma_Y$ and $\sigma_{cr} = \pi^2 \times E/(L/k)^2$ is no longer applicable. For mild steel, this point occurs at a slenderness ratio of approximately 100.

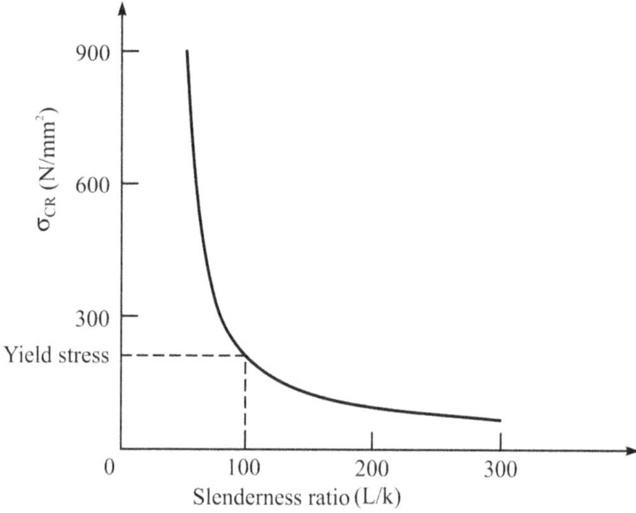

FIGURE 13.4 Critical stress Vs Slenderness ratio

13.3 COLUMN END FIXITY

The above relations are derived for the column with both ends hinged so that it can rotate freely. In most cases, however, compression members are connected in such a way that they are restrained against rotation at the ends. In such cases, the critical buckling load is determined from the following modified general equation by applying appropriate end conditions.

The general equation $d^2y/dx^2 + [P/(EI)] \times y = 0$ is rewritten as

$$d^4y/dx^4 + \lambda^2 (d^2y/dx^2) = 0 \qquad \text{where} \quad \lambda^2 = P/EI$$

The general solution is $y = C_1 \times \sin \lambda x + C_2 \times \cos \lambda x + C_3 \times x + C_4$

The constants are evaluated from the following end conditions for a column of length 'L' and critical buckling load is obtained as $P_{cr} = c \times \pi^2 \times EI / L^2$.

Equivalent length of the column with both ends hinged (c = 1) is obtained from

$$P_{cr} = \pi^2 \times EI / L_e^2 \quad \text{so that} \quad L_e^2 = L^2 / c = L^2 \qquad \qquad .....(13.3)$$

	Column end fixity	End conditions to be applied	Equivalent length, L_e
A	Hinged at both ends	$y = 0$ and $d^2y/dx^2 = 0$ at $x = 0$ & $x = L$	$L_e = L$
B	Fixed at both ends	$y = 0$ and $dy/dx = 0$ at $x = 0$ & $x = L$	$L_e = L/2$
C	Fixed at one end and hinged at the other end	$y = 0$ and $dy/dx = 0$ at $x = 0$ $y = 0$ and $d^2y/dx^2 = 0$ at $x = L$	$L_e = L/\sqrt{2}$ $= 0.707\,L$
D	Fixed at one end and free at the other end	$y = 0$ and $dy/dx = 0$ at $x = 0$ $y = \delta$ and $d^2y/dx^2 = 0$ at $x = L$	$L_e = 2L$

These four conditions are shown in Fig.13.5

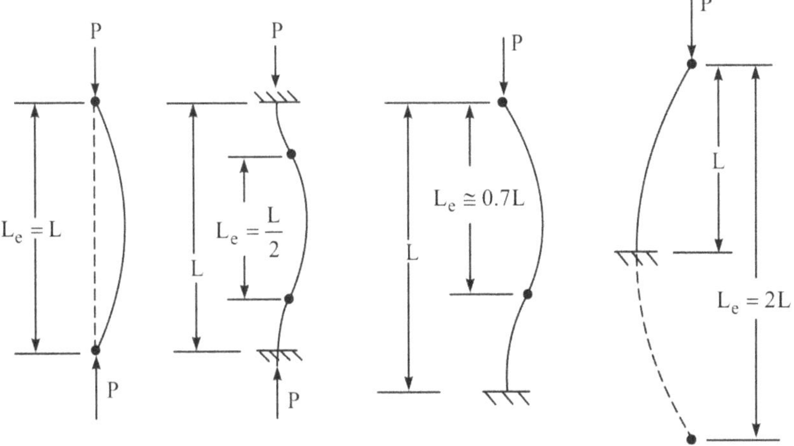

FIGURE 13.5 Effective length of column under buckling for different end conditions

Thus, the fixed end column will resist four times the load of a similar pin-ended column, if both are in the long column range. The same is not true for short columns because the critical load depends on E_t, which is smaller for small values of L_e.

For long columns ($L_e/k > 20$), Euler formula gives buckling stress. For short columns ($0 \leq L_e/k \leq 20$) beyond elastic limit, relation between the allowable column stress σ_c and L_e/k can not be expressed by a simple equation. Empirical formulae such as, ***Johnson's parabola formula***, matching with the experimental data, are commonly used. Buckling stress by ***Rankine formula***, based on bending and compressive stresses, can be used for both long columns and short columns.

Fixed end conditions, depend on infinite rigidity of the structure to which the column is attached. In practice, however, the structure provides finite rigidity which permit the column ends to rotate slightly. Columns have end-fixity conditions somewhere between hinged and fixed and it is difficult to determine end-fixity condition exactly. Tension members that connect to the ends of compression members provide greater restraint than similar compression members.

13.4 SHORT COLUMNS

Material of a column may exhibit a linear or non-linear stress-strain relationship, based on slenderness ratio L/k of the column. For the linear or straight portion, the constant slope is the modulus of elasticity 'E'. Above the elastic limit, the variable slope of non-linear portion is called ***tangent modulus of elasticity*** 'E_t' (Ref Fig. 13.6).

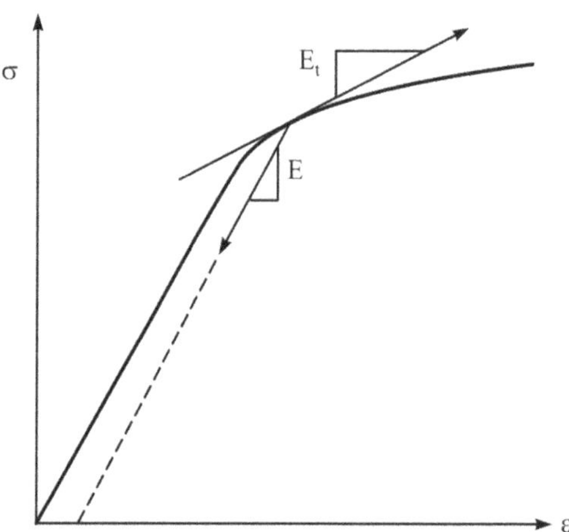

FIGURE 13.6 Elastic modulus

Ductile materials such as mild steel may have a zero or negative value of E_t near yield point. If E_t is positive at all points, a short column may remain perfectly straight when loaded to stresses beyond yield point. *In a short column, failure occurs mainly due to direct compressive stress and the role of bending stress is negligible*. If such a column has a slight lateral deformation, internal resisting moment $M_I = E_t I \times (d^2y/dx^2)$ is generated. If this internal moment is greater than the bending moment M_B produced by the load $P = \pi^2 \times E_t I / L^2$, the column will remain straight when loaded and is said to be in stable equilibrium. If $M_I < M_B$, the deflection will increase and the column will fail. Such a column is said to be in unstable equilibrium.

If a column is subjected to direct compressive load P, uniform compressive stress of P/A is seen across the cross section. However, when the column is loaded at a distance 'e' from its centroid (Ref Fig13.7), an additional bending stress is added over the direct compressive stress decreasing compressive stress on the convex side (σ_1) and increasing compressive stress on the concave side (σ_2). If σ_1 is less than elastic limit and σ_2 is more than elastic limit, then modulus of elasticity changes from E on the convex side to E_t on the concave side and the column behaves as if it were made of non-homogeneous material.

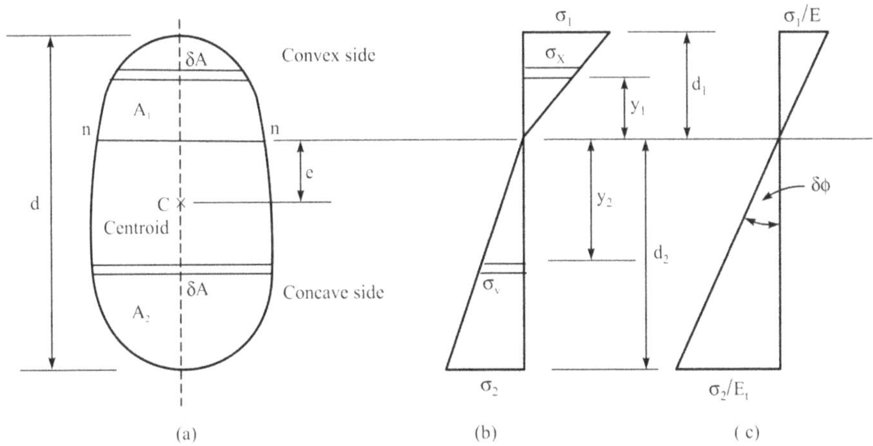

FIGURE 13.7 Determination of reduced elastic modulus

Then, axial force

$$P = \int_0^{d1} \sigma_X \, dA = \int_0^{d2} \sigma_Y \, dA$$

and bending moment

$$M = -P \times e = \int_0^{d1} \sigma_X \times (y_1 + e) \, dA + \int_0^{d2} \sigma_Y \times (y_2 - e) \, dA$$

Solving them, we get,

$$- P \times e = (d^2y/dz^2) \times (E \times I_1 + E_t \times I_2)$$
$$= (d^2y/dz^2) \times (E_r \times I)$$

where,

$$I_1 = \int_0^{d1} y_1{}^2 \, dA; \quad I_2 = \int_0^{d2} y_2{}^2 \, dA; \quad I = \int_{-d/2}^{+d/2} y^2 \, dA$$

and

$$E_r \times I = E \times I_1 + E_t \times I_2$$

Here, E_r is known as the ***reduced modulus***.

Critical buckling load is given by $P_{cr} = \pi^2 \times E_r \, I / L_e{}^2$ .....(13.4)

and the corresponding buckling stress is given by $\sigma_{cr} = \pi^2 \times E_r / (L_e/k)^2$

This method of predicting critical buckling loads and stresses outside the elastic range is known as the ***reduced modulus theory***.

In general, $E > E_t$. Therefore, $E_r > E_t$. It can be easily proved that $I_1 + I_2 > I$

Hence, P_{cr} predicted by reduced modulus theory is greater than P_{cr} predicted by tangent modulus theory. For general purposes, buckling load of columns is given most accurately by the tangent modulus theory.

13.4.1 RANKINE'S FORMULA

Critical buckling load 'P', based on compressive as well as bending stresses, suggested by Rankine and applicable for long or short columns is given by

$$1/P = 1/P_E + 1/P_C \qquad \qquad .....(13.5)$$

where, P_E is the Euler's critical load for long columns

$P_C = f_C \times A$ is the maximum allowable crushing load for column

f_C is the maximum allowable compressive stress for the material

and   A is the area of cross section of the column

For long columns, P_E is very small and $1/P_E$ is very large compared to $1/P_C$

Therefore, $P \approx P_E$

For short columns, P_E is very large and $1/P_E$ is very small compared to $1/P_C$

Therefore, $P \approx P_C$

Rankine's formula can also be expressed as

$$P = P_C \times P_E / (P_E + P_C) = P_C / [1 + (P_C/P_E)] \qquad .....(13.6)$$

$$P = \frac{f_c \times A}{1 + \left(f_c / \pi^2 E\right) \times \left(L_{eq} / k\right)^2} \quad \text{since, } P_E = \pi^2 \times E\,I / L_{eq}^2 \text{ and } I = A \times k^2$$

13.4.2 JOHNSON'S PARABOLA FORMULA

Euler buckling load was based on two assumptions – the column was perfectly straight before the application of load and the load was aligned perfectly along its axis. These columns fail at average stress within the linearly elastic range. Perfect column assumption is unrealistic. It is a coincidence that the ultimate load carrying capacity of an imperfect column can be predicted by the linear buckling theory for perfect columns

i.e., critical buckling stress, $\sigma_{cr} = \pi^2 \times E / (L_e/k)^2 \qquad .....(13.7)$

Bending displacements of imperfect columns increase rapidly as the axial force approaches the Euler load and its stress exceeds yield stress. Empirical relations are used to predict the critical loads of short columns, particularly in the case of columns which fail due to secondary instability where the tangent-modulus theory fails. Most commonly used empirical relationship uses a simple power law of the form

$$\sigma_{cr} = \sigma_c - \beta \times (L_e / k)^n \qquad \ldots\ldots(13.8)$$

where, σ_c = column yield stress, a function of compressive yield stress

n = parameter that establishes shape of the empirical curve

Equating it with the Euler equation and evaluating β at the common point of the two curves, critical slenderness ratio which divides long-column and short-column ranges is found to be

$$(L_e/k)_{cr} = \pi \times [\,(E/\sigma_c) \times (1 + 2/n)\,]^{1/2}$$

If $(L_e/k) \le (L_e/k)_{cr}$, the stress for failure can be computed from eq. (13.8), otherwise it is calculated from Euler equation (13.7).

The parameters σ_c and n are chosen so that the empirical equation fits the test data. To simplify the equation n is usually taken as an integer. The most commonly used values are n=1 which gives a straight line and n=2 which gives a parabola.

With n = 2, $\quad \sigma_{cr} = \sigma_c \,[1 - \{\,\sigma_c \times (L_e/k)^2\} / (4\,\pi^2 \times E)\,] \qquad \ldots\ldots(13.9)$

$$\text{applicable for} \quad L_e/k \le \pi \times (2E/\sigma_c)^{1/2}$$

Equation 13.9 is known as ***Johnson's parabola*** short-column equation

With n = 1, $\quad \sigma_{cr} = \sigma_c \,[\,1 - \{0.385 \times (L_e/k)\} / \{\pi \times (E/\sigma_c)^{1/2}\}\,] \qquad \ldots\ldots(13.10)$

$$\text{applicable for} \quad L_e/k \le \pi \times (3E/\sigma_c)^{1/2}$$

The straight line equation (13.10) will give a better approximation than the Johnson's parabola, for stable sections such as round or rectangular bars or sheet sections with relatively heavy walls and for most aluminium alloys.

Example 13.1

A square bar is made from PVC plastic that has a modulus of elasticity of E = 9 GPa and a yield strain of 0.001 mm/mm. Determine its smallest cross sectional dimension 'a' so that it does not fail from elastic buckling. It is pinned at its ends and has a length of 1250 mm.

Solution

Yield stress, $\sigma_Y = E \times \varepsilon_Y = 9\text{GPa} \times 0.001 = 9$ MPa ; $\quad L_e = L = 1.25$ m

Area of cross section, $A = a \times a = a^2$;   Moment of inertia, $I = a \times a^3 /12 = a^4/12$

$$\sigma_Y = \pi^2 \times EI / (L_e^2 A) = \pi^2 \times 9 \times 10^9 \times (a^4/12) / (1.25^2 \times a^2) = 9 \times 10^6$$

$$\Rightarrow a^2 = (12 \times 1.25^2)/(\pi^2 \times 10^3) = 18.982 \times 10^{-4} ;\quad a = 4.36 \times 10^{-2}\,m = 4.36\ cm$$

Example 13.2

A steel column has a length of 5m and is fixed at both ends. If the column has a hollow rectangular cross section of outer dimensions 100 mm × 50 mm and thickness 10 mm, determine the critical load. Assume $E = 200$ GPa

Solution

Effective length, $L_e = L/2 = 5000/2 = 2500$ mm ; $E = 200 \times 10^3\ N/mm^2$

Min moment of inertia, $I = (1/12)\,[100 \times 50^3 - 80 \times 30^3] = 86.17 \times 10^4\ mm^4$

Critical buckling load, $P_{cr} = \pi^2\,EI / L_e^2 = \pi^2 \times (200 \times 10^3) \times (86.17 \times 10^4)/ 2500^2$

$$= 2.72 \times 10^7\ N = 2.72 \times 10^4\ kN$$

Example 13.3

A solid square bar AB of side 40mm is pin connected at its ends. Determine the maximum allowable load P that can be applied to the frame. Use $E = 210$ GPa ; yield stress $= 250$ MPa and a factor of safety for buckling of 2.

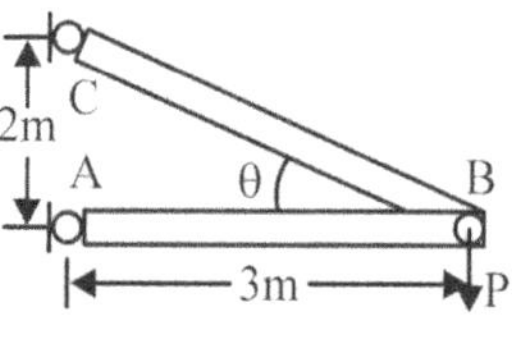

Solution

Forces in members BC and BA can be calculated from static equilibrium at B as, $F_{BC} = P / \sin\theta = P / (2/\sqrt{13})$ and $F_{BA} = F_{BC} \times \cos\theta = P / \tan\theta = P / (2/3)$

where, θ is the angle between members BA and BC at B

Force in BA is compressive and BA can fail by direct compression or buckling.

Moment of inertia, $I = 40^4/12 = 21.33 \times 10^4\ mm^4$

Area of cross section, $A = 40^2 = 1600\ mm^2$

Effective length of member AB (hinged-hinged), $L_e = L = 3000$ mm

(a) For direct compression, $P_1 = (2/3) \times F_{BA} = (2/3) \times (\sigma_Y \times A)$

$$= (2/3) \times 250\ N/mm^2 \times 1600\ mm^2$$

$$= 267\ kN$$

(b) For buckling, $P_2 = (2/3) \times (F_{BA})_{cr} / f_s = [\, (2/3) \times \pi^2 \times EI / L_e^2 \,] / f_s$

$$= [\, (2/3) \times \pi^2 \times (210 \times 10^3) \times 21.33 \times 10^4 / 3000^2 \,] / 2$$

$$= 16.385 \text{ kN}$$

Maximum allowable load, $P = \min(P_1, P_2) = 16.385$ kN

Example 13.4

Two columns are made of same material and have same length, same area and same end conditions. One of them is circular in section of diameter 10cm and the other is square in section. Determine which column will be stronger and by how many times?

Solution

$E_{square} = E_{circle} = E$ and $L_{square} = L_{circle} = L$

Let 's' be the side of square cross section.

$A_{square} = A_{circle} \Rightarrow s^2 = (\pi/4) \times d^2$

$I_{square} = (1/12)\, s^4 = (1/12) \times [(\pi/4) \times d^2]^2$; $I_{circle} = (\pi/64) \times d^4$

$(P_{cr})_{square} = \pi^2 EI_{square} / (L)^2$; $(P_{cr})_{circle} = \pi^2 E\, I_{circle} / (L)^2$

$(P_{cr})_{square} / (P_{cr})_{circle} = I_{square} / I_{circle} = (1/12) \times [(\pi/4) \times d^2]^2 / [(\pi/64) \times d^4]$

$$= \pi / 3 = 1.05$$

Square section is stronger

Example 13.5

A 8 m long steel column AB is fixed at its ends. Its load carrying capacity is increased by bracing it about the y-y (weak) axis using pin connected struts at C, mid-height. Determine critical load, if E = 200 GPa, yield strength = 410 MPa, Area of cross section is 30 cm², $I_{XX} = 1340$ cm⁴ and $I_{YY} = 183$ cm⁴

Solution

In X-Z plane, the column behaves as fixed-fixed column

$(P_{cr})_{XZ} = \pi^2 \times E\, I_{XX} / (L_e)_X^2 = \pi^2 \times 20000 \times 1340 / (800/2)^2 = 1653$ kN

In Y-Z plane, each half of the column AB behaves as fixed-hinged column

$(P_{cr})_{YZ} = \pi^2 \times E\, I_{YY} / (L_e)_Y^2 = \pi^2 \times 20000 \times 183 / (400/\sqrt{2})^2 = 460$ kN

$\sigma_{cr} = (P_{cr})_{min} / A = 460 \text{ kN} / 30 \text{ cm}^2 = 15.3 \text{ kN/cm}^2$ or 153 MPa

Since $\sigma_{cr} < \sigma_Y$, buckling will occur before yielding

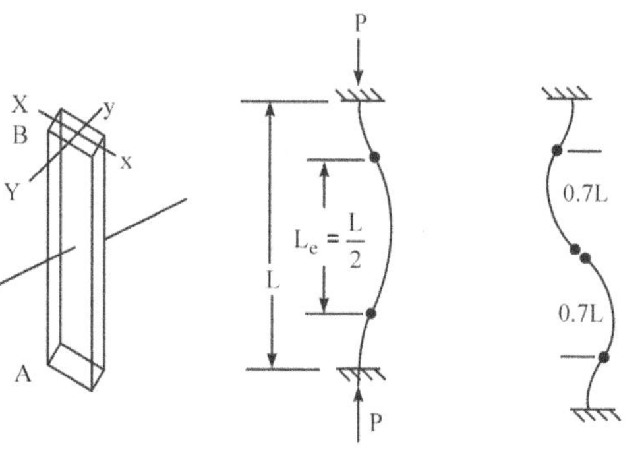

Buckling in X-Z plane   Buckling in Y-Z plane

Example 13.6

A 5m long column AB is fixed at its bottom and is braced at its top by cables so as to prevent movement along the X-axis, as shown. Determine the largest allowable load if modulus of elasticity is 70 GPa, yield stress is 215 MPa, factor of safety for buckling is 3, Area of cross section is 75 cm^2, I_{XX} = 6130 cm^4 and I_{YY} = 2320 cm^4.

Solution

In Y-Z plane, the column behaves as fixed-free column

$$(P_{cr})_{YZ} = \pi^2 \times E\, I_{YY} / (L_e)_Y^2 = \pi^2 \times 7000 \times 6130 / (2 \times 500)^2 = 424 \text{ kN}$$

In X-Z plane, the column behaves as fixed-hinged column

$$(P_{cr})_{XZ} = \pi^2 \times E\, I_{XX} / (L_e)_X^2 = \pi^2 \times 7000 \times 2320 / (500/\sqrt{2})^2 = 1310 \text{ kN}$$

$$P_{Allow} = (P_{cr})_{min} / f_s = 424 / 3 = 141.3 \text{ kN}$$

$$\sigma_{cr} = (P_{cr})_{min} / A = 424 / 75 = 56.5 \text{ MPa} \ < \ \sigma_Y \ (215 \text{ MPa})$$

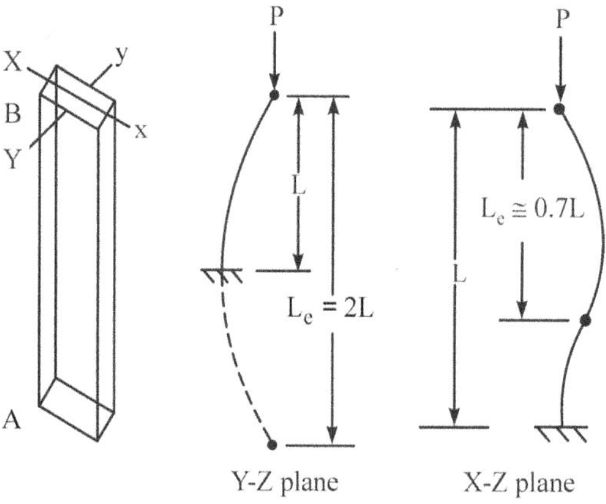

Y-Z plane   X-Z plane

Example 13.7

Four mild steel plates are welded to form a hollow square section of external size 15cm x 15cm. The section is used as a 6m long column, with both ends pinned to support an axial compressive load of 400 kN. If the working load is not to exceed 50% of the crippling load or crushing load, whichever is less, determine suitable thickness of plates. Assume E = 200 GPa, α = 1/7500 and max crushing strength as 220 MPa.

Solution

$$I_{Min} = (1/12)\ 15 \times 15^3 - (1/12) \times (15-2t) \times (15-2t)^3$$

$$P_{Euler} = \pi^2 \times E\ I\ /\ (L_e)^2 = 400\ /\ 0.5 \ \Rightarrow\ I = 1460\ cm^4 \ \Rightarrow\ t = 7.6\ mm$$

$$A = 15 \times 15 - (15-2t) \times (15-2t)\ ;\quad k^2 = I_{Min}\ /\ A$$

$$P_{Rankine} = (f_C \times A)\ /\ [1 + \alpha \times (L_e/k)^2\ ] = 400\ /\ 0.5 \qquad \Rightarrow\ t = 7.15\ mm$$

13.5 ECCENTRICALLY LOADED COLUMNS

In practice, a column is neither perfectly straight nor can be loaded exactly at the centroid of the cross section. Thus, if 'e' is the distance between the point of load application and the centroid or eccentricity (Ref Fig. 13.8), $y = C_1 \sin ax + C_2 \sin ax$ where the constants are now evaluated from the conditions, $y = e + \delta$ at x = 0 and dy/dx = 0 at x = 0, choosing the origin at the mid point of the column. Evaluating the constants and substituting them, we get

$y = (e + \delta) \times \cos (a \times x)$

δ can be found from the condition $y = e$ at $x = L/2$

Then, $e + \delta = e \times \sec (a \times L/2)$

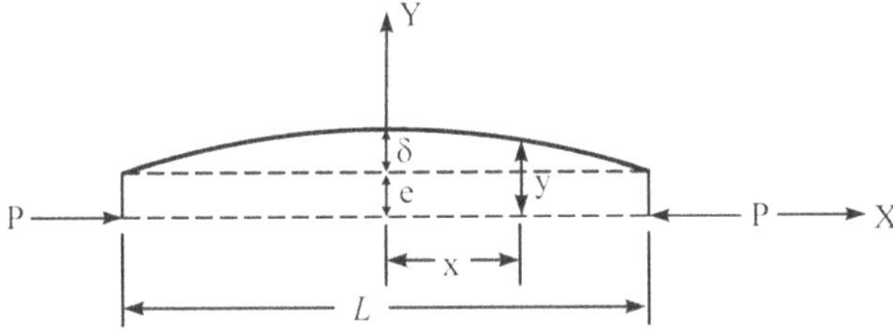

FIGURE 13.8 Buckling with eccentric load

As P approaches P_{cr}, $e + \delta = e \times \sec (\pi/2) = \infty$, which leads to $\delta = \infty$

Thus, for various values of eccentricity 'e', relationships between P and δ are asymptotic to the line $P = P_{cr}$, as shown in Fig. 13.9

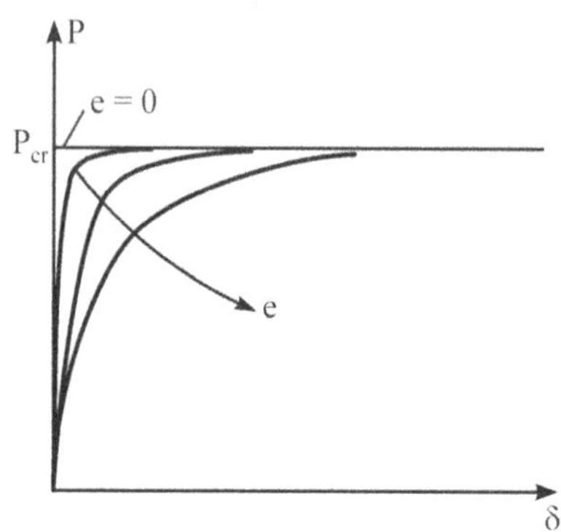

FIGURE 13.9 Variation of critical load with eccentricity

13.6 BEAM – COLUMNS

Stresses and deflections in a linearly elastic beam subjected to transverse loads, as predicted by simple beam theory, are directly proportional to the applied loads. This relationship is valid if the deflections are small so that the slight change in geometry has insignificant effect on the applied loads (Ref Fig. 13.10). This situation changes when axial loads act simultaneously with transverse loads. The internal moments, shear forces, stresses and deflections are dependent upon the magnitude of deflections as well as the magnitude of external loads. These are known as ***transversely loaded columns*** or ***beam-columns***.

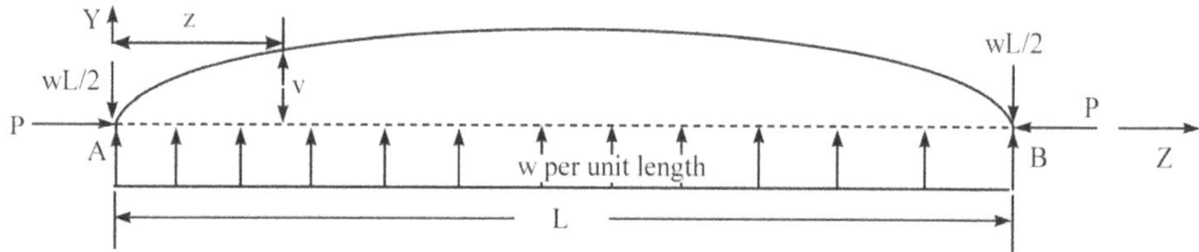

FIGURE 13.10 Column with uniformly distributed transverse load

13.6.1 BEAM WITH UNIFORMLY DISTRIBUTED LATERAL LOAD AND AXIAL COMPRESSIVE LOAD

In a beam carrying a uniformly distributed load of intensity 'w' per unit length and an axial compressive load 'P', bending moment 'M' at any section 'Z' is given by

$$M = P \times v + (w \times L/2) \times z - (w \times z) \times (z/2) = - E I \times (d^2v/dz^2)$$
$$\Rightarrow \quad (d^2v/dz^2) + (P/EI) \times v = [w/(2EI)] \times (z^2 - L \times z)$$

Standard solution of this differential equation is

$$v = A \times \cos(\lambda \times z) + B \times \sin(\lambda \times z) + (w/2P) \times (z^2 - Lz - 2/\lambda^2)$$

where,   A and B are unknown constants, to be evaluated from boundary conditions

and $\lambda^2 = P / (EI)$

In a pin-ended beam, boundary conditions are $v = 0$ at $z = 0$ and at $z = L$ so that

$$v = [w/(\lambda^2 P)] \times [\cos \lambda z + \{(1 - \cos \lambda L)/\sin \lambda L\} \times \sin \lambda z]$$
$$+ (w/2P) \times (z^2 - Lz - 2/\lambda^2)$$

We are usually concerned with maximum value, which occurs at the center ($z = L/2$) due to symmetry of loading and boundary conditions and is given by

$$v_{max} = [w/(\lambda^2 P)] \times [\sec(\lambda L/2) - 1] - wL^2/8P \qquad .....(13.11)$$

Corresponding maximum bending moment is

$$M_{max} = -P \times v_{max} - (wL^2/8) = (w/\lambda^2) \times [1 - \sec(\lambda L/2)]$$

It can be rewritten, in terms of Euler buckling load $P_{cr} = \pi^2 EI/L^2$ for a pin-ended column, as

$$M_{max} = (wL^2/\pi^2) \times (P_{cr}/P) \times [1 - \sec\{(\pi/2) \times \sqrt{(P/P_{cr})}\}].....(13.12)$$

As P approaches P_{cr}, the bending moment (and deflection) approaches infinity.

This theory is based on the assumption of small deflections. Otherwise, (d^2v/dz^2) would not be a close approximation for curvature. In reality, large deflections will be produced by the presence of axial compressive load, irrespective of the magnitude of the transverse load.

13.6.2   BEAM WITH CONCENTRATED LATERAL LOAD AND AXIAL COMPRESSIVE LOAD

Let us consider a beam-column with hinged ends carrying a concentrated lateral load W at a distance 'a' from the right-hand support and an axial compressive load 'P' (Fig. 13.11). Governing differential equation for equilibrium at a section 'z' is

$$EI \times (d^2v/dz^2) = -M = -P \times v - (W \times a/L) \times z$$

$$\text{for } z \leq L - a \qquad .....(13.13)$$

and $$EI \times (d^2v/dz^2) = -M = -P \times v - \{W \times (L-a)/L\} \times (L-z)$$

$$\text{for } z > L - a \qquad .....(13.14)$$

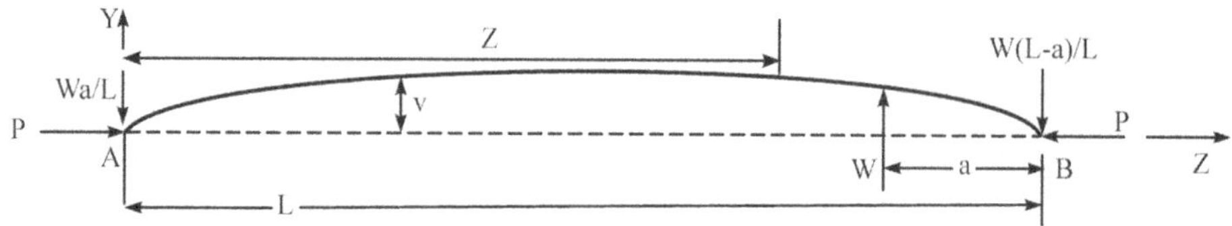

FIGURE 13.11 Column with a concentrated transverse load

With $\lambda^2 = P/EI$, general solution for eq (13.13) can be written, as

$$v = A \times \cos(\lambda \times z) + B \times \sin(\lambda \times z) - [W \times a/(P \times L)] \times z \quad .....(13.15)$$

and general solution for eq (13.14) can be written, as

$$v = C \times \cos(\lambda \times z) + D \times \sin(\lambda \times z) - (W/PL) \times (L - a) \times (L - z)$$

$$.....(13.16)$$

Constants A, B, C and D are found from the boundary conditions

when $z = 0$, $v = 0$ $\Rightarrow$ $A = 0$ from eq (13.15)

when $z = L$, $v = 0$ $\Rightarrow$ $C = - D \times \tan(\lambda \times L)$ from eq (13.16)

Deflection and slope from eq (13.15) and eq (13.16) should be the same at the point of application of load 'W' (i.e., at $z = L - a$)

Therefore, equating deflections,

$$v = B \times \sin\{\lambda \times (L - a)\} - [W \times a / (P \times L)] \times (L - a)$$

$$= D \times \tan(\lambda \times L) \times \cos\{\lambda \times (L - a)\} + D \times \sin\{\lambda \times (L - a)\}$$

$$- [W / (P \times L)] \times (L - a) \times a$$

and equating slopes,

$$dv/dz = B \times \lambda \times \cos\{\lambda \times (L - a)\} - [W \times a / (P \times L)]$$

$$= D \times \lambda \times \tan(\lambda \times L) \times \sin\{\lambda \times (L - a)\} + D \times \lambda \times \cos\{\lambda \times (L - a)\}$$

$$- [W / (P \times L)] \times (L - a)$$

Solving these two equations for B and D and substituting in eq (13.15) and (13.16), we get

$$v = [W \times \sin(\lambda \times a) / \{P \times \lambda \times \sin(\lambda \times L)\}] \sin(\lambda \times z) - [W \times a / (P \times L)] \times z$$
$$\text{for } z \leq L - a \quad .....(13.17)$$

$$v = [W \times \sin\{\lambda \times (L - a)\} / \{P \times \lambda \times \sin(\lambda \times L)\}] \times \sin\{\lambda \times (L - z)\}$$
$$- \{W / (P \times L)\} \times (L - a) \times (L - z) \quad \text{for } z > L - a \quad .....(13.18)$$

When the load W is applied at the center ($a = L/2$), then the deflection curve is symmetrical with maximum deflection at the mid point and is given by

$$v_{max} = [W / (2P \times \lambda)] \times \tan(\lambda \times L/2) - W \times L / (4P) \quad .....(13.19)$$

13.6.3 BEAM WITH END MOMENTS AND AXIAL COMPRESSIVE LOAD

Solution to this problem (Ref Fig. 13.12) can be found by superposing the results of the previous case. If the concentrated load W moves towards B and its magnitude increases simultaneously so that the product $M_B = W \times a$ is constant, then in the limit as $a \to 0$, we have M_B applied at B. Since '$\lambda \times a$' is now very small, $\sin(\lambda \times a) \approx \lambda \times a$, and the deflection curve given by eq (13.17) reduces to

$$v = (M_B/P) \times [\ \{\sin(\lambda \times z)\ /\ \sin(\lambda \times L)\} - z/L\ ] \qquad \ldots\ldots(13.20)$$

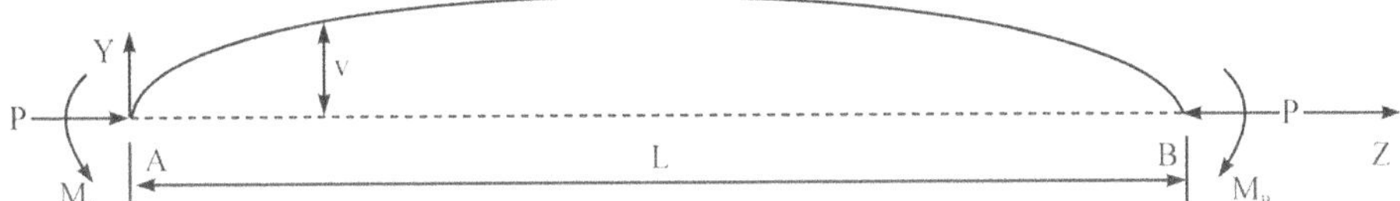

FIGURE 13.12 Column with end moments and axial compressive load

In a similar way, we find the deflection curve corresponding to M_A acting alone by substituting $W \times (L - a) = M_A$ and $\sin\{\lambda \times (L - a)\} \approx \lambda \times (L - a)$ in eq (13.18) as

$$v = (M_A/P) \times [\ \{\sin \lambda \times (L - z)\ /\ \sin(\lambda \times L)\} - (L - z)/L\ ] \qquad \ldots..(13.21)$$

The effect of both the moments acting simultaneously is obtained by superposition of the results of eq (13.20) and eq (13.21) as

$$v = \frac{M_B}{P}\left[\frac{\sin \lambda z}{\sin \lambda L} - \frac{z}{L}\right] + \frac{M_A}{P}\left[\frac{\sin \lambda z}{\sin \lambda L} - \frac{z}{L}\right] \qquad \ldots..(13.22)$$

Eq (13.22) is similar to the deflected form of a beam column subjected to axial compressive loads at the two ends with eccentricities e_A and e_B.

Then, $M_A = P \times e_A$ and $M_B = P \times e_B$ and the eq (13.22) changes to

$$v = \left(e_A + e_B\right) \times \left[\frac{\sin \lambda z}{\sin \lambda L} - \frac{z}{L}\right] \qquad \ldots..(13.23)$$

13.7   FAILURE OF COLUMNS WITH THIN WALLS

Stiffeners or other members used to carry compression loads are frequently made by forming sheet into channels, U, J or Z sections and the design of these members is a problem of their strength as columns. Such sections are subject to three major types of failure

- If the material is sufficiently thick, they fall into the class of columns having a stable cross section and can be analysed by methods described earlier

- If the material is relatively thin, portions of the cross section may be subject to local buckling and the crushing strength of the column will largely depend upon the stability properties of the various parts of the cross section

- Open sections are subject to a torsional type of failure in which induced shearing stresses tend to cause the column to fail by twisting

13.7.1 BENT-UP SHEET ANGLE SECTIONS

(a) Let us first consider simplest of bent-up sections, namely, an equal-legged angle. If the cross section has heavy walls and short outstanding legs, failing stress will correspond to the appropriate (long / short) columns curves.

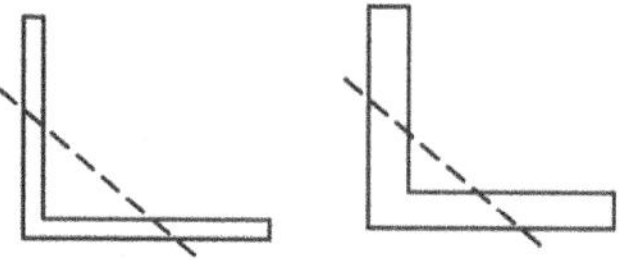

(b) If the cross section has long thin legs, it is possible for some value of L/k, the allowable stress given by appropriate (long / short) column curves will be higher than the stress which will cause the legs to buckle. For such sections, lower of the two stresses – stress calculated by column action and the buckling stress of legs (assuming three sides simply supported and the fourth free) – will be in close agreement with the experimentally measured buckling stress.

13.7.2 CHANNEL AND EQUAL-LEGGED Z-SECTIONS

These are the next most complex sections, subjected to many possible types of failure depending upon their cross sectional dimensions and their length –

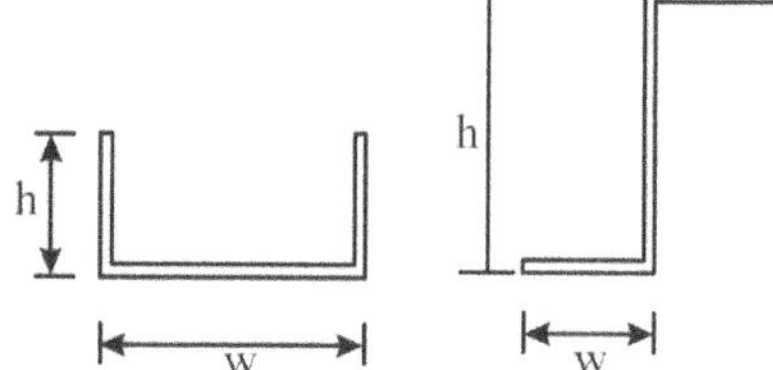

(a) Column failure about X-X axis and

(b) Column failure about Y-Y axis, which were already discussed.

(c) Local failure due to plate buckling : Critical buckling stress is given by empirical relation $\sigma_{cr} = K_F \times (f \times E) \times (t_h/h)^2$

where, K_F is a function of (w/h) and (t_w/t_h)

w, h, t_w and t_h are the width and thickness of web and flange respectively

and f is a factor,

$= 1$ if σ_{cr} is less than the proportional limit for the material

< 1, if σ_{cr} exceeds the proportional limit to take into account

reduced modulus of elasticity

This method of calculation is not applicable for channels and Z-sections, whose length is less than 2 to 3 times the width of the largest cross sectional dimension.

13.7.3 SQUARE AND RECTANGULAR TUBES WITH THIN WALLS

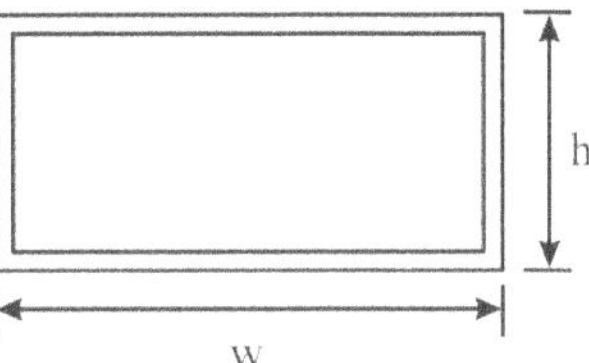

Considering interaction of the two walls of the tube on each other during buckling, critical buckling stress is given by empirical relation

$$\sigma_{cr} = K_F \times [\pi^2 \times (f \times E) / (1 - v^2)] \times (t_w/w)^2$$

where, w is the longer side of the cross section, of thickness t_w

It can be seen that the stress at which buckling occurs is not the stress at which the tube loses its load-carrying ability. During buckling, the corners carry a stress close to the yield point of the material and the problem becomes one of effective width of sheet acting with the corners.

13.7.4 GENERAL THIN-WALLED SHAPES

If certain flat elements buckle before the weakest curved section becomes unstable, then the total load carried by the section will be sum of the buckling loads of the flat portions which have buckled, plus the critical buckling stress of the weakest curved portion times the remaining unbuckled area of the cross section.

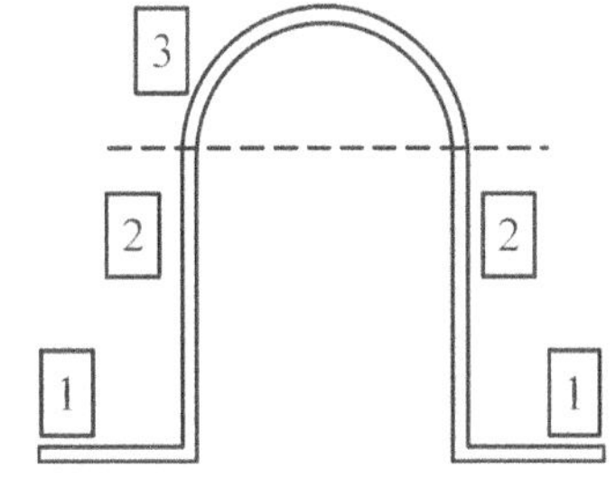

For example, in the section shown in figure, critical buckling stress for section-1, σ_{cr1}, is calculated on the basis of a plate simply supported on three sides with the fourth side free; critical buckling stress for section-2, σ_{cr2}, is calculated on the basis of a plate simply supported on all the four sides; and critical buckling stress for section-3, σ_{cr3}, is calculated on the basis of a curved plate simply supported on all the four sides. Then, critical crushing stress σ_{cc} of the column is given by

If $\sigma_{cr3} < \sigma_{cr1}$ and $\sigma_{cr3} < \sigma_{cr2}$,

$$\sigma_{cc} = \sigma_{cr3} \times (2A_1 + 2A_2 + A_3) / (2A_1 + 2A_2 + A_3) = \sigma_{cr3}$$

If $\sigma_{cr1} < \sigma_{cr3}$ and $\sigma_{cr3} < \sigma_{cr2}$,

$$\sigma_{cc} = [2\sigma_{cr1} \times A_1 + \sigma_{cr3} \times (2A_2 + A_3)] / (2A_1 + 2A_2 + A_3)$$

If $\sigma_{cr1} < \sigma_{cr3}$ and $\sigma_{cr2} < \sigma_{cr3}$,

$$\sigma_{cc} = (2\sigma_{cr1} \times A_1 + 2\sigma_{cr2} \times A_2 + \sigma_{cr3} \times A_3) / (2A_1 + 2A_2 + A_3)$$

13.8   TORSIONAL INSTABILITY OF COLUMNS

In all the above cases, it has been assumed that failure of the column was due to bending instability. However, certain sections with low torsional rigidity may fail due to torsional instability – twisting of the central portion of the column relative to the two ends. Twisting failure occurs by a rotation of the sections of the column about the shear center of the cross section. For column sections attached to plates as stiffening elements, the center of rotation may not be at the shear center of the section due to restraints introduced by the sheet. Critical stress for a concentrically loaded column subject to twisting failure is given by

$$\sigma_{ct} = [G \times J + (\pi/L_e)^2 \times E \times C_{bt}] / I_z$$

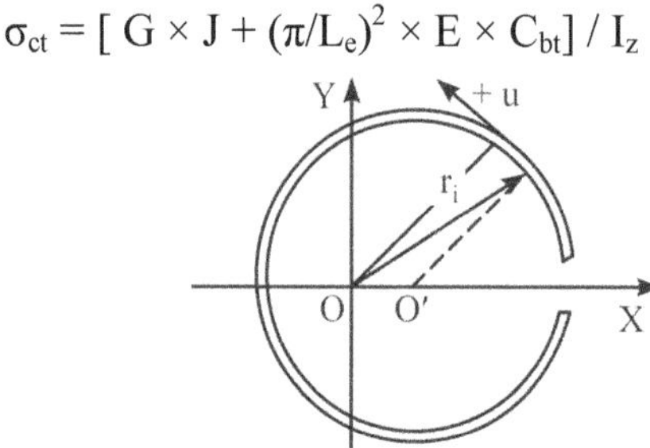

where, I_z is the polar moment of inertia of the section about the axis of rotation

and C_{bt} is the ***torsion-bending constant***, which is dependent on the dimensions of the cross section

The value of torsion-bending constant C_{bt} can be calculated using the expression

$$C_{bt} = \int w^2 \, dA - (1/A) [\int w \, dA]^2$$

where,   w is the circumferential warping given by $\int r_i \, du$

and   u is the circumferential coordinate, positive in the direction of rotation

For column sections attached to plates, critical torsional buckling stress is calculated for a number of assumed locations for the center of rotation, starting with some point in the plane of the sheet as a first approximation, and to take that location giving the lowest critical stress as the actual center of rotation. In each case, the origin of the rectangular coordinates X and Y must be taken at the assumed center of rotation.

The procedure is explained through the following example of a simple C-channel.

(a) Let us take O as the origin of u. We will calculate C_{bt} about the point O taking it as the center of rotation.

Then, from 1-2, $r_i = 0$; $w = 0$ and

from 1-4, $r_i = 0$; $w = 0$

from 2-3, $r_i = +a/2$; du is $-$ve   and,

therefore, $w = - a \times u/2$

from 4-5, $r_i = +a/2$; du is $+$ve   and,

therefore, $w = +a \times u/2$

$$\int_1^2 w \times dA = \int_1^3 w \times dA + \int_2^4 w \times dA + \int_1^5 w \times dA + \int_4 w \times dA$$

$$= 0 + \int_0^b (- a \times u / 2) \times (t \times du) + 0 + \int_0^b (+a \times u / 2) \times (t \times du) = 0$$

Hence, $C_{bt} = \int w^2 dA - (1 / A) \left[\int w \times dA \right]^2$

$$= \int_1^2 w^2 dA + \int_2^3 w^2 dA + \int_1^4 w^2 dA + \int_4^5 w^2 dA - (1 / A) \times (0)$$

$$= 0 + \int_0^b (-a \times u / 2)^2 \times (t \times du) + 0 + \int_0^b (+a \times u / 2)^2 \times (t \times du)$$

$$= (a^2 \times t/4) \times [(b^3/3) + (b^3/3)] = a^2 \times b^3 \times t/6$$

(b) We will now calculate C_{bt} about the point O' taking point O again as the origin of u. Then,

from 1-2, $r_i = c$;   du is $-$ve   and, therefore, $w = - c \times u$

from 2-3, $r_i = + a/2$;   du is $-$ve   and, therefore, $w = - a \times u/2 - a \times c/2$

from 1-4, $r_i = c$;   du is $+$ve   and, therefore, $w = + c \times u$

from 4-5, $r_i = + a/2$;   du is $+$ve   and, therefore, $w = + a \times u/2 + a \times c/2$

$$\int w \times dA = \int_1^2 w \times dA + \int_2^3 w \times dA + \int_1^4 w \times dA + \int_4^5 w \times dA$$

$$= \int_1^{a/2} (-c \times u) \times (t \times du) + \int_0^b (-a/2) \times (u+c) \times (t \times du)$$

$$+ \int_0^{a/2} (+c \times u) \times (t \times du) + \int_0^b (+a/2) \times (u+c) \times (t \times du) = 0$$

Hence, $C_{bt} = \int w^2 dA - (1/A) \left[\int w \times dA \right]^2$

$$= \int_1^2 w^2 dA + \int_2^3 w^2 dA + \int_1^4 w^2 dA + \int_4^5 w^2 dA - (1/A) \times (0)$$

$$= \int_1^{a/2} (-c \times u)^2 \times (t \times du)^2 + \int_0^b (-a/2)^2 \times (u+c)^2 \times (t \times du)$$

$$+ \int_0^{a/2} (+c \times u)^2 \times (t \times du) + \int_0^b (+a/2)^2 \times (u+c)^2 \times (t \times du)$$

$$= 2[\{(c^2 t/3) \times (a^3/8)\} + \{(a^2 t/4) \times (b^3/3 + b^2 c/2 + b\, c^2)\}\,]$$

$$= (a^2 \times t/12) \times [c^2 \times (a + 6b) + 2b^2 \times (b+3c)\,]$$

$\int w\, dA$ is not always equal to zero if, for example, the origin of u is taken at point 2.

Additional Problems for Practice

1. A straight uniform column of length 'L' and bending stiffness 'EI' is subjected to uniform lateral loading w/unit length. The end attachments do not restrict rotation of the column ends. The longitudinal; compressive force 'P' has eccentricity 'e' from the centroids of the end sections and is placed so as to oppose the bending effect of the lateral loading. The eccentricity e can be varied and is to be adjusted to the value which, for given values of P and w, will result in the least maximum bending moment of the column. Show that $e = (w/Pa^2) \times \tan^2(aL/4)$ where, $a = P/EI$

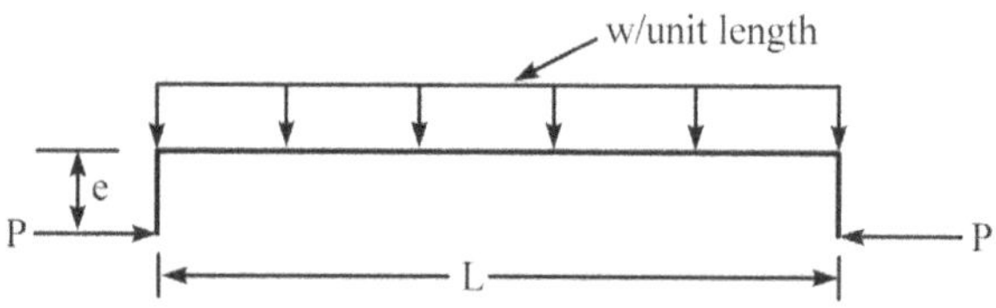

2. A thin-walled column of length '2L' with doubly symmetric cross-section shown in figure is compressed, preventing warping and twisting of its ends. Determine the average compressive stress at which the column first buckles in torsion. Take L = 500 mm, b = 25 mm, t = 2.5 mm and E = 70 GPa.

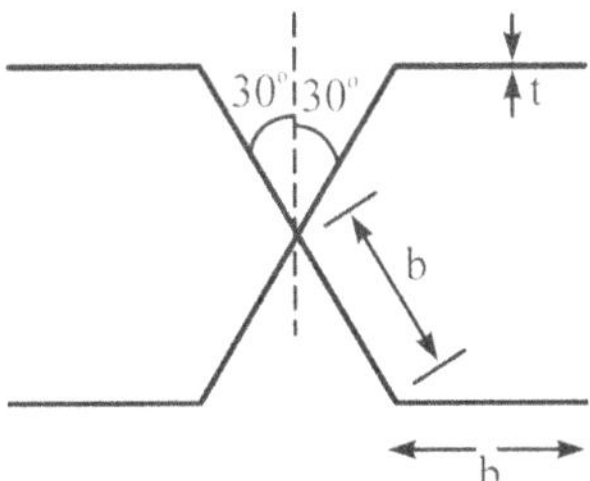

3. A thin-walled pin-ended column is 2 m long and has the section as shown. If the ends of the column are free to warp, determine the lowest value of flexural-torsional buckling load. Take E = 75 GPa and G = 30 GPa.

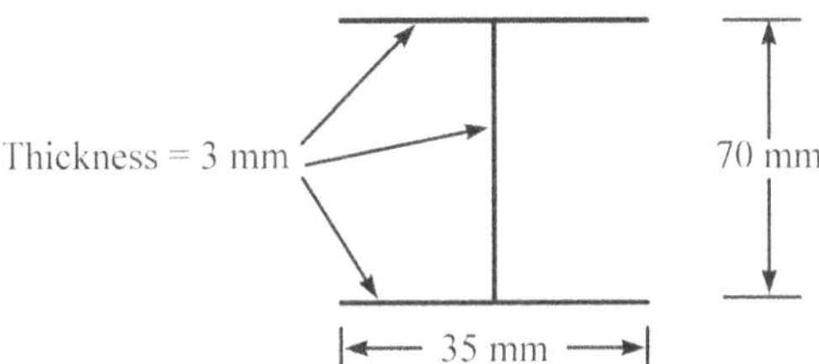

4. A column of length 1 m has the cross section shown. If the ends of the column are pinned and free to warp, calculate the lowest value of flexural-torsional buckling load. Take E = 70 GPa and G = 28 GPa

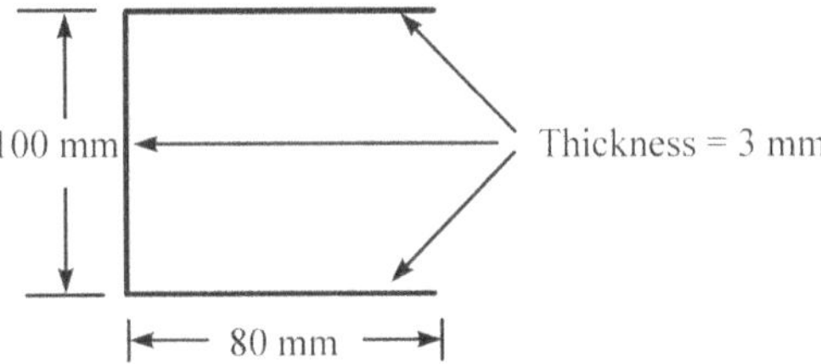

5. Figure shows the cross section of a thin walled column with rigidly fixed ends. Find an expression, in terms of section dimensions and Poisson's ratio, for the column length for which the purely flexural and purely torsional modes of instability would occur at the same axial load

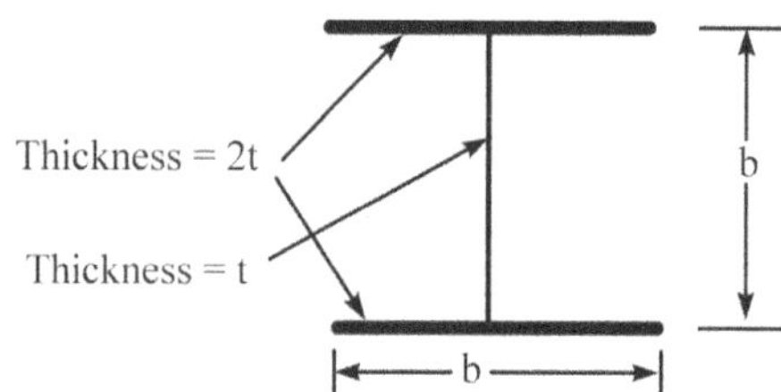

6. A truss ABC supports a vertical load P at joint B. Each member is a slender hollow circular steel pipe of 100mm outer dia and 6mm wall thickness. Assuming E = 200 GPa, determine critical value of P

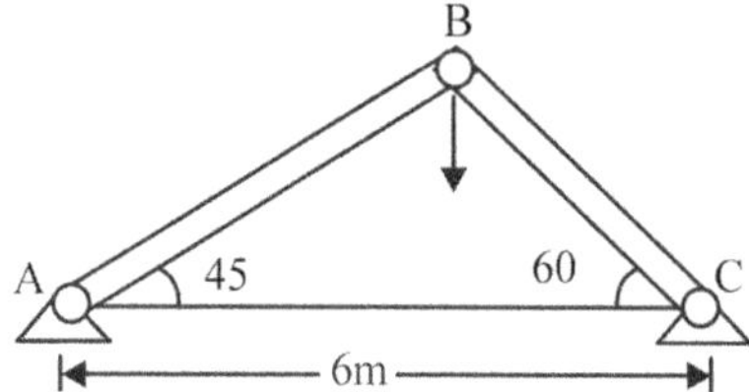

CHAPTER 14

BENDING AND BUCKLING OF THIN PLATES

Thin plate is a sheet of material whose thickness is very small compared to other dimensions. It is capable of resisting bending in addition to membrane forces. A thin plate such as stressed skin bounded by adjacent stringers and ribs, forms one basic part of an aeroplane structure. It can be analysed by solving a differential equation using an exact theory or by energy method. The latter approach is also used subsequently to determine buckling loads for unstiffened and stiffened panels.

14.1   ANALYSIS OF ISOTROPIC FLAT PLATES IN TENSION

Let us consider a thin flat plate of thickness 't' in X-Y plane subjected to in-plane tensile loads along X-direction and Y-direction on the two parallel edges. The stresses on a small elemental area of size $\delta x \times \delta y$ are as given in Fig.14.1. It is customary to represent stresses in a thin plate as forces 'N' per unit length, where $N = \sigma.t$

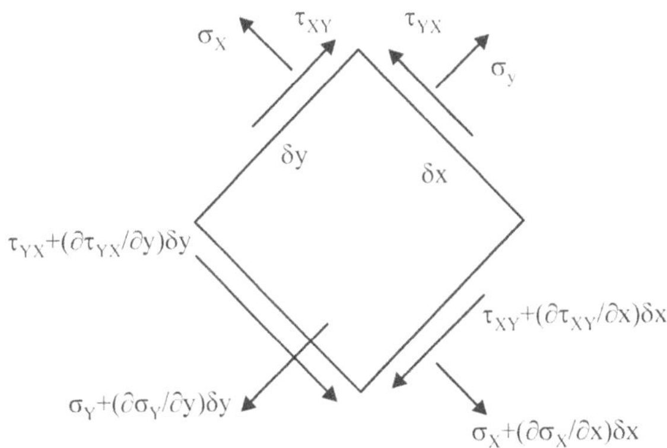

FIGURE 14.1 Stresses in a thin plate due to in-plane loads

279

For equilibrium in X-direction,

$$\left(\sigma_X + \frac{\partial \sigma_X}{\partial x}\delta x\right) \times t \times \delta y - \sigma_X \times t \times \delta y + \left(\tau_{YX} + \frac{\partial \tau_{YX}}{\partial y}\delta y\right) \times t \times \delta x - \tau_{YX} \times t \times \delta x = 0$$

Dividing by (t × δx × δy), we get $\dfrac{\partial \sigma_X}{\partial x} + \dfrac{\partial \tau_{YX}}{\partial y} = 0$ .....(14.1)

Similarly, for equilibrium in Y-direction,

$$\left(\sigma_Y + \frac{\partial \sigma_Y}{\partial y}\delta y\right) \times t \times \delta x - \sigma_Y \times t \times \delta x + \left(\tau_{XY} + \frac{\partial \tau_{XY}}{\partial x}\delta x\right) \times t \times \delta y - \tau_{YX} \times t \times \delta y = 0$$

Dividing by (t×δx×δy), we get $\dfrac{\partial \sigma_Y}{\partial y} + \dfrac{\partial \tau_{XY}}{\partial x} = 0$ .....(14.2)

For equilibrium about Z-axis, taking moments at the centre of the element,

$$\sigma_{XY} \times t \times \delta y \times (\delta x / 2) + \left(\tau_{XY} + \frac{\partial \tau_{XY}}{\partial x}\delta x\right) \times t \times \delta y \times (\delta x / 2) - \tau_{YX} \times t \times \delta x \times (\delta y / 2)$$

$$- \left(\tau_{YX} + \frac{\partial \tau_{YX}}{\partial y}\delta y\right) \times t \times \delta x \times (\delta y / 2) = 0$$

Neglecting higher order terms and dividing by (t × δx × δy), we get

$$\tau_{XY} \times \delta y \times (\delta x / 2) + \tau_{XY} \times \delta y \times (\delta x / 2) - \tau_{YX} \times \delta x \times (\delta y / 2) - \tau_{YX} \times \delta x \times (\delta y / 2) = 0$$

Dividing throughout by (δx×δy), $\tau_{XY} = \tau_{YX}$ .....(14.3)

From eq (14.1), if $\sigma_X = 0$,

$\partial \tau_{YX}/\partial y = 0$ or τ_{YX} does not vary along Y-direction

Similarly, from eq (14.2), if $\sigma_Y = 0$,

$\partial \tau_{XY}/\partial x = 0$ or τ_{XY} does not vary along X-direction

In-plane loads produce displacements in X and Y directions only and w = 0

14.2   PURE BENDING OF THIN PLATES

A thin rectangular plate is subjected to bending moments of intensity M_X (along edges parallel to Y-axis) and M_Y (along edges parallel to X-axis), as shown in Fig. 14.2. As in the case of simple beam theory, assuming plane section before bending remains plane after bending, a middle plane (also called neutral plane) does not deform. If the radii of curvature of the neutral plane are R_X and R_Y in the X-Z plane and Y-Z plane respectively, direct strains ε_X and ε_Y corresponding to direct stresses σ_X and σ_Y at a distance 'z' from the neutral plane (Fig. 14.3a) are given by $\varepsilon_X = z\,/\,R_X$ and $\varepsilon_Y = z\,/\,R_Y$ (Fig. 14.3b).

Using the stress-strain relations, $\varepsilon_X = (\sigma_X - \nu\,\sigma_Y)\,/\,E$ and $\varepsilon_Y = (\sigma_Y - \nu\,\sigma_X)\,/\,E$,

$$\text{we get } \sigma_X = \frac{E \times z}{(1-\nu^2)}\left[\frac{1}{R_X} + \nu\frac{1}{R_Y}\right] \text{ and } \sigma_Y = \frac{E \times z}{(1-\nu^2)}\left[\frac{1}{R_Y} + \nu\frac{1}{R_X}\right]$$

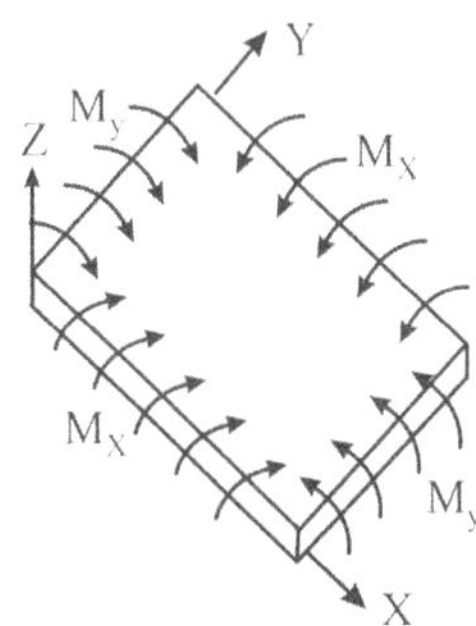

FIGURE 14.2 Pure bending of a thin plate

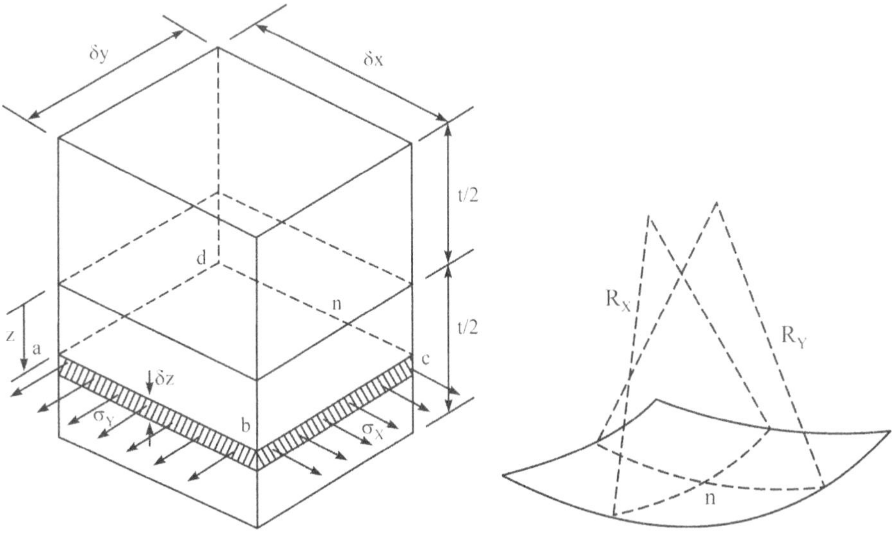

(a) Normal stresses in a plane   (b) Bending of neutral plane

FIGURE 14.3 Bending stresses in a thin plate

The moments can be expressed in terms of stresses as

$$M_X = \int \sigma_X \times z \times dz = \int \frac{E \times z^2}{(1-v^2)} \left[\frac{1}{R_X} + v\frac{1}{R_Y} \right] dz \quad \text{in the limits } -t/2 \le z \le t/2$$

$$= \frac{E \times t^3}{12(1-v^2)} \left[\frac{1}{R_X} + v\frac{1}{R_Y} \right] = D \left[\frac{1}{R_X} + v\frac{1}{R_Y} \right]$$

$$\text{and } M_Y = \int \sigma_Y \times z \times dz = \int \frac{E \times z^2}{(1-v^2)} \left[\frac{1}{R_Y} + v\frac{1}{R_X} \right] dz \quad \text{in the limits } -t/2 \le z \le t/2$$

$$= \frac{E \times t^3}{12(1-v^2)} \left[\frac{1}{R_Y} + v\frac{1}{R_X} \right] = D \left[\frac{1}{R_Y} + v\frac{1}{R_X} \right] \qquad \dots(14.4)$$

where, $D = \dfrac{E \times t^3}{12(1-v^2)}$ is called **flexural rigidity of plates**

If w is the deflection of a point on the neutral plane of the plate, then

$$1/R_X = -\partial^2 w/\partial x^2 \quad \text{and} \quad 1/R_Y = -\partial^2 w/\partial y^2 \qquad \dots(14.5)$$

Substituting these values,

$$M_X = -D \times \left[\frac{\partial^2 w}{\partial x^2} + v\frac{\partial^2 w}{\partial y^2} \right] \quad \text{and} \quad M_Y = -D \times \left[\frac{\partial^2 w}{\partial y^2} + v\frac{\partial^2 w}{\partial x^2} \right]$$

If $M_X = 0$, $\partial^2 w/\partial x^2 = -v\,\partial^2 w/\partial y^2$ and if $M_Y = 0$, $\partial^2 w/\partial y^2 = -v\,\partial^2 w/\partial x^2$

Thus, if M_X or $M_Y = 0$, the plate has curvatures of opposite sign and such a surface is called ___anticlastic surface___ (Ref Fig. 14.4).

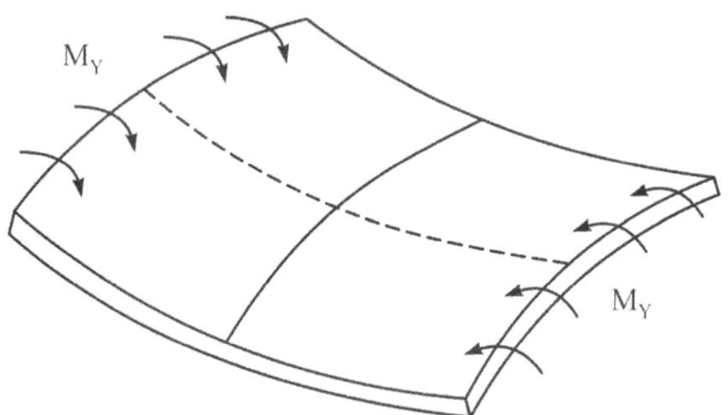

FIGURE 14.4 Anticlastic surface

If $M_X = M_Y = M$, then, $1/R_X = 1/R_Y = 1/R$ which means that the deformed shape is spherical and is of radius $R = D \times (1+v) / M$ $\dots(14.6)$

Example 14.1

A plate 10mmthick is subjected to bending moments M_X equal to 10 Nm/mm and M_Y equal to 5 Nm/mm. Calculate the maximum direct stresses in the plate.

Solution

In-plane stresses σ_X and σ_Y are given by

$$\sigma_X = \frac{E \times z}{(1-v^2)}\left[\frac{1}{R_X} + v\frac{1}{R_Y}\right] = \frac{E \times z}{(1-v^2)} \times \frac{M_X}{D}$$

and $\quad \sigma_Y = \frac{E \times z}{(1-v^2)}\left[\frac{1}{R_Y} + v\frac{1}{R_X}\right] = \frac{E \times z}{(1-v^2)} \times \frac{M_Y}{D}$

Substituting, $\quad D = E \times t^3 / 12(1-v^2)$, we get

$$\sigma_X = 12 \times z \times M_X / t^3$$

$$\sigma_Y = 12 \times z \times M_Y / t^3$$

The maximum values of σ_X and σ_Y will occur when $z = \pm\, t/2$.

Hence, $\quad \sigma_X (max) = \pm\, 6\, M_X / t^2 = \pm\, 6 \times 10 \times 10^3 / 10^2 = \pm\, 600 \text{N/mm}^2$

$$\sigma_Y (max) = \pm\, 6\, M_Y / t^2 = \pm\, 6 \times 5 \times 10^3 / 10^2 = \pm\, 300 \text{N/mm}^2$$

14.3 PLATE SUBJECTED TO BENDING AND TWISTING

In general, a bending moment may not be applied in planes perpendicular to the edges of a plate. Such moments can be resolved into components along (or tangential) and perpendicular to the edges. While the perpendicular component produces bending of the plate, tangential component (M_{XY} or M_{YX}) produces twisting of the plate. The twisting moment produces horizontal shear stresses (τ_{XY}) in the plate (Ref Fig. 14.5). M_{XY} being the twisting moment per unit length has the dimensions of force. For equilibrium of any small element of the plate, shear stresses on adjacent perpendicular edges act in pairs in opposite directions (complementing each other) or $\tau_{XY} = \tau_{XY}$.

Thus, $M_{XY} = - M_{YX}$

Similar to pure bending of a thin plate,

$$M_{XY} = -\int \tau_{XY} \times z\, dz = -\int G \times \gamma_{XY} \times z\, dz \qquad \text{in the limits} - t/2 \le z \le t/2$$

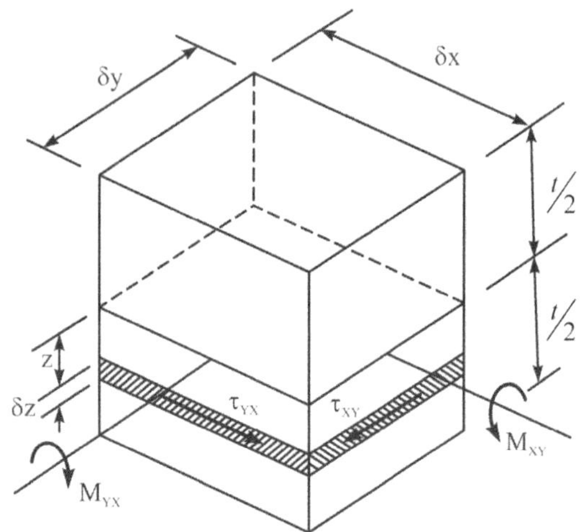

FIGURE 14.5 Stresses in a plate due to bending

Substituting, $\quad \gamma_{XY} = \dfrac{\partial v}{\partial x} + \dfrac{\partial u}{\partial y} = -2z\dfrac{\partial^2 w}{\partial x \partial y}$

since $\qquad u = -(\partial w/\partial x)\,z \quad$ and $\quad v = -(\partial w/\partial y)\,z$

we get $\qquad M_{XY} = \dfrac{G \times t^3}{6} \times \dfrac{\partial^2 w}{\partial x \times \partial y} = \dfrac{E \times t^3}{12 \times (1+v)} \times \dfrac{\partial^2 w}{\partial x \times \partial y}$

$$= D(1-v) \times \dfrac{\partial^2 w}{\partial x \times \partial y} \qquad\qquad \ldots\ldots(14.7)$$

14.4   PLATE SUBJECTED TO A DISTRIBUTED LATERAL LOAD

A plate may, in general, be subjected to distributed transverse load q which varies from point to point (function of x and y coordinates of the point). The transverse load produces shear stresses τ_{XZ} and τ_{YZ}. In order that the assumption that plane sections before bending remain plane after bending is valid, we ignore shear strains $\gamma_{XZ} = \tau_{XZ}/G$ and $\gamma_{YZ} = \tau_{YZ}/G$ as negligible. From the previous sections, we have moments

$$M_X = \int \sigma_X \times z \times dz; \quad M_Y = \int \sigma_Y \times z \times dz \quad \text{and} \quad M_{XY} = \int \tau_{XY} \times z \times dz$$

In a similar way, vertical shear forces are given by

$$Q_X = \int \tau_{XZ} \times dz \quad \text{and} \quad Q_Y = \int \tau_{YZ} \times dz$$

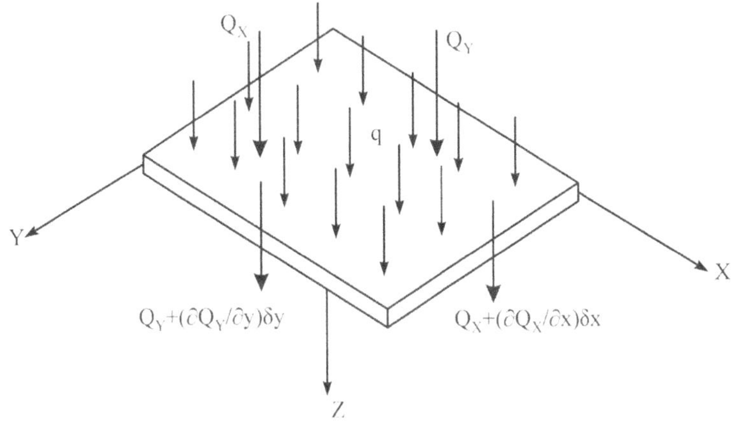

FIGURE 14.6 Thin plate with distributed lateral load

For equilibrium of a small element parallel to X-Y plane, we get

$$\partial Q_X/\partial x + \partial Q_Y/\partial y + q = 0 \qquad \qquad .....(14.8)$$

Taking moments about X-axis and simplifying, we get

$$\partial M_{XY}/\partial x - \partial M_Y/\partial y + Q_Y = 0 \qquad \qquad .....(14.9)$$

Similarly, by taking moments about Y-axis and simplifying,

we get $\quad \partial M_{XY}/\partial y - \partial M_X/\partial x + Q_X = 0 \qquad \qquad .....(14.10)$

Substituting for Q_X and Q_Y from eq (14.9) and (14.10) in eq (14.8), we get

$$\frac{\partial^2 M_X}{\partial x^2} - 2\frac{\partial^2 M_{XY}}{\partial x \partial y} + \frac{\partial^2 M_Y}{\partial y^2} = -q \qquad \qquad .....(14.11)$$

Substituting $\quad M_X = -D \times \left[\frac{\partial^2 w}{\partial x^2} + v\frac{\partial^2 w}{\partial y^2}\right] ; \quad M_Y = -D \times \left[\frac{\partial^2 w}{\partial y^2} + v\frac{\partial^2 w}{\partial x^2}\right]$

and $\quad M_{XY} = D \times (1-v)\frac{\partial^2 w}{\partial x \partial y}$

eq 14.11 can be written as $\quad \dfrac{\partial^4 w}{\partial x^4} + 2\dfrac{\partial^4 w}{\partial x^2 \partial y^2} + \dfrac{\partial^4 w}{\partial y^4} = \dfrac{q}{D}$

$$\Rightarrow \left(\frac{\partial^2}{\partial x^2} + \frac{\partial^2}{\partial y^2}\right)^2 w = \left(\nabla^2\right)^2 w = \frac{q}{D} \qquad \qquad .....(14.12)$$

where, $\quad \dfrac{\partial^2}{\partial x^2} + \dfrac{\partial^2}{\partial y^2} = \nabla^2 \quad$ is the Laplace operator

Solution to eq. (14.12) for a given transverse load q depends on the support conditions on the edges of the plate.

- **Simply supported edge** along x = 0: Along this edge, the plate is free to rotate but not to deflect. i.e., $[w]_{x=0} = 0 \Rightarrow \partial w/\partial y = \partial^2 w/\partial y^2 = 0$

   Therefore, $(M_X)_{x=0} = -D[(\partial^2 w/\partial x^2) + v \times (\partial^2 w/\partial y^2)]_{x=0} = 0$

   From the above two equations, we get $[\partial^2 w/\partial x^2]_{x=0} = 0$

- **Built-in edge** along x = 0: Along this edge, the plate is firmly clamped so that it can neither rotate nor deflect. i.e.,

   $$[w]_{x=0} = 0 \quad \text{and} \quad [\partial w/\partial x]_{x=0} = 0$$

- **Free edge** along x = 0; Along this edge, there are no bending moments, twisting moments or vertical shearing forces i.e.,

   $$(M_X)_{x=0} = 0 \; ; \; (M_{XY})_{x=0} = 0 \; \text{and} \; (Q_X)_{x=0} = 0$$

   Since the twisting moment along an edge can be replaced by a vertical force system, the last two conditions may be replaced by a single equivalent condition,

   $$\left[Q_X - \frac{\partial M_{XY}}{\partial y}\right]_{X=0} = 0 \quad \text{or} \quad \left[\frac{\partial^3 w}{\partial x^3} + (2-v) \times \frac{\partial^3 w}{\partial x \partial y^2}\right]_{X=0} = 0$$

- **Special case of a thin rectangular plate of sides 'a' (along X) and 'b' (along Y) with all four edges simply supported:** As explained above, simply supported condition along an edge results in $\partial^2 w/\partial x^2$ or $M_{XY} = 0$. Replacement of the twisting moment M_{XY} along the edges x = 0 and x = a by a vertical force distribution results in concentrated forces M_{XY} at the corners, as shown in Fig. 14.7.

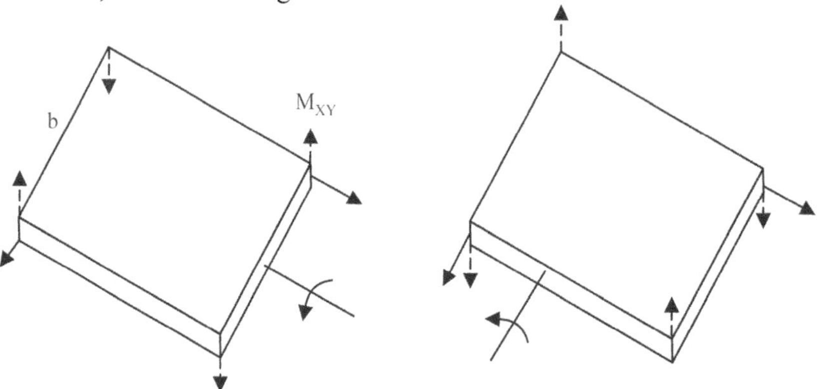

FIGURE 14.7 Corner forces in a simply supported plate with transverse load

Similarly, replacement of twisting moment M_{YX} by a vertical force distribution results in concentrated forces M_{YX} at the corners. Since $M_{XY} = -M_{YX}$, resultant forces of $2M_{XY}$ act at each corner and must be restrained, if the corners are not expected to move. In the thin plate simply supported along the four edges and subjected to uniformly distributed normal load 'q', slope $\partial w/\partial x$ is positive and numerically increasing with increasing 'y' near the corner $x = 0$ and $y = 0$. Hence, M_{XY} or $\partial^2 w/(\partial x\,\partial y)$ is positive and M_{YX} is negative, producing a resultant downward force of $2M_{XY}$. By symmetry, force at each of the remaining three corners is also $2M_{XY}$ downwards. Thus, ***all corners of the plate have a tendency to rise***.

Example 14.2

A thin rectangular plate $a \times b \times t$ is simply supported along its four edges and carries a uniformly distributed normal load q. Determine displacement w of the deflected shape and distribution of bending moment. Deduce the result if $E = 70\,GPa$, $t = 2\,cm$, $v = 0.3$, $q = 10\,N/cm^2$ and $a = b = 0.5\,m$

Solution

Deflection of the plate is the solution for the differential equation

$$\frac{\partial^4 w}{\partial x^4} + 2\frac{\partial^4 w}{\partial x^2 \partial y^2} + \frac{\partial^4 w}{\partial y^4} = \frac{q(x,y)}{D}$$

with the boundary conditions $(w)_{X=0,a} = 0$; $(\partial^2 w/\partial x^2)_{X=0,a} = 0$

and $(w)_{Y=0,b} = 0$; $(\partial^2 w/\partial y^2)_{Y=0,b} = 0$

These conditions are satisfied by representing the deflection w as an infinite trigonometric series, as

$$w = \sum_{m=1}^{\infty}\sum_{n=1}^{\infty} a_{mn} \times \sin\frac{m\pi x}{a} \times \sin\frac{n\pi y}{b}$$

where,   m = number of half waves in X-direction

and   n = number of half waves in Y-direction

If the applied load q(x,y) is also represented by a Fourier series

$$q(x,y) = \sum_{m=1}^{\infty}\sum_{n=1}^{\infty} a_{mn} \times \sin\frac{m\pi x}{a} \times \sin\frac{n\pi y}{b}$$

then, $$a_{mn} = \frac{4}{ab}\int_0^a\int_0^b q(x,y) \times \sin\frac{m\pi x}{a} \times \sin\frac{n\pi y}{b} \times dx \times dy$$

For uniform load q, we get

$$a_{mn} = \frac{16 \times q}{\pi^2 \times m \times n} \quad \text{for odd integer values of m and n}$$

and $\quad a_{mn} = 0 \qquad$ for even integer values of m and n

Then, $\quad w = \dfrac{1}{\pi^4 D} \displaystyle\sum_{m=1,3,5,..}^{\infty} \sum_{n=1,3,5..}^{\infty} \dfrac{a_{mn}}{\left[(m^2/a^2)+(n^2/b^2)\right]^2}$

$$= \frac{16q}{\pi^6 D} \sum_{m=1,3,5,..}^{\infty} \sum_{n=1,3,5..}^{\infty} \frac{\sin(m\pi x/a) \times \sin(n\pi y/b)}{\left[(m^2/a^2)+(n^2/b^2)\right]^2} \qquad .....(14.13)$$

Maximum deflection occurs at the centre of the plate ($x = a/2$ and $y = b/2$) and is given by

$$w_{max} = \frac{16q}{\pi^6 D} \sum_{m=1,3,5,..}^{\infty} \sum_{n=1,3,5..}^{\infty} \frac{\sin(m\pi/2) \times \sin(n\pi/2)}{\left[(m^2/a^2)+(n^2/b^2)\right]^2} \qquad .....(14.14)$$

The bending moments can now be obtained from

$$M_X = \frac{16q}{\pi^4} \sum_{m=1,3,5,..}^{\infty} \sum_{n=1,3,5..}^{\infty} \frac{\left[(m^2/a^2)+v(n^2/b^2)\right]}{mn\left[(m^2/a^2)+(n^2/b^2)\right]^2} \times \sin\frac{m\pi x}{a} \times \sin\frac{n\pi y}{b}$$

and $\quad M_Y = \dfrac{16q}{\pi^4} \displaystyle\sum_{m=1,3,5,..}^{\infty} \sum_{n=1,3,5..}^{\infty} \dfrac{\left[v(m^2/a^2)+(n^2/b^2)\right]}{mn\left[(m^2/a^2)+(n^2/b^2)\right]^2} \times \sin\dfrac{m\pi x}{a} \times \sin\dfrac{n\pi y}{b}$

Special case: If $a = b = 0.5m$, $t = 2cm$, $E = 70GPa$, $v = 0.3$ and $q = 10N/cm^2$

$$D = \frac{Et^3}{12(1-v^2)} = 70 \times 10^5 \times 2^3 / [12 \times (1 - 0.3^2)] = 51.28 \times 10^5 \text{ N/cm}^2$$

With $m = 1$ and $n = 1$,

$$w_{max} = \frac{16q}{\pi^6 D} \times \frac{\sin(\pi/2).\sin(\pi/2)}{\left[(1/a^2)+(1/b^2)\right]^2} = \frac{16 \times 10}{\pi^6 \times 51.28 \times 10^5} \times \frac{1}{\left[(1/50^2)+(1/50^2)\right]^2}$$

$$= 0.05086 \text{ cm}$$

Example 14.3

A thin square plate of side 'a' is simply supported along its four edges and carries a uniformly distributed normal load 'q'. With origin at the centre of the plate, show that deflection 'w' of the plate can be represented by

$$w = \frac{q}{96(1-v)D}\left[2(x^4+y^4) - 3a^2(1-v)(x^2+y^2) - 12vx^2y^2 + A\right]$$

Calculate the value of A and deflection at the origin.

Solution:

The deflection is zero at $x = \pm a/2$ and $y = \pm a/2$.

Then, from the deflection equation,

$$0 = (a^4/4) - (3/2)\,a^4 \times (1-v) - (3/4)\,a^4 \times v + A$$

Hence, $A = (a^4/4) \times (5 - 3v)$

The central deflection, i.e. at $x = 0$, $y = 0$ is then

$$w_{max} = [\,q/\{\,96\times(1-v)\times D\}\,] \times [\,(a^4/4)\times(5-3v)\,]$$

$$= [\,q\times a^4/(384\times D)\,] \times [\,(5-3v)/(1-v)\,]$$

Example 14.4

Show that the deflection function $w = A\times(x^2y^2 - b\times x^2y - a\times xy^2 + a\times b\times xy)$ is valid for a rectangular plate of sides a and b, built in on all four edges and subjected to a uniformly distributed load of intensity q. If the material of the plate has a Young's modulus E and is of thickness t, determine the distribution of bending moments along the edges of the plate

Solution

Differentiating the deflection function gives

$$\partial^4 w/\partial x^4 = 0 \;\; ; \;\; \partial^4 w/\partial y^4 = 0 \;\; \text{and} \;\; \partial^4 w/\partial x^2 \partial y^2 = 4A$$

Substituting in $$\frac{\partial^4 w}{\partial x^4} + 2\frac{\partial^4 w}{\partial x^2 \partial y^2} + \frac{\partial^4 w}{\partial y^4} = \frac{q}{D}$$

we get $0 + 2 \times 4A + 0 = \text{constant} = q/D$

The deflection function is therefore valid   and $A = q/8D$

Therefore, $w = (q/8D) \times (x^2y^2 - b \times x^2y - a \times xy^2 + a \times b \times xy)$

The bending moment distributions are given by

$$M_X = -D \times \left[\frac{\partial^2 w}{\partial x^2} + v \frac{\partial^2 w}{\partial y^2} \right] = -(q/4) \times [y^2 - b \times y + v \times (x^2 - ax)]$$

$$M_Y = -D \times \left[\frac{\partial^2 w}{\partial y^2} + v \frac{\partial^2 w}{\partial x^2} \right] = -(q/4) \times [x^2 - a \times x + v \times (y^2 - by)]$$

For the edges $x = 0$ and $x = a$, $M_x = -(q/4) \times (y^2 - by)$; $M_y = -(vq/4) \times (y^2 - by)$

For the edges $y = 0$ and $y = b$, $M_x = -(vq/4) \times (x^2 - ax)$; $M_y = -(q/4) \times (x^2 - ax)$

14.5 PLATE SUBJECTED TO LATERAL LOAD AND IN-PLANE LOAD

Determination of the contribution of shear stress due to in-plane loads to the equilibrium in the Z-direction of any small element is complicated since the element has curvature in both X-Z and Y-Z planes. Component of shear stress τ_{XY} in Z-direction is obtained from Fig 14.8 as

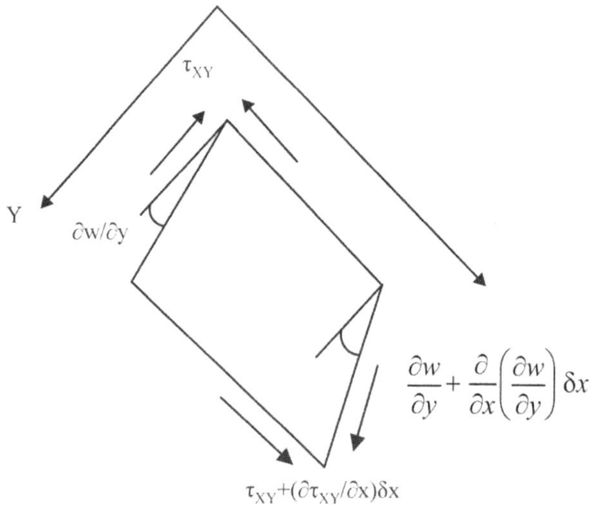

FIGURE 14.8 Z-component of shear stress in X-Y plane

$$(\tau_{XY})_Z = \left(\tau_{XY} + \frac{\partial \tau_{XY}}{\partial x} \times \delta x \right) \times \left(\frac{\partial w}{\partial y} + \frac{\partial^2 w}{\partial x \partial y} \times \delta x \right) \times \delta y - \tau_{XY} \times \frac{\partial w}{\partial y} \times \delta y$$

which gets simplified, by neglecting higher order derivatives, as

$$(\tau_{XY})_Z = \tau_{XY} \times \frac{\partial^2 w}{\partial x \partial y} \times \delta x \times \delta y + \frac{\partial \tau_{XY}}{\partial x} \times \frac{\partial w}{\partial y} \times \delta x \times \delta y$$

Similarly, contribution of τ_{YX} in Z-direction is $\quad \tau_{YX}\dfrac{\partial^2 w}{\partial x \partial y}\delta x \delta y + \dfrac{\partial \tau_{YX}}{\partial y}\dfrac{\partial w}{\partial x}\delta x \delta y$

The component of σ_X in Z-direction is obtained from Fig 14.9 as

$$(\sigma_X)_Z = \left(\sigma_X + \frac{\partial \sigma_X}{\partial x}\times \delta x\right)\times \left(\frac{\partial w}{\partial y} + \frac{\partial^2 w}{\partial x \partial y}\times \delta x\right)\times \delta y - \sigma_X \times \frac{\partial w}{\partial y}\times \delta y$$

which gets simplified, by neglecting higher order derivatives, as

$$(\sigma_X)_Z = \sigma_X \times \frac{\partial^2 w}{\partial x^2}\times \delta x \times \delta y + \frac{\partial \sigma_X}{\partial x}\times \frac{\partial w}{\partial x}\times \delta x \times \delta y$$

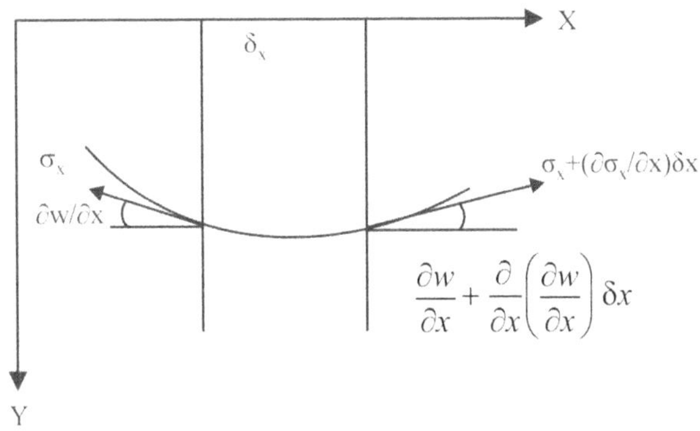

FIGURE 14.9 Z-component of normal stress in X-Y plane

Similarly, contribution of σ_Y in Z-direction is

$$(\sigma_Y)_Z = \sigma_Y \times \frac{\partial^2 w}{\partial y^2}\times \delta x \delta y + \frac{\partial \sigma_Y}{\partial y}\times \frac{\partial w}{\partial y}\times \delta x \delta y$$

The total force from all these components, including applied transverse load q, in Z-direction is given by their summation as

$$q' = q + \sigma_X \frac{\partial^2 w}{\partial x^2}\delta x \delta y + \frac{\partial \sigma_X}{\partial x}\frac{\partial w}{\partial x}\delta x \delta y + \sigma_Y \frac{\partial^2 w}{\partial y^2}\delta x \delta y + \frac{\partial \sigma_Y}{\partial y}\frac{\partial w}{\partial y}\delta x \delta y$$

$$+ \tau_{XY}\frac{\partial^2 w}{\partial x \partial y}\delta x \delta y + \frac{\partial \tau_{XY}}{\partial x}\frac{\partial w}{\partial y}\delta x \delta y + \tau_{YX}\frac{\partial^2 w}{\partial x \partial y}\delta x \delta y + \frac{\partial \tau_{YX}}{\partial y}\frac{\partial w}{\partial x}\delta x \delta y$$

which can be simplified, by using the eq 14.1, 14.2 and 14.3, as

$$q' = q + \left(\sigma_X \frac{\partial^2 w}{\partial x^2} + 2\tau_{XY}\frac{\partial^2 w}{\partial x \partial y} + \sigma_Y \frac{\partial^2 w}{\partial y^2}\right)\delta x \delta y$$

The vertical deflection w is obtained from Eq (14.12)

$$\frac{\partial^4 w}{\partial x^4} + 2\frac{\partial^4 w}{\partial x^2 \partial y^2} + \frac{\partial^4 w}{\partial y^4} = \left(\nabla^2\right)^2 w = \frac{q'}{D} \qquad\qquad ....(14.15)$$

using the appropriate boundary conditions to evaluate integration constants.

14.6   ENERGY METHOD FOR THE BENDING OF THIN PLATES

Principle of stationary value of total potential energy (sum of strain energy U due to induced stresses and potential energy V due to the applied loads) can also be used to calculate deflection w of a thin plate. If the deflection w can be assumed to match with the actual deformation, which is usually very rare, we get exact solution. A more common approach, Rayleigh-Ritz method, assumes the deflection in the form of a finite series satisfying boundary conditions and obtains an approximate solution.

Considering a small element of size $\delta x \times \delta y$,

Strain energy due to $M_X = \dfrac{1}{2} M_X \delta y \left[-\dfrac{\partial^2 w}{\partial x^2} \delta x \right]$ (Ref Fig 14.9)

Similarly, Strain energy due to $M_Y = \dfrac{1}{2} M_Y \delta x \left[-\dfrac{\partial^2 w}{\partial y^2} \delta y \right]$

Strain energy due to $M_{XY} = \dfrac{1}{2} M_{XY} \delta y \left[-\dfrac{\partial^2 w}{\partial x \partial y} \delta x \right]$ (Ref Fig 14.8)

Strain energy due to $M_{YX} = \dfrac{1}{2} M_{YX} \delta x \left[-\dfrac{\partial^2 w}{\partial y \partial x} \delta y \right]$

Substituting $M_X = -D \times \left[\dfrac{\partial^2 w}{\partial x^2} + v \times \dfrac{\partial^2 w}{\partial y^2} \right]$; $M_Y = -D \times \left[\dfrac{\partial^2 w}{\partial y^2} + v \times \dfrac{\partial^2 w}{\partial x^2} \right]$

and $M_{XY} = M_{YX} = D \times (1-v) \times \dfrac{\partial^2 w}{\partial x \partial y}$

we get, strain energy for the element as

$$U_e = \frac{D}{2} \times \left[\left(\frac{\partial^2 w}{\partial x^2} \right)^2 + \left(\frac{\partial^2 w}{\partial y^2} \right)^2 + 2v \times \frac{\partial^2 w}{\partial x^2} \times \frac{\partial^2 w}{\partial y^2} + 2(1-v) \times \left(\frac{\partial^2 w}{\partial x \partial y} \right)^2 \right] \times \delta x \times \delta y$$

Total strain energy for the entire plate is now obtained by integration as

$$U = \frac{D}{2} \times \int_0^a \int_0^b \left[\left(\frac{\partial^2 w}{\partial x^2} \right)^2 + \left(\frac{\partial^2 w}{\partial y^2} \right)^2 + 2\nu \times \frac{\partial^2 w}{\partial x^2} \times \frac{\partial^2 w}{\partial y^2} + 2(1-\nu) \times \left(\frac{\partial^2 w}{\partial x \partial y} \right)^2 \right] \times dx \times dy$$

$$\ldots.(14.16)$$

If the plate is subjected to pure bending only, $M_{XY} = 0$.

Therefore, $U = \dfrac{D}{2} \times \int_0^a \int_0^b \left[\left(\dfrac{\partial^2 w}{\partial x^2} \right)^2 + \left(\dfrac{\partial^2 w}{\partial y^2} \right)^2 + 2\nu \times \dfrac{\partial^2 w}{\partial x^2} \times \dfrac{\partial^2 w}{\partial y^2} \right] \times dx \times dy$

Potential energy due to transverse load, $V_t = -\int_0^a \int_0^b w \times q \times dx \times dy$

If the in-plane loads per unit length are N_X, N_Y and N_{XY}, potential energy due to the in-plane loads is given by

$$V_i = -\frac{1}{2} \int_0^a \int_0^b N_X \left(\frac{\partial w}{\partial x} \right)^2 dxdy - \frac{1}{2} \int_0^a \int_0^b N_Y \left(\frac{\partial w}{\partial y} \right)^2 dxdy - \frac{1}{2} \int_0^a \int_0^b N_{XY} \left(\frac{\partial w}{\partial x} \frac{\partial w}{\partial y} \right) dxdy$$

Assuming deflection of the plate as $w = A_1 f_1(x,y) + A_2 f_2(x,y) + A_3 f_3(x,y)$, where A_1, A_2 and A_3 are unknown coefficients, stationary potential energy $V = V_t + V_i$ gives

$$\frac{\partial(U+V)}{\partial A_1} = 0 \;;\quad \frac{\partial(U+V)}{\partial A_2} = 0 \;\text{ and }\; \frac{\partial(U+V)}{\partial A_3} = 0$$

As an ***example***, let us assume $w = A_{11} \times \sin\dfrac{\pi x}{a} \times \sin\dfrac{\pi y}{b}$. Then,

$$U + V = \int_0^a \int_0^b \left[\frac{DA_{11}^2}{2} \left\{ \begin{array}{l} \dfrac{\pi^4}{\left(a^2 b^2\right)^2}\left(a^2 + b^2\right)\sin^2\dfrac{\pi x}{a}\sin^2\dfrac{\pi y}{b} \\[2ex] -2(1-\nu)\times\left[\dfrac{\pi^4}{a^2 b^2}\sin^2\dfrac{\pi x}{a}\sin^2\dfrac{\pi y}{b} - \dfrac{\pi^4}{a^2 b^2}\cos^2\dfrac{\pi x}{a}\cos^2\dfrac{\pi y}{b} \right] \end{array} \right\} - qA_{11}\sin\dfrac{\pi x}{a}\sin\dfrac{\pi y}{b} \right] dxdy$$

$$= \frac{DA_{11}^2}{2} \times \frac{\pi^4}{4a^3 b^3} \times \left(a^2 + b^2\right)^2 - q \times A_{11} \times \frac{4ab}{\pi^2}$$

and $\dfrac{\partial(U+V)}{\partial A_{11}} = \dfrac{D \times A_{11} \times \pi^4}{4 \times a^3 \times b^3} \times \left(a^2 + b^2\right)^2 - q \times \dfrac{4a \times b}{\pi^2} = 0$

gives $$A_{11} = \frac{16q \times a^4 \times b^4}{\pi^6 \times D \times \left(a^2 + b^2\right)^2}$$

and $$w = \frac{16q \times a^4 \times b^4}{\pi^6 \times D \times \left(a^2 + b^2\right)^2} \times \sin\frac{\pi x}{a} \times \sin\frac{\pi y}{b}$$

At the centre of the plate, $$w_{max} = \frac{16q \times a^4 \times b^4}{\pi^6 \times D \times \left(a^2 + b^2\right)^2}$$

The solution matches with the result of example 14.2 for $m = 1$ and $n = 1$.

14.7   BENDING OF PLATES HAVING A SMALL INITIAL CURVATURE

Bending moments at any point on the plate depend on the change of curvature and produce lateral deflection. Thus, an initial deflection w_0 at any point of the mid-plane of a thin plate is equivalent to a transverse load of intensity q_0, given by

$$q_0 = \sigma_X \times \frac{\partial^2 w_0}{\partial x^2} + 2\tau_{XY} \times \frac{\partial^2 w_0}{\partial x \partial y} + \sigma_Y \times \frac{\partial^2 w_0}{\partial y^2} \qquad .....(14.17)$$

If lateral load q and in-plane loads are also applied, producing an additional deflection w_1, total deflection w $(= w_0 + w_1)$ of the mid-plane at any point is given by

$$q + \sigma_X \frac{\partial^2 \left(w_0 + w_1\right)}{\partial x^2} + 2\tau_{XY} \frac{\partial^2 \left(w_0 + w_1\right)}{\partial x \partial y} + \sigma_Y \frac{\partial^2 \left(w_0 + w_1\right)}{\partial y^2}$$

$$= D\left[\frac{\partial^4 w_1}{\partial x^4} + 2\frac{\partial^4 w_1}{\partial x^2 \partial y^2} + \frac{\partial^4 w_1}{\partial y^4}\right] \qquad ....(14.18)$$

Thus, in a plate with initial curvature, in-plane loads alone can produce bending.

If the shape of the plate with initial curvature, before applying transverse load is given by

$$w_0 = \sum_{m=1}^{\infty}\sum_{n=1}^{\infty} A_{mn} \times \sin\frac{m\pi x}{a} \times \sin\frac{n\pi y}{b} ,$$

for the particular case of compressive σ_X, we get from eq (14.18),

$$\sigma_Y = 0 \text{ and } \tau_{XY} = 0.$$

Then the additional deflection is given by

$$w_1 = \sum_{m=1}^{\infty}\sum_{n=1}^{\infty} B_{mn} \times \sin\frac{m\pi x}{a} \times \sin\frac{n\pi y}{b}$$

For a square plate, $m = 1$ and $n = 1$ give a close estimate of the maximum deflection at the centre ($x = y = a/2$).

$$w_1 = \frac{A_{11} \times \sigma_X}{\sigma_{X,CR} - \sigma_X} \quad \text{or} \quad w_1 = \sigma_{X,CR} \times \frac{w_1}{\sigma_X} - A_{11}$$

A graph of σ_X plotted against w_1/σ_X will have a slope, in the region of critical load, equal to $\sigma_{X,CR}$, and the intercept equals A_{11}, similar to the ***Southwell plot*** of a column.

14.8   ELASTIC BUCKLING OF ISOTROPIC FLAT PLATES IN COMPRESSION

A flat plate in X-Y plane, in which the thickness is small compared to the other dimensions and is subjected to compressive load along x-direction on two parallel edges, can be approximated by a beam subjected to axial compressive load. If a small rectangular element of unit length is considered, it will have elongations δ_X and δ_Y given by

$$\delta_X = (\sigma_X/E) - v \times (\sigma_Y/E) \quad \text{and} \quad \delta_Y = (\sigma_Y/E) - v \times (\sigma_X/E)$$

where v is Poisson's ratio

If the plate is assumed to have no curvature in the Y direction, $\varepsilon_Y = 0$. Then, $\sigma_Y = v \times \sigma_X$ and $\varepsilon_X = \sigma_X \times (1 - v^2) / E$ and hence, the curvature resulting from an equivalent bending moment will be smaller by the ratio $1 - v^2$.

Euler formula for a flat plate of thickness 't', whose unloaded edges (length 'a') are free and loaded edges (length 'b') are simply supported, may thus be obtained as $P_{cr} = \pi^2 \times E \times I / \{(1 - v^2) \times L^2\}$

Substituting $I = b \times t^3/12$, $L = a$ and $P_{cr} = \sigma_{cr} \times t \times b$ gives

$$\sigma_{cr} = \left[\frac{\pi^2 \times E}{12 \times (1 - v^2)}\right] \times \left(\frac{t}{a}\right)^2 \qquad \qquad \text{.....(14.19)}$$

The buckling strength for a flat sheet with a large ratio of length to width 'b', with one side simply supported and the other free can be obtained as

$$\sigma_{cr} = 0.385 \; E \times (t/b)^2$$

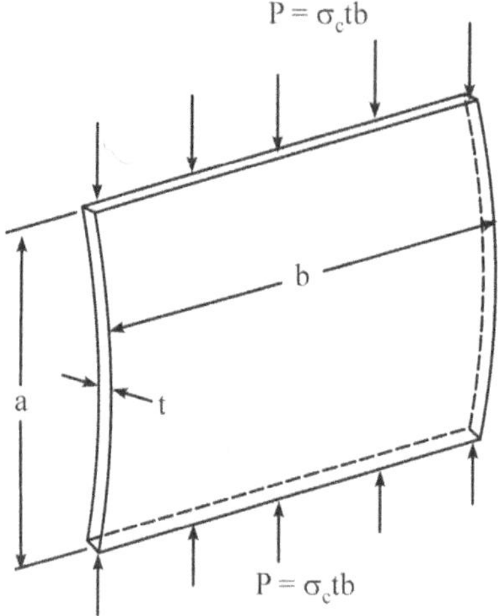

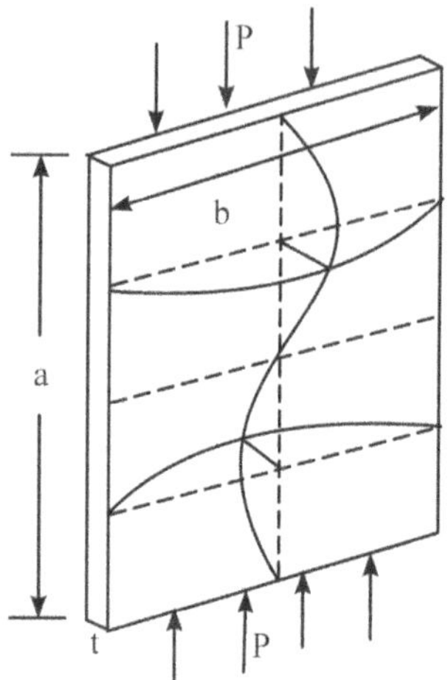

(a) Loaded edges simply supported   b) All four edges simply supported
and Unloaded edges free

FIGURE 14.10 Buckling of a flat sheet

A plate simply supported along loaded edges and free on unloaded edges buckles like a beam with hinged ends (Ref Fig 14.10a). For a plate simply supported on all four edges, the buckling compressive load is considerably higher. As the plate deflects, both vertical and horizontal strips must bend. The supporting effect of the horizontal strips may be sufficient to cause a vertical strip to deflect into two or more waves (Ref Fig. 14.10 b) and the buckling stress is given by **Bryan equation**

$$\sigma_{cr} = \frac{\pi^2 \times E}{12(1-v^2)} \times \left[\frac{bm}{a} + \frac{a}{bm}\right]^2 \times \left(\frac{t}{b}\right)^2 = K \times E \times \left(\frac{t}{b}\right)^2 \qquad .....(14.20)$$

where, m = number of half waves and is a function of a/b

$$K = [\ \pi^2 / \{12 \times (1 - v^2)\}\ ] \times C \quad \text{and is a function of a/b,}$$

and $C = [\ (bm/a) + (a/bm)\ ]^2$ is known as **plate buckling coefficient**

A few buckling mode shapes of a rectangular plate, with all four sides simply supported, obtained by an FEM software are shown in Fig 14.11.

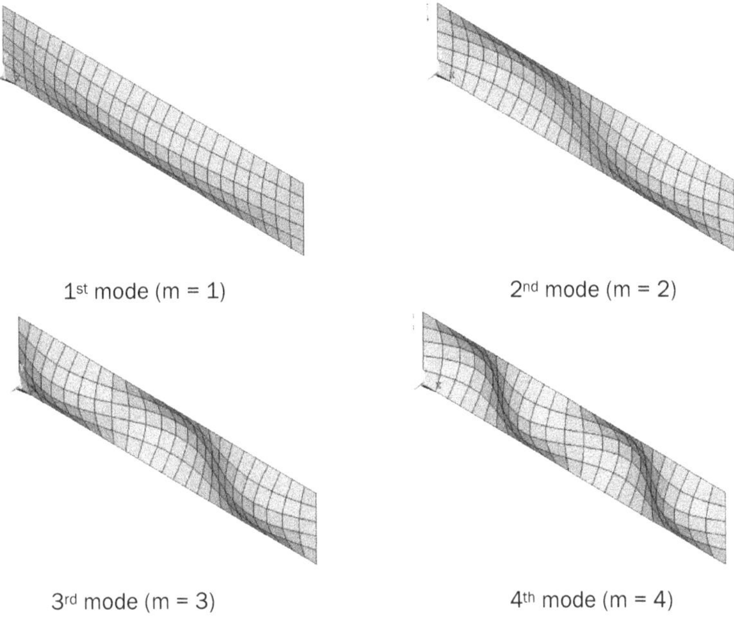

FIGURE 14.11 Buckling mode shapes of a rectangular plate, with all four sides simply supported

It is seen that wavelength of the buckles in the longitudinal (or loaded) direction reduces with load while it is same and equal to width 'b' in the transverse (or unloaded) direction in all modes. The minimum value of K, irrespective of the number of waves, for a rectangular plate with all four sides simply supported is 3.62, as seen from fig 14.12.

The buckling stress for a square plate with four edges simply supported is 4 times that obtained for a plate with two unloaded edges free and two parallel loaded edges simply supported. The buckling loads for rectangular plates with other edge conditions also can be found by using the correct values of K.

The true edge-fixity conditions for flat plates in an airplane structure can not be calculated in most cases. It is necessary to estimate the edge fixity after the supporting structure has been considered, in a manner similar to that for estimating column end-fixity conditions. K is found to be largely independent of the restraint on the loaded edges when a/b > 3, which is common in aircrafts. In most structures, value of K is used to represent a conservative mean between simply supported and clamped edge conditions.

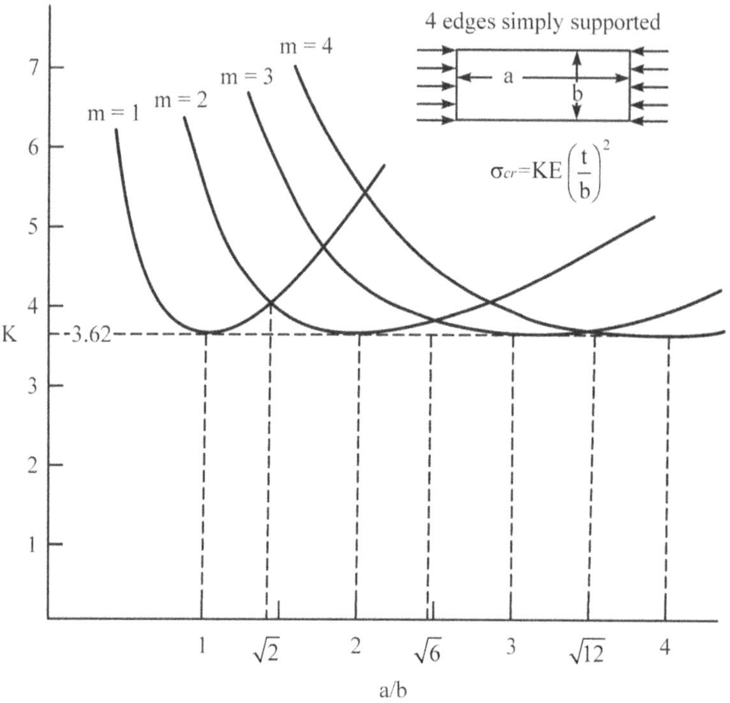

FIGURE 14.12 K Vs a/b for different values of m

The general expression for the shearing stress at the onset of buckling has the same form as eq (14.20) i.e., $\tau_{cr} = KE \times (t/b)^2$. The value of constant K for shear buckling will be different from that of buckling due to compression.

14.9 ELASTIC BUCKLING OF PLATES DUE TO SHEAR AND BENDING STRESSES

In addition to direct compression stress, other types of stresses such as shear stresses and bending stresses also may produce elastic buckling of thin plates. Critical shear stress can be expressed by the relation, $\tau_{cr} = KE \times (t/b)^2$. The value of constant K for buckling due to shear and bending will be different from that of buckling due to compression. Values of K for different conditions are given in Table 14.1.

Since the plate in shear is loaded on all four sides, the dimension 'b' is considered as the smaller of the two plate dimensions. The critical shearing stress is uniformly distributed along all four sides of the plate. The rectangular plate, which is loaded in pure shear, has principal tensile and compressive stresses at 45^0 to the edges, equal in magnitude to the shear stresses. The diagonal compressive stresses cause the sheet to buckle and wrinkles form at approximately 45^0 to the edges.

Critical buckling stresses in a thin plate loaded in bending is given by

$$(\sigma_b)_{cr} = K \times E \times (t/b)^2$$

TABLE 14.1 Values of K for long (a/b > 3) rectangular plates

Loading	Edge support	K
Compression	All edges simply supported	3.62
	All edges clamped	6.35
	3 edges simply supported, 1 unloaded edge free	0.385
Shear	All edges simply supported	5.35
	All edges clamped	8.98
Bending	All edges simply supported	23.9
	All edges clamped	41.8

14.10 PLASTIC BUCKLING OF FLAT SHEETS

Elastic buckling of flat sheets is similar to the elastic buckling of long columns in that the modulus of elasticity (E) is the only significant material property. In the case of sheets for which the thickness is greater (or b/t is small), the compressive stress (σ_{cr}) will exceed the elastic limit before buckling will occur, as is the case for short columns. Buckling stress is then expressed, replacing modulus of elasticity (E) with the tangent modulus (E_t), by the relation,

$$\sigma_{cr} = K \times E_t \times (t/b)^2 = K \times E_t / (b/t)^2$$

This equation can be compared with $P_c = \pi^2 E_t I / L^2$ or $\sigma_c = \pi^2 E_t / (L/\rho)^2$ for a short column and the critical buckling stress for a flat plate can be expressed as $\quad \sigma_{cr} = \pi^2 \times E_t / (L'/K)^2$

where, $L'/K = (\pi / \sqrt{K}) \times (b/t)$ is the slenderness ratio of an equivalent short column

It can also be expressed as $\sigma_{cr} = \eta \times K \times E \times (t/b)^2$ where η is ***plasticity correction factor***

One common application of plastic buckling is the buckling of compressive skin between rivets attaching the skin to the stringers or spars, also called ***inter-rivet buckling***. If the rivets have a uniform spacing 's' as shown in Fig. 14.13, the skin element of length 's' and infinite width has clamped ends and free sides. The element therefore resists four times the load of a similar element with hinged ends.

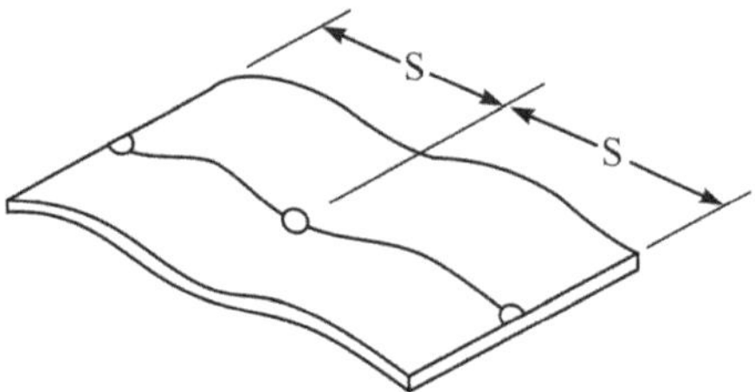

FIGURE 14.13 Buckling of riveted skin

The compressive stress is given by,

$$\sigma_{cr} = [\pi^2 E_t / 3(1 - v^2)] \times (t/s)^2 = 3.62\ E_t / (s/t)^2 \quad \text{for} \quad v = 0.3$$

$$\text{Equivalent}\quad L'/\rho = (\pi/\sqrt{3.62}) \times (s/t) = 1.65 \times (s/t)$$

Panels built-up by riveting stiffeners to the skin may experience ***inter-rivet buckling*** or ***wrinkling*** failure (also called ***forced crippling***) at a stress lower than the monolithic (stiffeners integral with the skin) crippling stress. In inter-rivet buckling (Ref Fig. 14.14a), wavelength of the buckle is equal to the pitch (or spacing) of the rivets. In wrinkling failure (Ref Fig. 14.14b), the wavelength of the buckles and the interaction with the stiffeners are different. In wrinkling, stiffener provides an elastic line support.

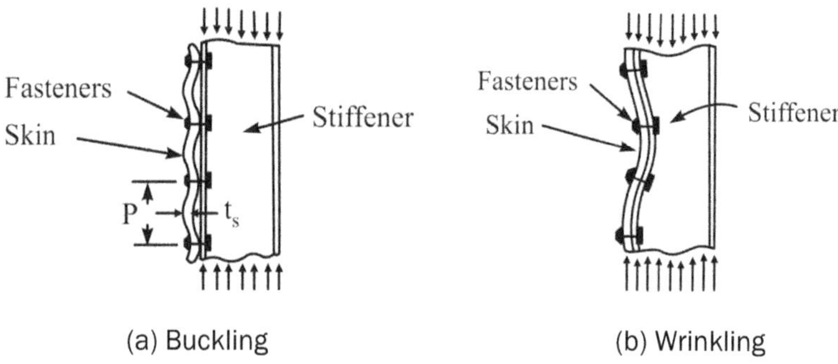

FIGURE 14.14 Comparison of buckling and wrinkling

Example 14.5

Find the compression buckling stress for a sheet 10 cm × 10 cm × 3 mm with all four edges simply supported, assuming $E_t = 55$ GPa

Solution

For each leg, one side is considered free and other edges simply supported, and so taking K = 3.62,

Equivalent $L' / k = (\pi / \sqrt{K}) \times (b/t) = (\pi / \sqrt{3.62}) \times (100/3) = 55.06$

Then, $\sigma_{cr} = \pi^2 \times E_t / (L'/k)^2 = \pi^2 \times 55000 / 55.06^2$

$= 179$ MPa or 179 N/mm^2

Example 14.6

The angle extrusion shown in Fig 14.15 is loaded in compression. Each leg of the angle buckles as a plate simply supported on the ends and on one side and free on the other side. Find the stress, at which buckling occurs. If a sheet of 1mm thick is riveted to the extrusion by rivets spaced 25mm apart, find also the compression stress in the extrusion, which produces buckling of the sheet between rivets. Assume $E_t = 55$ GPa for the extrusion material and 37.5 GPa for the sheet material; K = 3.62 for the plate with 2 sides simply supported & 2 sides free; K = 0.385 for the extrusion with 3 sides simply supported & 1 side free

Solution

For each leg, with 3 sides simply supported and 1 side free

Equivalent $L' / K = (\pi / \sqrt{K}) \times (b/t) = (\pi / \sqrt{0.385}) \times (25/1.5) = 84.42$

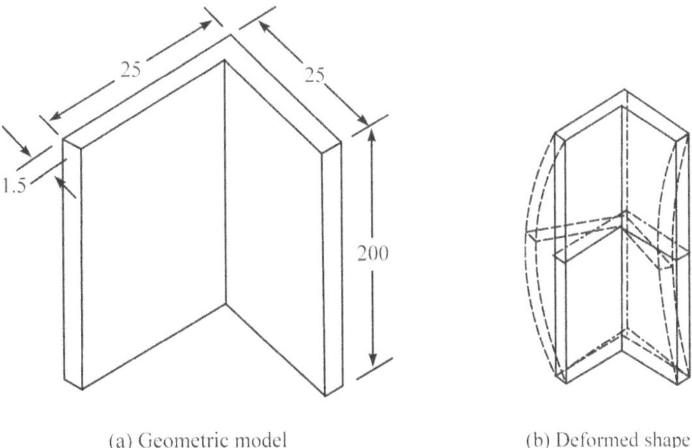

FIGURE 14.15 Angle extrusion

Then, $\sigma_{cr} = \pi^2 \, E_t \, / \, (L' / K)^2 = \pi^2 \times 55000 \, / \, 84.42^2$

$$= 76.23 \text{ MPa or } 76.23 \text{ N/mm}^2$$

For the plate with clamped ends (riveted lines) and free sides,

Equivalent $L' / K = (\pi / \sqrt{K}) \times (s / t) = (\pi / \sqrt{3.62}) \times (25/1) = 41.2$

Then, $\sigma_{cr} = \pi^2 \, E_t \, / \, (L' / k)^2 = \pi^2 \times 37500 \, / \, 41.2^2 = 218 \text{ MPa}$

14.11   COLUMNS SUBJECTED TO LOCAL CRIPPLING FAILURE

The column equations previously derived are applicable to closed tubular sections with comparatively thick walls or to other cross sections which are not subject to local crippling failure. Many of the columns used in semi-monocoque light vehicle structures are made of extruded sections or of bent sheet sections and fail by *local crippling* (also called *secondary instability*). As a result, the flange and web elements of the cross section buckle like plates and cross section of column is deformed. Local instability occurs when the weakest plate element reaches the buckling stress, usually for $L_e/k < 20$.

An approximate value of the crippling stress may be derived by finding the sum of the plastic buckling strengths of the rectangular elements (plates) of the cross sections. These plate elements are assumed to be simply supported on both sides (sections shown in Fig. 14.16 with width b), or free on one side and restrained on the other side (sections shown in Fig. 14.16 with width b'), depending on the way these plate elements are assembled to form the member. A few examples of extruded sections are shown in Fig. 14.16.

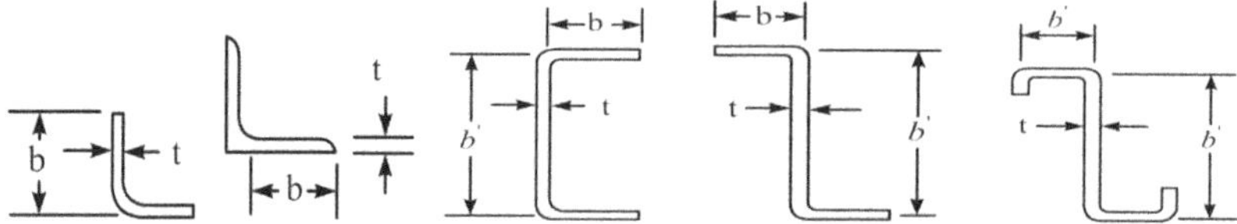

FIGURE 14.16 Different extruded column sections

The legs of the channel, shown in Fig 14.17, buckle into the same number of half waves as the back of the channel which buckles into square panels. Thus, the initial buckling stress of the plates may be smaller than the stress at which collapse of the member occurs, since the corner resists load after the initial buckling.

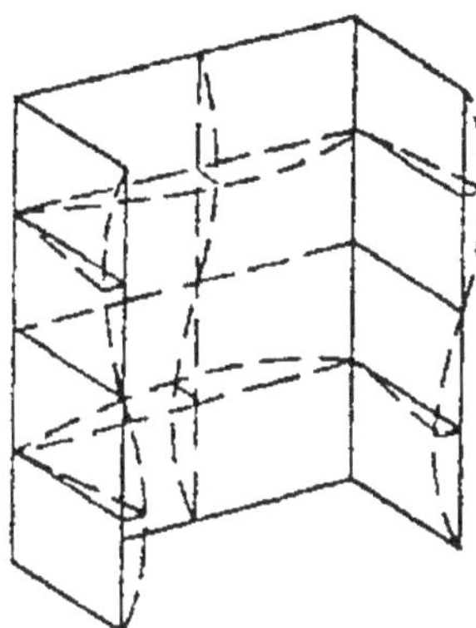

FIGURE 14.17 Buckling of an extruded channel section

This effect is considered empirically by assuming the effective width 'b' to be less than the total width. The extrusions resist a greater load at the corners than the bent sheet sections, as indicated by the widths b. If the areas have dimensions (b_1,t_1), (b_2,t_2), (b_3,t_3), .. and buckling stresses σ_1, σ_2, σ_3,.. , the average crippling stress on the cross section is found from

$$\sigma_{cc} = (\sigma_1 \times b_1 \times t_1 + \sigma_2 \times b_2 \times t_2 + \sigma_3 \times b_3 \times t_3 + \ldots) /$$
$$(b_1 \times t_1 + b_2 \times t_2 + b_3 \times t_3 + \ldots)$$

$$= (\Sigma\, \sigma \times b \times t) / (\Sigma\, b \times t) \qquad\qquad \ldots\ldots(14.21)$$

Total critical buckling load on the column,

$$P_{CR} = (P_{CR})_1 + (P_{CR})_2 + (P_{CR})_3 + \ldots$$

The denominator may not be equal to the total area because the corner areas are not included. If thickness of flanges and web is equal, $\sigma_{cc} = (\Sigma\, \sigma \times b) / \Sigma\, b$

The short column curve for aluminium alloy materials closely approximates a second-degree parabola and is tangent to the Euler curve. It is represented by

$$\sigma_c = \sigma_{cc} \times [\, 1 - \sigma_{cc} \times (L'/K)^2 / (4\pi^2 \times E)\,]$$

Example 14.7

Estimate the crippling load of a channel section of length 1m, width of flanges 200mm, depth of web 300mm and thickness of web and flanges 2mm. The channel is made of aluminium alloy with E = 69 GPa and v = 0.3.

Solution

The channel is assumed to be of an assembly of three flat plates. The edges along the junctions between elements are approximated as simply supported edges and open ends of flanges are obviously free.

The two flanges are simply supported along three sides and free on one side. The aspect ratio a/b = 5, which gives buckling coefficient K = 0.385

Local buckling load for flange (plates),

$$(P_{cr})_F = K \times E \times t^3 / b = 0.385 \times 69 \times 10^9 \times 0.002^3 / 0.2 = 1062.6 \text{ N}$$

For the web element, the plate is simply supported along the four edges. The aspect ratio a/b = 3.33, which gives buckling coefficient K = 3.62

Local buckling load for the web,

$$(P_{cr})_W = K \times E \times t^3 / b = 3.62 \times 69 \times 10^9 \times 0.002^3 / 0.3 = 6660.8 \text{ N}$$

Total crippling load for the channel section,

$$P_{cr} = 2 \times (P_{cr})_F + (P_{cr})_W = 2 \times 1062.6 + 6660.8 = 8786 \text{ N}$$

14.11.1 NEEDHAM METHOD

In this method, the member section is divided into angle elements. Through extensive tests, Needham arrived at the following semi-empirical equation for the crippling stress of angle sections.

$$\sigma_c = k_e \times \sqrt{\left(E_c / \sigma_{cy}\right)} / [(a + b)/2t]^{0.75} \qquad \ldots\ldots(14.22)$$

where, E_c = compressive modulus of elasticity

σ_{cy} = compressive yield stress

and k_e = constant, depending on support condition of angle edges

= 0.366 with no edge free

= 0.342 with one edge free

= 0.316 with two edges free

and the crippling stress for the actual section is obtained from

$$\sigma_{cc} = (\Sigma \, \sigma_{ci} \times A_i) / \Sigma \, A_i \qquad \ldots\ldots(14.23)$$

14.11.2 GERARD METHOD

It is a generalization of Needham's method and gives three semi-empirical equations for various shapes of structural members.

(i) For sections with straight unloaded edges such as plates, Tee, cruciform and H sections,

$$\sigma_c = 0.67\,\sigma_{cy} \times [\,(g\,t^2\,/\,A) \times \sqrt{(E_c\,/\,\sigma_{cy})}\,]^{0.85} < 0.8\,\sigma_{cy} \quad ....(14.24)$$

(ii) For sections with distorted unloaded edges such as tubes, angles, stiffened panels and multi-corner sections,

$$\sigma_c = 0.56\,\sigma_{cy} \times [\,(g\,t^2\,/\,A) \times \sqrt{(E_c\,/\,\sigma_{cy})}\,]^{0.4} < 0.7\,\sigma_{cy} \quad .....(14.25)$$

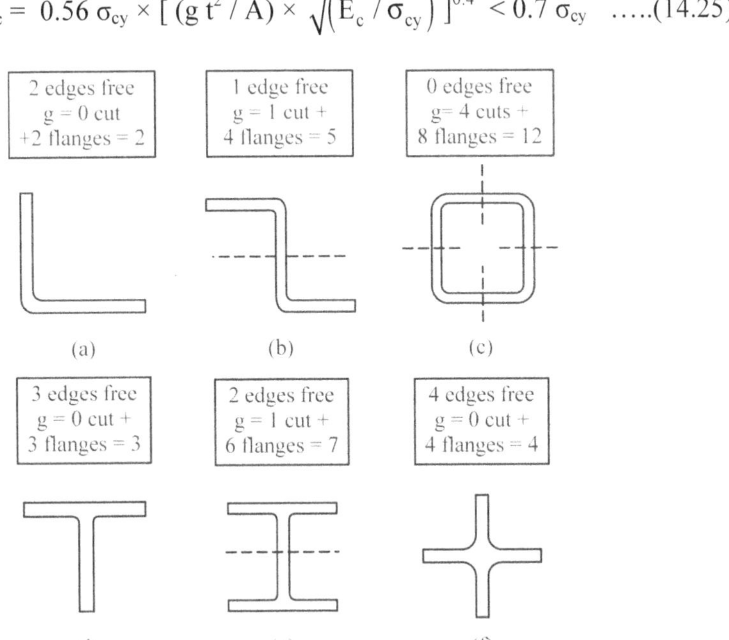

FIGURE 14.18   Idealisation of sections for estimating crippling strength

(iii) For sections such as 2-corner sections, J, Z and channel sections,

$$\sigma_c = 3.2\,\sigma_{cy} \times [\,(t^2\,/\,A)\,/\,(E_c\,/\,\sigma_{cy})^{1/3}\,]^{0.75} < 0.9\,\sigma_{cy} \qquad .....(14.26)$$

where g = number of flanges which make up the section
$\qquad\qquad$ + number of cuts required to divide the section into a number of flanges (Ref Fig 14.18)

Example 14.8

Calculate crippling stress for the given extrusion section. Assume E = 75 GPa, Thickness of web is 1.5 mm and thickness of flanges is 2 mm

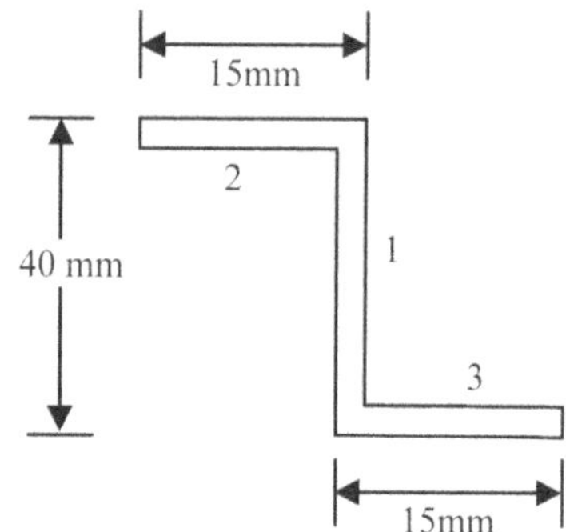

Solution

For the 4 sides simply supported case, taking K = 3.62,

Area No.	b/t	Equivalent L'/K $= (\pi/\sqrt{K})(b/t)$	Crippling stress, σ $= \pi^2 E/(L'/K)^2$
1	40/1.5 =26.7	$(\pi/\sqrt{3.62}) \times 26.7 = 44.1$	$\pi^2 \times 75000/(44.1)^2 = 380.92$MPa
2	15/2 = 7.5	$(\pi/\sqrt{3.62}) \times 7.5 = 12.9$	$\pi^2 \times 75000/(12.9)^2 = 4451.75$MPa
3	15/2 = 7.5	$(\pi/\sqrt{3.62}) \times 7.5 = 12.9$	$\pi^2 \times 75000/(12.9)^2 = 4451.75$MPa

Then,

$$\sigma_{cc} = \frac{\Sigma\,\sigma\,b\,t}{\Sigma\,b\,t} = \frac{380.92 \times 40 \times 1.5 + 4451.75 \times 15 \times 2 + 4451.75 \times 15 \times 2}{40 \times 1.5 + 15 \times 2 + 15 \times 2}$$

$$= 2416.34 \text{ MPa}$$

The short column curve is now obtained as

$$\sigma_c = \sigma_{cc} \times [\,1 - \sigma_{cc} \times (L'/K)^2/(4\pi^2 E)\,]$$

$$= 2416.34 \times [\,1 - 2416.34 \times (L'/K)^2/(4\pi^2 \times 75000)\,] \text{ MPa}$$

14.12 ELASTIC BUCKLING OF CURVED RECTANGULAR PLATES

A large part of the structure of an airplane consists of outer shell or skin, usually curved to provide the necessary aerodynamic shape and it must resist tension, compression, shear and bending stresses. Since wrinkles can adversely affect aerodynamic performance, the skin must be designed so that it will not wrinkle under normal flight conditions. The buckling stress for a curved plate in shear is higher than the buckling stress for a flat plate with corresponding dimensions. Theoretical shear buckling stress for curved plates, with all four edges simply supported, is given by

$$\tau_{cr} = K_s \times E \times (t/b)^2$$

Here, K_s is a function of (a/b) as well as (b^2/Rt) where 'R' is the radius of the curved plate. Theoretical shear buckling stresses for curved plates are higher than values obtained experimentally.

Shear buckling stress for curved plates, with all four edges simply supported, can also be expressed as

$$\tau_{cr} = K \times E \times (t/b)^2 + K_1 \times E \times (t/R)$$

where, the first term represents buckling stress for a flat plate and
the last term represents additional stress because of curvature

Thus, $K_s = K + K_1 \times (b^2/Rt)$ ($K_1 = 0.10$ is recommended)

Example 14.9

For the wing part shown in Fig. 14.19, Radius of curvature (R) = 1250 mm, sheet thickness (t) = 1.5 mm, span between stringers (b) = 150 mm and the rib spacing (L) = 450 mm. Find the compressive stress in the skin at which buckling occurs, if E = 70 GPa.

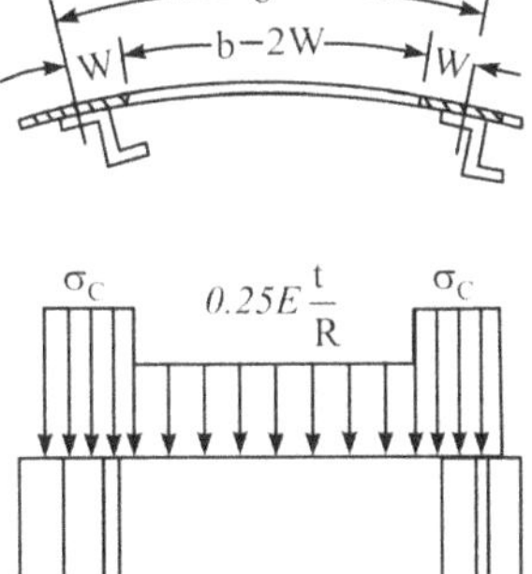

FIGURE 14.19 Wing skin with stiffeners

Solution

The buckling stress (σ_c) is obtained as the sum of the buckling stress for a flat sheet simply supported on four sides (σ_{c1}) and the buckling stress for a cylinder (σ_{c2}).

Taking K = 3.62, buckling stress for a flat plate with four edges simply supported,

$$\sigma_{c1} = K \times E \times (t/b)^2 = 3.62 \times 70000 \times (1.5/150)^2$$
$$= 25.34 \text{ MPa or } 25.34 \text{ N/mm}^2$$

Buckling stress for a cylinder is given by

$$\sigma_{c2} / E = 9 \, (t/R)^{1.6} + 0.16 \, (t/L)^{1.2} = 9 \times (1.5/1250)^{1.6} + 0.16 \times (1.5/450)^{1.2}$$
$$= 1.9 \times 10^{-4} + 1.7 \times 10^{-4} = 3.6 \times 10^{-4}$$
$$\sigma_{c2} = 70000 \times 3.6 \times 10^{-4} = 25.2 \text{ MPa}$$

Therefore, $\sigma_c = \sigma_{c1} + \sigma_{c2} = 25.34 + 25.2 = 50.54$ MPa   or   50.54 N/mm^2

14.13 STIFFENED PANELS

A stiffened panel, shown in Fig 14.20, is a plate whose buckling stress is increased by longitudinal and/or transverse stiffeners. The spacing and size of stiffeners are usually selected so that the skin and the stiffeners buckle at about the same stress level. Buckling may occur by primary instability with a wavelength of the order of panel length or by local (secondary) instability with a wavelength of the order of panel width between stiffeners. The equation for critical buckling stress has the same form as that for flat plate

i.e., $\sigma_{cr} = K \times [\pi^2 E / \{12(1 - v^2)\}] \times (t/b)^2$

except that K is a function of not only the dimensions of the plate but is also dependent upon the number and the bending rigidities of the stiffeners.

The stress in the sheet across the width is not uniform, when the sheet buckles. It is therefore customary to deal with effective width 'w', which is a function of the edge condition. Thus, $P = 2 \times t \times w \times \sigma_c$

where, $w = 0.85 \times t \sqrt{(E / \sigma_c)}$ when all the four edges are simply supported

and $w_1 = 0.60 \times t \sqrt{(E / \sigma_c)}$ when two edges and one side are simply supported and one side is free

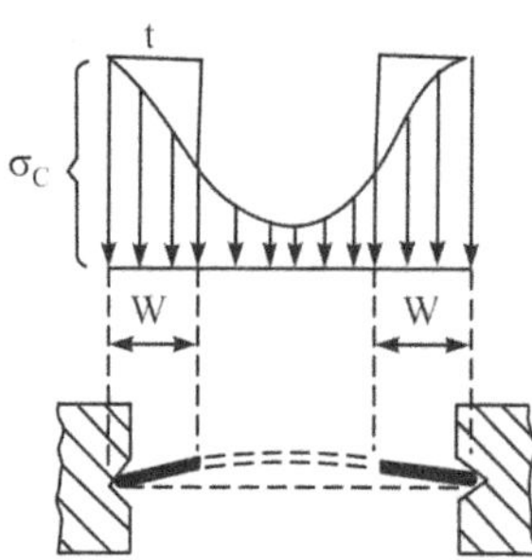

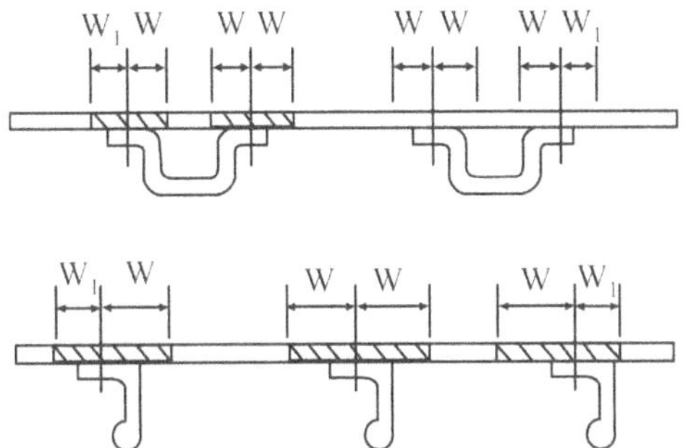

FIGURE 14.20 Stiffened panel & its deflection pattern

Example 14.10

The sheet stringer panel, consisting of 3 stringers of 25cm long located with 10cm spacing and sheet of 30 cm × 25 cm × 1 mm as shown in Fig. 14.21, is

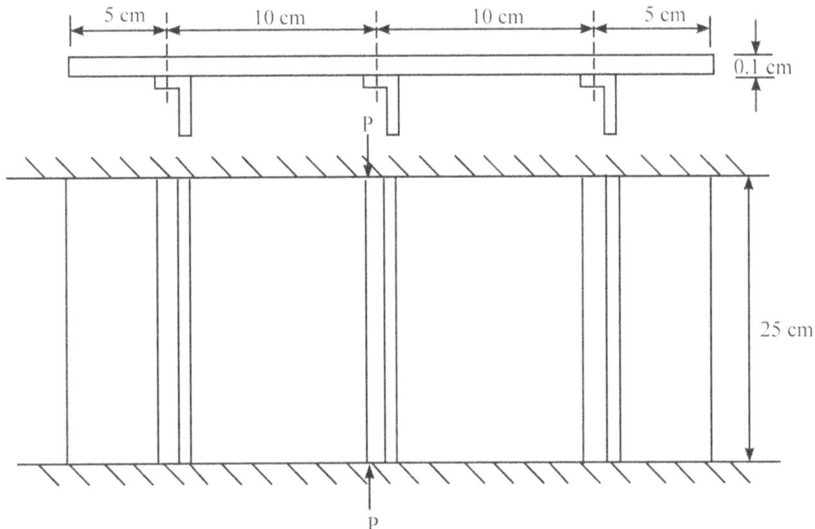

FIGURE 14.21 Buckling of riveted skin

loaded in compression by means of rigid members. The sheet is assumed to be simply supported at the loaded ends and at the rivet lines and to be free at the sides. Each stringer has an area of 50mm². Assume E = 72.5 GPa for the sheet and stringers. Find the total compressive load P,

(a) when the sheet first buckles

(b) when the stringer stress $\sigma_c = 70$ MPa

(c) when the stringer stress $\sigma_c = 210$ MPa

Solution

(a)   The sheet between stringers is simply supported on all four edges.
With K=3.62, minimum buckling stress of sheet between stringers,

$\sigma_1 = K \times E \times (t/b)^2 = 3.62 \times 72.5 \times (0.1/10)^2$ GPa

$\quad = 26.2$ MPa or 26.2 N/mm^2

The edge of the sheet is simply supported on three edges and free on the fourth edge. Therefore, with K= 0.385, minimum buckling stress of edge sheet,

$\sigma_2 = K \times E \times (t/b)^2 = 0.385 \times 72.5 \times (0.1/10)^2$ GPa

$\quad = 38.4$ MPa or 38.4 N/mm^2

Since $\sigma_1 < \sigma_2$, the sheet buckles initially between the stringers. The total area of the sheet is assumed to be effective before buckling occurs. Therefore, total load before buckling occurs is,

$P = \sigma \times A = 26.2 \times (3 \times 50 + 300 \times 1) = 11{,}790$ N

(b)   Effective sheet width w for the two middle plates, with all edges simply supported,

$$w = 0.85 \times t \sqrt{(E / \sigma_c)} = 0.85 \times 1 \times \sqrt{(72.5 \times 10^3 / 70)} = 27.4 \text{ mm}$$

Effective sheet width w_1 for the two end plates, with three edges simply supported and one edge free,

$$w_1 = 0.60 \times t \sqrt{(E / \sigma_c)} = 0.60 \times 1 \times \sqrt{(72.5 \times 10^3 / 70)} = 19.3 \text{ mm}$$

Effective sheet cross sectional area,

$A_1 = (4w + 2w_1) \times t = (4 \times 27.4 + 2 \times 19.3) \times 1 = 148.2$ mm^2

The total compressive load, $P = \sigma_c \times A = 70 \times (3 \times 50 + 148.2) = 20874$ N

(c)   Effective sheet width w for the 2 middle plates, with all edges simply supported,

$$w = 0.85 \times t \sqrt{(E / \sigma_c)} = 0.85 \times 1 \times \sqrt{(72.5 \times 10^3 / 210)} = 15.8 \text{ mm}$$

Effective sheet width w_1 for the 2 end plates, with one edge free,

$$w_1 = 0.60 \times t \sqrt{(E / \sigma_c)} = 0.6 \times 1 \times \sqrt{(72.5 \times 10^3 / 210)} = 11.14 \text{ mm}$$

Effective sheet cross sectional area,

$A_1 = (4w + 2w_1) \times t = (4 \times 15.8 + 2 \times 11.14) \times 1 = 85.48$ mm^2

The total compressive load, $P = \sigma_c \times A = 210 \times (3 \times 50 + 85.48)$

$$= 49{,}450 \text{ N} \quad \text{or} \quad 49.45 \text{ kN}$$

14.14 PURE OR COMPLETE TENSION FIELD BEAM (WAGNER'S THEORY)

The spans of aircraft wings usually comprise an upper and a lower flange connected by thin stiffened webs. Pure tension field beam is one in which the web buckles when the shearing forces are initially applied. The flange areas are assumed to resist the entire beam bending moments. The beam web has thickness 't' and depth 'h' between centroids of flanges (Ref Fig. 14.22). The shear force V is constant for all cross sections and is equal to P_S. The shear flow at all points in the web is therefore equal to V/h and shear stress $\tau = V/(t \times h)$. Vertical stiffeners are spaced uniformly at a distance 'd' along the span.

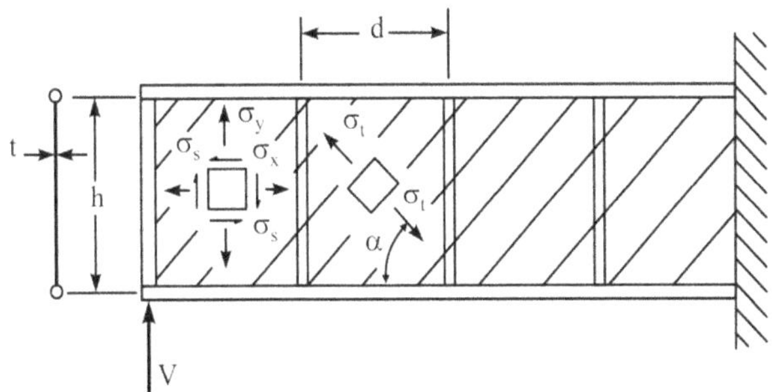

FIGURE 14.22 Shear flow in a pure tension field beam

A web element in Y-Z plane for a pure tension field web is as shown in Fig.14.23. The vertical and horizontal faces will then have normal stresses σ_z and σ_y also, in addition to the shear stress. For equilibrium,

$$\tau \times FD \times t = \sigma_t \times CD \times t \sin \alpha \quad \text{or} \quad \tau = \sigma_t \times \sin \alpha \times \cos \alpha = \sigma_t \sin \alpha \times \frac{CD}{FD}$$

Similarly, $\sigma_z = \tau / \tan \alpha = \sigma_t \times \cos^2 \alpha$ and $\sigma_y = \tau \times \tan \alpha = \sigma_t \times \sin^2 \alpha$

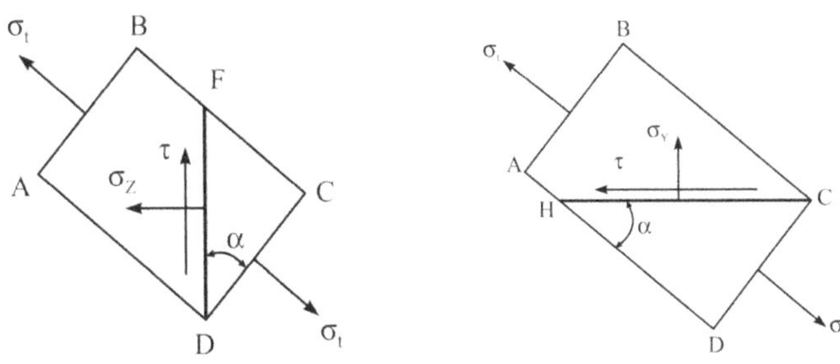

FIGURE 14.23 Stresses in a pure tension field on vertical and horizontal faces

If the load P is sufficiently high, the stiffeners will buckle as columns of equivalent length

$$L_e = h / \sqrt{4 - 2d/h} \qquad \text{for} \qquad d < 1.5\,h$$

$$\text{or} \qquad L_e = h \qquad \text{for} \qquad d > 1.5\,h$$

In a pure tension field beam, the vertical web tension stresses σ_y tend to pull the beam flanges together, which must be resisted by compression forces in the stiffeners given by

$$P_S = \sigma_Y \times t \times d = (V \times d / h) \times \tan \alpha$$

In addition to causing direct compression in the stiffeners, the direct stress σ_y produces bending of the beam flanges between stiffeners. Each flange acts as a continuous beam supported by the stiffeners and carrying a uniformly distributed load of intensity $(\sigma_Y \times t)$. The maximum bending moment M_{max} for a continuous beam, with ends fixed against rotation, occurs at a stiffener, as shown in the fig 14.24, and is given by

$$M_{max} = \frac{\sigma_Y \times I}{d} = \frac{\sigma_Y \times t \times d^2}{12} = \frac{P \times d^2 \times \tan \alpha}{12h}$$

Midway between the stiffeners, this bending moment reduces to $\dfrac{P \times d^2 \times \tan \alpha}{24h}$

In practice, both flanges and stiffeners deform so that 'α' is slightly less than 45^0, usually of the order of 40^0 and rarely below 38^0.

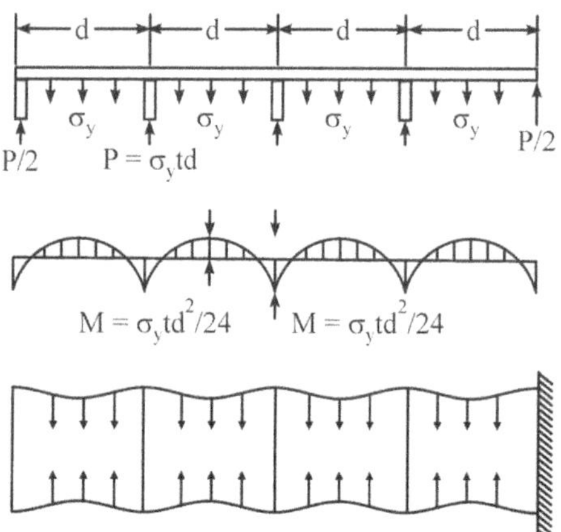

FIGURE 14.24 Bending moment distribution and bending of flange

For beams having all components made of the same material, 'α' is given by

$$\tan^2 \alpha = (\sigma_t + \sigma_F) / (\sigma_t + \sigma_S)$$

where, σ_F and σ_S are the uniform direct compressive stresses in the flanges and stiffeners respectively, induced by the diagonal tension

Substituting $\sigma_F = P / (2\, A_F \times \tan \alpha)$ and $\sigma_S = P \times d \times \tan \alpha / (A_S \times h)$,

we get $$\tan^4 \alpha = \frac{1 + t \times h / 2A_F}{1 + t \times d / A_S}$$

If the web is **shear-resistant** (Ref Fig. 14.25a), the element resists only the shear stress τ and no normal stresses on the vertical and horizontal faces X and Y (Ref Fig. 14.25b). The principal stresses σ_t and σ_c occur along the two diagonals, σ_t along the diagonal represented by solid line and σ_c along the diagonal represented by dotted line (Ref Fig. 14.25c), at angle α (equal to 45° to the horizontal in a square web), and from Mohr's circle construction, $\sigma_t = \sigma_c = \tau$. If the web is assumed to be extremely flexible, it will not be able to resist the diagonal compressive stress and acts as a group of parallel wires, inclined in the direction of tension diagonal (Ref Fig. 14.25d), at an angle α, where $\tan \alpha = h/d$ (approximately 45°).

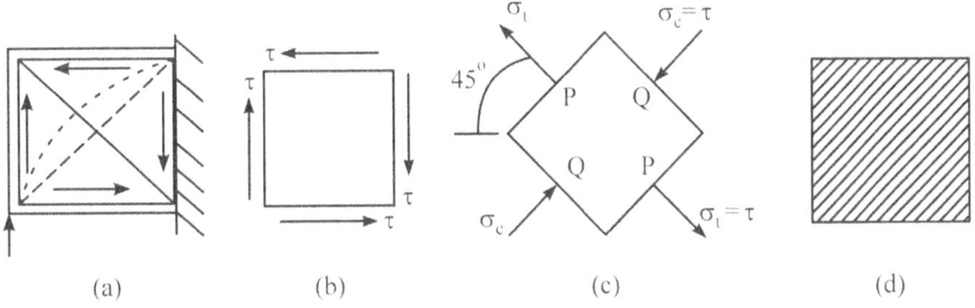

FIGURE 14.25   Principal normal stresses in a shear resistant tension-field beam

In a shear-resistant beam, the vertical stiffeners resist no compression load; they only divide the web into smaller unsupported rectangles and thus increase the web buckling stress.

Example 14.11

The beam shown in Fig 14.26 is assumed to have a pure tension field web. Draw free body diagrams for the stiffeners and flanges and plot the axial loads in the stiffeners and flanges. Assume $\alpha = 45^0$. All dimensions are in mm.

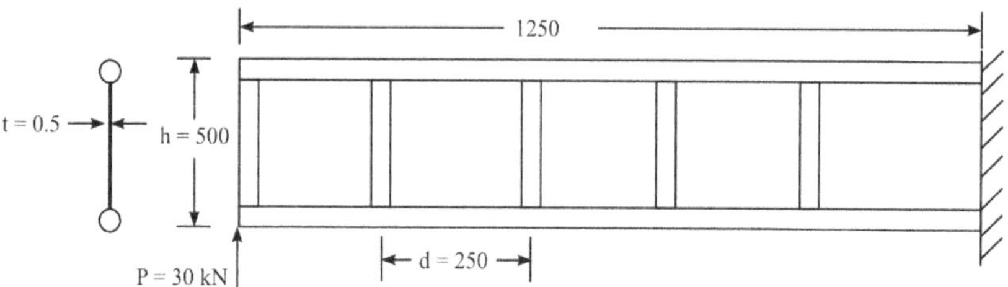

FIGURE 14.26

Solution

The shearing stress on a horizontal or vertical plane of a web element is

$$\tau = V / (t \times h) = 30000 / (0.5 \times 500) = 120 \text{ N/mm}^2$$

The running shear, $q = \tau \times t = 120 \times 0.5 = 60$ N/mm

The tensile stresses σ_X and σ_Y on these planes and the tension loads per mm also equal τ and q. The compressive load on an intermediate stiffener,

$$P_S = V \times d / h = 30 \times 250 / 500 = 15 \text{ kN}$$

The stiffener at the left end has a compressive load of $P_S/2$ and an additional compressive force of 30 kN applied at the lower end. Both beam flanges have compressive loads of $P_S/2 = 15$ kN at the left end. The beam flange loads vary linearly along the span. At the support, the flange loads from beam bending moment are $M/h = (P_S \times L) / h = 30 \times 1250 / 500 = 75$ kN

The compression flange resists a load of $- M/h - P/2 = - 75 - 15 = - 90$ kN and the tension flange resists a load of $+ M/h - P/2 = + 75 - 15 = + 60$ kN

The free body diagrams are shown in Fig 14.27.

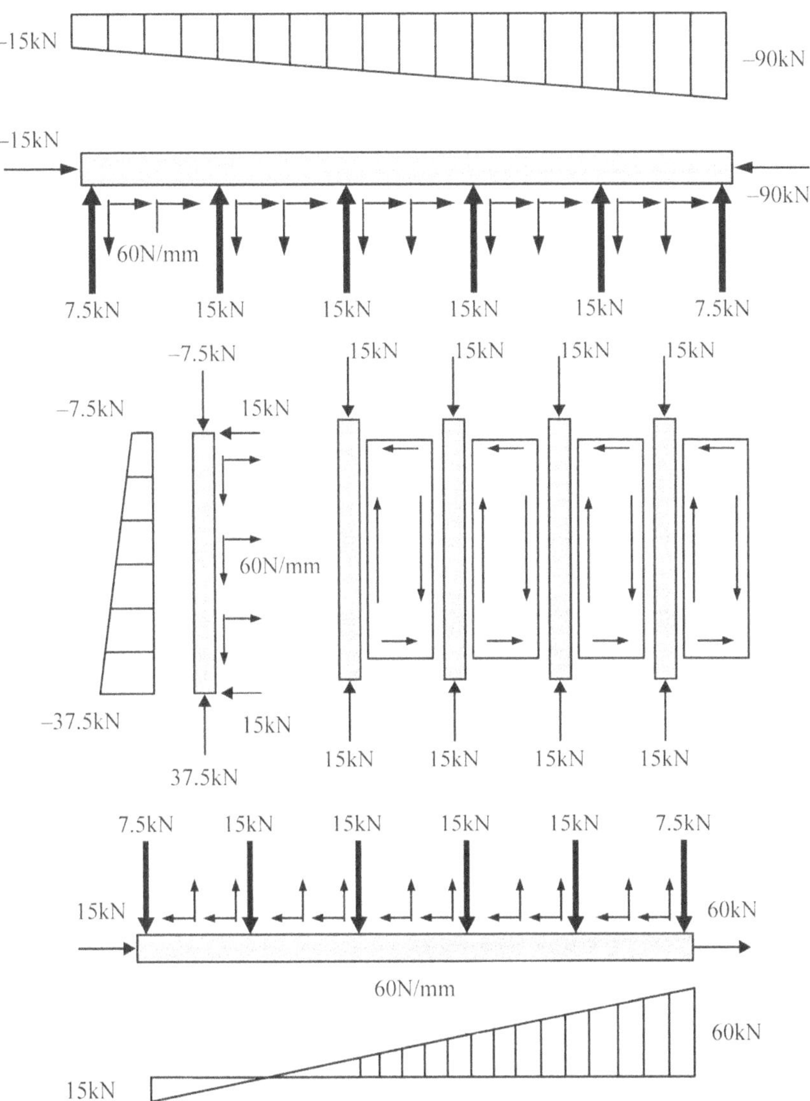

FIGURE 14.27 Free body diagrams of stiffeners and flanges

Example 14.12

The beam shown in Fig. 14.28 is assumed to have a complete tension field web. If the cross sectional areas of the flanges and stiffeners are 350mm^2 and 300mm^2 respectively and the section modulus of each flange is 750mm^3, determine the maximum stress in a flange and also whether or not the stiffeners will buckle. Thickness of the web is 2mm and second moment of area of a stiffener about an axis in the plane of the web is 2000mm^4; E = 70kN/mm^2. All dimensions are in mm.

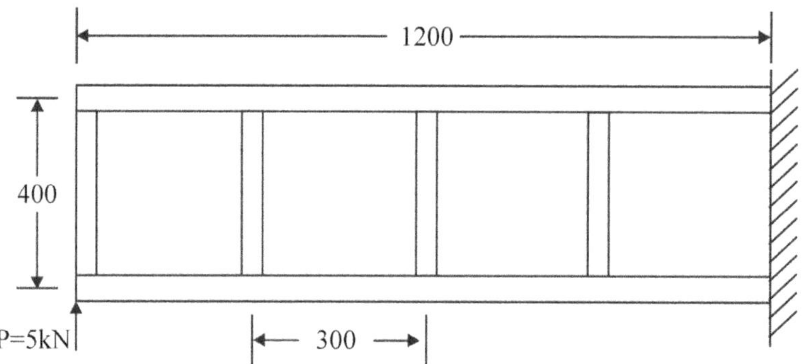

Solution

$$\tan^4 \alpha = \frac{1+t\times h / 2A_F}{1+t\times d / A_S} = \frac{1+2\times 400 / (2\times 350)}{1+2\times 300 / 300} = 0.7143 \quad \text{or} \quad \alpha = 42.6^0$$

The maximum flange stress will occur in the top flange at the fixed end where the bending moment on the beam is greatest and the stresses due to bending and diagonal tension are additive. Thus,

$$F_T = \frac{5000\times 1200}{400} + \frac{5000}{2\tan 42.6^0} = 17700\text{N}$$

Direct stress in the top flange produced by the externally applied bending moment and the diagonal tension is $17700 / 350 = 50.7 \text{ N/mm}^2$

Local bending moment in the top flange at the fixed end is calculated from

$$M_{Max} = \frac{5000\times 300^2 \times \tan 42.6}{12\times 400} = 86000\text{Nmm}$$

Maximum compressive stress, corresponding to this bending moment, at the lower surface of the top flange is $M/Z = 86000 / 750 = 114.9 \text{ N/mm}^2$

Thus, the maximum stress in a flange occurs at the built-in end, is compressive and equal to

$$114.9 + 50.7 = 165.6 \text{ N/mm}^2 \quad \text{or} \quad 0.1656 \text{ kN/mm}^2$$

Compressive load in a stiffener is

$$P_S = 5000 \times 300 \times \tan 42.6^0 / 400 = 3400 \text{ N}$$

With the given values of b = 300mm and d = 400 mm, b < 1.5 d and, therefore $L_e = 400 / \sqrt{4 - 2\times 300 / 400} = 253\text{mm}$

Critical buckling load of the stiffener,

$$P_{Cr} = \pi^2 \times 70000 \times 2000 / 253^2 = 22 \text{ kN}$$

Applied load, $P = 5000 \text{ N}$ and $P < P_{Cr}$

Therefore, the stiffener will not buckle.

14.15   INCOMPLETE DIAGONAL TENSION-FIELD BEAM

In modern aircraft structures, beams having extremely thin webs are rare. The web can resist some diagonal compressive stress after buckling. Stress in the web is somewhere between that of pure tension field web and shear resistant web. Such a beam is described as an incomplete diagonal tension field beam or semi-tension field beam and may be analysed by semi-empirical theory as follows. If 'k' is the diagonal tension factor (k=0 for unbuckled web and k=1 for a web in complete diagonal tension), the nominal web shear τ (=S/th) may be divided into

$$\text{True shear component, } \tau_S = (1 - k) \times \tau$$

and diagonal tension component, $\tau_{DT} = k \times \tau$

The compressive stresses in the flanges and stiffeners can be rewritten in terms of the applied shear stress τ as

$$\sigma_F = k \times \tau \times \cot \alpha / [\{2A_F/(t \times h)\} + 0.5(1 - k)]$$

and $\sigma_S = k \times \tau \times \tan \alpha / [\{A_S/(t \times h)\} + 0.5(1 - k)]$

The web stress σ_t consists of two direct stress components,

$$\sigma_1 = (2 \, k \times \tau / \sin 2\alpha) + \tau \times (1 - k) \times \sin 2\alpha \qquad \text{along the direction of } \alpha$$

and $\sigma_2 = -\tau \times (1 - k) \times \sin 2\alpha$ \qquad perpendicular to the direction of α

The secondary bending moment is given by $M_{max} = k \times P \times d^2 \times \tan \alpha / (12 \, h)$

and the effective length for the calculation of stiffener buckling loads becomes

$$L_C = h_S / [\, 1 + k^2 \times (3 - 2d/h_S)]^{0.5} \qquad \text{for } d < 1.5 \, h$$

or $L_C = h_S$ \qquad for $d > 1.5 \, h$

where h_S is the actual stiffener height, as opposed to the effective height h of the web

14.16   TAPERED TENSION-FIELD BEAM

In many cases beams taper along their lengths, as shown in Fig. 14.28. In that case, the flange loads are no longer horizontal but have vertical components, which reduce the shear load carried by the web. If 'h' is the height of the section considered,

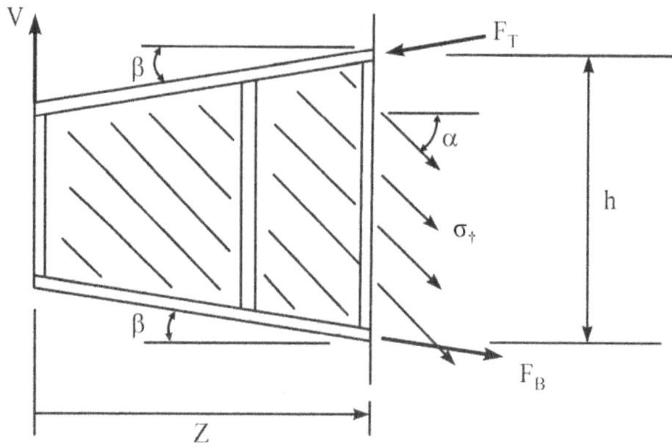

Fig 14.28 Free body diagram of a tapered tension field beam

For vertical equilibrium,

$$V - (F_T + F_B) \times \sin \beta - \sigma_t \times (h \cos \alpha) \times \sin \alpha = 0$$

For horizontal equilibrium,

$$(F_T - F_B) \times \cos \beta - \sigma_t \times t \times h \times \cos^2 \alpha = 0$$

Taking moments about B,

$$V \times z - F_T \times h \times \cos \beta + \sigma_t \times t \times h^2 \times \cos^2 \alpha = 0$$

Solving these equations will result in

$$F_T = \frac{V}{h \times \cos \beta} \left[z + \frac{h \times \cot \alpha}{2} \left\{ 1 - \frac{2z \times \tan \beta}{h} \right\} \right]$$

$$F_B = \frac{V}{h \times \cos \beta} \left[z - \frac{h \times \cot \alpha}{2} \left\{ 1 - \frac{2z \times \tan \beta}{h} \right\} \right]$$

and $$\sigma_t = \frac{2V}{t \times h \times \sin 2\alpha} \left(1 - \frac{2z \times \tan \beta}{h} \right)$$

Then, compressive load in the stiffener, $$P = \frac{V \times d}{h} \times \tan \alpha \times \left(1 - \frac{2z}{h} \times \tan \beta \right)$$

and shear force at any section of the beam, $S = V - (F_T + F_B) \times \sin \beta$

$$= V \left[1 - (2z \times \tan \beta / h) \right]$$

Additional Problems and Practice

1. (a) A uniform plate of thickness 't' has a width 'b' in the y-direction and length 'a' in the x-direction. The edges parallel to the x-axis are clamped and those parallel to the y-axis are simply supported. A uniform compressive stress σ is applied in the x-direction along the edges parallel to y-axis. Using an energy method, find an approximate expression for the magnitude of the stress σ which causes the plate to buckle, assuming that the deflected shape of the plate is given by

$$w_0 = \delta \times \sin(m\pi x/a) \times \sin^2(\pi y/b)$$

   (b) For the particular case a = 2b, find the number of half waves 'm' (to the nearest integer) corresponding to the lowest critical stress. Determine also the lowest critical stress.

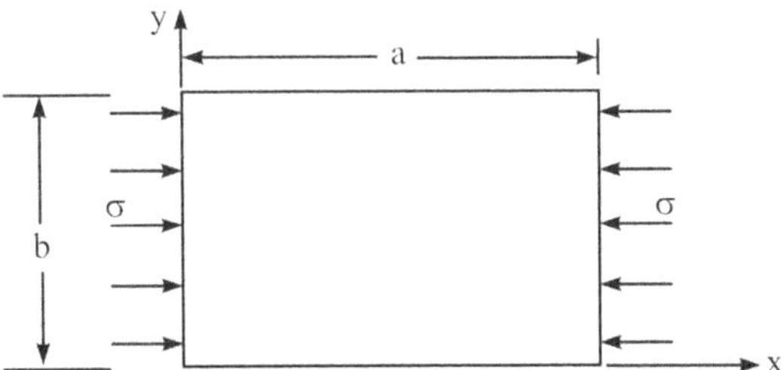

2. The spar of an aircraft is to be designed as an incomplete diagonal tension beam, the flanges being parallel. The stiffener spacing will be 250 mm, the effective depth of web will be 750 mm and the depth between web-to-flange attachments is 75 mm. The spar is to carry ultimate shear force of 100 kN. The max permissible shear stress is 165 MPa; but it is also required that the shear stress should not exceed 15 times the critical shear stress for the web panel. Assuming angle of diagonal tension, measured from the spanwise axis of the beam is 40^0, calculate the stiffener end load.

3. A simply supported beam has a span of 2.4 m and carries a central concentrated load of 10 kN. The flanges of the beam each have a cross-sectional area of $300\,mm^2$ while that of the vertical web stiffeners is $280\ mm^2$. Depth of the beam, measured between the centroids of area of the flanges, is 350 mm and the stiffeners are symmetrically arranged

about the web and spaced at 300 mm intervals. Determine the maximum axial load in a flange and the compressive load in a stiffener. Assume that the beam web, of thickness 1.5 mm, is capable of resisting diagonal tension only.

4. Find crippling stress for the angles shown in figure, using Gerard's method

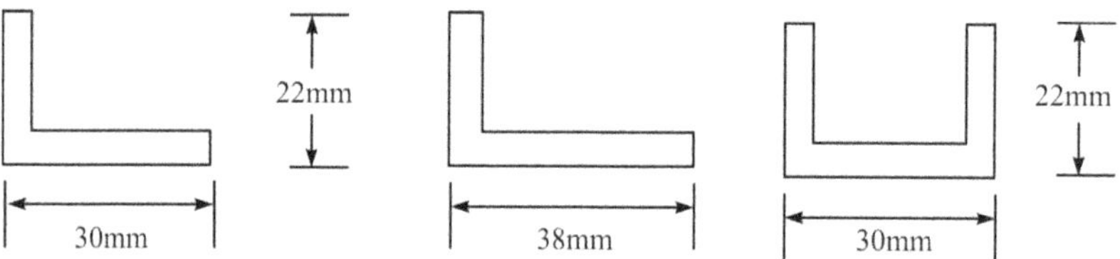

5. Find crippling stress of rectangular tubes shown in figure using Nedham's method, when formed from aluminium. Uniform thickness, t = 1.5mm

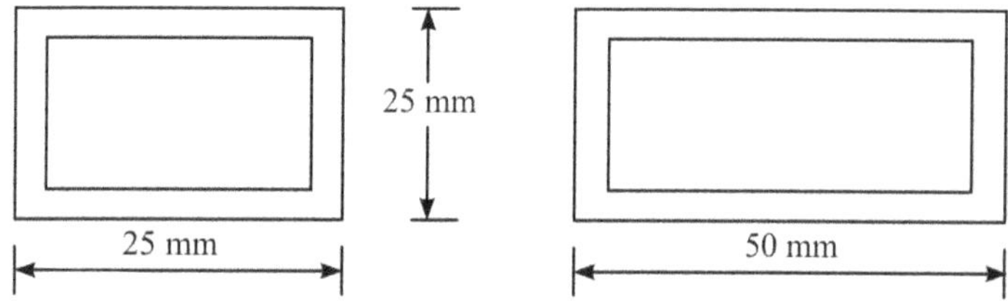

6. Determine the buckling strength of a panel, comprising flat sheet and uniformly spaced Z-section stringers, a part of whose cross section is shown in figure, under uniform compressive loads.

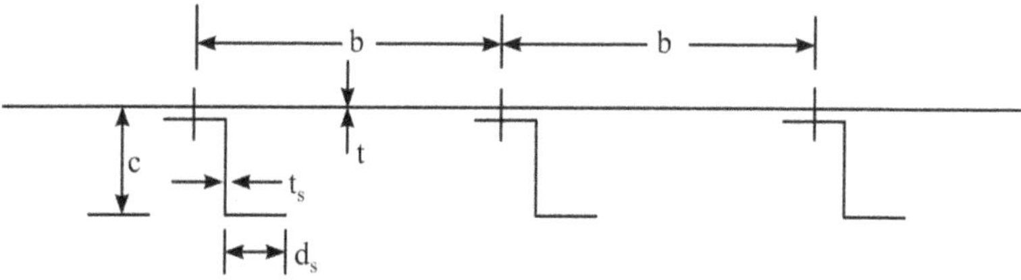

7. Determine the buckling strength of a panel, comprising flat sheet and uniformly spaced Z-section stringers, a part of whose cross section is shown in figure, under uniform compressive loads. Take E = 70 GPa and compressive stress = 300 MPa. All dimensions are in mm.

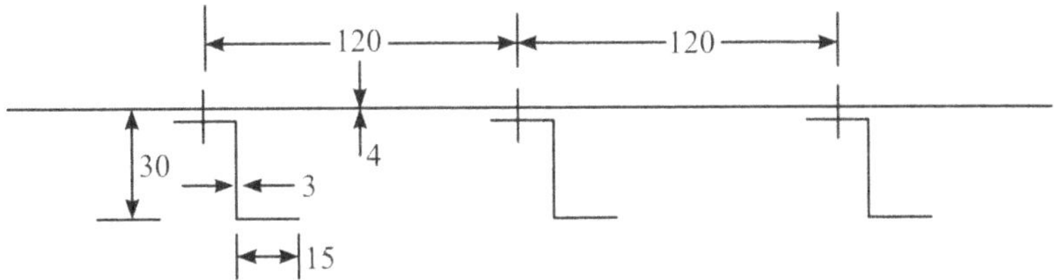

8. Determine the buckling strength of a panel, comprising flat sheet and uniformly spaced stringers, a part of whose cross section is shown in figure, under uniform compressive loads. Take E = 70 GPa and compressive stress = 300 MPa. All dimensions are in mm.

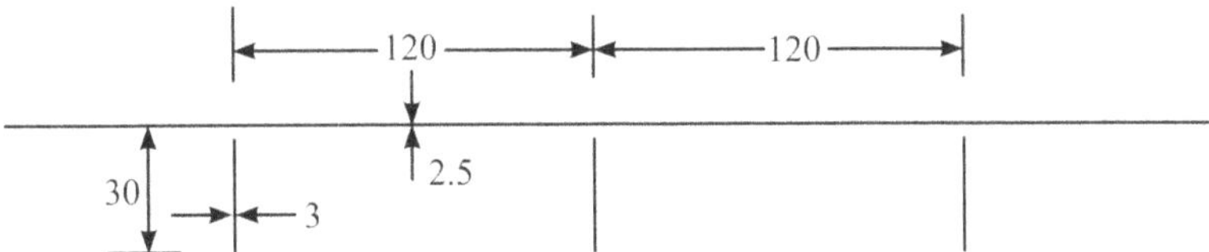

9. The spar of an aircraft is to be designed as an incomplete diagonal tension beam, the flanges being parallel. The stiffener spacing is 250 mm; effective width of web is 750 mm and depth between web-to-flange attachments is 725 mm. The spar is to carry an ultimate shear force of 100 kN. Max permissible shear stress is 165 N/mm^2. It is required that the shear stress should not exceed 15 times the critical shear stress for the web panel. Assuming angle of diagonal tension to be 40^0, calculate stiffener end load and secondary bending moment in the flanges.

10. Determine the crippling stress of the panel, formed with hat-section stiffeners, as shown. Take σ_{cy} = 470 MPa and E = 70 GPa for stiffeners while σ_{cy} = 280 MPa.

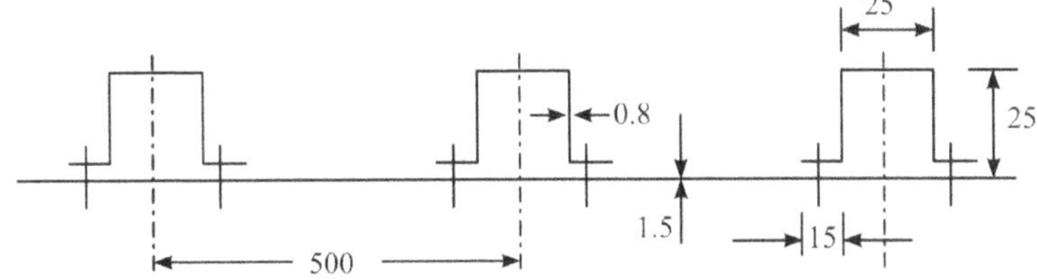

11. A simply supported panel is 600 mm wide and 1200 mm long. The 3mm thick panel is stiffened on both sides by angle section longitudinal stiffeners with 150 mm spacing, as shown. Determine the compressive stress which produces buckling of the panel. Take A = 50 mm^2,

I (through centroid) = 500 mm^4 and distance of centroid from the center of panel and z =3 mm for the stiffeners and E = 70 GPa.

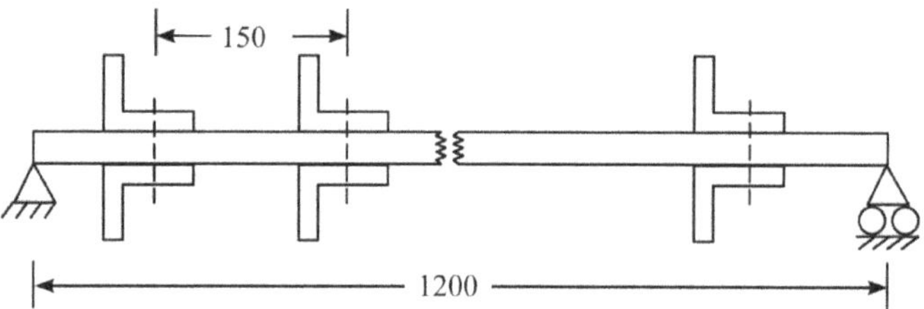

12. Derive an expression for buckling stress of a rectangular plate and deduce the relation for elastic buckling when a plate (ratio of sides, a/b>3) is simply supported on three sides and free on one edge.

13. The skin on a fuselage is supported by stringers which are spaced at 125mm and by rings spaced at 500mm. Assume E = 70GPa and an average between simply supported and clamped-edge conditions. Find the shear buckling stress for the flat sheet if   (i) t = 0.5mm and   (ii) t = 1mm

CHAPTER 15

SHEAR FLOW, SHEAR CENTER AND SHEAR LAG

15.1 SHEAR STRESSES FOR BENDING IN ONE PLANE

It is seen that strain in bending along the axis of the beam is a linear function of distance from the neutral plane, tensile on one side and compressive on the other side.

Within the elastic limit, since stress is proportional to strain, **bending (normal) stress** (σ_X) is also a linear function of distance from the neutral plane, tensile on one side and compressive on the other side. Different normal stresses along adjacent layers, parallel to neutral plane, induce shear force (S_Y) between the layers. The shear stress ($\tau_{YX} = S_Y/A$) is seen to be zero on the outermost layers and maximum along the plane where bending stress changes from tensile to compressive, giving a variable shear stress distribution across any beam section.

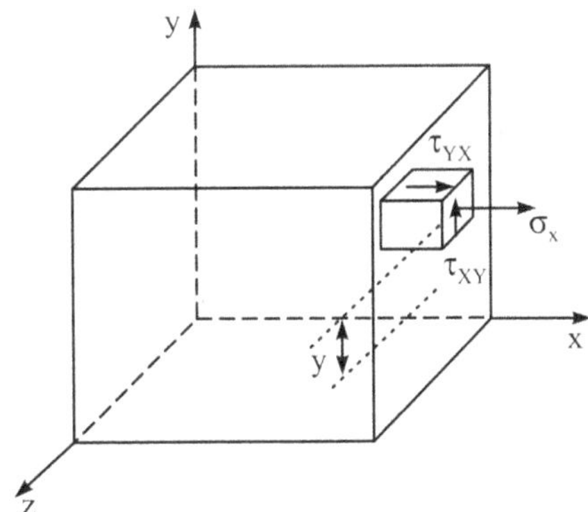

FIGURE 15.1 Complementary shear stress

The shear stress (τ_{YX}) produces complementary shear stress (τ_{XY}) on the perpendicular face, (Ref Fig. 15.1) whose sum over the cross sectional area is called shear force at that section. Similar shear stresses exist on the other faces of the infinitesimal element considered. The shear stresses τ_{ij} on the faces of the element are designated with the first subscript indicating the face (identified by the direction of its normal) on which the shear stress acts and the second subscript indicating the direction of shear stress.

The stresses τ_{XY} and σ_X are related to the shear force and bending moment at a section of the beam subjected to bending in X-Y plane (X being the axial direction), by

$$S_Y = \int \tau_{XY} \times dA \qquad \text{and} \qquad M_Z = \int \sigma_X \times y \times dA \qquad \text{.....(15.1)}$$

Similarly, for bending in X-Z plane, $S_Z = \int \tau_{XZ} \times dA$ and $M_Y = \int \sigma_X \times z \times dA$

Maximum compressive stress may not always be equal to the maximum tensile stress (as in unsymmetric T, L sections) except for beam sections which are symmetric w.r.t. the neutral plane (as in rectangular, circular, I sections). In an **unsymmetric section**, width of section on one side of neutral axis is more and smaller stress acting on the larger width will produce the same total force as larger stress on smaller width. Thus, stress is maximum on the outermost layer on the side of smaller width (or more depth) w.r.t. neutral axis, as can be seen from the Fig 15.2.

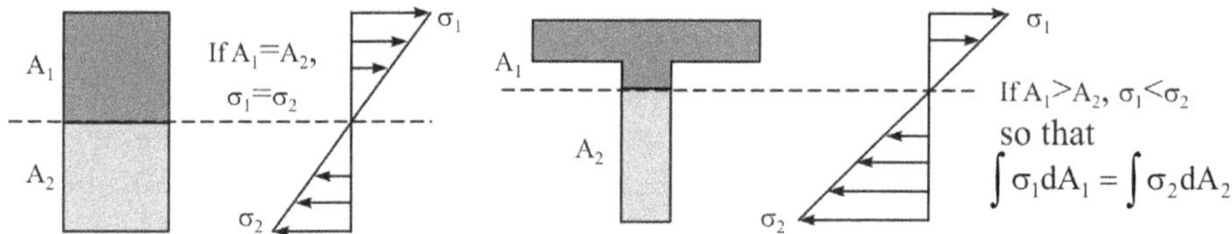

FIGURE 15.2 Bending stress distribution across different beam sections

For the moments about an axis through the center of the element in the Z-direction to be zero, $(\tau_{XY} \times dy \times dz) \times dx = (\tau_{YX} \times dx \times dz) \times dy$ or $\tau_{XY} = \tau_{YX}$

Similarly, equating moments about axes through the center of the element in the X and Y directions, $\tau_{YZ} = \tau_{ZY}$ and $\tau_{XZ} = \tau_{ZX}$

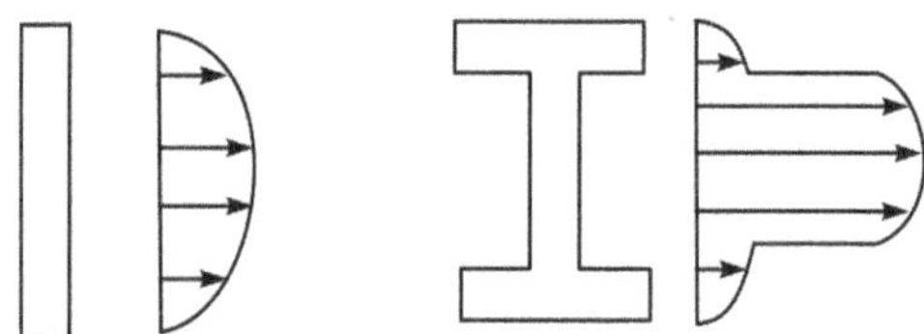

FIGURE 15.3 Shear stress distribution across different beam sections

The shear stress ($\tau_{YX} = S_Y/A$) produced across the cross section depends on thickness distribution of the cross section. The shear stress distribution for rectangular section and I-section are given in Fig. 15.3, to indicate how thickness of each layer affects shear stress distribution.

Since, there can be no abrupt changes in shear distribution, the shear stress must be zero at any sharp corner and may be very high at a sharp re-entrant angle. At the junction of the web and the flange, a radius is desirable in order to permit a good distribution of shear stress.

15.2   BENDING MOMENT APPLIED IN A LONGITUDINAL PLANE

Bending moment 'M' on a beam section may not be always be applied coinciding with the width and breadth directions of a section (M_Z or M_Y). If the neutral axis (N-A) of the plane of bending is inclined at angle 'θ' to the Z-axis (Ref Fig. 15.4), the bending moment on this plane can be resolved into two components

$$M_Y = M \times \cos\theta \quad \text{and} \quad M_Z = M \times \sin\theta \qquad \dots\dots(15.2)$$

Then, for a small area 'dA' at a distance 's' from the centroid,

$$\sigma_X = E \times s / R \qquad \text{where, } s = y \times \cos\theta + z \times \sin\theta$$

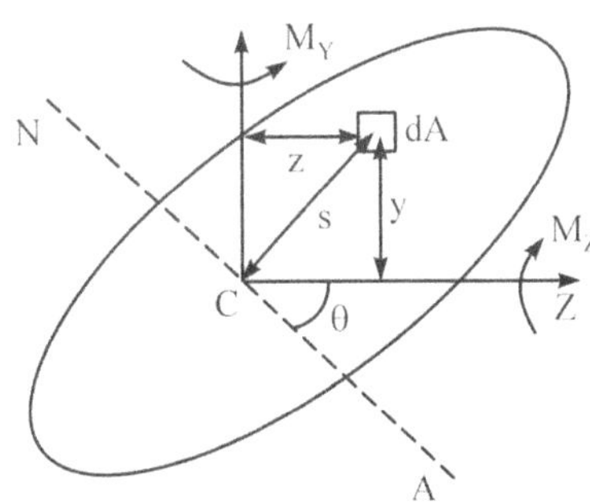

FIGURE 15.4 Resolution of bending moment

The bending stress distribution should also satisfy the conditions

$$\int \sigma_X \times dA = \int (E \times s / R) \times dA = 0$$

where, $\sigma_X = E \times s / R = (E/R) \times (y \times \cos\theta + z \times \sin\theta)$

$\Rightarrow \qquad \int s\, dA = 0 \qquad\qquad \text{since } E/R \text{ is a constant}$

and $M_Y = \int \sigma_X \times z \times dA = (E/R) \int (y \times \cos\theta + z \times \sin\theta) \times z \times dA$
$M_Z = \int \sigma_X \times y \times dA = (E/R) \int (y \times \cos\theta + z \times \sin\theta) \times y \times dA$

Using $I_{ZZ} = \int y^2 \times dA ; \quad I_{YY} = \int z^2 \times dA \quad \text{and} \quad I_{YZ} = \int y \times z \times dA$

$$\left\{ \begin{matrix} M_Y \\ M_Z \end{matrix} \right\} = \frac{E}{R} \begin{bmatrix} I_{YY} & I_{YZ} \\ I_{YZ} & I_{ZZ} \end{bmatrix} \left\{ \begin{matrix} \sin\theta \\ \cos\theta \end{matrix} \right\} \qquad \text{.....(15.3)}$$

Sin θ and cos θ can now be evaluated from

$$\frac{E}{R} \left\{ \begin{matrix} \sin\theta \\ \cos\theta \end{matrix} \right\} = \begin{bmatrix} I_{YY} & I_{YZ} \\ I_{YZ} & I_{ZZ} \end{bmatrix}^{-1} \left\{ \begin{matrix} M_Y \\ M_Z \end{matrix} \right\} = \frac{1}{I_{YY} \times I_{ZZ} - I_{YZ}^2} \begin{bmatrix} I_{ZZ} & -I_{YZ} \\ -I_{YZ} & I_{YY} \end{bmatrix} \left\{ \begin{matrix} M_Y \\ M_Z \end{matrix} \right\}$$

$$\text{.....(15.4)}$$

Substituting these values in the equation for σ_X, we get,

$$\sigma_X = \frac{(M_Z \times I_{YY} - M_Y \times I_{YZ}) \times y}{I_{ZZ} \times I_{YY} - I_{YZ}^2} + \frac{(M_Y \times I_{ZZ} - M_Z \times I_{YZ}) \times z}{I_{ZZ} \times I_{YY} - I_{YZ}^2} \quad \text{.....(15.5)}$$

If the terms are rearranged,

$$\sigma_X = \frac{M_Z(I_{YY} \times y - I_{YZ} \times z)}{I_{ZZ} \times I_{YY} - I_{YZ}^2} + \frac{M_Y(I_{ZZ} \times z - I_{YZ} \times y)}{I_{ZZ} \times I_{YY} - I_{YZ}^2} \qquad \text{......(15.6)}$$

The position of neutral axis depends on the geometry of the cross section while the orientation of neutral axis depends on the form of the applied loading as well as the geometrical properties of the cross section as seen from the eq. 15.6

For a symmetrical section, $I_{YZ} = 0$. There exists, in any unsymmetrical cross section, a centroidal set of axes (called principal axes) inclined at angle 'α' with y-axis for which $I_{YZ} = 0$. Then, the above equation simplifies to

$$\sigma_X = M_Z \times y / I_{ZZ} + M_Y \times z / I_{YY} \qquad \text{.....(15.7)}$$

Thus, bending stress in a symmetric section can be obtained by algebraic addition of bending stresses due to the two moments, as shown in the Fig. 15.5.

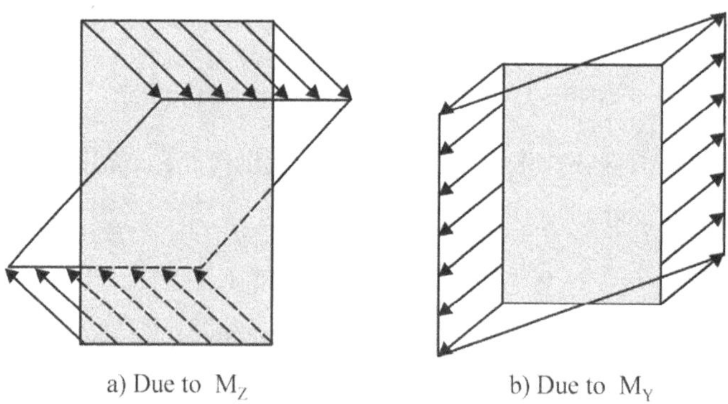

FIGURE 15.5 Bending stress distribution

If only one bending moment M_Z or M_Y is applied on the beam,

$$\sigma_X = M_Z \times y \,/\, I_{ZZ} \quad \text{or} \quad M_Y \times z \,/\, I_{YY} \qquad \qquad .....(15.8)$$

The angle of inclination (α) of principal axes with Y-axis is obtained from the condition that bending stress at any point (y', z') on the principal axis is zero.

$$\text{i.e.,} \qquad 0 = \frac{\left(M_Z \times I_{YY} - M_Y \times I_{YZ}\right) \times y'}{I_{ZZ} \times I_{YY} - I_{YZ}^{\,2}} + \frac{\left(M_Y \times I_{ZZ} - M_Z \times I_{YZ}\right) \times z'}{I_{ZZ} \times I_{YY} - I_{YZ}^{\,2}}$$

$$\text{or} \qquad \frac{z'}{y'} = \tan\alpha = -\frac{M_Z \times I_{YY} - M_Y \times I_{YZ}}{M_Y \times I_{ZZ} - M_Z \times I_{YZ}} \qquad \qquad .....(15.9)$$

If $M_Z = 0$, $\tan\alpha = I_{YZ}/I_{ZZ}$; If $M_Y = 0$, $\tan\alpha = I_{YY}/I_{YZ}$ and α is independent of magnitude of bending moment.

Example 15.1

An unsymmetrical T-section shown in figure is subjected to a bending moment of 1500Nm in a vertical plane. Calculate maximum bending stress and its location. All dimensions in the figure are in mm.

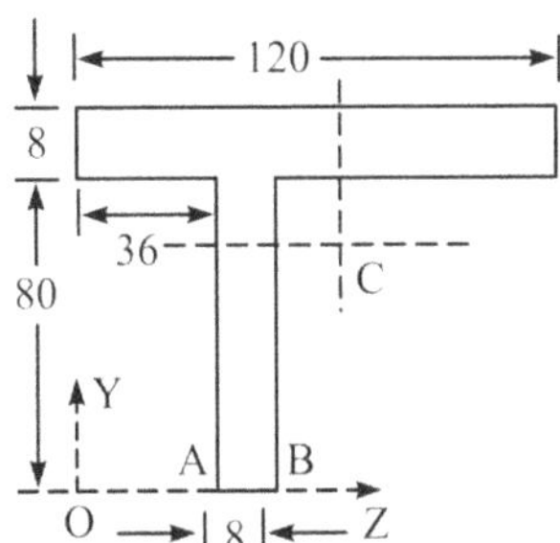

Solution

Position of the centroid of the section can be found by taking moments of areas about O.

$$(120 \times 8 + 80 \times 8) \times y_c = 120 \times 8 \times 84 + 80 \times 8 \times 40 \quad \Rightarrow \quad y_c = 66.4 \text{ mm}$$

$$(120 \times 8 + 80 \times 8) \times z_c = 120 \times 8 \times 60 + 80 \times 8 \times 40 \quad \Rightarrow \quad z_c = 52 \text{ mm}$$

w.r.t. the centroidal axes,

$$I_{ZZ} = [120 \times 8^3 \,/\, 12 + 120 \times 8 \times 17.6^2] + [8 \times 80^3 \,/\, 12 + 80 \times 8 \times 26.4^2]$$

$$= 1.09 \times 10^6 \text{ mm}^4$$

$$I_{YY} = [8 \times 120^3 \,/\, 12 + 120 \times 8 \times 8^2] + [80 \times 8^3 \,/\, 12 + 80 \times 8 \times 12^2]$$

$$= 1.31 \times 10^6 \text{ mm}^4$$

$$I_{YZ} = 120 \times 8 \times 8 \times 17.6 + 8 \times 80 \times (-12) \times (-26.4) = 0.34 \times 10^6 \text{ mm}^4$$

$$\sigma_X = M_Z \times (I_{YY} \times y - I_{YZ} \times z) / (I_{ZZ} \times I_{YY} - I_{YZ}^2)$$

$$= 1500 \times 1000 \times (1.31\,y - 0.34\,z) \times 10^6 / (1.09 \times 1.31 - 0.34^2) \times 10^{12}$$

$$= 1.497 \times y - 0.3886 \times z$$

Maximum bending stress in an unsymmetrical section occurs at the maximum distance from the centroidal (or neutral) axis, at A or B in this example (with max y). Since A is farther from neutral axis than B, maximum bending stress occurs at B

$$\sigma_{max} = 1.497 \times (-66.4) - 0.3886 \times (-8) = -96.292 \text{ N/mm}^2$$

Example 15.2

A symmetrical section shown in figure is subjected to a bending moment of 100kNm in a plane parallel to the longitudinal axis of the beam but inclined at 30^0 to the left of vertical. Determine the distribution of bending stress and orientation of neutral axis.

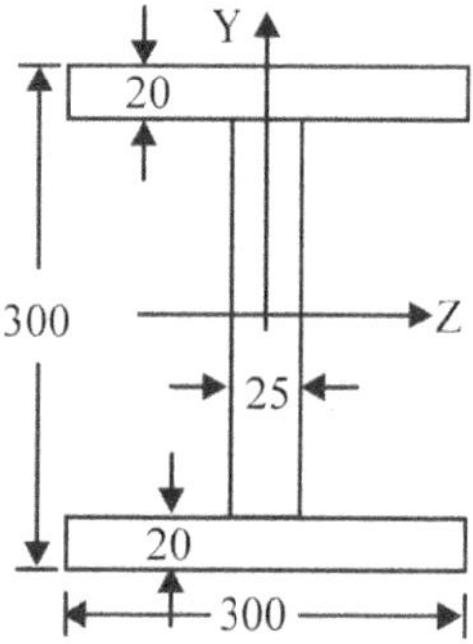

Solution

The components of bending moment are

$$M_Z = +100 \times \cos 30 = +86.6 \text{ kNm} = +86.6 \times 10^6 \text{ Nmm}$$

> since it produces tensile stress in the upper half of the beam, where y is positive

and $M_Y = -100 \times \sin 30 = -50 \text{ kNm} = -50 \times 10^6 \text{ Nmm}$

> since it produces tensile stress in the left half of the beam, where z is negative

Moments of inertia, $I_{ZZ} = (200 \times 300^3 - 175 \times 260^3) / 12 = 193.7 \times 10^6 \text{ mm}^4$

$$I_{YY} = (2 \times 20 \times 200^3 - 260 \times 25^3) / 12 = 27 \times 10^6 \text{ mm}^4$$

$$I_{YZ} = 0 \quad \text{since the section is symmetric}$$

Bending stress is given by

$$\sigma_X = M_Z \times y / I_{ZZ} + M_Y \times z / I_{YY}$$
$$= (+\, 86.6 \times 10^6 / 193.7 \times 10^6) \times y + (-50 \times 10^6 / 27 \times 10^6) \times z$$
$$= +\, 0.45 \times y - 1.85 \times z$$

On the upper edge of the top flange,

$$y = 150 \qquad \text{while} \qquad z \text{ varies from } -100 \text{ to } +100.$$

Maximum bending stress, therefore, occurs at the top left corner

$$\sigma_{Max} = +0.45 \times 150 - 1.85 \times (-100) \quad \text{or} \quad 252.5 \text{ N/mm}^2$$

Distribution of bending stress thus obtained at different points (y,z) over the entire cross section is shown in the figure. It can be noticed from the figure that the neutral plane does not pass through Y-axis or Z-axis, but passes through the centroid. If the neutral plane is inclined at α with Z-axis, the angle can be calculated from the condition $\sigma_X = 0$

i.e. , $\sigma_X = +0.45 \times y - 1.85 \times z = 0$

from which, we get $y/z = 1.85 / 0.45 = 4.11 = \tan \alpha$

$\Rightarrow \quad \alpha = 76.3^0$

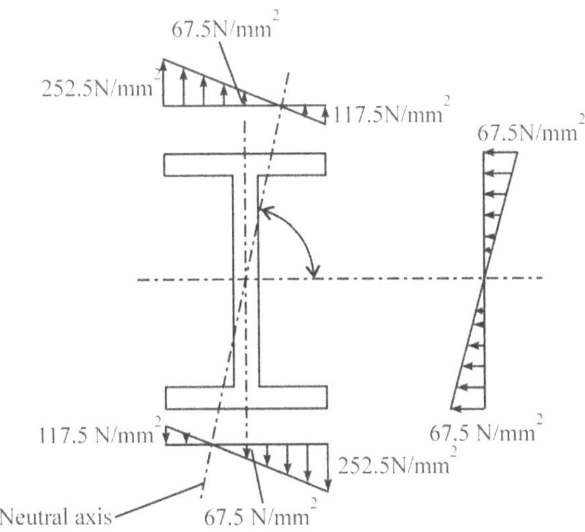

15.3 RELATIONSHIPS BETWEEN LOAD, DEFLECTION, SHEAR FORCE AND BENDING MOMENT

Consider free body diagram of a small element of size $\delta x \times \delta y \times t$, subjected to bending in X-Y plane due to distributed load $w_Y(x)$ varying over the element along the length (X-direction) of the plate.

$$\left(S_Y + \frac{\partial S_Y}{\partial x} \times \delta x \right) + p_Y \times \delta x - S_Y = 0$$

For force equilibrium in Y-direction, we get
which reduces to $p_Y = -(\partial S_Y / \partial x)$

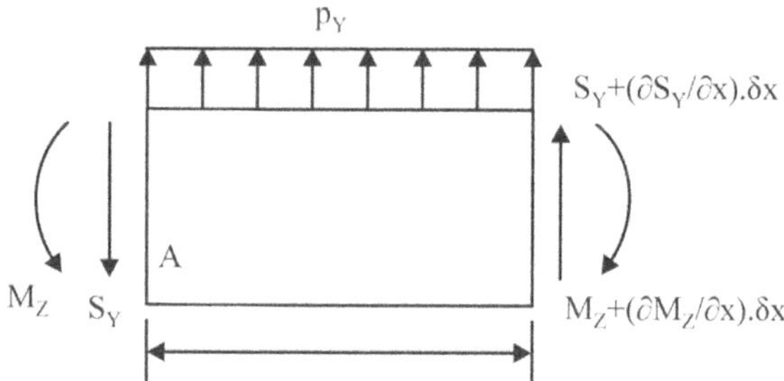

FIGURE 15.6 Shear force and bending moment over a small element

Taking moments about A, for equilibrium

$$\left(M_Z + \frac{\partial M_Z}{\partial x} \times \delta x \right) - \left(S_Y + \frac{\partial S_Y}{\partial x} \times \delta x \right) \times \delta x - w_Y \frac{(\delta x)^2}{2} - M_Z = 0$$

which reduces, by neglecting higher order small terms, to $S_Y \approx \partial M_Z / \partial x$

Combining these two relations, we get $-p_Y = \partial S_Y / \partial x = \partial^2 M_Z / \partial x^2$

Similarly, for load w_Z which causes bending in X-Z plane, we get

$$-p_Z = \partial S_Z / \partial x = \partial^2 M_Y / \partial x^2$$

From simple beam theory, bending moment M acting about an arbitrary direction produces deflection Ω of the plate in the plane of bending with a radius of curvature R given by $1/R = d^2\Omega/dx^2$

Components of Ω along Y and Z axes can be written as

$$w = -\Omega \sin \theta \quad \text{and} \quad v = -\Omega \cos \theta$$

where θ is the angle between Y-axis and neutral axis

Substituting these values in beam deflection equation,

$$\sin \theta / R = -d^2w/dx^2 = w'' \quad \text{and} \quad \cos \theta / R = -d^2v/dx^2 = v''$$

From eq (15.4)

$$\left\{ \begin{matrix} w'' \\ v'' \end{matrix} \right\} = -\frac{1}{R}\left\{ \begin{matrix} \sin\theta \\ \cos\theta \end{matrix} \right\} = -\frac{1}{E}\begin{bmatrix} I_{YZ} & I_{ZZ} \\ I_{YY} & I_{YZ} \end{bmatrix}^{-1}\left\{ \begin{matrix} M_Z \\ M_Y \end{matrix} \right\} = \frac{1}{I_{YY} \times I_{ZZ} - I_{YZ}^2}\begin{bmatrix} I_{YZ} & -I_{ZZ} \\ -I_{YY} & I_{YZ} \end{bmatrix}\left\{ \begin{matrix} M_Z \\ M_Y \end{matrix} \right\}$$

$$\text{or} \quad \begin{Bmatrix} M_Z \\ M_Y \end{Bmatrix} = -E \begin{bmatrix} I_{YZ} & I_{ZZ} \\ I_{YY} & I_{YZ} \end{bmatrix} \begin{Bmatrix} w'' \\ v'' \end{Bmatrix}$$

It can be seen that in an unsymmetric beam section, M_Z produces curvatures (or deflections) in both X-Y and X-Z planes, even if $M_Y = 0$. Similarly, for M_Y when $M_Z = 0$.

In a beam of symmetric cross section, $I_{YZ} = 0$. Hence,

$$w'' = -M_Y / EI_{YY} \qquad \text{and} \qquad v'' = -M_Z / EI_{ZZ}$$

which are the equations given by simple beam theory

15.4   MOMENTS OF INERTIA FOR THIN WALLED BEAMS

For optimizing weight, airplane structures are usually made of thin walled beam sections. In these sections, thickness 't' is assumed to be small in comparison to other cross-sectional dimensions. Hence, bending stresses on planes normal to the beam surface are constant across the thickness. Also, higher powers of 't' can be neglected so that moments of inertia are simplified as given here for a few simple cases.

(a)   If a singly symmetric c-channel is considered, $I_{YZ} = 0$ and

$$I_{ZZ} = 2\,[\,(b+t) \times t^3/12 + (b+t) \times t \times h^2\,] + t \times [2(h-t)]^3/12$$

$$\approx b \times t \times h^2 / 4 + t \times (2h)^3 / 12$$

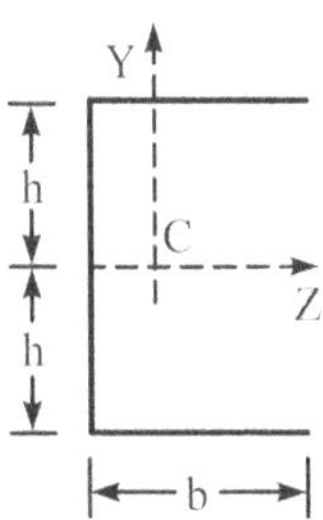

For a thin-walled semicircular section, $I_{YZ} = 0$ and

$$I_{YY} = I_{ZZ} = \int (t \times ds) \times y^2 = \int t \times (r \cos \theta)^2 \times (r \times d\theta) = \pi\, r^3 \times t / 2$$

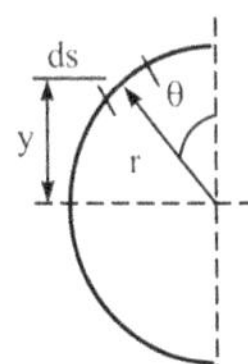

(b) For the antisymmetrical thin-walled Z-section,

$$I_{ZZ} = 2 \times (h \times t/2) \times (h/2)^2 + t \times h^3/12 \approx h^3 \times t / 3$$

$$I_{YY} \approx 2 \times (t/3) \times (h/2)^3 = h^3 \times t / 12$$

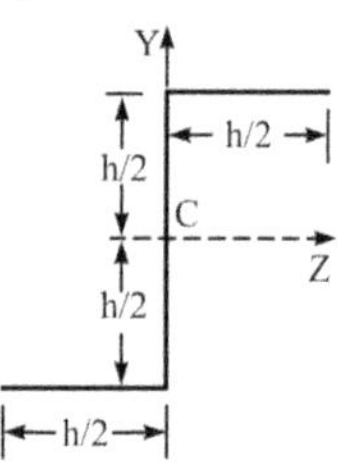

and $I_{YZ} = (h \times t/2) \times (h/4) \times (h/2) + (h \times t/2) \times (-h/4) \times (-h/2)$
$$= h^3 \times t / 8$$

(c) For a line of length L, inclined at angle θ to the horizontal (Z) axis,

$$I_{ZZ} = L^3 \times t \times \sin\theta / 12; \quad I_{YY} = L^3 \times t \times \cos\theta / 12;$$

$$I_{YZ} = L^3 \times t \times \sin 2\theta / 24$$

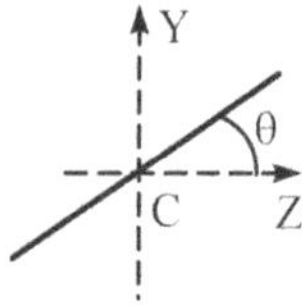

15.5 BENDING STRESS DISTRIBUTION IN A THIN WALLED Z-SECTION

Using the simplified expressions, derived earlier, for moments of inertia of a thin walled Z-section with each flange of width equal to half the height of web (b = h/2),

$$\sigma_X = \frac{M_Z \times (I_{YY} \times y - I_{YZ} \times z)}{I_{ZZ} \times I_{YY} - I_{YZ}^2} + \frac{M_Y \times (I_{ZZ} \times z - I_{YZ} \times y)}{I_{ZZ} \times I_{YY} - I_{YZ}^2}$$

Substituting the values of I_{YY}, I_{ZZ} and I_{YZ}, for $M_Y = 0$, the equation reduces to

$$\sigma_X = [M_Z /(h^3 \times t)] \times (6.86 \times y - 10.3 \times z) \qquad \dots(15.10)$$

For the top flange, y = h/2 and so,

At $z = + h/2$, $\sigma_X = [M_Z / (h^3 \times t)] \times [6.86 \times (h/2) - 10.3 \times (h/2)]$

$$= -1.72\, M_Z / (h^2 \times t)$$

At $z = 0$, $\sigma_X = [M_Z / (h^3 \times t)] \times [6.86 \times (h/2) - 10.3 \times 0]$

$$= 3.43\, M_Z / (h^2 \times t)$$

For the web, $z = 0$ and so,

$$\text{At } y = +h/2, \quad \sigma_X = [M_Z / (h^3 \times t)] \times [6.86 \times (h/2) - 10.3 \times 0]$$
$$= 3.43 \, M_z / (h^2 \times t)$$
$$\text{At } y = -h/2, \quad \sigma_X = [M_Z / (h^3 \times t)] \times [6.86 \times (-h/2) - 10.3 \times 0]$$
$$= -3.43 \, M_z / (h^2 \times t)$$

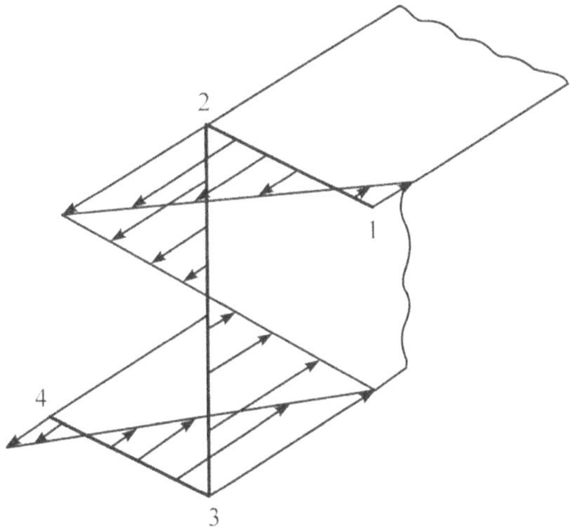

FIGURE 15.7 Bending stress distribution

For the bottom flange, $y = -h/2$ and so,

$$\text{At } z = -h/2, \quad \sigma_X = [M_Z / (h^3 \times t)] \times [6.86 \times (-h/2) - 10.3 \times (-h/2)]$$
$$= 1.72 \, M_z / (h^2 \times t)$$
$$\text{At } z = 0, \quad \sigma_X = [M_Z / (h^3 \times t)] \times [6.86 \times (-h/2) - 10.3 \times 0]$$
$$= -3.43 \, M_z / (h^2 \times t)$$

Stress in the bottom flange can also be directly estimated from the stress in the top flange, due to anti-symmetry of the cross section (Ref Fig 15.7).

15.6   SHEAR FLOW DISTRIBUTION DUE TO SHEAR FORCE

Shear problem is similar to the torsion problem, since in both cases a closed-form solution for an arbitrary cross section is not known. The shearing stresses due to lateral shear force are seldom of concern in the design of slender beams of compact cross section, for the longitudinal stresses are more critical. Aircraft structures are usually fabricated with thin skins and webs supported by longitudinal and transverse stiffening members. In these thin-walled structures, the shearing stresses are important, since they may produce buckling of the walls.

15.6.1 SHEAR FLOW IN A THIN WALLED OPEN SECTION BEAM

Thin walled beam sections are, in general, classified as open sections (such as L, T, Z, H,..) and closed sections (such as hollow circular, rectangular,..). Open sections are analysed by considering them as different segments in X-S plane, oriented parallel to the beam axis (X-direction) and inclined to the Y-Z plane (S-direction). In each of these thin walled segments, shear stresses normal to the beam surface may be neglected, since they are zero at the two surfaces and the wall is thin. It is often more convenient to work in terms of **shear flow** 'q' (defined as shear force per unit length) rather than in terms of shear stress (τ). Thus, $q = \tau \times t$. and $\tau_{XS} = \tau_{SX} = \tau$. Corresponding shear stress and shear flow distributions in a small element of size $\delta x \times \delta s \times t$ are given in Fig 15.8(a) and 15.8(b).

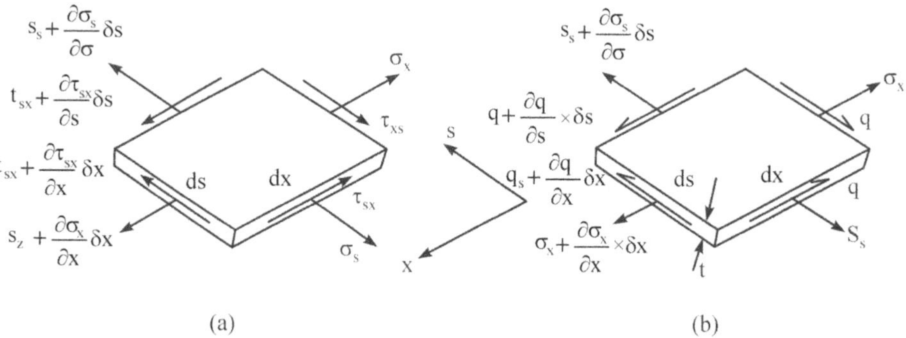

FIGURE 15.8 Shear stress and shear flow in a beam wall

For the equilibrium of the element in X-direction,

$$[\sigma_x + (\partial \sigma_x/\partial x)\, \delta x] \times t \times \delta s - \sigma_x \times t \times \delta s + [q + (\partial q/\partial s)\, \delta s] \times \delta x - q \times \delta x = 0$$

which reduces to $t \times (\partial \sigma_x/\partial x) + (\partial q/\partial s) = 0$

Similarly, equilibrium of the element in S-direction gives

$$t \times (\partial \sigma_s/\partial s) + (\partial q/\partial x) = 0$$

Thus, the shear flow q can now be evaluated from $\quad \partial q/\partial s = - t \times (\partial \sigma_x/\partial x)$

Substituting for σ_X in terms of M_Y and M_Z from eq 15.6,

$$\delta q/\delta s = -t \times \left[\frac{(\partial M_Z / \partial x) \times (I_{YY} \times y - I_{YZ} \times z)}{I_{ZZ} \times I_{YY} - I_{YZ}^2} + \frac{(\partial M_Y / \partial x) \times (I_{ZZ} \times z - I_{YZ} \times y)}{I_{ZZ} \times I_{YY} - I_{YZ}^2} \right]$$

$$= -t \times z \times \left[\frac{S_Z \times I_{ZZ} - S_Y \times I_{YZ}}{I_{ZZ} \times I_{YY} - I_{YZ}^2} \right] - t \times y \times \left[\frac{S_Y \times I_{YY} - S_Z \times I_{YZ}}{I_{ZZ} \times I_{YY} - I_{YZ}^2} \right]$$

and $q_s = \int (\partial q / \partial s) \times ds$

$$= -\left[\frac{S_Z \times I_{ZZ} - S_Y \times I_{YZ}}{I_{ZZ} \times I_{YY} - I_{YZ}^2} \right] \int t \times z \times ds - \left[\frac{S_Y \times I_{YY} - S_Z \times I_{YZ}}{I_{ZZ} \times I_{YY} - I_{YZ}^2} \right] \int t \times y \times ds$$

$$.....(15.11)$$

If Y-axis or Z-axis is an axis of symmetry, $I_{YZ} = 0$ and the shear flow equation reduces to

$$q = -\frac{S_Z}{I_{YY}} \int t \times z \times ds - \frac{S_Y}{I_{ZZ}} \int t \times y \times ds \qquad .....(15.12)$$

For the particular thin-walled Z-section with $S_Z = 0$,

$$q = \frac{1}{I_{ZZ} \times I_{YY} - I_{YZ}^2} \left[S_Y \times I_{YZ} \int t \times z \times ds - S_Y \times I_{YY} \int t \times y \times ds \right] \quad .....(15.13)$$

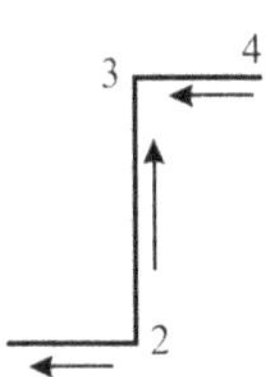

For *example*, in a Z-section,

$$I_{ZZ} = h^3 \times t / 3; \ I_{YY} = h^3 \times t / 12 \text{ and } I_{YZ} = h^3 \times t / 8,$$

$$q_s = (S_Y/h^3) \int (10.32 \times z - 6.84 \times y) \times ds$$

For the bottom flange 1-2, substituting

$$y = -h/2 \quad \text{and} \quad z = -h/2 + s_1 \qquad \text{where, } \ 0 \le s_1 \le h/2,$$

$$q_{12} = (S_Y/h^3) \int [10.32 \times (-h/2 + s_1) - 6.84 \times (-h/2)] \ ds_1$$

$$= (S_Y/h^3) \times [5.16 \times s_1^2 - 1.74 \times h \times s_1]$$

At 1, $\quad s_1 = 0 \quad$ and $\quad q_1 = 0$, being the free end of section

At 2, $\quad s_1 = h/2 \quad$ and $\quad q_2 = 0.42 \ S_Y/h$

It is seen that q_{12} is a quadratic expression, as shown in Fig 15.9, with $q = 0$

at $s_1 = 1.74 \times h/5.16$

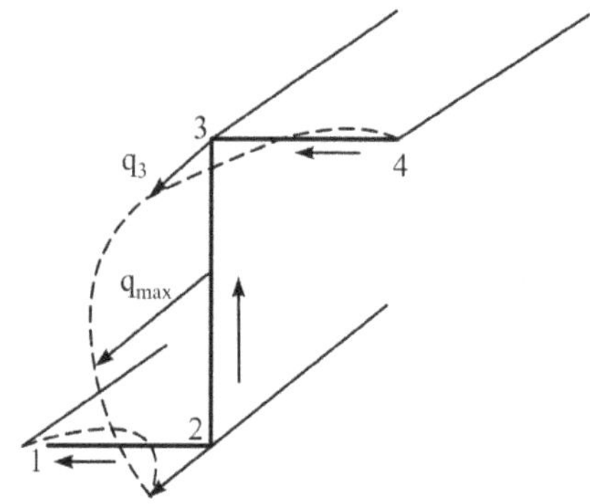

FIGURE 15.9 Shear stress distribution

For the web 2-3, substituting $z = 0$ and $y = -h/2 + s_2$ where $0 \le s_2 \le h$

$$q_{23} = (S_Y/h^3) \int [\, 10.32 \times (0) - 6.84 \times (-h/2 + s_2) \,] \, ds_2$$

$$= (S_Y / h^3) \times [\, 3.42 \times h \times s_2 - 3.42 \times s_2^2 \,] + q_2$$

At 3, $s_2 = +h$ or $y = +h/2$ and $q_3 = 0.42\, S_Y/h$

It is seen that $q_2 = q_3$, but q_{23} is quadratic and symmetric

Therefore, $q_{max} = 1.275\, S_Y/h$ at $s_2 = h/2$ or $y = 0$

In the top flange, $q_4 = 0$ at 4, another free end of the section

Since the section is anti-symmetric, q_{12} and q_{34} are similar

15.6.2 SHEAR FLOW IN A THIN WALLED CLOSED SECTION BEAM

Shear flow distribution in this case is calculated by superposition of shear flow in a open section as explained above, assuming shear flow in any one segment as zero and the additional shear flow needed to satisfy moment equilibrium. Thus, $(q_s)_{closed} = (q_s)_{open} + q_{s,0}$ .....(15.14)

15.7 SHEAR FLOW DISTRIBUTION DUE TO TORSION

15.7.1 SHEAR FLOW IN A THIN WALLED CLOSED SECTION BEAM

In the absence of an axial constraint, a closed section beam subjected to a pure torque T does not develop normal stresses (σ_Z and σ_S). It follows that the equilibrium equations derived earlier for a thin walled open section

$$T \times (\partial \sigma_Z/\partial z) + (\partial q/\partial s) = 0 \quad \text{and} \quad t \times (\partial \sigma_S/\partial s) + (\partial q/\partial z) = 0$$

reduce to $\partial q/\partial s = 0$ and $\partial q/\partial z = 0$ respectively. These relations can only be satisfied simultaneously by a constant value of 'q'. However, the shear stress τ may vary around the cross section if the wall thickness t is a function of 's'.

Since shear stresses at the free surface of a member are parallel to the surface, it will be sufficiently accurate to assume that the shear stresses in thin webs are parallel to the surfaces for the entire thickness of the web. Air loads normal to the surface must, of course, be resisted by shear stresses perpendicular to the web, but these stresses usually are negligible and are not considered here. In most design calculations, thickness of the web needs to be calculated. Hence, shear flow 'q', the product of shear stress and thickness, at different points along the web is more convenient to use than the shear stress. Consider the example of a curved web, representing the leading edge of a wing, shown in Fig.15.10. The diagonal tensile and compressive stresses σ_t and σ_c are shown on principal planes at 45° from the plane of maximum shear τ. From Mohr's circle, for a condition of pure torsion, it is seen that the diagonal compressive stress and diagonal tensile stress are both equal to the maximum shear stress τ.

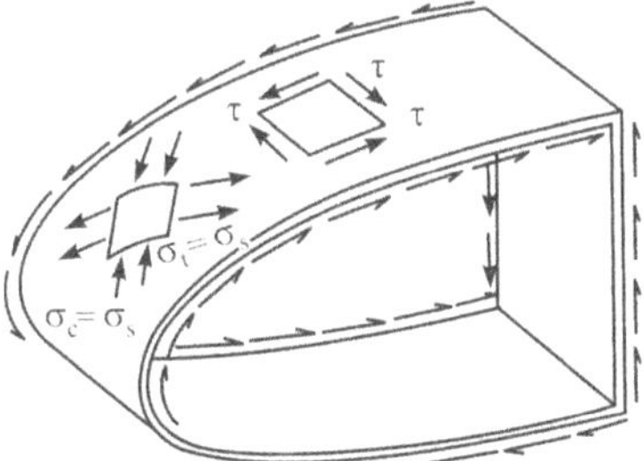

FIGURE 15.10 Shear flow on a curved web

The diagonal compression tends to bend the web to an increased curvature while the diagonal tension tends to bend the web to a reduced curvature. These two effects counteract each other. Consequently, the curved web resists high shear stress without deforming from its original curvature.

If the shear flow q is constant over the length of the web, the resultant forces in X and Y directions are given by

$$F_X = \int q\,dx = q \times x \quad \text{and} \quad F_Y = \int q\,dy = q \times y$$

It is obvious from these equations that the resultant horizontal or vertical force for a closed section is equal to zero, since $\int dx = \int dy = 0$ for an endless contour of a closed section.

At any point 'O' in the plane of the section, the shear flow over a small section 'ds' along the contour produces torsional moment 'dT', given by $dT = p \times q \times ds$ where 'p' is the normal distance of the segment ds from the reference point. The summation of these moments over the entire contour of the closed section gives

$$T = \int dT = \int p \times q \times ds = \int q \times 2dA = q \times 2A \qquad .....(15.15)$$

This is called **Bredt-Batho threory** and the relation $T = 2 \times q \times A$ is called Bredt-Batho equation.

15.7.2 SHEAR FLOW IN A THIN WALLED OPEN SECTION BEAM

The resultant force is $q \times L$ and is parallel to the straight line joining ends of the web, where 'L' is the length of the line. The resultant force induces torsional moment, whose magnitude depends on the shape of the web. The torque induced at a point, say 'O', is $r \times q \times ds$ and area of the triangle formed by the arc ds and the lines joining its ends with 'O' is $dA = (r \times ds)/2$. Then, the total torque induced along the entire web,

$$T = \int r \times q \, ds = \int 2q \times dA = 2 \, q \times A$$

where 'A' is the area enclosed by the web and the lines joining its ends with the point 'O' (Ref Fig 15.11). The distance 'e' of the resultant shear force from the point 'O' may be obtained from $\quad T = 2q \times A = (q \times L) \times e$

$$\text{or} \qquad e = 2 \, A \, / \, L \qquad\qquad .....(15.16)$$

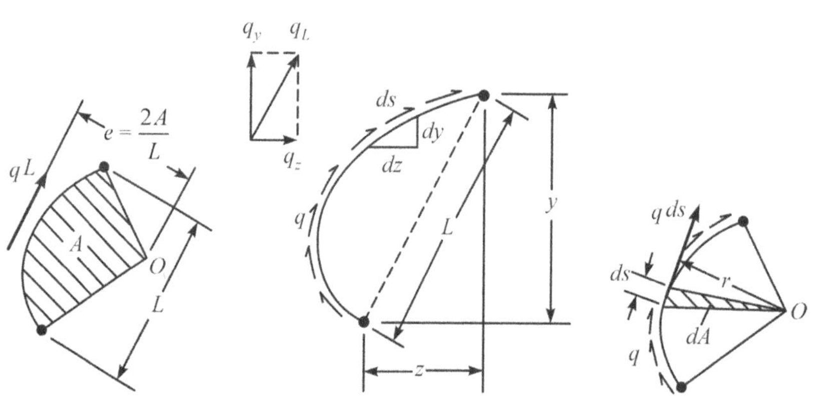

FIGURE 15.11 Shear flow due to torsion in an open section.

In a thick walled section, the shear stress varies across the thickness of the wall. In the simple case of a circular tube with a diameter-thickness ratio of 20:1, average shear stress given by $q = T/2A$ is about 5% lower than the maximum stress in the outer fiber, since stress in a fiber is directly proportional to the radius of that fiber.

The thin webs considered earlier are capable of resisting loads, which are applied at the shear center, but are not suitable for resisting torsion. In many structures, the resultant load acts at different positions for different loading conditions and consequently produces torsion. On an airplane wing, for example, the position of the resultant aerodynamic load changes with change in the angle of attack or when the flaps or ailerons are deflected. A closed box structure capable of resisting torsion is, therefore, used for airplane wings and similar structures. In some wings, two or more closed boxes may act together in resisting torsion. The skin forward and aft of the box in a wing is not used to resist bending loads.

15.8   SHEAR FLOW IN STIFFENED SHEAR WEBS

Let us consider a thin sheet, connected to stiffeners along its four edges, fixed at the two corners of one end and free at the other end as shown in (Fig 15.12). If the sheet is eliminated and the flanges pinned together, it would be an unstable truss. It becomes a statically determinate truss, if a diagonal bar is provided or if a thin sheet or panel, capable of carrying only shear load, is provided. The panel is called '*shear panel*' because it has only shear forces along its edges. Changes in shear flow in three different cases are explained here, for an understanding of the influence of shape of the stiffened web – rectangular, parallelogram and trapezoidal.

In Fig 15.12, a rectangular web is considered. Bars AB and CD are called flanges while bars BC and AD are called stiffeners. A small element 1-2-3-4 of size (dx × dy) at a distance x from A towards B is also shown.

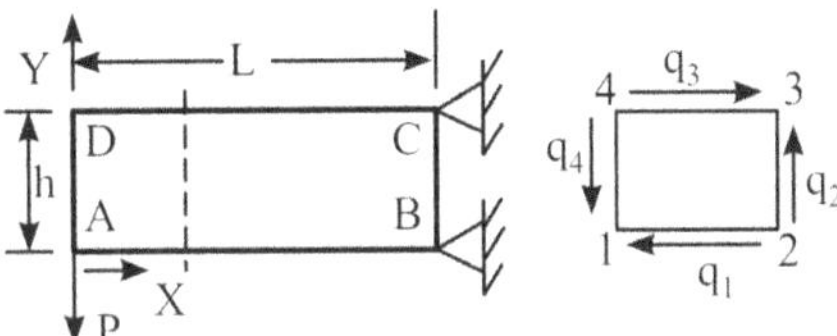

FIGURE 15.12 Shear flow in a rectangular web

At any section x, vertical shear force in Y-direction 'S' is equal to the applied load 'P' and the shear flow $q_2 = S/h$. Axial force in the flanges, tensile in CD and compressive in AB, is obtained from moment equilibrium. Taking moments about A, $F_{CD} = S \times x / h$

From force equilibrium condition of the shear web in X-direction,

$$F_{AB} = - F_{CD}$$

Considering a small element of this web and taking moments about 1,

$$q_3 = q_2 \times (dx/dy)$$

If a square element is chosen, dx = dy and, therefore, $q_3 = q_2$.

From force equilibrium conditions of the small element, $q_1 = q_2 = q_3 = q_4$

In Fig 15.13, a cantilevered parallelogram stiffened shear panel is considered. A small element 1-2-3-4 of sides 'a' and 'b' at a distance x from A towards B is also shown.

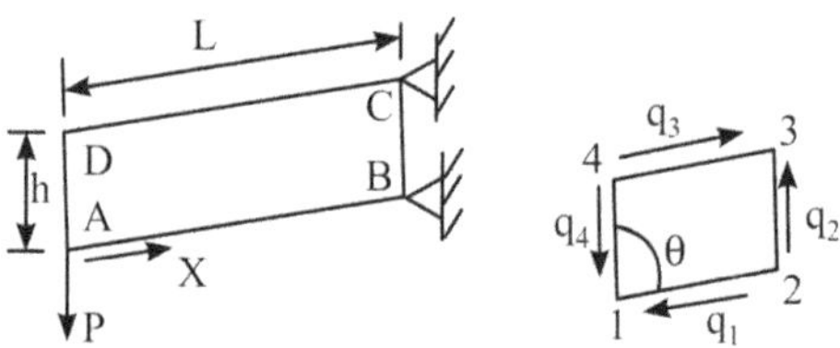

FIGURE 15.13 Shear flow in a parallelogram web

In this case also, at any section x along AB, shear force 'S' in the section parallel to AD is equal to the applied load 'P' and the shear flow $q_2 = S/h$. Axial force in the flanges, tensile in CD and compressive in AB, is obtained from moment equilibrium.

Taking moments about A, $F_{CD} = S \times (x \sin \theta) / (h \sin \theta) = S \times (x/h)$

From force equilibrium condition of the shear web in X-direction,

$$F_{AB} = - F_{CD}$$

Considering a small element of this web and taking moments about 1,

$$q_3 = q_2 \times (a \times \sin \theta) / (b \times \sin \theta) = q_2 \times (a/b)$$

If the element is chosen such that $a = b$, $q_3 = q_2$. From force equilibrium conditions of the small element, $q_1 = q_2 = q_3 = q_4$

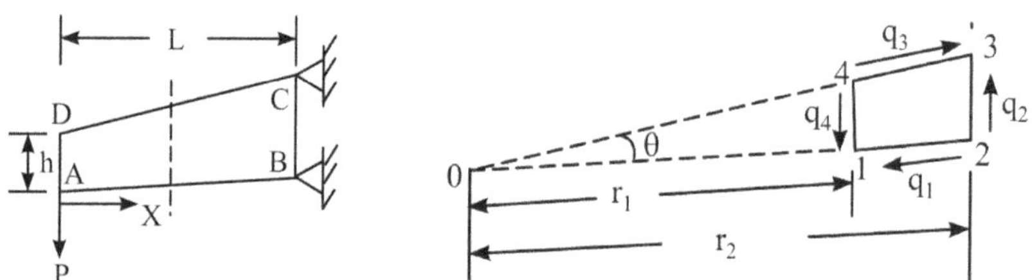

FIGURE 15.14 Shear flow in a tapered web   **FIGURE 15.15** A trapezoidal web

In Fig 15.14, a tapered and stiffened trapezoid-shaped shear panel is considered. A small element 1-2-3-4 at a distance x from A towards B is shown in Fig 15.15. Extend non-parallel lines 1-2 and 3-4 to meet at O. Let the edge 1-4 be at a distance r_1 from O and edge 2-3 at a distance r_2 from O.

Let S_{2-3} and S_{1-4} be the shear forces along the edges 2-3 and 1-4. Taking moments about O, for equilibrium of the small element, $S_{2-3} \times r_2 = S_{1-4} \times r_1$.

From similar triangles O-1-4 and O-2-3, $r_1 / r_2 = L_{1-4} / L_{2-3}$

where, L_{1-4} and L_{2-3} are the lengths of the two parallel sides of the element.

From the two relations, we get $S_{2-3} = S_{1-4} \times (r_1 / r_2) = S_{1-4} \times (L_{1-4} / L_{2-3})$

The shear flows are given by,

$$q_2 = S_{2-3} / L_{2-3} = [S_{1-4} \times (L_{1-4} / L_{2-3})] / L_{2-3}$$

$$= q_4 \times (L_{1-4} / L_{2-3})^2$$

$$\Rightarrow q_2 = q_4 \times (r_1/r_2)^2 \quad \text{or} \quad q_2 \times (r_2)^2 = q_4 \times (r_1)^2$$

Thus, shear flow on the parallel faces of a trapezoidal shear panel is inversely proportional to the square of the distance from the vertex O.

i.e., $q = k/r^2$ .....(15.17)

where, k is a constant which depends on the boundary conditions

The shear flow is uniform along the panel's parallel edges, since r is constant on each of them. Shear force along an inclined edge of length L is obtained from

$$S = \int_0^L q \times ds = \int_{r_1}^{r_2} q \times (dr / \sin\theta) = \int_{r_1}^{r_2} (k / r^2) \times (dr / \sin\theta) = \left[(r_2 - r_1) / \sin\theta\right] \times \left[k / (r_1 . r_2)\right]$$

Since $(r_2 - r_1)/\sin\theta = L$,   average shear flow, $q_{av} = S/L = k/(r_1 \times r_2)$ .....(15.18)

Substituting for k from eq (15.18) in eq (15.17) in terms of q_{av},

$$q = q_{av} \times (r_1 \times r_2) / r^2 \qquad .....(15.19)$$

15.9 SHEAR FLOW IN WEB WITH STRINGERS

Stringers that may be attached to the tube, subjected to pure torsion, would be unstressed because stresses in such stiffeners could be produced only by a variable shear flow. However, if bending load is applied, axial loads are induced in the stringers and the shear flow in the thin web varies at each stringer. Thus, in a thin web with a number of stringers attached to it, as shown in Fig. 15.16, the shear flow is expressed by $q_n = q_{n-1} + P_{n-1}$

or $q_n = q_0 + \sum P_i$ for $1 \leq i \leq n$

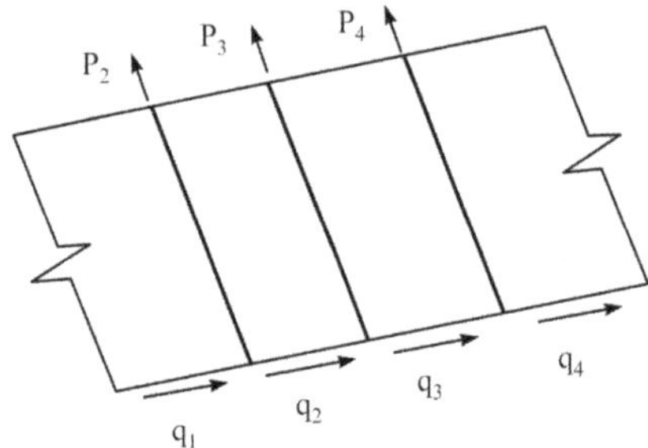

FIGURE 15.16 Shear flow in web with stringers

15.10 SHEAR FLOW DUE TO BENDING LOAD AND TORSIONAL MOMENT

Linear superposition of shear flow can be carried when bending load as well as torsional moment are applied on a thin web with stringers. A simple example of one web with two stringers, representing leading edge of an airplane wing is shown in Fig. 15.17. In this figure, $q = q_1 - q_2$

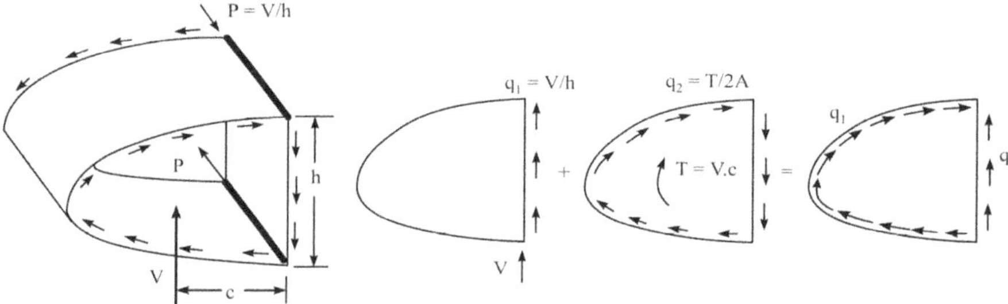

FIGURE 15.17 Shear flow in web due to combined bending and torsion stringers

15.11 SHEAR CENTER

When a load is applied on a beam through its plane of symmetry or plane of bending, it does not induce any torsional moment on the section and, hence, does not produce any twist. Such *a point in the cross section through which any applied load does not twist the section is called **shear center** (S)*. However, if the load is applied through some other point, it produces twist in the section. For closed sections symmetric in two perpendicular planes (Fig. 15.18 a), shear center always lies at the intersection of symmetry planes. For angle sections (Fig. 15.18 b,c,d), the shear center is located at the intersection of the sides since the resultant internal shear loads all pass through these points. For unsymmetrical sections and open sections (Fig. 5.18e), calculation of shear center is essential to avoid twisting of the section by any bending load. Location of shear center in a thin walled section beam depends on shear flow distribution and is detailed later. Elastic axis of a beam is a straight line which passes through the shear centers of the cross sections. A shear force at any point on this axis produces bending without any twisting.

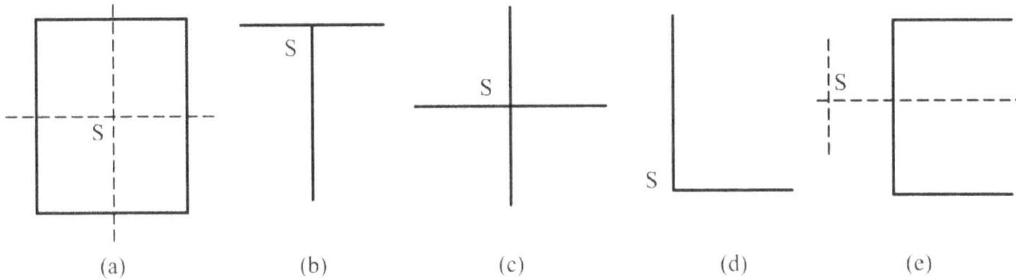

FIGURE 15.18 Shear center in different beam sections

15.12 SHEAR CENTER LOCATION FOR OPEN SECTIONS

Shear center location can be calculated, using the expression for shear flow in a thin-walled section, so as to produce no torque when the shear force is applied through the shear center. For example, in a thin-walled singly symmetrical C-section with $S_Z = 0$,

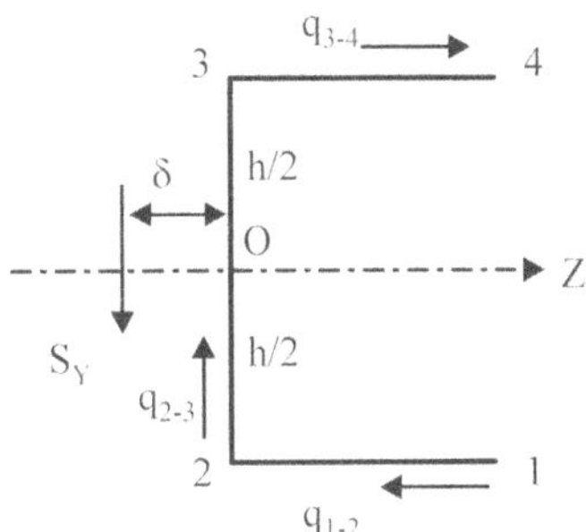

$$q_S = -(S_Y / I_{ZZ}) \times \int t \times y \times ds \qquad \qquad .....(15.20)$$

where, $I_{ZZ} = 2\,b \times t \times (h/2)^2 + t \times h^3/12 = h^3 \times t \times (1 + 6b/h) / 12$

The shear flow along 1-2-3-4 is given by q_{1-2}, q_{2-3} and q_{3-4}. The direction of shear flow depends on the direction of shear force. If the force in the figure is upwards, the shear flow direction will reverse. Due to symmetry of the section,

$$q_{1-2} = q_{3-4}$$

$$= -(S_Y / I_{ZZ}) \times \int t \times (-h/2) \times ds_1 = (S_Y / I_{ZZ}) \times t \times (h/2) \times s_1 \quad 0 \le s_1 \le b$$

For $s_1 = 0$ at 1 (free end), $q_1 = 0$

and for $s_1 = b$ at 2, $q_2 = (S_Y / I_{ZZ}) \times t \times (h/2) \times b$

Thus, variation of shear flow in flanges 1–2 and 3– 4 is linear

If 'S' is the shear center, located at a distance 'δ' from O, assuming the shear force S_Y acts through S along y-direction and taking moments about O,

$$S_Y \times \delta = 2 \times (q_{1-2} \times s_1) \times h/2$$

$$= - (S_Y \times h \times t / I_{ZZ}) \times \int (h/2) \times s_1 \times ds_1 \qquad \text{since, } y = h/2$$

$$= - [12\, S_Y / \{h^2 (1 + 6b/h)\}\,] \times (h/2) \times (s_1^2/2) \qquad \text{where, } s_1 = b$$

$$\Rightarrow \qquad \delta = 3\, b^2 / [h \times (1+6b/h)] = 3\, b^2 / (h + 6b) \qquad\qquad .....(15.21)$$

Shear flow in web can be calculated from

$$q_{2-3} = - (S_Y / I_{ZZ}) \times \int t \times (- h/2 + s_2) \times ds_2 + q_2 \qquad \text{for} \qquad 0 \le s_2 \le h$$

$$= - (S_Y / I_{ZZ}) \times t \times [(-h/2) \times s_2 + (s_2)^2/2] + q_2 \qquad\qquad .....(15.22)$$

Variation of shear flow in the web is quadratic with

$$q_3 = (S_Y / I_{ZZ}) \times t \times (h/2) \times b = q_2 \qquad \text{at} \quad s_2 = h$$

and maximum value,

$$q_{max} = (S_Y / I_{ZZ}) \times t \times [(h^2/8) + (h/2) \times b]$$

$$= (S_Y / I_{ZZ}) \times t \times h \times [h + 4b] / 8 \qquad \text{at} \quad s_2 = h/2$$

Note: Assume vertical shear force S_Y is downwards and horizontal shear force $S_Z = 0$ in all the following problems, with cross section in Z-Y plane

Example 15.3

Cross section of a slit rectangular tube of constant thickness is shown in figure. Show that shear center is located at a distance of $\delta = b \times (2h + 3b)/[2(h + 3b)]$

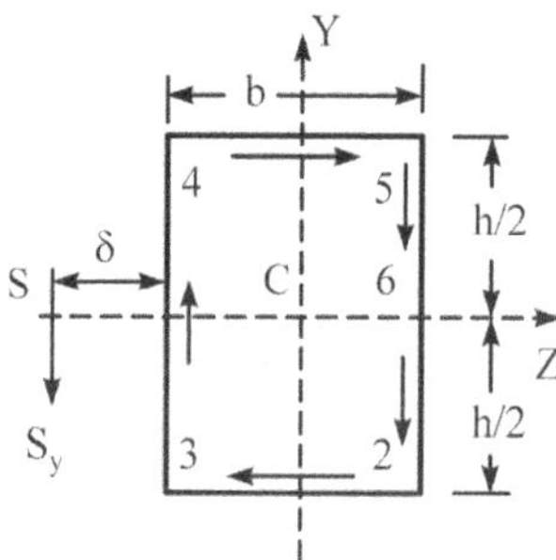

Solution

$$I_{ZZ} = 2\, [t \times h^3/12 + b \times t \times (h/2)^2] = t \times h^2 \times (h + 3b) / 6$$

Due to symmetry, $I_{YZ} = 0$

Therefore, $\quad q_s = - (S_Y/I_{ZZ}) \int t \times y \times ds$

At the free end, $q_1 = 0$

$$q_{1-2} = - (S_Y/I_{ZZ}) \int t \times (0 - s_1) \times ds_1 + q_1 \qquad \text{for} \quad 0 \le s_1 \le h/2$$

$$= - [6S_Y / \{t \times h^2 \times (h + 3b)\}] \times [- t \times (s_1^2/2)]$$

At $s_1 = h/2$, $q_2 = 3S_Y / 4(h + 3b)$

$$q_{2-3} = -(S_Y/I_{ZZ}) \int t \times (-h/2) \times ds_2 + q_2 \qquad \text{for} \quad 0 \le s_2 \le b$$

$$= -[6S_Y /\{t \times h^2 \times (h + 3b)\}] \{t \times h^2 \times (h + 3b)\}$$

$$\times (-t \times h/2) \times s_2 + 3S_Y / \{4(h + 3b)\}$$

At $s_2 = b$, $q_3 = 3S_Y \times (4b + h) / \{4h \times (h + 3b)\}$

$$q_{3-4} = -(S_Y/I_{ZZ}) \int t \times (-h/2 + s_3) \times ds_3 + q_3 \quad \text{for} \quad 0 \le s_3 \le h$$

$$= -[6S_Y / \{t \times h^2 \times (h + 3b)\}] \times t \times [(-h/2) \times s_3 + s_3^2/2]$$

$$+ 3S_Y \times (4b + h) / \{4 h \times (h + 3b)\}$$

At $s_3 = h$, $q_4 = q_3 = 3S_Y \times (4b + h) / \{4h \times (h + 3b)\}$

$$q_{4-5} = -(S_Y/I_{ZZ}) \int t \times (h/2) \times ds_4 + q_4 \qquad \text{for} \quad 0 \le s_4 \le b$$

$$= -[6S_Y /\{t \times h^2 \times (h + 3b)\}] \times t \times (h/2) \times s_4$$

$$+ 3S_Y \times (4b + h) / \{4 h \times (h + 3b)\}$$

At $s_4 = b$, $q_5 = 3S_Y / \{4(h + 3b)\} = q_2$

$$q_{5-6} = -(S_Y/I_{ZZ}) \int t \times (h/2 - s_5) \times ds_5 + q_5 \quad \text{for} \quad 0 \le s_5 \le +h/2$$

$$= -[6S_Y /\{t \times h^2 \times (h + 3b)\}] \times t \times [(h/2) \times s_5 - s_5^2/2]$$

$$+ 3S_Y /\{4(h + 3b)\}$$

At $s_5 = h/2$, $q_6 = 0 = q_1$, being another free end

Taking moments about 1,

$$M_1 = S_Y(b + \delta) = \int q_s \times y \times ds$$

$$= \int q_{1-2} \times 0 \times ds_1 + \int q_{2-3} \times (h/2) \times ds_2 + \int q_{3-4} \times b \times ds_3$$

$$+ \int q_{4-5} \times (h/2) \times ds_4 + \int q_{5-6} \times 0 \times ds_5$$

$$= (h/2) \times \int [-(S_Y/I_{ZZ}) \times (-t \times h/2) \times s_2 + q_2] ds_2$$

$$+ b \times \int [(-S_Y/I_{ZZ}) \times \{t \times (-h/2) \times s_3 + t \times (s_3^2/2)\}$$

$$+ q_3] ds_3 + (h/2) \times \int [-(S_Y/I_{ZZ}) \times t \times (h/2) \times s_4 + q_4] ds_4$$

$$= (h/2) \times [-(S_Y/I_{ZZ}) \times (-t \times h/2) \times (s_2^2/2) + q_2 \times s_2]$$

$$+ b \times [-(S_Y/I_{ZZ}) \times \{t \times (-h/2) \times (s_3^2/2) + t \times (s_3^3/6)\} + q_3 \times s_3]$$

$$+ (h/2) \times [-(S_Y/I_{ZZ}) \times t \times (h/2) \times (s_4^2/2) + q_4 \times s_4]$$

$$= (h/2) \times [-(S_Y/I_{ZZ}) \times (-t \times h \times b^2/4) + q_2 \times b]$$

$$+ b \times [-(S_Y/I_{ZZ}) \times \{t \times (-h^3/4) + t \times (h^3/6)\} + q_3 \times h]$$

$$+ (h/2) \times [-(S_Y/I_{ZZ}) \times t \times h \times (b^2/4) + q_4 \times b]$$

$$= - (S_Y/I_{ZZ}) \times \{[(-t \times h \times b^2/4) \times (h/2)] + [t \times (-h^3/4) + t \times (h^3/6)] \times b$$
$$+ [t \times h \times (b^2/4) \times (h/2)] \} + [q_2 \times b \times (h/2) + q_3 \times h \times b + q_4 \times b \times (h/2)]$$
$$= - [6S_Y/t \times h^2 \times (h + 3b)] \times (t \times h^2 \times b) \times \{ [-b/8 - h/4 + h/6 + b/8]$$
$$+ [3S_Y \times b \times h / \{4h \times (h + 3b)\}] \times [h/2 + (4b + h) + (4b + h)/2]$$
$$= [- 6S_Y b/\{24(h + 3b)\}] \times (-3b - 6h + 4h + 3b)$$
$$+ [3S_Y \times b / \{8(h + 3b)\}] \times [h + 2(4b + h) + (4b + h)]$$
$$= [- S_Y b / \{4(h + 3b)\}] \times (-2h) + [3S_Y \times b / \{8(h + 3b)\}] \times (4h + 12b)$$
$$= [S_Y b / \{8(h + 3b)\}] \times [4h + 3(4h + 12b)]$$
$$= S_Y \times b \times (4h + 9b) / [2(h + 3b)]$$

Therefore, $\delta = [b \times (4h + 9b) / \{2(h + 3b)\}] - b$
$$= b \times [(4h + 9b) - 2(h + 3b)] / [2(h + 3b)]$$
$$\boldsymbol{\delta = b \times (2h + 3b) / [2(h + 3b)]}$$

Example 15.4

Determine the location of shear center for the cross section shown in figure, assuming uniform thickness of 3 mm. All dimensions are in mm

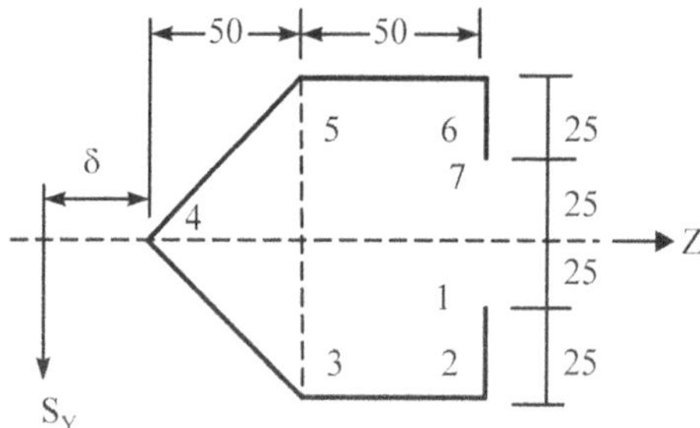

Solution

$$I_{ZZ} = 2 [\{3 \times 25^3/12 + 3 \times 25 \times 37.5^2\} + 3 \times 50 \times 50^2$$
$$+ \{3 \times (50\sqrt{2})^3 \times \sin^2 45/12\}] = 105.7 \times 10^4 \text{ mm}^4$$

Due to symmetry, $I_{YZ} = 0$ Therefore, $q_s = - (S_Y/I_{ZZ}) \int t \times y \times ds$

At the free end, $q_1 = 0$

$$q_{1-2} = - (S_Y/I_{ZZ}) \int 3 \times (-25 - s_1) \times ds_1 + q_1 \quad \text{for} \quad 0 \le s_1 \le 25$$
$$= - (S_Y/I_{ZZ}) \times [- 3 \times 25 s_1 - 3 \times (s_1^2/2)]$$

At $s_1 = 25$, $q_2 = 3 \times 25^2 S_Y /(2 \times I_{ZZ})$

$$q_{2-3} = - (S_Y/I_{ZZ}) \int 3 \times (-50) \times ds_2 + q_2 \quad \text{for} \quad 0 \le s_2 \le 50$$
$$= - (S_Y/I_{ZZ}) \times [3 \times (-50) \times s_2] + 3 \times 3 \times 25^2 S_Y / (2 \times I_{ZZ})$$

At $s_2 = 50$,

$$q_3 = 3 \times 25^2 S_Y (4 + 3/2) / (2 \times I_{ZZ}) = 11 \times 3 \times 25^2 S_Y / (2 \times I_{ZZ})$$

$$q_{3\text{-}4} = -(S_Y/I_{ZZ}) \int 3 \times (-50 + s_3/\sqrt{2}) \times ds_3 + q_3 \quad \text{for} \quad 0 \le s_3 \le 50\sqrt{2}$$

$$= -(S_Y/I_{ZZ}) \times 3 \times (-50\, s_3 + s_3^2/2\sqrt{2}) + 11 \times 3 \times 25^2 S_Y / (2 \times I_{ZZ})$$

At $s_3 = 50\sqrt{2}$,

$$q_4 = 3 \times 25^2 S_Y (4\sqrt{2} - 4/\sqrt{2} + 11/2) / I_{ZZ}$$

$$= 3 \times 25^2 S_Y (4\sqrt{2} + 11) / (2 \times I_{ZZ})$$

Due to symmetry, $q_5 = q_3$; $q_6 = q_2$ and $q_7 = q_1$

However, these values are evaluated here for verifying the above.

$$q_{4\text{-}5} = -(S_Y/I_{ZZ}) \int 3 \times (s_4/\sqrt{2}) \times ds_4 + q_4 \quad \text{for} \quad 0 \le s_4 \le 50\sqrt{2}$$

$$= -(S_Y/I_{ZZ}) \times 3 \times (s_4^2/2\sqrt{2}) + 3 \times 25^2 S_Y (4\sqrt{2} + 11) / (2 \times I_{ZZ})$$

At $s_4 = 50\sqrt{2}$,

$$q_5 = 3 \times 25^2 S_Y (-4/\sqrt{2} + 2\sqrt{2} + 11/2) / I_{ZZ}$$

$$= 11 \times 3 \times 25^2 S_Y / (2.I_{ZZ}) = q_3$$

$$q_{5\text{-}6} = -(S_Y/I_{ZZ}) \int 3 \times 50.ds_5 + q_5 \qquad \text{for} \quad 0 \le s_5 \le 50$$

$$= -(S_Y/I_{ZZ}) \times 3 \times 50\, s_5 + 11 \times 3 \times 25^2 S_Y / (2 \times I_{ZZ})$$

At $s_5 = 50$,

$$q_6 = 3 \times 25^2 S_Y (-4 + 11/2) / I_{ZZ} = 3 \times 3 \times 25^2 S_Y / 2.I_{ZZ} = q_2$$

$$q_{6\text{-}7} = -(S_Y/I_{ZZ}) \int 3 \times (50 - s_6) \times ds_6 + q_6 \quad \text{for} \quad 0 \le s_1 \le 25$$

$$= -(S_Y/I_{ZZ}) \times [3 \times 50 \times s_6 - 3 \times (s_6^2/2)] + 3 \times 3 \times 25^2 S_Y / (2 \times I_{ZZ})$$

At $s_6 = 25$,

$$q_7 = 3 \times 25^2 S_Y (-2 + 1/2 + 3/2) / I_{ZZ} = 0$$

$$= q_1, \text{ being another free end}$$

Taking moments about 4,

$$\begin{aligned}
M_4 &= S_y \times \delta = \int q_s \times y \times ds = \int q_{1\text{-}2} \times 100 \times ds_1 + \int q_{2\text{-}3} \times 50 \times ds_2 \\
&\quad + \int q_{3\text{-}4} \times 0 \times ds_3 + \int q_{4\text{-}5} \times 0 \times ds_4 \\
&\quad + \int q_{5\text{-}6} \times 50 \times ds_5 + \int q_{6\text{-}7} \times 100 \times ds_6 \\
&= 2 \left[\int q_{1\text{-}2} \times 100 \times ds_1 + \int q_{2\text{-}3} \times 50 \times ds_2 \right] \\
&= 2 \int (-S_Y/I_{ZZ}) \times [-3 \times 25 \times s_1 - 3 \times (s_1^2/2)] \times 100\, ds_1 \\
&\quad + 2 \int [(-S_Y/I_{ZZ}) \times 3 \times (-50) \times s_2 \\
&\quad + 3 \times 3 \times 25^2 S_Y / (2\, I_{ZZ})] \times 50\, ds_2
\end{aligned}$$

$$= 2\ (-S_Y/I_{ZZ}) \times [-3 \times 25(s_1{}^2/2) - 3 \times (s_1{}^3/6)] \times 100$$
$$+\ 2\ [(-S_Y/I_{ZZ}) \times 3 \times (-50) \times (s_2{}^2/2)$$
$$+\ 3 \times 3 \times 25^2 \times s_2 \times S_Y / (2\ I_{ZZ})] \times 50$$

$$= 2\ (-S_Y/I_{ZZ}) \times [-3 \times 25(25^2/2) - 3 \times (25^3/6)] \times 100$$
$$+\ 2\ [(-S_Y/I_{ZZ}) \times 3 \times (-50) \times (50^2/2)$$
$$+\ 3 \times 3 \times 25^2 \times 50 \times S_Y / (2\ I_{ZZ})] \times 50$$

$$= 2 \times 25^3 \times 50\ (S_Y/I_{ZZ}) \times [(3/2 + 1/2).2 + (3 \times 8/2 + 9)]$$

$$= 2 \times 25^3 \times 50\ (S_Y/I_{ZZ}) \times [4 + 21] = 4 \times 25^5\ (S_Y/I_{ZZ})$$

Therefore, $\delta = 4 \times 25^5 / I_{ZZ} = 4 \times 25^5 / (105.7 \times 10^4) = 36.95$ mm

Example 15.5

Determine location of shear center for the thin walled section shown

Solution

Due to symmetry, location of centroid is along X-axis

$$I_{ZZ} = 2 \times [\ t \times (r/2)^3/12 + t \times (r/2) \times (r + r/4)^2] + \pi\ r^3 t\ /2$$
$$= r^3 t \times (19 + 6\ \pi)\ /\ 12$$

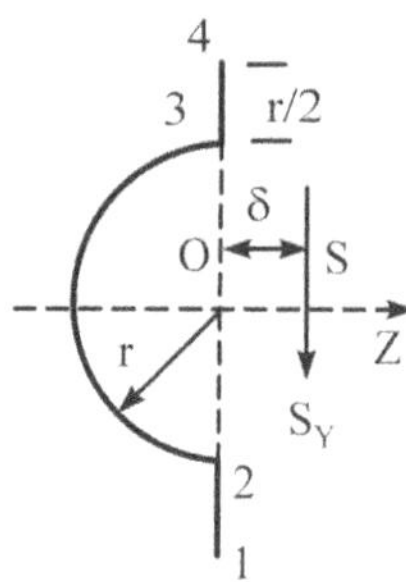

Due to symmetry, $I_{XY} = 0$

Therefore,

$$q_s \quad = -(S_Y/I_{ZZ}) \int t \times y \times ds$$

At the free end, $q_1 = 0$

$$q_{1-2} = -(S_Y/I_{ZZ}) \int t \times (-3r/2 + s_1) \times ds_1 + q_1 \text{ for } 0 \le s_1 \le r/2$$
$$= -(S_Y/I_{ZZ}) \times [-t \times (3r/2) \times s_1 + t \times (s_1{}^2/2)]$$

At $s_1 = r/2$,

$$q_2 \quad = -[S_Y/\ I_{ZZ}] \times [-t \times (3r/2) \times (r/2) + t \times (r^2/8)] = 5\ t\ r^2\ S_Y / (8\ I_{ZZ})$$
$$q_{2-3} = -(S_Y/I_{ZZ}) \int t \times (-r \cos \theta) \times (r\ d\theta) + q_2 \text{ for } 0 \le \theta \le \pi$$
$$= -(S_Y/I_{ZZ}) \times [-t\ r^2 \sin \theta] + 5\ t\ r^2\ S_Y / (8\ I_{ZZ})$$

At $\theta = \pi$,

$$q_3 = - (S_Y/I_{ZZ}) \times [- t \, r^2 \sin \pi] + 5 \, t \, r^2 \, S_Y / (8 \, I_{ZZ})$$

$$= 5 \, t \, r^2 \, S_Y/(8 \, I_{ZZ}) = q_2$$

$$q_{3-4} = - (S_Y/I_{ZZ}) \int t \times (r + s_3) \times ds_3 + q_3 \qquad \text{for } 0 \leq s_3 \leq r/2$$

$$= - (S_Y/I_{ZZ}) \times t \times (r \, s_3 + s_3^2/2) + 5 \, t \, r^2 \, S_Y / (8 \, I_{ZZ})$$

At $s_3 = r/2$, $q_4 = - (S_Y/I_{ZZ}) \times t \times (r \times r/2 + r^2/8) + 5 \, t \, r^2 \, S_Y / (8 \, I_{ZZ}) = 0 = q_1$

It can also be seen from symmetry, that $q_2 = q_3$ and $q_1 = q_4$

Taking moments about O,

$$M_O = S_Y \times \delta = \int q_s \times y \times ds$$

$$= \int (q_{1-2} \times ds_1) \times 0 + \int (q_{2-3} \times r \, d\theta) \times (r \sin \theta) + \int (q_{3-4} \times ds_3) \times 0$$

$$= \int (q_{2-3} \times r \, d\theta) \times (r \sin \theta)$$

$$= \int [- (S_Y/I_{ZZ}) \times (-t \, r^2 \sin \theta) + 5 \, t \, r^2 \, S_Y / (8 \, I_{ZZ})] \times (r \, d\theta) \times (r \sin \theta)$$

$$= t \, r^4 \, (- S_Y/I_{ZZ}) \times [\int (- \sin^2 \theta) \, d\theta - \int (5/8) \times \sin \theta \, d\theta]$$

$$= t \, r^4 \, (- S_Y/I_{ZZ}) \times [\int \{-(1-\cos 2\theta)/2\} \, d\theta - \int (5/8) \times \sin \theta \, d\theta]$$

$$\text{for } 0 \leq \theta \leq \pi$$

$$= t \, r^4 \, (- S_Y/I_{ZZ}) \times [- (\theta - \sin 2\theta/2)/2 - (5/8) \times (- \cos \theta)]$$

$$= t \, r^4 \, (- S_Y/I_{ZZ}) \times [- (\pi - 0)/2 - (5/8) \times (-2)]$$

$$= t \, r^4 \, (S_Y/I_{ZZ}) \times [\pi/2 - 5/4] = t \, r^4 \times (2\pi - 5) \times S_Y / (4 \, I_{ZZ})$$

Therefore, $\delta = t \, r^4 \times (2\pi - 5) / 4 I_{ZZ}$

$$= t \, r^4 \times (2\pi - 5) / [4 \, r^3 t \times (19 + 6 \, \pi) / 12]$$

$$\delta = 3 \, r \times (2\pi - 5) / (6 \, \pi + 19)$$

Example 15.6

Determine the location of shear center for the thin walled section of uniform thickness shown in figure.

Solution

Due to symmetry, location of centroid is along Z-axis

$$I_{ZZ} = [\pi \times r^3 \times t /2] + 2 \times [(t \times 2r) \times r^2)]$$

$$= r^3 \times t \times (\pi/2 + 4)$$

Due to symmetry, $I_{YZ} = 0$ Therefore, $q_s = -(S_Y/I_{ZZ}) \int t \times y \times ds$

At the free end, $q_1 = 0$

$$q_{1-2} = -(S_Y/I_{ZZ}) \int t \times (-r) \times ds_1 + q_1 \qquad \text{for} \quad 0 \le s_1 \le 2r$$
$$= -(S_Y/I_{ZZ}) \times [-t \times r \times s_1] + 0$$

At $s_1 = 2r$, $q_2 = (S_Y/I_{ZZ}) \times [t \times r \times 2r] = 2\,t\,r^2\,S_Y/I_{ZZ}$

$$q_{2-3} = -(S_Y/I_{ZZ}) \int t \times (-r \cos \theta) \times (r\,d\theta) + q_2 \quad \text{for} \quad 0 \le \theta \le \pi$$
$$= -(S_Y/I_{ZZ}) \times [-t\,r^2 \sin \theta] + 2\,t\,r^2\,S_Y/I_{ZZ}$$
$$= (S_Y/I_{ZZ}) \times t\,r^2 \times (\sin \theta - 2)$$

At $\theta = \pi$, $q_3 = (S_Y/I_{ZZ}) \times [0] + 2\,t\,r^2\,S_Y/(8\,I_{ZZ}) = +2\,t\,r^2\,S_Y/I_{ZZ} = q_2$

$$q_{3-4} = -S_Y/I_{ZZ}) \int t \times r \times ds_3 + q_3 \qquad \text{for} \quad 0 \le s_3 \le 2\,r$$
$$= -(S_Y/I_{ZZ}) \times t \times r\,s_3 + 2\,t\,r^2\,S_Y/I_{XX}$$
$$= -(S_Y/I_{ZZ}) \times t \times r \times (s_3 - 2\,r)$$

At $s_3 = 2r$, $q_4 = -(S_Y/I_{ZZ}) \times t \times r \times (2r - 2r) = 0 = q_1$

It can also be seen from symmetry, that $\quad q_2 = q_3 \quad$ and $\quad q_1 = q_4$

Taking moments about O,

$$M_O = S_y \times \delta = \int q_s \times y \times ds$$
$$= \int (q_{1-2} \times ds_1) \times r + \int (q_{2-3} \times r\,d\theta) \times (r \sin \theta) + \int (q_{3-4} \times ds_3) \times r$$
$$= \int -(S_Y/I_{ZZ}) \times (t \times r \times s_1)\,r\,ds_1$$
$$\qquad + \int (S_Y/I_{ZZ}) \times t\,r^2 \times (\sin \theta - 2)] \times (r\,d\theta) \times (r \sin \theta)$$
$$\qquad + \int [-(S_Y/I_{ZZ}) \times t \times r \times (s_3 - 2\,r)] \times r \times ds_3$$
$$= (S_Y/I_{ZZ}) \times [-(t \times r^2 \times s_1^2/2) + t\,r^4 \times (\theta - \sin 2\theta/2)/2$$
$$\qquad + 2t\,r^4 \times \cos \theta - t\,r^2 \times (s_3^2/2 - 2\,r\,s_3)\,]$$
$$= (S_Y/I_{ZZ}) \times t\,r^2 \times [-4\,r^2/2 + r^2 \times (\pi)/2 - 4\,r^2$$
$$\qquad - (4\,r^2/2 - 2\,r \times 2\,r)]$$
$$= (S_Y/I_{ZZ}) \times t\,r^4 \times (\pi/2 - 4)$$

Therefore, $\delta = t\,r^4 \times (\pi/2 - 4)/I_{ZZ} = [t\,r^4 \times (\pi/2 - 4)]/[r^3\,t \times (\pi/2 + 4)]$
$$= r \times (\pi/2 - 4)/(\pi/2 + 4) \quad \text{or} \quad r \times (\pi - 8)/(\pi + 8)$$

Example 15.7

Determine the location of shear center for the section of uniform thickness shown in figure.

Solution

$$I_{ZZ} = \pi\,r^3\,t/2 + t \times (2r)^3/12 = t \times r^3(3\,\pi + 4)/6$$

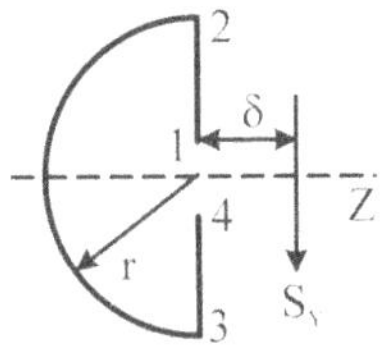

Since the cross section is symmetric, $I_{XY} = 0$

Therefore, $q_S = -(S_Y/I_{ZZ}) \int t \times y \times ds$

At the free end, $q_1 = 0$

$$q_{1-2} = -(S_Y/I_{ZZ}) \int t \times s_1 \times ds_1 + q_1 \qquad \text{for} \quad 0 \leq s_1 \leq r$$

$$= -(S_Y/I_{ZZ}) \times [t \times (s_1^2/2) + 0]$$

At $s_1 = r$, $q_2 = (S_Y/I_{ZZ}) \times t \times (r^2/2) = -t\, r^2\, S_Y / (2\, I_{ZZ})$

$$q_{2-3} = -(S_Y/I_{ZZ}) \int t \times (r \cos \theta) \times (r\, d\theta) + q_2 \qquad \text{for} \quad 0 \leq \theta \leq \pi$$

$$= -(S_Y/I_{ZZ}) \times (t\, r^2 \sin \theta) - t\, r^2\, S_Y / (2\, I_{ZZ})$$

$$= -(S_Y/I_{ZZ}) \times t\, r^2 \times (\sin \theta + 1/2)$$

At $\theta = \pi$, $q_3 = -(S_Y/I_{ZZ}) \times t\, r^2 \times (0 + 1/2) = -t\, r^2\, S_Y / (2\, I_{ZZ}) = q_2$

$$q_{3-4} = -(S_Y/I_{ZZ}) \int t \times (-r + s_3) \times ds_3 + q_3 \qquad \text{for} \quad 0 \leq s_3 \leq r$$

$$= -(S_Y/I_{ZZ}) \times [-t \times r\, s_3 + t \times (s_3^2/2) - t \times r^2 /2]$$

$$= -(S_Y/2I_{ZZ}) \times t \times (s_3 - r)^2$$

At $s_3 = r$, $q_4 = -(S_Y/2I_{ZZ}) \times t \times (r - r)^2 = 0 = q_1$ since 4 is another free end

It can also be seen from symmetry, that $q_2 = q_3$ and $q_1 = q_4$

Taking moments about O,

$$M_O = S_Y \times \delta = \int q_s \times y \times ds$$

$$= \int (q_{1-2} \times ds_1) \times 0 + \int (q_{2-3} \times r\, d\theta) \times (r \sin \theta)$$

$$+ \int (q_{3-4} \times ds_3) \times 0$$

$$= 0 + \int [-(S_Y/I_{ZZ}) \times t\, r^2 \times (\sin \theta + 1/2)] \times (r\, d\theta) \times (r \sin \theta) + 0$$

$$= -(S_Y/I_{ZZ}) \times t\, r^4 \times [(\theta - \sin 2\theta/2)/2 - (t\, r^4 \times \cos \theta)/2\,]$$

$$= -(S_Y/I_{ZZ}) \times t\, r^4 \times [\pi - 0 - (-2)\,]/2$$

$$= (S_Y/I_{ZZ}) \times t\, r^4 \times (\pi + 2)/2$$

Therefore, $\delta = t\, r^4 \times (\pi + 2)/(2\, I_{ZZ}) = [t \times r^4 \times (\pi + 2)/2] / [t \times r^3 \times (3\pi + 4)/6]$

$$\delta = 3\, r \times (\pi + 2)/(3\pi + 4)$$

Example 15.8

Determine shear center for a circular section of radius 'R', thickness 't', having a narrow slit

Solution

For a circular ring of radius 'R' and thickness 'T', $I_{ZZ} = \pi R^3 T$

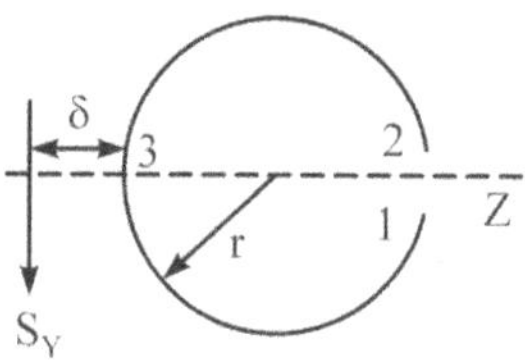

At the free end, $q_1 = 0$

$$q_{1-2} = -(S_Y/I_{ZZ}) \times \int T \times y \times ds + q_1 = -(S_Y/I_{ZZ}) \times \int T \times (R \sin \theta) \times (R \, d\theta)$$

$$= -(S_Y/I_{ZZ}) \times T \times R^2 \times \cos \theta \quad \text{for} \quad 0 \le \theta \le 2\pi$$

At 2, $\theta = 2\pi$ and $q_2 = 0$ since 2 is also a free boundary point|

At 3, $\theta = \pi$ and $q_3 = q_{max} = (S_Y/I_{ZZ}) \times T \times R^2$

Taking moments about 3,

$$M_3 = S_Y \times \delta = \int q_{1-2} \times ds \times y$$

$$= \int \{-(S_Y/I_{ZZ}) \times T \times R^2 \times \cos \theta\} \times (R \, d\theta) \times (R + R \cos \theta)$$

$$\therefore \qquad \delta = -T \times R^4 \left[\int \{\cos \theta + (1 + \cos 2\theta)/2\} \, d\theta \right]/I_{XX} \quad \text{for} \quad 0 \le \theta \le 2\pi$$

$$= T \times R^4 \int \{\sin \theta + (\theta/2) + (\sin 2\theta)/4\} / (\pi R^3 T) = T \times R^4 (\pi) / (\pi R^3 T)$$

Thus, $\delta = R$

15.13 SHEAR CENTER LOCATION FOR SINGLE-CELL CLOSED SECTIONS

It is difficult to find the value of shear flow in a closed section, since there is no reference point or free boundary where shear flow is zero, based on which the shear flow at other points is evaluated.

Shear stresses produced by torsion in a closed section beam has exactly the same form as the shear stresses produced by shear force while the shear stresses due to torsion and shear are different in an open section. Hence, shear loads may be applied through points in the cross section other than shear center so that torsional and shear stresses are included. Shear loads S_y and S_z cause bending stresses and shear flows. Assuming that hoop stresses and body forces are absent, shear flows are related by the equilibrium equation,

$$t \times (\partial \sigma_x / \partial x) + (\partial q / \partial s) = 0$$

$\Rightarrow \int (\partial q / \partial s) \times ds = q_s - q_{s,o}$

$$= -\left[\frac{(S_Z \times I_{ZZ} - S_Y \times I_{YZ})}{I_{ZZ} \times I_{YY} - I_{YZ}{}^2}\right] \int t \times x \times ds - \left[\frac{(S_Y \times I_{YY} - S_Z \times I_{YZ})}{I_{ZZ} \times I_{YY} - I_{YZ}{}^2}\right] \int t \times y \times d$$

$$....(15.23)$$

where, $q_{s,o}$ is the shear flow value at a chosen origin for contour s and is the basic difference from the shear flow in a open section

Representing 'open' section or 'basic' shear flow as q_b,

we can write $q_S = q_b + q_{s,0}$

Shear loads which are not applied through the shear center of a closed section beam cause cross section to twist and warp i.e., in addition to rotation, they have out of plane axial displacement. We can express q_S in terms of warping and tangential displacements w and v_t of a point in the beam wall as

$$q_S = \tau \times t = G \times t \times \gamma = G \times t \times (\varphi_1 + \varphi_2) = G \times t \,[\, \partial w/\partial s + \partial v_t/\partial x \,]$$

Aircraft structures of thin shell type are stiffened by ribs or frames at frequent intervals along their length. It is reasonable to assume that the cross section moves as a rigid body in its own plane but is flexible normal to its plane offering no resistance to axial displacement 'w'

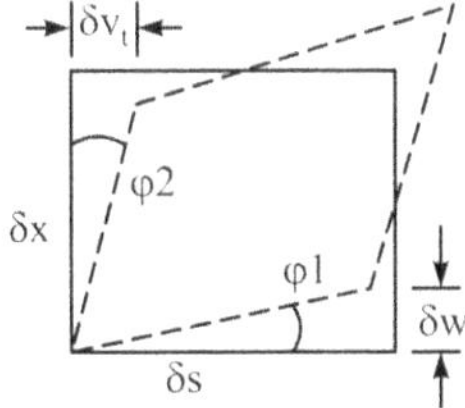

Along the contour of the cross section,

$$\oint (q_s / Gt)ds = \oint \left\{ \left(q_b + q_{s,0}\right) / Gt \right\} ds = 0$$

$\Rightarrow \qquad q_{s,0} = -\oint q_0 ds / \oint ds$ $\qquad\qquad$ ......(15.24)

Example 15.9

A thin-walled closed section beam has the singly symmetrical cross section shown in figure. Each wall of the section is flat and has the same thickness t and shear modulus G. Calculate the distance of shear center from 4.

Solution

Let 'δ' be the distance of the shear center 'S' from 4 and S_Y the shear force acting through 'S'. Since S lies on the axis of symmetry, product of inertia $I_{YZ} = 0$. If shear force along Z-axis, $S_Z = 0$,

$$q_s = - (S_Y/I_{ZZ}) \int t \times y\, ds + q_{s,o}$$

where, $I_{ZZ} = 2\, [\int t \times (8s_1/10)^2\, ds_1 + \int t \times (8s_2/17)^2\, ds_2]$

$$\text{for } 0 \le s_1 \le L_{4-1}; \quad 0 \le s_2 \le L_{1-2}$$

with length $L_{4-1} = \sqrt{(6a)^2 + (8a)^2} = 10a$; $L_{1-2} = \sqrt{(15a)^2 + (8a)^2} = 17a$

$$I_{ZZ} = 1152\, a^3 t$$

Since the section is symmetric about Z-axis, in the wall $4-1$, with $q_4 = q_{s,0}$ as the reference value

$$q_{4-1} = -(S_Y/I_{ZZ}) \times \int t \times y \times ds + q_4 = q_{b,41} + q_{s,0}$$

$$q_{b,41} = -[S_Y/(1152\, a^3 t)] \times \int t \times (8s_1/10)\, ds_1 \qquad \text{for} \quad 0 \le s_1 \le L_{4-1}$$

$$= -[S_Y/(1152\, a^3)] \times (2s_1^2/5)$$

At 1, $s_1 = L_{4-1} = 10a$ and $q_{b,1} = -(S_Y/1152\, a^3) \times (40a^2)$

Similarly, in the wall 1-2,

$$q_{1-2} = -(S_Y / I_{ZZ}) \oint t \times y \times ds + q_1 = (q_{b,12} + q_{b,1}) + q_{s,0}$$

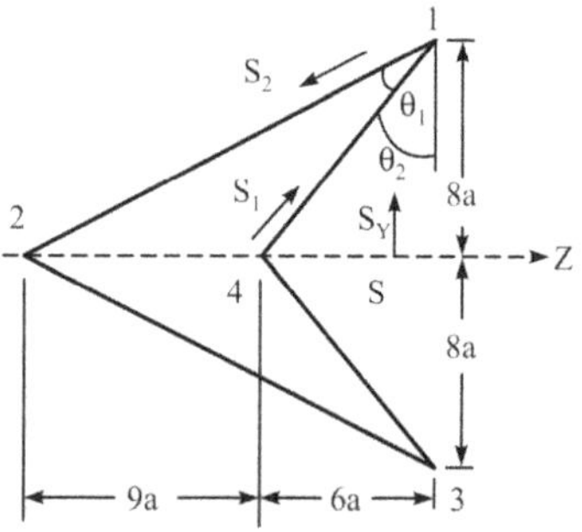

$$q_{b,12} = -[S_Y/(1152\, a^3 t)] \times \int t \times [8 \times (17a - s_2)/17 + 40\, a^2]\, ds_2$$

$$= -[S_Y/(1152\, a^3)] \times [-4\, s_2^2/17 + 8a\, s_2 + 40\, a^2] \text{ for } 0 \le s_2 \le L_{1-2} \text{ or } 17a$$

The shear flow at the reference point '4', $q_{s,0}$ is calculated from

$$q_{s,0} = \oint q_{b,s}\, ds / \oint ds \qquad\qquad \text{along periphery}$$

$$\oint ds = 2\, [L_{4-1} + L_{1-2}] = 2\left(\sqrt{(6a)^2 + (8a)^2} + \sqrt{(15a)^2 + (8a)^2}\right)$$

$$= 2 \times (10a + 17a) = 54a$$

$$q_{s,0} = 2\{ q_{b,41} \times ds_1 + \int q_{b,12} \times ds_2 \} / (54a)$$

$$= 2\, [\{S_Y/(1152\, a^3 t)\} \times \{\int (2\, s_1^2/5)\, ds_1$$

$$+ \int (-4s_2^2/17 + 8a\, s_2 + 40\, a^2 \times ds_2\} \,] / (54a)$$

$$= [S_Y/(1152\, a^3)] \times 58.7\, a^2$$

Taking moments about 2,

$$S_Y \times (\delta + 9a) = 2 \int q_{4-1} \{17a \times \sin(\theta_1 - \theta_2)\}\, ds_1$$

since moment arm for q_{4-1} w.r.t. 2 is $17a \times \sin(\theta_1 - \theta_2)$

$$\sin(\theta_1 - \theta_2) = \sin\theta_1 \times \cos\theta_2 - \cos\theta_1 \times \sin\theta_2$$
$$= (15a/17a) \times (8a/10a) - (8a/17a) \times (6a/10a) = 72/170$$

$$S_Y \times (\delta + 9a) = 2 \int (q_{b,41} + q_{s,0}) \times (17a \times \sin\theta)\, ds_1$$
$$= 2\,[S_Y/(1152\,a^3)] \times \int(-2s_1^2/5 + 58.7a^2) \times (17a \times \sin\theta)\, ds_1$$
$$= 2 \times S_Y\,[-2s_1^3/15 + 58.7\,a^2 \times s_1] \times (17a \times \sin\theta)\,/1152\,a^3$$
$$\text{for} \quad 0 \le s_1 \le L_{4-1} \text{ or } 10a$$

or $\quad \delta + 9a = 2 \times [-2000a^3/15 + 58.7 \times 10\,a^3] \times (17a \times \sin\theta)\,/1152\,a^3$
$$= 5.65\,a$$

$\Rightarrow \quad \delta = -3.35\,a$

Negative sign indicates that the shear center lies to the left of 4, between 2 and 4

Example 15.10

Determine the location of shear center for the box beam shown in figure, assuming uniform thickness of 3 mm. All dimensions are in mm

Solution

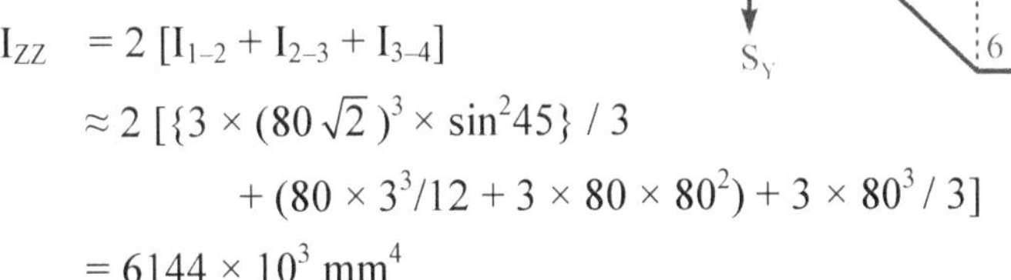

$$I_{ZZ} = 2\,[I_{1-2} + I_{2-3} + I_{3-4}]$$
$$\approx 2\,[\{3 \times (80\sqrt{2})^3 \times \sin^2 45\}\,/\,3$$
$$+ (80 \times 3^3/12 + 3 \times 80 \times 80^2) + 3 \times 80^3\,/\,3]$$
$$= 6144 \times 10^3 \text{ mm}^4$$

Since the section is symmetric about z-axis, shear flow in the wall 1-2, with $q_1 = q_{s,0}$ as the reference value, $q_{1-2} = -(S_Y/I_{ZZ}) \times \int t \times y \times ds + q_1 = q_{b,12} + q_{s,0}$

$$q_{b,12} = -(S_Y/I_{ZZ}) \int t \times (s_1/\sqrt{2})\, ds_1 \quad \text{for} \quad 0 \le s_1 \le L_{1-2} \text{ or } 80\sqrt{2}$$
$$= -(S_Y/I_{ZZ}) \times (3\,s_1^2/2\sqrt{2})$$

At 2, $\quad s_1 = L_{1-2} = 80\sqrt{2} \quad$ and $\quad q_{b,2} = -(S_Y/I_{ZZ}) \times (3 \times 80^2/\sqrt{2})$

Similarly, in the wall 2-3, $q_{2-3} = -(S_Y/I_{ZZ}) \times \int t \times y \times ds + q_2 = q_{b,23} + q_{s,0}$

$$q_{b,23} = -(S_Y/I_{ZZ}) \times [3 \times 80\,s_2 + 3 \times 80^2/\sqrt{2}] \quad \text{for} \quad 0 \le s_2 \le L_{2-3} \text{ or } 80$$

At 3, $\quad s_2 = 80 \quad$ and

$$q_{b,3} = -(S_Y/I_{ZZ}) \times 3 \times 80^2 \times (1 + 1/\sqrt{2}) = -1.707 \times 3 \times 80^2 \times (S_Y/I_{ZZ})$$

Similarly, in the wall 3-4, $\quad q_{3-4} = -(S_Y/I_{ZZ}) \int t \times y\, ds + q_3 = q_{b,34} + q_{s,0}$

$$q_{b,34} = -(S_Y/I_{ZZ}) \times [\int 3 \times (180 - s_3)\, ds_3 + 1.707 \times 3 \times 80^2]$$
$$\text{for} \quad 0 \le s_3 \le L_{3-4} \text{ or } 80$$
$$= -(S_Y/I_{ZZ}) \times [3 \times (80 \times s_3 - s_3^2/2) + 1.707 \times 3 \times 80^2]$$

At 4, $s_3 = 80$ and $q_{b,4} = -(S_Y/I_{ZZ}) \times 3 \times 80^2 \times (1/2 + 1.707)$

$$= 2.207 \times 3 \times 80^2 \times (S_Y/I_{ZZ})$$

The shear flow at the reference point '1', $q_{s,0}$ is calculated from

$$q_{s,0} = -\oint q_{b,s}\, ds \,/ \oint ds \quad \text{along the contour of the thin closed section}$$

$$\oint ds = 2\left[L_{1-2} + L_{2-3} + L_{3-4}\right] = 2 \times (80\sqrt{2} + 80 + 80) = 546.3 \text{ mm}$$

$$q_{s,0} = -2 \left\{ \int q_{b,12}.ds_1 + \int q_{b,23} \times ds_2 + \int q_{b,34}.ds_3 \right\} / 546.3$$

$$= -2 \left\{(-S_Y/I_{ZZ}) \left[\int (t.s_1^2/2\sqrt{2})\, ds_1 + \int (3 \times 80\, s_2 + 3 \times 80^2/\sqrt{2})\, ds_2 \right.\right.$$

$$\left.\left. + \int (3 \times (80 \times s_3 - s_3^2/2) + 1.707 \times 3 \times 80^2)ds_3\right]\right\} / 546.3$$

$$= 2 \times (S_Y/I_{ZZ}) \left[(t \times s_1^3/6\sqrt{2}) + (3 \times 80\, s_2^2/2 + 3 \times 80^2 \times s_2/\sqrt{2})\right.$$

$$\left. + \{3 \times (80 \times s_3^2/2 - s_3^3/6) + 1.707 \times 3 \times 80^2 \times s_3\}\right] / 546.3$$

$$= 2 \times (S_Y/I_{ZZ}) \left[3 \times (80\sqrt{2})^3/6\sqrt{2} + (3 \times 80 \times 80^2/2 + 3 \times 80^2 \times 80/\sqrt{2})\right.$$

$$\left. + \{3 \times (80 \times 80^2/2 - 80^3/6) + 1.707 \times 3 \times 80^2 \times 80\}\right] / 546.3$$

$$= 2 \times (S_Y/I_{ZZ}) \times 3 \times 80^3 \times [1/3 + 1/2 + 1/\sqrt{2} + 1/2 - 1/6$$

$$+ 1.707] / 546.3$$

$$= 20.136 \times 10^3 \times (S_Y/I_{ZZ})$$

Taking moments about 1,

$$S_Y \times \delta = \int q_s\, z \times ds$$

$$= 2 \left[\int (q_{12} \times 0)\, ds_1 + \int(q_{23} \times 80)\, ds_2 + \int(q_{34} \times 160)\, ds_3\right]$$

$$= 2 \left[0 + \int (q_{b,23} + q_{s,0}) \times 80\, ds_2 + \int (q_{b,34} + q_{s,0}) \times 160\, ds_3\right]$$

$$= -2(S_Y/I_{ZZ}) \times \left[\int(3 \times 80\, s_2 + 3 \times 80^2/\sqrt{2}\right.$$

$$- 20.136 \times 10^3) \times 80\, ds_2 + \int\{3 \times (80 \times s_3 - s_3^2/2)$$

$$+ 1.707 \times 3 \times 80^2 - 20.136 \times 10^3\} \times 160\, ds_3]$$

$$= -2(S_Y/I_{ZZ}) \left[3 \times 80 \times 80 \times (s_2^2/2)\right.$$

$$+ (3 \times 80^2/\sqrt{2} - 20.136 \times 10^3) \times 80\, s_2$$

$$+ 3 \times (80 \times s_3^2/2 - s_3^3/6) \times 160$$

$$+ (1.707 \times 3 \times 80^2 - 20.136 \times 10^3) \times 160\, s_3]$$

$$= -2(S_Y/I_{ZZ}) \times [6144 \times 10^4 - 4199.4 \times 10^4 + 8192 \times 10^4$$

$$+ 16177.15 \times 10^4]$$

$$= -52627.2 \times 10^4 \, (S_Y / I_{ZZ})$$

Therefore, $\delta = -52627.2 \times 10^4 / I_{ZZ} = -52627.2 \times 10^4 / 6144 \times 10^3$

$$= -85.66 \text{ mm}$$

Negative sign indicates that the shear center lies to the right of 1, between 1 and 4

Example 15.11

Determine the shear center for an aircraft box beam shown in figure, assuming uniform thickness of 5mm. All dimensions are in mm

Solution

$$I_{ZZ} \approx \pi \times 250^3 \times 5/2 + 2 \times [5 \times 300 \times 250^2$$

$$+ 5 \times 250^3/3] = 3.6235 \times 10^8 \ mm^4$$

Since the section is symmetric, $I_{YZ} = 0$ and, therefore,

$$q_{b,12} = -(S_Y/I_{ZZ}) \times \int t \times s_1 \ ds_1 = -(S_Y/I_{ZZ}) \times 5 \times s_1^2/2 \quad for \quad 0 \le s_1 \le 250$$

At 2, $s_1 = 250$ and $\quad q_{b,2} = -(S_Y/I_{XX}) \times 5 \times 250^2/2 = -156250 \ (S_Y/I_{ZZ})$

$$q_{b,23} = -(S_Y/I_{ZZ}) \times \int t \times h \times ds_2 + q_{b,2}$$

$$= -(S_Y/I_{ZZ}) \times 5 \times 250 \times s_2 - 156250 \ (S_Y/I_{ZZ}) \quad for \quad 0 \le s_2 \le 300$$

At 3, $s_2 = 300$ and

$$q_{b,3} = -(S_Y/I_{ZZ}) \times (5 \times 250 \times 300 + 156250) = -531250 \ (S_Y/I_{ZZ})$$

$$q_{b,34} = -(S_Y/I_{ZZ}) \int t \times (250 - 250 \cos \theta) \times (250d\theta) + q_{b,3} \quad for \quad 0 \le \theta \le \pi/2$$

$$= -(S_Y/I_{ZZ}) [5 \times (250 \times 250 \ \theta - 250 \times 250 \sin \theta) + 531250]$$

At 4, $\theta = \pi/2$ and

$$q_{b,4} = -(S_Y/I_{ZZ}) \times [5 \times (250^2 \times \pi/2 - 250^2 \times 1) + 531250]$$

$$= -267 \times 10^4 \ (S_Y/I_{ZZ})$$

The reference shear flow, q_1 or $q_{s,0}$ is given by

$$q_{s,0} = \oint qds / \oint ds \quad \text{with integration along the periphery}$$

$$= 2 \ [\int q_{b,12} \ ds_1 + \int q_{b,23} \ ds_2 + \int q_{b,34} \ ds_3 \ ] / \{2 \ [\pi \times 250/2 + 300 + 250]\}$$

$$= -2(S_Y/I_{ZZ}) \ [\int 5 \times (s_1^2/2) \ ds_1 + \int (5 \times 250 \times s_2 + 156250) \ ds_2$$

$$+ \int \{5 \times (250 \times 250 \ \theta - 250 \times 250 \sin \theta) + 531250\} d\theta] / (2 \times 942.9)$$

$$= -2(S_Y/I_{ZZ}) \ [5 \times (250^3/6) + 5 \times 250 \times (300^2/2) + 156250 \times 300$$

$$+ 5 \times 250 \times 250 \times (\pi^2/8) - 5 \times 250 \times 250 \times (1 - 0)$$

$$+ 531250 \times (\pi/2)] / 1885.8$$

$$= -124140 \ (S_Y/I_{ZZ})$$

Location of shear center is now obtained by taking moments about any point, say 1,

$$M_1 = S_Y \times \delta = \int q \times y \times ds$$

$$= 2 \left[\int q_{12} \times 0 \, ds_1 + \int q_{23} \times 250 \, ds_2 + \int q_{34} \times (300 + r \sin\theta) \, r \, d\theta \right]$$

$$= 2 \left[0 + \int (q_{b,23} + q_{s,0}) \times 250 \, ds_2 \right.$$

$$\left. + \int (q_{b,34} + q_{s,0}) \times (300 + 250 \sin\theta) \times 250 \, d\theta \right]$$

$$= -2 \, (S_Y/I_{ZZ}) \left[\int (5 \times 250 \times s_2 + 156250 + 124140) \times 250 \, ds_2 \right.$$

$$+ \int \{ 5 \times (250 \times 250 \, \theta - 250 \times 250 \sin\theta)$$

$$\left. + 531250 + 124140 \} \times (300 + r \sin\theta) \times 250 \, d\theta \right]$$

$$= -2 \, (S_Y/I_{ZZ}) \left[\{ 5 \times 250 \times (300^2/2) + 280390 \times 300 \} \times 250 \right.$$

$$+ \{ 5 \times 250 \times 250 \times (\pi/2) - 5 \times 250 \times 250 \times (1-0)$$

$$\left. + 655390 \} \times \{ 300 \times (\pi/2) + 250 \times (1-0) \} \times 250 \right]$$

$$= -2 \, (S_Y/I_{ZZ}) \left[(56.25 + 84.117) \times 250 + \{ (0.491 - 0.31 \right.$$

$$\left. + 0.6554) \times (471.43 + 250) \} \times 250 \right] \times 10^6$$

$$= -37.1886 \times 10^{10} \, (S_Y/I_{ZZ})$$

$$\therefore \delta = \int q \times y \times ds \, / \, S_Y = -37.1886 \times 10^{10} / I_{ZZ}$$

$$= -37.1886 \times 10^{10} / (3.6235 \times 10^8) = -1026.32 \text{ mm}$$

Negative sign indicates that the shear center lies to the left of 4, along X-axis

Example 15.12

Determine the location of shear center for the rectangular section 200mm × 300 mm, shown in figure. Thickness of sides 6-2, 2-3 and 5-6 is 10mm while thickness of side 3-5 is 20 mm.

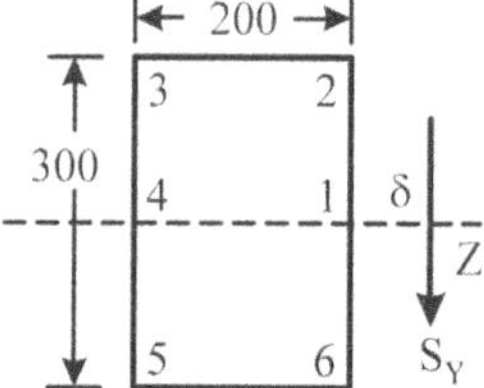

Solution

$$I_{ZZ} \approx 2 \times [10 \times 150^3/3 + 10 \times 200 \times 150^2 + 20 \times 150^3/3]$$

$$= 1.575 \times 10^8 \text{ mm}^4$$

Since the section is symmetric, $I_{YZ} = 0$ and, therefore,

$$q_{b,12} = -(S_Y/I_{ZZ}) \int t \, s_1 \, ds_1 = -(S_Y/I_{ZZ}) \, 10 \times s_1^2/2 \quad \text{for} \quad 0 \le s_1 \le 150$$

At 2, $s_1 = 150$ and $q_{b,2} = -(S_Y/I_{ZZ}) \times 10 \times 150^2/2 = -112500 \, (S_Y/I_{ZZ})$

$$q_{b,23} = -(S_Y/I_{ZZ}) \int t \, h \, ds_2 + q_{b,2}$$

$$= -(S_Y/I_{ZZ}) \, 10 \text{x} 150 \text{x} s_2 - 112500 \, (S_Y/I_{ZZ}) \qquad \text{for} \quad 0 \le s_2 \le 200$$

$$q_{b,2} = -(S_Y/I_{ZZ}) \, 10 \times 150 \times s_2 - 112500 \, (S_Y/I_{ZZ}) \quad \text{for} \quad 0 \le s_2 \le 200$$

At 3, $s_2 = 200$ and

$$q_{b,3} = -(S_Y/I_{ZZ})(10 \times 150 \times 200 + 112500) = -412500 \times (S_Y/I_{ZZ})$$

$$q_{b,34} = -(S_Y/I_{ZZ}) \int t \times (150 - s_3)\, ds_3 + q_{b,3} \qquad \text{for} \quad 0 \le s_3 \le 150$$

$$= -(S_Y/I_{ZZ})[20 \times 150 \times s_3 - 20 \times (s_3^2/2) + 412500]$$

At 4, $s_3 = 150$ and

$$q_{b,4} = -(S_Y/I_{ZZ}) \times [20 \times 150 \times 150 - 20 \times (150^2/2) + 412500]$$

$$= -637500 \times (S_Y/I_{ZZ})$$

The reference shear flow, q_1 or $q_{s,0}$ is given by

$$q_{s,0} = \oint q\,ds / \oint ds \qquad \text{with integration along the periphery}$$

$$= 2\,[\int q_{b,12}\,ds_1 + \int q_{b,23}\,ds_2 + \int q_{b,34}\,ds_3\,]/2\,[150 + 200 + 150]$$

$$= -2\,(S_Y/I_{ZZ})\,[\int 10 \times (s_1^2/2)\,ds_1 + \int (10 \times 150 \times s_2 + 112500)\,ds_2$$

$$+ \int (20 \times 150 \times s_3 - 20 \times (s_3^2/2) + 412500)\,ds_3\,] / (2 \times 500)$$

$$= -2\,(S_Y/I_{ZZ})\,[10 \times (150^3/6) + 10 \times 150 \times (200^2/2) + 112500 \times 200$$

$$+ 20 \times 150 \times (150^2/2) - 20 \times (150^3/6) + 412500 \times 150]/1000$$

$$= -193125 \times (S_Y/I_{ZZ})$$

Location of shear center is now obtained by taking moments about any point, say 1,

$$M_1 = S_Y \times \delta = \oint q \times y \times ds$$

$$= 2\,[\int q_{12} \times 0\, ds_1 + \int q_{23} \times 150\, ds_2 + \int q_{34} \times 200\, ds_3]$$

$$= 2\,[0 + \int (q_{b,23} + q_{s.0}) \times 150\, ds_2 + \int (q_{b,34} + q_{s.0}) \times 200\, ds_3\,]$$

$$= -2\,(S_Y/I_{ZZ})\,[\int (10 \times 150 \times s_2 + 112500 + 193125) \times 150\, ds_2$$

$$+ \int (20 \times 150 \times s_3 - 20 \times (s_3^2/2) + 412500 + 193125) \times 200\, ds_3]$$

$$= -2\,(S_Y/I_{ZZ})\,[\{10 \times 150 \times (200^2/2) + 307625 \times 200\} \times 150$$

$$+ \{20 \times 150 \times (150^2/2) - 20 \times (150^3/6) + 605625\} \times 200]$$

$$= -2\,(S_Y/I_{ZZ})\,[(30 + 61.525) \times 150$$

$$+ (33.75 - 11.25 + 12.1125) \times 200] \times 10^6$$

$$= -413 \times 10^8 \times (S_Y/I_{ZZ})$$

Therefore, $\delta = -413 \times 10^8 / I_{ZZ} = -413 \times 10^8 / 1.575 \times 10^8 = -262.24$ mm

Negative sign indicates that the shear center lies to the left of 4, along Z-axis

Example 15.13

A thin walled box beam of an equilateral triangular section of side 'L' and thickness 't', oriented as shown in the figure, is used in the support structure of an aeroplane wing. Determine the location of shear center.

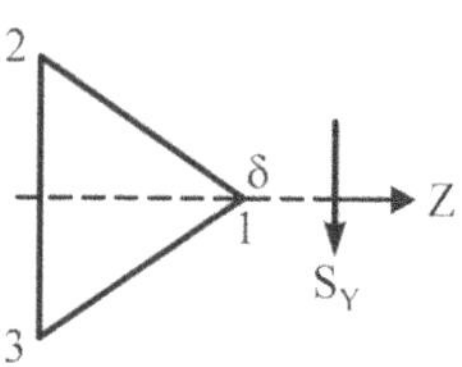

Solution

$$I_{ZZ} = [t \times L^3 \times \sin^2 30 /3 + t \times L^3/12 + t \times L^3 \times \sin^2 30 /3] = t.L^3/4$$

Since the section is symmetric, $I_{XY} = 0$ and, therefore, with $q_{b,1} = 0$

$$q_{b,12} = - (S_Y/I_{ZZ}) \int t \times s_1 \sin 30 \, ds_1 = - (S_Y/I_{ZZ}) \, t \times s_1^2/4 \quad \text{for} \quad 0 \le s_1 \le L$$

At 2, $s_1 = L$ and $q_{b,2} = - (S_Y/I_{ZZ}) \, t \times L^2/4$

$$q_{b,23} = - (S_Y/I_{ZZ}) \int t \, (L/2 - s_2) \, ds_2 + q_{b.2}$$

$$= - (S_Y/I_{ZZ}) \, [t \times (L.s_2/2 - s_2^2/2) + t \times L^2/4 \,] \quad \text{for} \quad 0 \le s_2 \le L$$

At 3, $s_2 = L$ and

$$q_{b,3} = - (S_Y/I_{ZZ}) \times t \times [L \times L/2 - L^2/2 + L^2/4] = - (S_Y/I_{ZZ}) \times t \times L^2/4$$

$$= q_{b,2} \quad \text{by symmetry}$$

$$q_{b,31} = - (S_Y/I_{ZZ}) \int t \, (-L/2 + s_3 \sin 30) \, ds_3 + q_{b,3} \quad \text{for} \quad 0 \le s_3 \le L$$

$$= - (S_Y/I_{ZZ}) \, t \times [- L \times s_3/2 + s_3^2/4 + L^2/4]$$

At 1, $s_1 = L$ and $q_{b,4} = - (S_Y/I_{ZZ}) \times t \times [- L \times L/2 + L^2/4 + L^2/4] = 0$

$$= q_{b,1} \quad \text{as expected}$$

The reference shear flow, q_1 or $q_{s,0}$ is given by

$$q_{s,0} = \oint q ds / \oint ds \qquad \text{with integration along the periphery}$$

$$= [\int q_{b,12} \, ds_1 + \int q_{b,23} \, ds_2 + \int q_{b,31} \, ds_3 \,] / [L + L + L]$$

$$= - (S_Y/I_{ZZ}) [\int (t \times s_1^2/4) \, ds_1 + \int t \times (L \times s_2/2 - s_2^2/2 + L^2/4) \, ds_2$$

$$+ \int t \times (-L \times s_3/2 + s_3^2/4 + L^2/4) \, ds_3 \,] / 3L$$

$$= - (S_Y/I_{ZZ}) \times t \, [\, s_1^3/12 + L \times s_2^2/4 - s_2^3/6 + s_2 \, L^2/4$$

$$- L \times s_3^2/4 + s_3^3/12 + s_3 \, L^2/4 \,] / 3L$$

$$= - (S_Y/I_{ZZ}) \times t \, [\, L^3/12 + L \times L^2/4 - L^3/6 + L \times L^2/4$$

$$- L \times L^2/4 + L^3/12 + L \times L^2/4 \,] / 3L$$

$$= - (S_Y/I_{ZZ}) \times t \times L^3 \, (1 + 3 - 2 + 3 - 3 + 1 + 3) / (3 \times 12 \, L)$$

$$= - (t \, L^2 / 6) \times (S_Y/I_{ZZ})$$

Location of shear center is now obtained by taking moments about any point, say 1,

$$M_1 = S_Y \times \delta = \int q \times y \times ds$$

$$= \int q_{12} \times 0 \, ds_1 + \int q_{23} \times (L \cos 30) \, ds_2 + \int q_{34} \times 0 \, ds_3$$

$$= 0 + \int (q_{b,23} + q_{s,0}) \times (L \cos 30) \, ds_2 + 0$$

$$= -(S_Y/I_{ZZ}) \times [\int [t \times (L \times s_2/2 - s_2^2/2) + t \times L^2/4 + t \, L^2/6]$$
$$\times (L \cos 30) \, ds_2$$

$$= -(S_Y/I_{ZZ}) \times (t \times L \times s_2^2/4 - t \times s_2^3/6 + 5 \times s_2 \times t \times L^2/12)$$
$$\times (\sqrt{3} \, L/2) \quad \text{for} \quad 0 \le s_2 \le L$$

$$= -(S_Y/I_{ZZ}) \times (t \times L \times L^2/4 - t \times L^3/6 + 5 \, t \times L \times L^2/12) \times (\sqrt{3} \times L/2)$$

$$= -(\sqrt{3} \, t \times L^4 / 4) \times (S_Y/I_{ZZ})$$

Therefore, $\delta = -\sqrt{3} \, t \times L^4 / (4 \times t \times L^3/4) = -\sqrt{3} \, L$

Negative sign indicates that the shear center lies to the left of 1, along X-axis

Example 15.14

100mmx200mm rectangular box beam of uniform thickness 3mm in Y-Z plane is subjected to a vertical shear force of 20kN in Y-direction. Determine the variation of shear flow throughout the cross section.

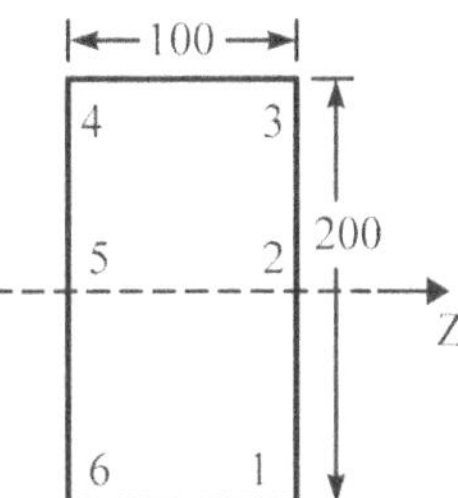

Solution

$$I_{ZZ} \approx 2 \times [3 \times 200^3 / 12 + 3 \times 100 \times 100^2] = 10^7 \, mm^4$$

Since the section is symmetric, $I_{YZ} = 0$ and, therefore, with $q_{b,1}=0$,

$$q_{b,13} = -(S_Y/I_{ZZ}) \times \int t \times (-100 + s_1) \, ds_1$$
$$= -(S_Y/I_{ZZ}) \times (-3 \times 100 \, s_1 + 3 \times s_1^2/2) \qquad \text{for} \quad 0 \le s_1 \le 200$$

At 3, $s_1 = 200$ and $q_{b,3} = -(S_Y/I_{ZZ}) \times (-100 \times 3 \times 200 + 3 \times 200^2 / 2) = 0$

Maximum shear flow along 1-3 occurs at 2 for $s_1 = 100$ and

$$q_{b,2} = -(S_Y/I_{ZZ}) \times (-100 \times 3 \times 100 + 3 \times 100^2/2)$$
$$= +1.5 \times 10^4 \, (S_Y/I_{XX})$$

$$q_{b,34} = -(S_Y/I_{ZZ}) \times \int t \times h \times ds_2 + q_{b,3}$$
$$= -(S_Y/I_{ZZ}) \times 3 \times 100 \times s_2 \qquad \text{for} \quad 0 \le s_2 \le 100$$

At 4, $s_2 = 100$ and $q_{b,4} = -(S_Y/I_{XX}) \times 3 \times 100 \times 100 = -3 \times 10^4 \, (S_Y/I_{XX})$

$$q_{b,46} = -(S_Y/I_{ZZ}) \times \int t \times (100 - s_3) \, ds_3 + q_{b,4} \qquad \text{for} \quad 0 \le s_3 \le 200$$
$$= -(S_Y/I_{ZZ}) \times [3 \times (100 \, s_3 - (s_3^2/2) + 3 \times 10^4]$$

At 6, $s_3 = 200$ and $q_{b,6} = -(S_Y/I_{ZZ}) \times [3 \times 100 \times 200 - 3 \times (200^2/2) + 3 \times 10^4]$

$$= -3 \times 10^4 \, (S_Y/I_{ZZ}) = q_{b,4}$$

Maximum shear flow along 4-6 occurs at 5 for $s_3 = 100$ and

$q_{b,5} = -(S_Y/I_{ZZ}) \times (3 \times 100 \times 100 - 3 \times (100^2/2) + 3 \times 10^4)$

$$= -4.5 \times 10^4 \, (S_Y/I_{ZZ})$$

$q_{b,61} = -(S_Y/I_{ZZ}) \times \int t \times (-100) \, ds_4 + q_{b,6}$ for $\;\; 0 \le s_4 \le 100$

$$= -(S_Y/I_{ZZ}) \times [3 \times 100 \, s_4 + 3 \times 10^4]$$

At 1, $s_1 = 100$ and $q_{b,1} = -(S_Y/I_{ZZ}) \times [-3 \times 100 \times 100 + 3 \times 10^4] = 0$

It can also be concluded, due to symmetry of section, that

$$q_{b,6} = q_{b,4} \quad \text{and} \quad q_{b,3} = q_{b,1}$$

The reference shear flow, q_1 or $q_{s,0}$ is given by

$q_{s,0} = \oint q \, ds / \oint ds$ with integration along the periphery

$$= [\int q_{b,13} \, ds_1 + \int q_{b,34} \, ds_2 + \int q_{b,46} \, ds_3 + \int q_{b,61} \, ds_4 \,] / [\, 2 \, (100 + 100 + 100) \,]$$

$$= -(S_Y/I_{ZZ}) [\int (-3 \times 100 \, s_1^2 + 3 \times s_1^2/2) \, ds_1 + \int (3 \times 100 \times s_2) \, ds_2$$

$$+ \int \{3 \times 100 \times s_3 - 3 \times (s_3^2/2)\} \, ds_3$$

$$+ \int (3 \times 100 \times s_4 + 3 \times 10^4) \, ds_4 \,] / (2 \times 300)$$

$$= -(S_Y/I_{ZZ}) [-3 \times 100 \times (s_1^2/2) + 3 \times (s_1^3/6) + 3 \times 100 \times (s_2^2/2)$$

$$+ \{3 \times 100 \times (s_3^2/2) - 3 \times (s_3^3/6) + 3 \times 10^4 \, s_3\} + \{3 \times 100 \, (s_4^2/2)$$

$$+ 3 \times 10^4 \, s_4 \,\} \,] / 600$$

$$= -(S_Y/I_{ZZ}) [-3 \times 100 \times 200^2/2 + 3 \times 200^3/6 + 3 \times 100 \times (100^2/2)$$

$$+ \{3 \times 100 \times (100^2/2) - 3 \times (100^3/6) + 3 \times 10^4 \times 100\}$$

$$+ \{3 \times 100 \times 100^2/2 + 3 \times 10^4 \times 100\}\,] / 600$$

$$= -10^6 \, (-6 + 4 + 1.5 + 1.5 - 0.5 + 3 + 1.5 + 3) \times S_Y / (600 \, I_{ZZ})$$

$$= -(8/6) \times 10^4 \, (S_Y/I_{ZZ}) = -1.333 \times 10^4 \, (S_Y/I_{ZZ})$$

$$= -1.333 \times 10^4 \times 20 \times 10^3 / 10^7$$

$$= -26.67 \text{ N/mm}$$

Therefore, $q_1 = q_3 = q_{s,0} = -26.67$ N/mm

$q_2 = q_{b,2} + q_{s,0} = (1.5 \times 10^4 - 1.333 \times 10^4) \times (S_Y/I_{ZZ})$

   $= 3.34$ N/mm

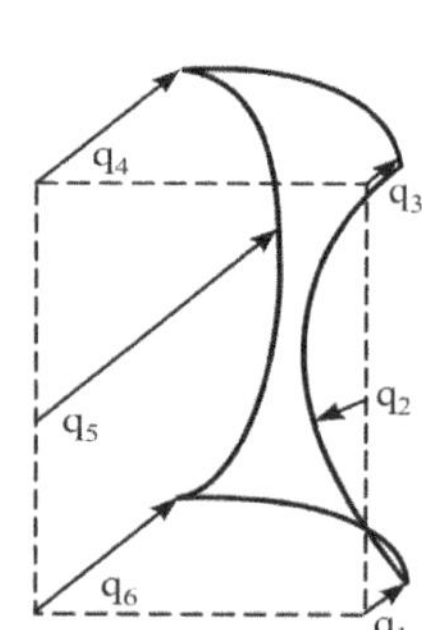

$$q_4 = q_6 = q_{b,4} + q_{s,0}$$

$$= - (3 \times 10^4 + 1.333 \times 10^4) \times (S_Y/I_{ZZ}) = - 86.67 \text{ N/mm}$$

$$\text{and } q_5 = q_{b,5} + q_{s,0} = - (4.5 \times 10^4 + 1.333 \times 10^4) \times (S_Y/I_{ZZ})$$

$$= - 116.7 \text{ N/mm}$$

The distribution of shear flow is shown graphically.

15.14 SHEAR FLOW IN THIN WEBS, ENCLOSED IN A FRAME OF STIFFENERS

Sheet metal, necessary for covering modern aircrafts, is also utilised for resisting loads. The thin sheets are very efficient in resisting shear or tension loads on the planes of the webs, but must be stiffened to resist compression loads and loads normal to the web. If they are used to resist all loads without stiffening, the construction is called *monocoque* or *full monocoque*. Such a construction is heavy since the skin should be thick enough to resist buckling due to compression loads. Hence, most airplane structures use thin sheet metal webs for resisting shear and tension while stiffeners resist compression loads and loads normal to the webs. Such a construction is called *semi-monocoque*. Shear flow calculation in semi-monocoque structures is explained through the following examples.

The stiffeners resist compression loads and transmit them to the webs as shear loads in the plane of the webs. Since the concentrated loads may have components along three mutually perpendicular axes, it is necessary to provide webs in different planes at the point of applying concentrated loads.

In this section, thin webs are assumed to resist pure shear. In actual structures, the thin webs may wrinkle due to tension field stresses. These effects are discussed in the subsequent sections and the stresses can be superimposed.

Example 15.15

Let us consider a simple example of a beam of length 'L' and depth 'd' with a concentrated load 'P' acting at a distance 'a' from its left end. From equations of equilibrium, the reactions are found to be inversely proportional to the horizontal distance between the load and the support. Thus, with L = a+b,

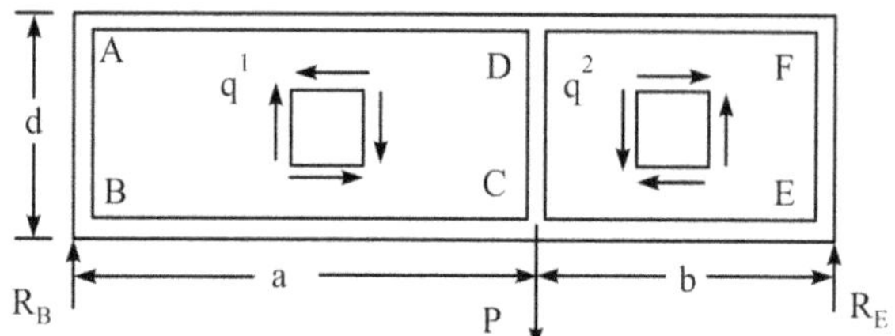

$$R_B = P \times b / L \quad \text{and} \quad R_E = P \times a/L$$

Shear flow in the web ABCD, $q_1 = R_B/d$

and in the web CEFD, $q_2 = R_E/d$

Force variation in the stiffener CD and in the webs along AB, EF and BCE are shown here.

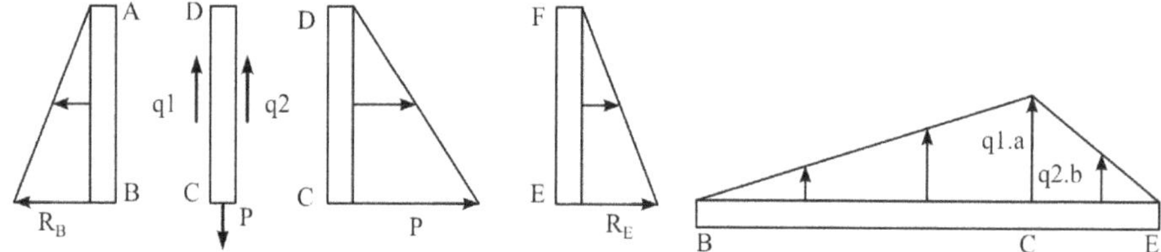

Example 15.16

Let us consider one more example, where a concentrated load 'P' is applied at B at an angle such that it has a vertical component 'V' and a horizontal component 'H'. As explained earlier, stiffeners AB and CBD have to be placed at the point of load application to resist the two components of force. Length of stiffener AB depends on the shear flow such that horizontal force at A is zero. Since stiffener AB can not end with a web, a vertical stiffener EAF is provided.

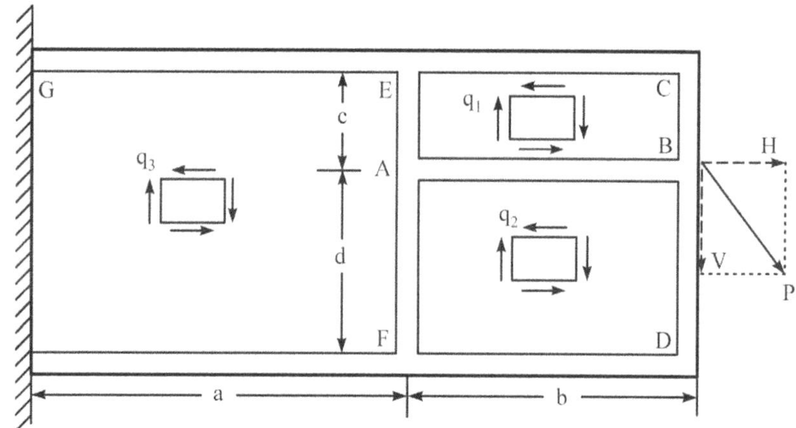

The shear flows q_1 and q_2 can be obtained from the equilibrium condition of the stiffeners AB and CBD as

$$c \times q_1 + d \times q_2 = V \quad \text{and} \quad b \times q_1 - b \times q_2 = H,$$

since q_1 and q_2 are in the opposite directions on the two sides of AB, as is evident from the 1st equation. Solving these two simultaneous equations, q_1 and q_2 can be evaluated. These values can also be derived by analyzing the beam separately for the two components of load and superimposing the results.

Vertical component of force results in uniform shear force

$$q_1' = q_2' = q_3 = V/(c + d)$$

while the horizontal component produces +ve q_1'' and –ve q_2''

such that $(q_1'' + q_2'') \times b = H$. Distribution of axial force along AB and shear force in the webs along CBD and GEC are shown below.

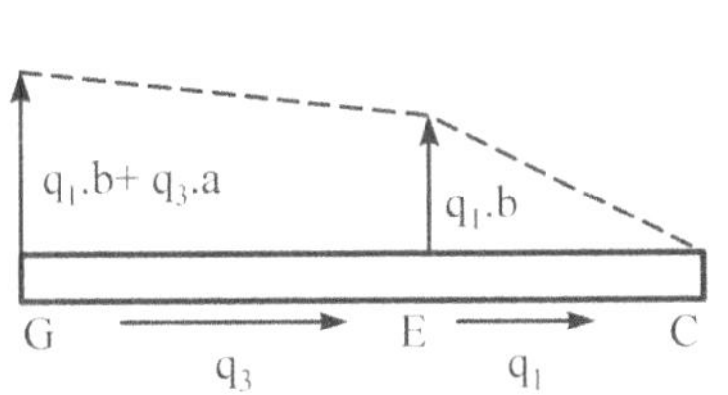

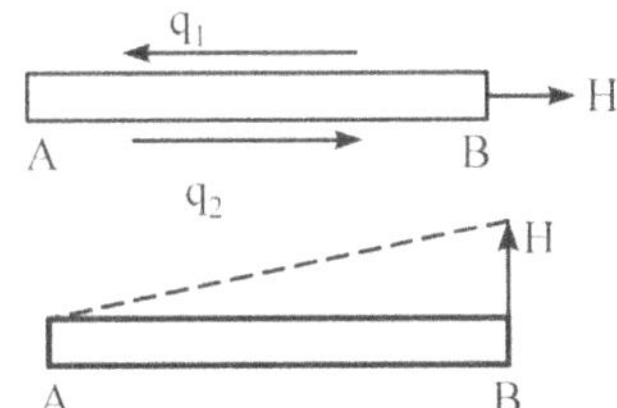

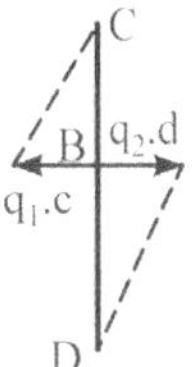

15.15 SHEAR LAG

In simple beam theory, the two assumptions made – plane sections before bending remain plane after bending and bending stresses are proportional to the distance from neutral axis – are less accurate for semi-monocoque structures. Unlike in heavy sections, shearing deformations in thin webs are not always negligible. Shear strains in the thin walls of beams cause cross sections to distort. The effect of shearing deformations in redistributing the bending stresses in a box beam is commonly known as ***shear lag***.

In a cantilever box beam, symmetrical about a central vertical plane, with a concentrated vertical load on the plane of symmetry at the tip section so that no twist is produced, simple beam theory predicts uniform bending stress at any straight section $a' - e'$. However, shear strains at the section cause distortion to take the form $e' - b' - c' - d' - e'$ as shown in Fig. 15.19. Hence, different stiffeners on the top plane carry different bending loads, a significant variation from the simple beam theory.

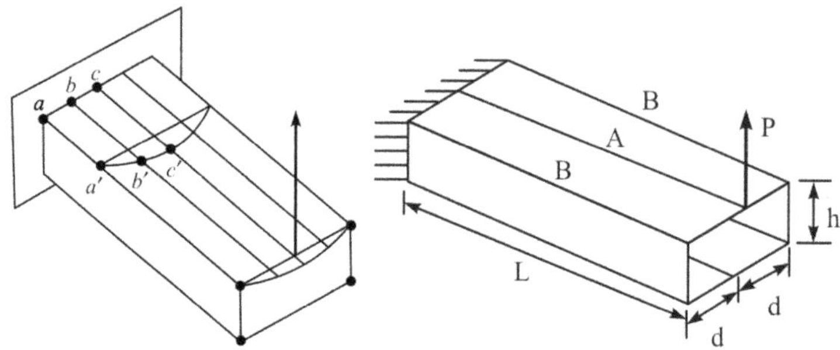

FIGURE 15.19 Shear lag in a cantilever beam

At the support, the section is restrained from warping out of its original plane and line a-b-c-d-e remains straight. Since the distance c–c' is greater than a– a', the stringer at c resists a smaller compressive stress than the stringer at a. Thus, the bending stress at a must be greater than that calculated by simple theory and the bending stress at c must be less than that calculated by simple theory. Hence, the end stiffeners have to be of larger section than those inside. At some distance from the support, all the stringers have approximately the same bending stress and strain. The shear lag effect is greatest at the support and is a local effect. Generally, shear lag becomes significant in wide, relatively shallow, thin walled beams such as wings.

Additional Problems for Practice

1. Find the shear flow in each web of the beam shown in the figure. Plot the distribution of axial load along each stiffening member when $P_1 = 25$kN, $P_2 = 15$kN All dimensions are in cm.

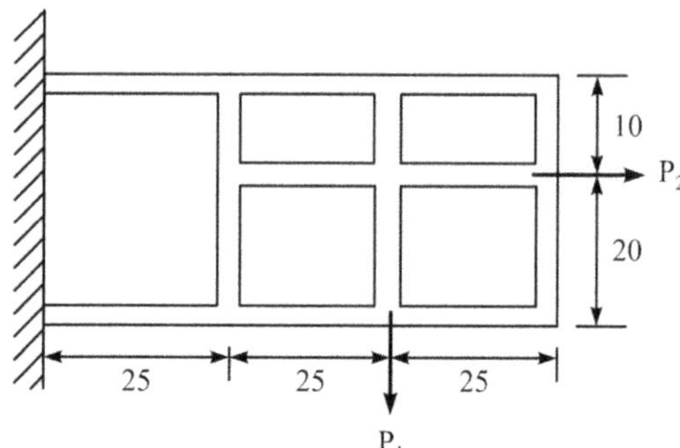

2. Find the shear flow in each web of the beam shown in the figure. Plot the distribution of axial load along each stiffening member when $P_1 = 20$kN and $P_2 = 10$kN. All dimensions are in cm.

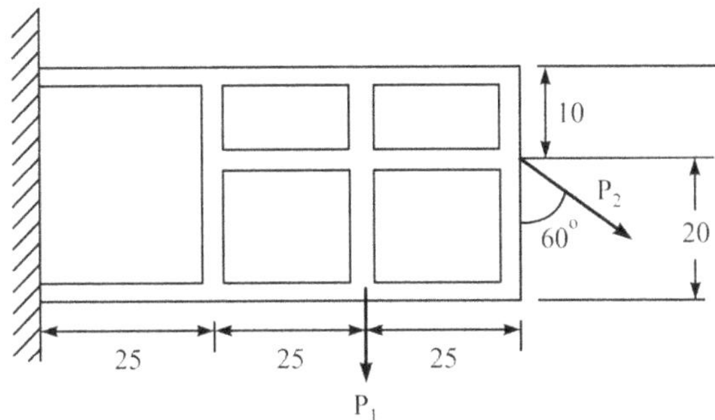

3. Find the shear flow in each web of the beam shown in the figure. Plot the distribution of axial load along each stiffening member when $P_1 = 20kN$, $P_2 = 15kN$ and $P_3 = 10kN$. All dimensions are in cm.

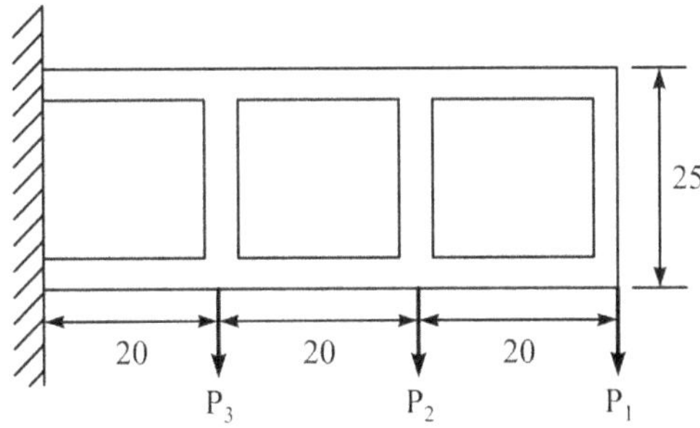

4. Find the shear flow in each web of the beam shown in the figure. Plot the distribution of axial load along each stiffening member when $P_1 = 20kN$ and $P_2 = 15kN$. All dimensions are in cm.

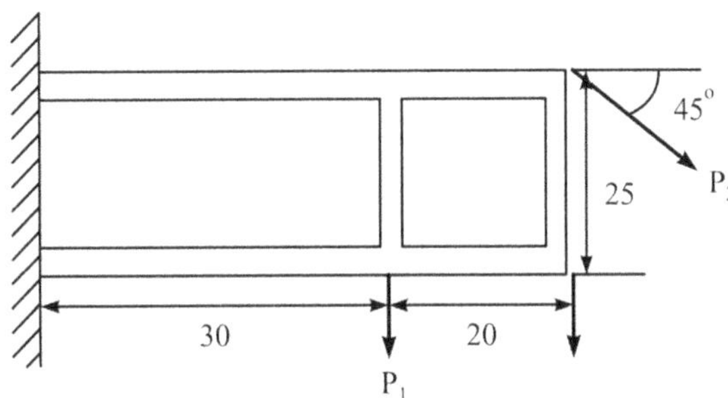

5. (a)   What are monocoque and semi-monocoque structures ? Explain briefly with suitable examples

   (b)   Find the shear flow in each web of the beam shown in the figure. Plot the distribution of axial load along each stiffening member when $P_1 = 20kN$, $P_2 = 15kN$ and $P_3 = 10kN$. All dimensions are in cm.

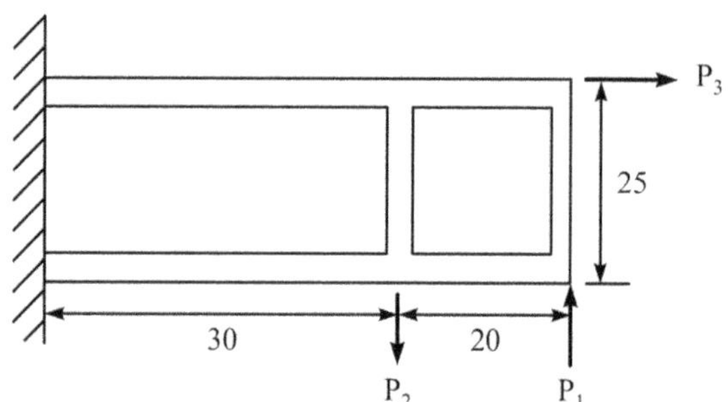

6. Find the shear flow in each web of the beam shown in the figure. Plot the distribution of axial load along each stiffening member when $P_1 = 20$kN, $P_2 = 15$kN and $P_3 = 10$kN. All dimensions are in cm.

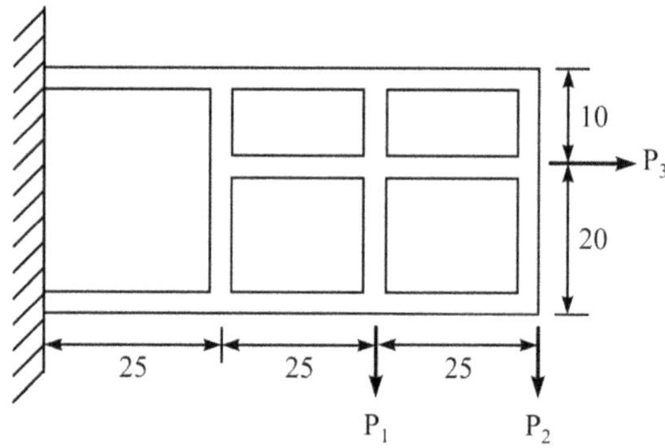

CHAPTER 16

ANALYSIS OF AEROPLANE FUSELAGE AND WING

Basic theory for the analysis of open and closed section thin-walled beams and stiffened panels subjected to bending, shear and torsion are explained in the earlier chapters. Analysis of actual aircraft components, such as wings and fuselage, using these methods is presented in this chapter. Analysis of box beams, etc. which form essential part of analysis of wing and fuselage are first discussed. Detailed analysis is usually carried out by computer using Finite Element Method, but cheaper approximate methods explained here are used in the preliminary stages when several possible structural alternatives are being investigated. They also provide an insight into physical behaviour of structures, which are incorporated into the particular computer software.

16.1 ANALYSIS OF BOX BEAMS – STRUCTURAL IDEALISATION

Practical airplane structures differ from the simpler models analysed in the earlier sections. For a preliminary design, these *complex structures are idealized into simpler forms*, which behave, under the given loading conditions, in very nearly the same way as the actual structure, so that a simpler solution is obtained quickly. A final solution is then obtained as accurately as possible by improving the analytical model.

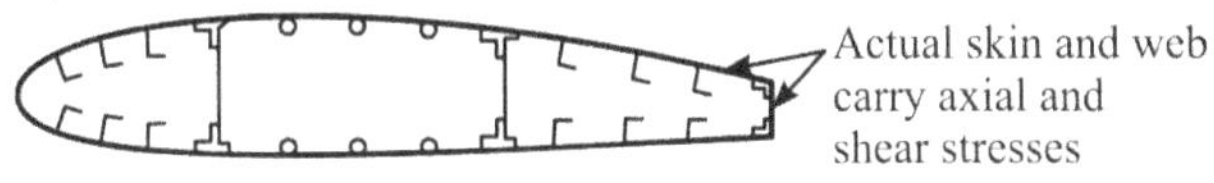

FIGURE 16.1 Typical wing section

A typical 2-spar wing section is shown in Fig 16.1, in which Z-section stringers are used to stiffen the thin skin while the angle sections form the spar flanges.

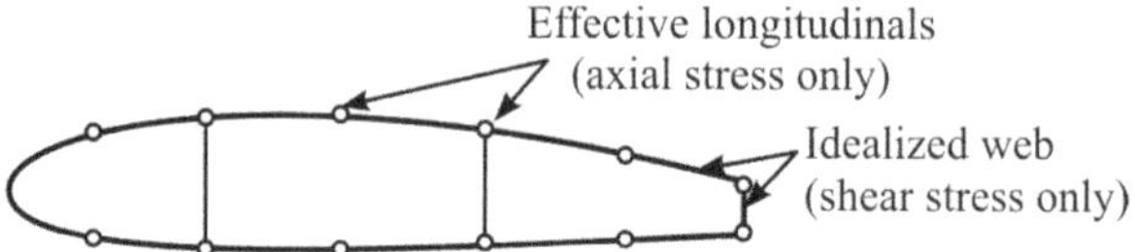

FIGURE 16.2 Idealisation of a wing section

It would be reasonable, as shown in Fig. 16.2, to assume that –

- the variation of bending stress over the small section of the stringers is negligibly small
- the distance between the centroids of stringers and angles forming spar flanges and the center-line of the skin is so small that the stringers and flanges can be considered as concentrated booms along the center-line of the thin skin

Stringers and spar flanges carry most of the direct stresses while the skin is mainly effective in resisting shear stresses. The panel between stringers is idealised by concentrations of area, known as **booms**, at the two ends to resist direct loads and skin of zero thickness which is effective only in shear. Areas of stringers at the ends of the panels are added to the areas of the booms. Suppose the direct stress distribution of the skin panel of thickness 't' varies from σ_1 at one end to σ_2 at the other end, it can be represented by a shear panel with boom areas B_1 and B_2 at its ends (shown in Fig.16.3).

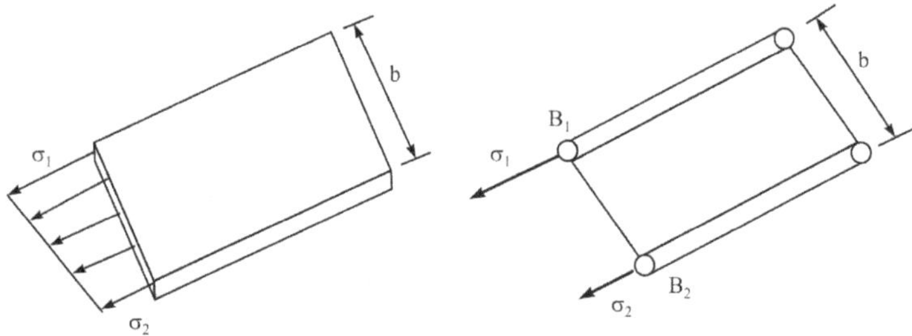

FIGURE 16.3 Actual panel and idealized panel

Thus, taking moments about the right hand edge of the panel, considering the stress variation as a combination of uniform stress σ_2 and a varying component stress of 0 to σ_1-σ_2

$$(\sigma_2 \times b \times t) \times (b/2) + \{(\sigma_1 - \sigma_2)/2\} \times t \times (b/2) \times (2b/3)$$
$$= \sigma_1 \times B_1 \times b$$

$$\Rightarrow B_1 = (t \times b/6) \times \{2 + (\sigma_2/\sigma_1)\} \qquad\qquad .....(16.1)$$

Equating the total direct force, $\{(\sigma_1 + \sigma_2)/2\} \times b \times t = \sigma_1 \times B_1 + \sigma_2 \times B_2$

$$\Rightarrow B_2 = (t \times b/6) \times \{2 + (\sigma_1/\sigma_2)\}$$

This can also be obtained by taking moments about other edge

Here, σ_1 and σ_2 are caused by a combination of axial load and bending moment.

If only axial load acts, $\sigma_1 = \sigma_2$ and, therefore, $B_1 = B_2 = t \times b / 2$

For a pure bending moment, $\sigma_1/\sigma_2 = -1$ since direct stress is tensile on one side and compressive on the other side and, therefore, $B_1 = B_2 = t \times b / 6$

Example 16.1

Part of a wing section is in the form of a two-cell box shown below, in which vertical spars are connected to the wing skin through angle sections, all having cross sectional area 'A'. Idealise the section into an arrangement of direct stress carrying booms and shear stress only carrying panels suitable for resisting bending moments in a vertical plane.

Solution

Since the section resists bending moment, direct stress at any point in the wing is directly proportional to its distance from the horizontal axis of symmetry. Also, the distribution of direct stress in all the panels will be linear i.e.,

$$\sigma_2/\sigma_1 = d/c \qquad \text{and} \qquad \sigma_3/\sigma_2 = e/d$$

The boom areas include areas of the existing angle sections, considered as spar flanges.

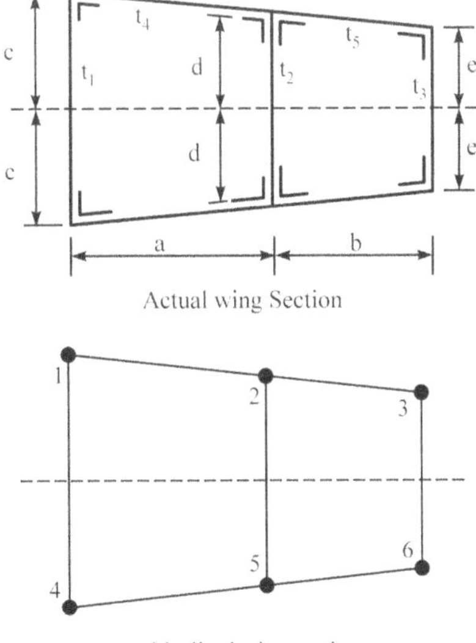

Actual wing Section

Idealised wing section

Area of boom at 1 has contributions from angle at 1 and panels 1-2 and 1-4

$$B_1 = A + (a \times t_4 / 6) \times \{2 + (\sigma_2/\sigma_1)\} + (c \times t_1 / 6) \times \{2 + (\sigma_4/\sigma_1)\}$$
$$= A + (a \times t_4 / 6) \times \{2 + (d/c)\} + (c \times t_1 / 6) \times \{2 + (-1)\}$$

Area of boom at 2 has contributions from two angles at 2 and panels 1-2, 2-3 and 2-5

$$B_2 = 2A + (a \times t_4 / 6) \times \{2 + (\sigma_1/\sigma_2)\} + (b \times t_5 / 6) \times \{2 + (\sigma_3/\sigma_2)\}$$
$$+ (d \times t_2 / 6) \times \{2 + (\sigma_5/\sigma_2)\}$$

$$= 2A + (a \times t_4 / 6) \times \{2 + (c/d)\} + (b \times t_5 / 6) \times \{2 + (e/d)\}$$
$$+ (d \times t_2 / 6) \times \{2 + (-1)\}$$

Area of boom at 3 has contributions from angle at 3, panels 2-3 and 3-6

$$B_3 = A + (b \times t_5 / 6) \times \{2 + (\sigma_2/\sigma_3)\} + (e \times t_3 / 6) \times \{2 + (\sigma_6/\sigma_3)\}$$
$$= A + (b \times t_5 / 6) \times \{2 + (d/e)\} + (e \times t_3 / 6) \times \{2 + (-1)\}$$

By symmetry, $\quad B_4 = B_1; \quad B_5 = B_2 \quad$ and $\quad B_6 = B_3$

Example 16.2

A thin walled beam with stiffeners at the junctions of two panels is shown in figure. Calculate boom areas of the idealised structure. Area of stiffeners at 1,3,4 and 6, $A_1 = 300\text{mm}^2$; Area of stiffeners at 2 and 5, $A_2 = 200\text{mm}^2$; $t_{1-2} = t_{5-6} = t_{2-3} = t_{4-5} = 1.2$ mm; $t_{1-6} = t_{3-4} = t_{2-5} = 2$ mm; a = 500 mm; b = 1000 mm; c = 150 mm; d = 250 mm; e = 100 mm

Solution

Direct stress at any point due to S_Y is proportional to its distance from Z-axis. Thus,

$\sigma_2/\sigma_1 = 250/100 = 2.5$

$\sigma_1/\sigma_2 = 1/2.5 = 0.4$

$\sigma_3/\sigma_2 = 150/250 = 0.6$

$\sigma_2/\sigma_3 = 1/0.6 = 1.67$

$\sigma_1/\sigma_6 = \sigma_2/\sigma_5 = \sigma_3/\sigma_4 = -1$

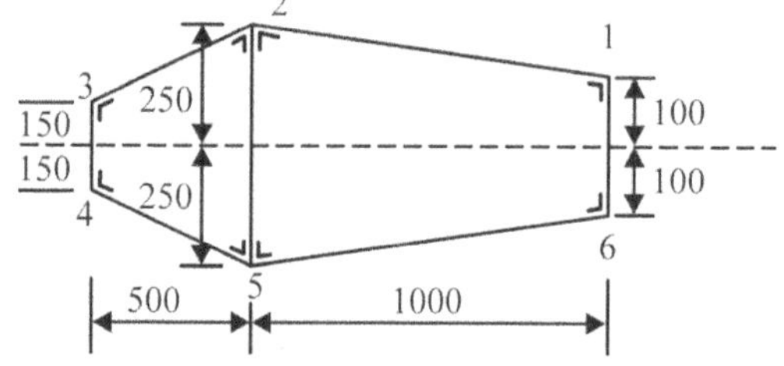

The boom areas include areas of the existing angle sections, considered as spar flanges

Area of boom at 1 has contributions from angle at 1 and panels 1-2 and 1-6

$$B_1 = A_1 + (b \times t_{2-3} / 6) \times \{2 + (\sigma_2/\sigma_1)\} + (e \times t_{1-6} / 6) \times \{2 + (\sigma_6/\sigma_1)\}$$

$$= 300 + (1000 \times 1.2/6) \times (2 + 2.5) + (100 \times 2/6) \times \{2 + (-1)\}$$

$$= 1233.3 \text{ mm}^2$$

Area of boom at 2 has contributions from two angles at 2 and panels 1-2, 2-3 and 2-5

$$B_2 = 2A_2 + (a \times t_{2\text{-}3} / 6) \times \{2 + (\sigma_3/\sigma_2)\} + (b \times t_{2\text{-}1}/6) \times \{2 + (\sigma_1/\sigma_2)\}$$
$$+ (d.t_{2\text{-}5} / 6) \times \{2 + (\sigma_5/\sigma_2)\}$$

$$= 2 \times 200 + (500 \times 1.2/6) \times (2 + 0.6) + (1000 \times 1.2/6) \times (2 + 0.4)$$
$$+ (250 \times 2/6) \times \{2 + (-1)\}$$

$$= 1223.3 \text{ mm}^2$$

Area of boom at 3 has contributions from angle at 3, panels 2-3 and 3-4

$$B_3 = A_1 + (a \times t_{2\text{-}3} / 6) \times \{2 + (\sigma_2/\sigma_3)\} + (c \times t_{3\text{-}4} / 6) \times \{2 + (\sigma_4/\sigma_3)\}$$

$$= 300 + (500 \times 1.2/6) \times (2 + 1.67) + (150 \times 2/6) \times \{2 + (-1)\} = 717 \text{ mm}^2$$

By symmetry, $B_4 = B_3$; $B_5 = B_2$ and $B_6 = B_1$

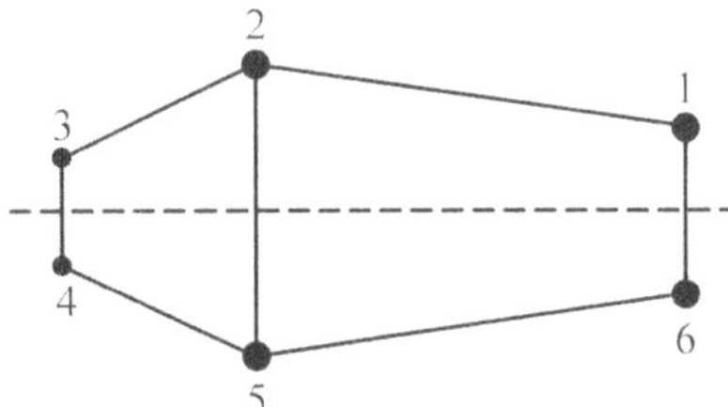

16.1.1 EFFECT OF BOOMS ON SHEAR FLOW DISTRIBUTION

Booms cause discontinuities in the skin and interrupt shear flow. The shear flow equation given below (derived earlier), therefore, needs to be changed

$$q_s = -\left[\frac{S_Z \times I_{ZZ} - S_Y \times I_{YZ}}{I_{ZZ} \times I_{YY} - I_{YZ}^2}\right] \int t \times z \times ds - \left[\frac{S_Y \times I_{YY} - S_Z \times I_{YZ}}{I_{ZZ} \times I_{YY} - I_{YZ}^2}\right] \int t \times y \times ds$$

If we consider a small segment δx of boom of area B with normal stresses σ_X and $(\partial\sigma_X/\partial x)\ \delta x$ at its ends and shear flows q_1 and q_2 on either side, for equilibrium,

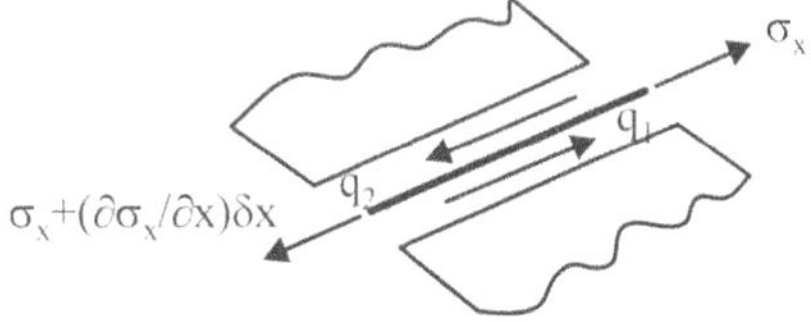

$$[\ \sigma_X + (\partial\sigma_X/\partial x)\ \delta x\ ] \times B - \sigma_X \times B + q_2 \times \delta x - q_1 \times \delta x = 0$$

$$\Rightarrow \quad q_2 - q_1 = -\ (\partial\sigma_X/\partial x) \times B$$

Thus, over every boom, there is a change in shear flow by

$$q_2 - q_1 = -\ (\partial\sigma_X/\partial x) \times B$$

Substituting for σ_X, increment in shear flow over r^{th} boom,

$$q_2 - q_1 = -\left[\frac{S_Z \times I_{ZZ} - S_Y \times I_{YZ}}{I_{ZZ} \times I_{YY} - I_{YZ}^2}\right] \times B_r \times z_r - \left[\frac{S_Y \times I_{YY} - S_Z \times I_{YZ}}{I_{ZZ} \times I_{YY} - I_{YZ}^2}\right] \times B_r \times y_r$$

After passing through 'n' booms around a open section beam, shear flow

$$q_s = -\left[\frac{S_Z \times I_{ZZ} - S_Y \times I_{YZ}}{I_{ZZ} \times I_{YY} - I_{YZ}^2}\right] \times \left(\int t \times z \times ds + \sum_{r=1}^{n} B_r \times z_r\right)$$

$$-\left[\frac{S_Y \times I_{YY} - S_Z \times I_{YZ}}{I_{ZZ} \times I_{YY} - I_{YZ}^2}\right] \times \left(\int t \times y \times ds + \sum_{r=1}^{n} B_r \times y_r\right)$$

Alternative Method of Calculating Shear Flow

Change in shear flow across r^{th} boom $q_2 - q_1 = -(\partial\sigma_X/\partial x) \times B_r$ can also be expressed as $\quad q_2 - q_1 = -(\partial P_r/\partial x) = -\Delta P_r$

A vertical load produces bending such that booms in the top flange are in compression and booms in the bottom flange are in tension or vice versa, depending on the direction of applied load.

This method is explained through the following example.

Example 16.3

Calculate the shear flow distribution in the open channel section shown, produced by a vertical shear load of 4.8 kN acting through its shear centre. Assume that the walls of the section are only effective in resisting shear stresses while the booms, each of area 300 mm^2, carry all the direct stresses.

Solution

Effective direct stress carrying thickness of walls is zero.

Section properties refer to the booms only.

Thus, $\quad I_{ZZ} = 4\,A \times y^2 = 4 \times 300 \times 200^2$

$$= 48 \times 10^6 \text{ mm}^4$$

Since Z-axis is the axis of symmetry,

$$I_{YZ} = 0$$

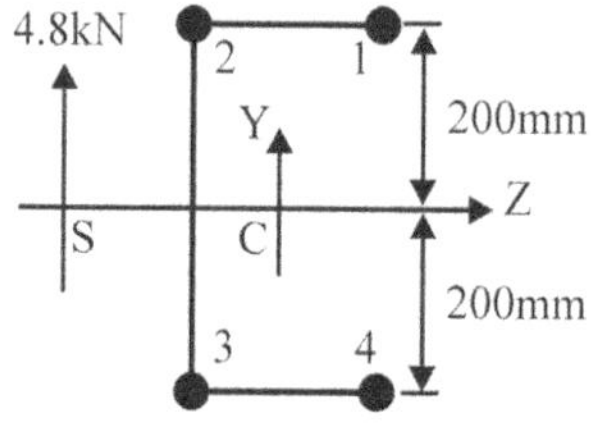

Shear flow at any section is given by

$$q_s = -\left[\frac{S_Z \times I_{ZZ} - S_Y \times I_{YZ}}{I_{ZZ} \times I_{YY} - I_{YZ}^2}\right] \times \left(\int t \times z \times ds + \sum_{r=1}^{n} B_r \times z_r\right)$$

$$-\left[\frac{S_Y \times I_{YY} - S_Z \times I_{YZ}}{I_{ZZ} \times I_{YY} - I_{YZ}^2}\right] \times \left(\int t \times y \times ds + \sum_{r=1}^{n} B_r \times y_r\right)$$

$$= -\,(S_Y/I_{ZZ}) \times \Sigma B_r \times y_r$$

$$= -\,[(4.8 \times 10^3)/(48 \times 10^6)] \times (\Sigma B_r \times y_r)$$

$$= -\,(1 \times 10^{-4}) \times (\Sigma B_r \times y_r)$$

Outside boom-1, shear flow $q_{1\text{-}0} = 0$

Shear flow between different booms is

$$q_{2\text{-}1} = -\,10^{-4} \times (300 \times 200) \qquad\qquad = -\,6 \text{ N/mm}$$

$$q_{3\text{-}2} = -\,6 - 10^{-4} \times (300 \times 200) \qquad = -\,12 \text{ N/mm}$$

$$q_{4\text{-}3} = -\,12 - 10^{-4} \times [300 \times (-200)] = -\,6 \text{ N/mm}$$

Outside boom-4, $q_{4\text{-}0} = -\,6 - 10^{-4} \times [300 \times (-200)] = 0$

Actual shear flow along the beam section varies quadratically. This method gives *average value of shear flow* between booms.

Alternative Method

Considering a beam of length 1mm,

Applied bending moment = 4.8kN × 1mm

Internal bending moment = 4 × (P × 200) Nmm

Equating the two, we get   P = 6 N

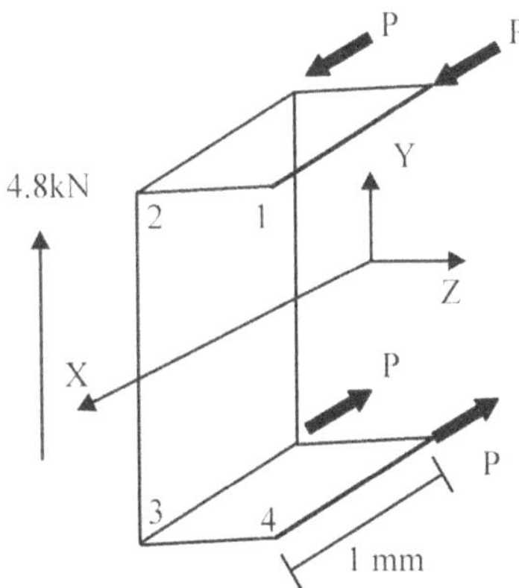

Change in load in booms across 1 mm length, $\Delta P = 6$ N/mm

With $q_{1\text{-}0} = 0$, shear flow between booms is then obtained as

$$q_{2\text{-}1} = q_{1\text{-}0} - \Delta P_1 = 0 - 6 \qquad = -\,6 \text{ N/mm}$$

$$q_{3\text{-}2} = q_{2\text{-}1} - \Delta P_2 = -\,6 - 6 \qquad = -\,12 \text{ N/mm}$$

$$q_{4\text{-}3} = q_{3\text{-}2} - \Delta P_3 = -12 + 6 = -6 \text{ N/mm}$$

$$q_{4\text{-}0} = q_{4\text{-}3} - \Delta P_4 = -6 + 6 = 0 \text{ N/mm}$$

Example 16.4

The thin-walled single cell closed section beam, shown in figure, has been idealized into a combination of direct stress carrying booms and shear stress only carrying walls. If the section supports a vertical shear load of 10 kN acting in a vertical plane through booms 3 and 6, calculate the distribution of shear flow around the section. Boom areas: $B_1 = B_8 = 200 \text{ mm}^2$, $B_2 = B_7 = 250 \text{ mm}^2$, $B_3 = B_6 = 400 \text{ mm}^2$, $B_4 = B_5 = 100 \text{ mm}^2$.

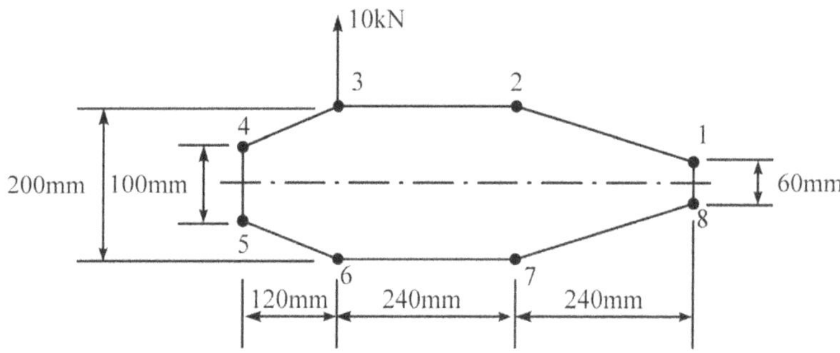

Solution

The cross section is symmetric about the horizontal axis.

Hence, $I_{YZ} = 0$; Also, $S_Z = 0$

Shear flow in any wall, $q_s = -(S_Y/I_{ZZ}) \times \Sigma B_r \times y_r + q_{s,0}$, where the first term represents shear flow of an open section (with an imaginary cut) and the second term represents constant shear flow throughout the beam to balance moments.

$$I_{ZZ} = 2 \times [200 \times 30^2 + 250 \times 100^2 + 400 \times 100^2 + 100 \times 50^2]$$

$$= 13.86 \times 10^6 \text{ mm}^4$$

Then, $q_s = -[(10 \times 10^3)/(13.86 \times 10^6)] \times \Sigma B_r \times y_r + q_{s,0}$

$$= -7.22 \times 10^{-4} \times \Sigma B_r \times y_r + q_{s,0}$$

With an imaginary cut in the wall in section 8-1,

$$q_{b,81} = 0$$

$$q_{b,12} = q_{b,81} + q_s = \quad 0 - 7.22 \times 10^{-4} \times (200 \times 30) = -4.33 \text{ N/mm}$$

$$q_{b,23} = q_{b,12} + q_s = -4.33 - 7.22 \times 10^{-4} \times (250 \times 100) = -22.38 \text{ N/mm}$$

$$q_{b,34} = q_{b,23} + q_s = -22.38 - 7.22 \times 10^{-4} \times (400 \times 100) = -51.26 \text{ N/mm}$$

$$q_{b,45} = q_{b,34} + q_s = -51.26 - 7.22 \times 10^{-4} \times (100 \times 50) = -54.87 \text{ N/mm}$$

By symmetry, $q_{b,56} = q_{b,34}$; $q_{b,67} = q_{b,23}$; $q_{b,78} = q_{b,12}$

Taking moments about C, we get

$$0 = [q_{b,81} \times 60 \times 480 + 2q_{b,12} \times (240 \times 100 + 70 \times 240)$$

$$+ 2q_{b,23} \times 240 \times 100 + 2q_{b,34} \times 120 \times 100 + q_{b,45} \times 100 \times 120]$$

$$+ 2 \{600 \times 200 - 2 \times (120 \times 50/2) - 2 \times (240 \times 70/2)\}\, q_{s,0}$$

$$= [0 \times 2.88 - 4.332 \times 8.16 - 22.382 \times 4.8 - 51.262 \times 2.4$$

$$- 54.872 \times 1.2] \times 10^4 + 19.44 \times 10^4 \times q_{s,0}$$

=> $q_{s,0} = +17.06$ N/mm

Total shear flows are given by

$$q_{8\text{-}1} = q_{b,81} + q_{s,0} \qquad = 0 + 17.06 \qquad = +17.06 \ \text{N/mm}$$

$$q_{7\text{-}8} = q_{1\text{-}2} = q_{b,12} + q_{s,0} = -4.33 \ + 17.06 = +12.73 \ \text{N/mm}$$

$$q_{6\text{-}7} = q_{2\text{-}3} = q_{b,23} + q_{s,0} = -22.38 + 17.06 = -5.32 \ \text{N/mm}$$

$$q_{5\text{-}6} = q_{3\text{-}4} = q_{b,34} + q_{s,0} = -51.26 + 17.06 = -34.2 \ \text{N/mm}$$

$$q_{4\text{-}5} = q_{b,45} + q_{s,0} \qquad = -54.87 + 17.06 = -37.81 \ \text{N/mm}$$

These values are plotted in the figure, with negative values represented in clockwise direction and positive values represented in counter-clockwise directions respectively.

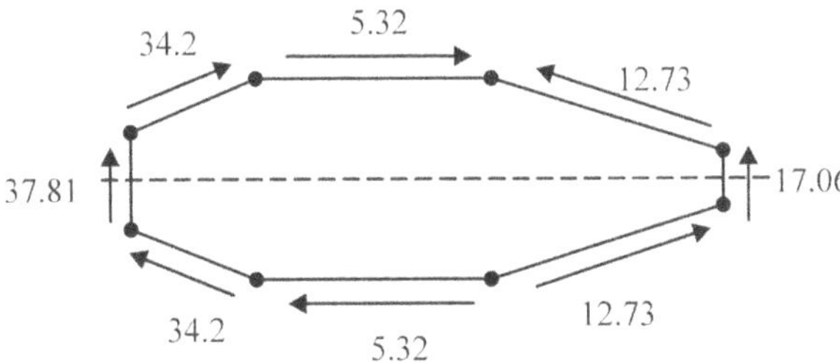

Example 16.5

A thin walled beam, shown in figure, is subjected to a vertical shear force of 10kN. Calculate shear flows in the skin panels. The boom areas of the idealised structure are: $B_1 = B_6 = 300$ mm^2 ; $B_2 = B_5 = 400$ mm^2 ; $B_3 = B_4 = 200$ mm^2

Solution

$$I_{ZZ} = 2 \times [B_1 \times y_1^2 + B_2 \times y_2^2 + B_3 \times y_3^2]$$
$$= 744 \times 10^4 \ \text{mm}^4$$

Imagining a cut in 1-6,

$$q_{b,61} = 0$$

Then,

$$q_s = - [S_Y / I_{ZZ}] \times (\Sigma B_r \times y_r) + q_{s,0}$$
$$= -13.44 \times 10^{-4} \times (\Sigma B_r \times y_r) + q_{s,0}$$

$$q_{b,12} = q_{b,61} + q_s$$
$$= 0 - 13.44 \times 10^{-4} \times (300 \times 50)$$
$$= -20.16 \text{ N/mm}$$

$$q_{b,23} = q_{b,12} + q_s = -20.16 - 13.44 \times 10^{-4} \times (400 \times 75) = -60.48 \text{ N/mm}$$

$$q_{b,34} = q_{b,23} + q_s = -60.48 - 13.44 \times 10^{-4} \times (200 \times 60) = -76.61 \text{ N/mm}$$

By symmetry, $q_{b,45} = q_{b,23}$; $q_{b,56} = q_{b,12}$

Shear flow over the cut portion can be obtained by taking moments about C.

$$0 = [\, q_{b,61} \times 100 \times 600 + 2q_{b,12} \times 600 \times 50$$
$$+ 2q_{b,23} \times 80 \times 60 + q_{b,34} \times 120 \times 80 \,]$$
$$+ 2 \{680 \times 150 - 2 \times (600 \times 25/2) - 2 \times (80 \times 15/2) \} \, q_{s,0}$$

$$= [\, 0 - 20.16 \times 6 - 60.48 \times 0.96 - 76.61 \times 0.96 \,] \times 10^4 + 23.64 \times 10^4 \times q_{s,0}$$

$$\Rightarrow q_{s,0} = +10.68 \text{ N/mm}$$

Total shear flows are given by

$$q_{6\text{-}1} = q_{b,61} + q_{s,0} \qquad = 0 + 10.68 \qquad = +10.68 \text{ N/mm}$$
$$q_{1\text{-}2} = q_{5\text{-}6} = q_{b,12} + q_{s,0} = -20.16 + 10.68 = -9.48 \text{ N/mm}$$
$$q_{2\text{-}3} = q_{4\text{-}5} = q_{b,23} + q_{s,0} = -60.48 + 10.68 = -49.80 \text{ N/mm}$$
$$q_{3\text{-}4} = q_{b,34} + q_{s,0} \qquad = -76.61 + 10.68 = -65.93 \text{ N/mm}$$

These values are plotted with positive values represented in the counter-clockwise direction and negative values represented in the clockwise direction.

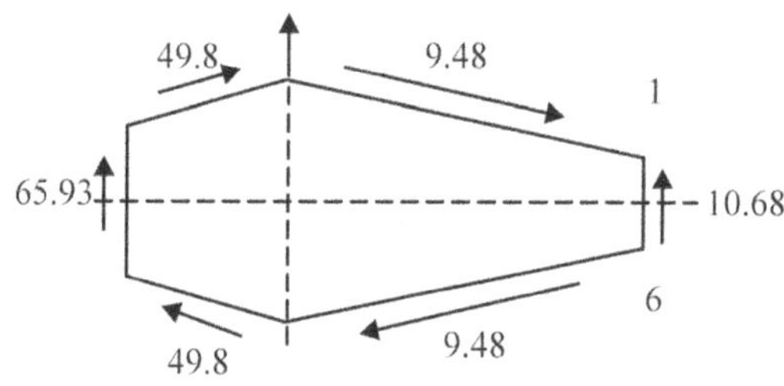

Check : Resultant of vertical components of shear forces,

$$\Sigma F_Y = 10.68 \times 100 + 2 \times 49.8 \times 15 + 65.93 \times 120 = 10.47 \text{kN} \approx 10 \text{kN} = S_Y$$

Example 16.6

A hollow thin walled beam in the form of an isosceles triangle of h = 300 mm and d = 400 mm, as shown, is subjected to a vertical shear load S_Y of 15 kN. Calculate shear flows around the section. Assume uniform thickness, t =1.5 mm

Solution

Shear flow at any point on the beam is given by

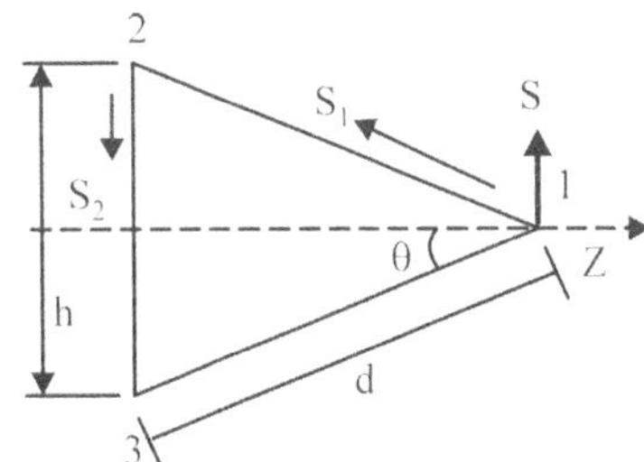

$$q_s = -\frac{S_Y}{I_{ZZ}}\int_0^s t \times y \times ds + q_{s,0}$$

$$I_{ZZ} = \frac{t \times h^3}{12} + 2 \times \frac{d^3 \times t \times \sin^2\theta}{3} = \frac{t \times h^2}{12}(h + 2d)$$

since $\sin\theta = (h/2)/d$

With an imaginary cut at 1, $q_1 = 0$

$$q_{b,12} = -\frac{S_Y}{I_{ZZ}}\int t \times (s_1 \times \sin\theta) \times ds_1 = -\frac{S_Y \times t \times \sin\theta}{2I_{ZZ}} \times s_1^2 = \frac{3S_Y}{h \times d \times (h + 2d)} \times s_1^2$$

Then, at $s_1 = d$, $q_{b,2} = \frac{3S_Y \times d}{h \times (h + 2d)}$

$$q_{b,23} = -\frac{S_Y}{I_{ZZ}}\int_0^{s_2} t \times \left(\frac{h}{2} - s_2\right) \times ds_2 + q_{b,2} = -\frac{6S_Y}{h.(h + 2d)} \times \left(s_2 - \frac{s_2^2}{h} + \frac{d}{2}\right)$$

At $s_2 = h$, $q_{b,3} = \frac{3S_Y \times d}{h \times (h + 2d)}$ (Due to symmetry also, $q_{b,2} = q_{b,3}$)

Now taking moments about the point 1,

$$0 = \int_0^h q_{b,23} \times (d \times \cos\theta) \times ds_2 + h \times q_{s,0} \times (d \times \cos\theta)$$

$$q_{s,0} = -\frac{S_Y \times (h + 3d)}{h \times (h + 2d)}$$

=> $q_1 = q_{s,0} = -15000 \times (300 + 3 \times 400) / [300 \times (300 + 2 \times 400)]$

       $= -68.18 \text{N/mm}$

$$q_2 = -\frac{S_Y}{(h + 2d)} = -15000 / (300 + 2 \times 400) = -13.64 \text{ N/mm}$$

$$q_{23} = \frac{6S_Y}{h \times (h+2d)}\left(s_2 - \frac{s_2^2}{h} + \frac{d}{2}\right) - \frac{S_Y \times (h+3d)}{h \times (h+2d)} = \frac{S_Y}{h \times (h+2d)}\left(6s_2 - \frac{6s_2^2}{h} - h\right)$$

At $s_2 = h$, $\quad q_3 = -\dfrac{S_Y}{(h+2d)} = -13.64$ N/mm

The quadratic shear flow variation between 2 and 3 with -13.64 N/mm at both ends reverses to a maximum value of $S_Y/[2(h+2d)]$ at $s_2 = h/2$ which gives $+6.818$ N/mm, for the particular data. Obviously, shear flow is zero at two points between 2 and 3, which can be obtained from $q_{23} = 0$

$\Rightarrow \quad 6s_2 - (6s_2^2/h) - h = 0$

i.e., $\quad s_2^2 - s_2 \times h + (h^2/6) = 0 \quad$ or $\quad s_2 = h/2 \pm h/\sqrt{12}$

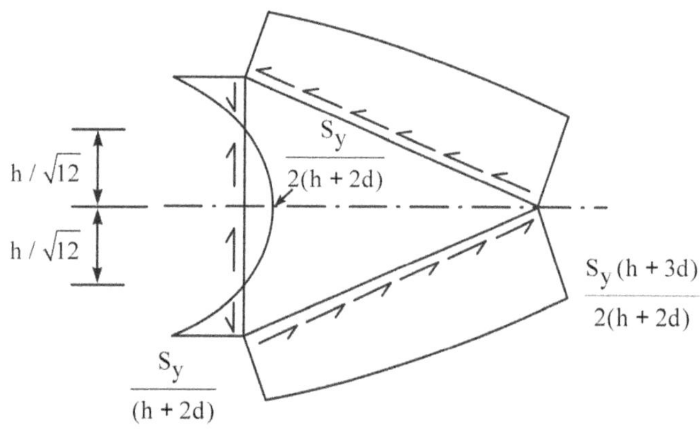

Alternative Method – *Structural Idealization with Booms*

Direct stress at any point due to S_Y is proportional to its distance from Z-axis.

Thus, $\sigma_1/\sigma_2 = 0 / (h/2) = 0$; $\sigma_3/\sigma_2 = -1$

$\theta = \sin^{-1}(h/2d) = 22.02^0$

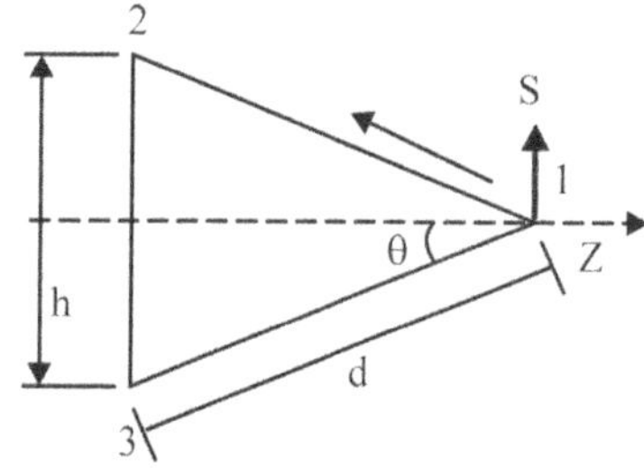

Area of boom at 2 has contributions from panels 1-2 and 2-3

$$B_2 = \{(d\cos\theta) \times t/6\} \times \{2 + (\sigma_1/\sigma_2)\}$$

$$+ \{(h/2) \times t/6\} \times \{2 + (\sigma_3/\sigma_2)\}$$

$$= \{(400 \times \cos 22.02^0) \times 1.5/6\} \times (2+0) + (150 \times 1.5/6) \times \{2 + (-1)\}$$

$$= 222.9 \text{ mm}^2$$

Due to symmetry, $B_3 = B_2$

$$I_{ZZ} = 2 \times [B_2 \times y_2{}^2] + B_1 \times y_1$$
$$= 1003 \times 10^4 \text{ mm}^4$$

Imagining a cut in 1-2, $q_{b,12} = 0$

Then, $q_s = - [S_Y/I_{ZZ}] \times (\Sigma B_r \times y_r) + q_{s,0}$
$$= -14.96 \times 10^{-4} \times (\Sigma B_r \times y_r) + q_{s,0}$$

$q_{b,23} = q_{b,12} + q_s$
$$= 0 - 14.96 \times 10^{-4} \times (222.9 \times 150) = -50 \text{ N/mm}$$

$q_{b,31} = q_{b,23} + q_s = -50 - 14.96 \times 10^{-4} \times \{222.9 \times (-150)\} = 0 \text{ N/mm}$

Taking moments about 1, we get

$$0 = q_{b,23} \times 300 \times (400 \cos 22.02^0) + 2 \times (400 \cos 22.02^0 \times 300 / 2) \, q_{s,0}$$
$$= -50 \times 11.125 \times 10^4 + 11.125 \times 10^4 \times q_{s,0}$$

$=> q_{s,0} = +50 \text{ N/mm}$

Then,

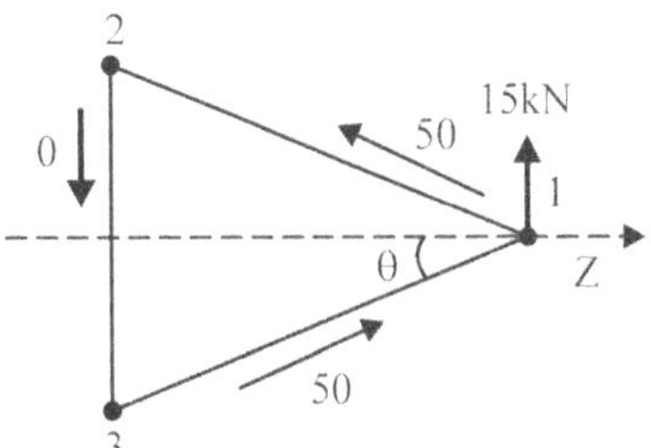

$q_{12} = q_{b,12} + q_{s,0} = \quad 0 + 50 = 50 \text{ N/mm}$

$q_{23} = q_{b,23} + q_{s,0} = -50 + 50 = \quad 0 \text{ N/mm}$

$q_{31} = q_{b,31} + q_{s,0} = \quad 0 + 50 = 50 \text{ N/mm}$

Check : Resultant of vertical components of shear forces,

$$\Sigma F_Y = 2 \times (50 \text{ N/mm} \times 400 \text{ mm}) \times \sin 20.22^0 = 15000 \text{ N or } 15 \text{ kN} = S_Y$$

Note **:** Actual shear flow varied from –68.18 N/mm at 1 to –13.64 N/mm at 2 as a quadratic function (first method) while the approximate method of structural idealization gave uniform shear flow of 50 N/mm. Similarly, actual flow varied from –13.64 N/mm at 2 to –13.64 N/mm at 3 as a quadratic function with a reversal to a maximum value of +6.818 N/mm (first method) while the approximate method gave uniform shear flow of 0 N/mm

16.2 AXIAL LOADS IN BOOMS

For a better understanding of the changes in axial loads in different stiffeners, let us consider an idealised six-boom beam (Ref Fig. 16.4) of width '2d', height 'h' and length 'L', built-in at one end and carrying a shear load at the other end, in the middle without producing any twisting moment in the section. Let A and B be the areas of cross section of central and corner booms.

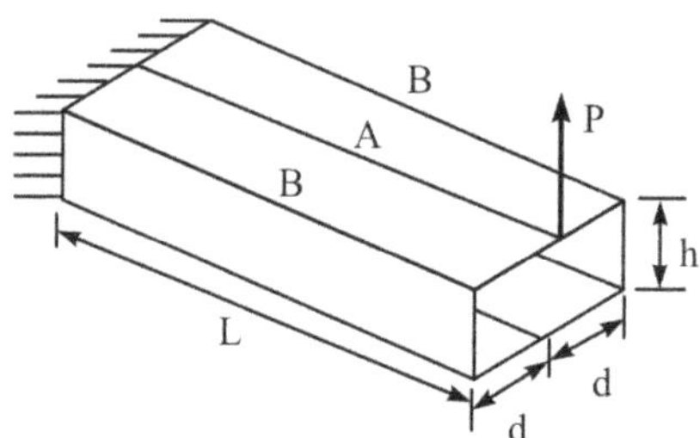

FIGURE 16.4 An Idealised box beam

Forces acting in the stiffeners and panels over a small length 'dx' are shown in the Fig 16.5. Due to symmetry of shear force P, each flange takes a load of P/2. Shear flow q_1 in the flange is given by $q_1 = (P/2)/h$

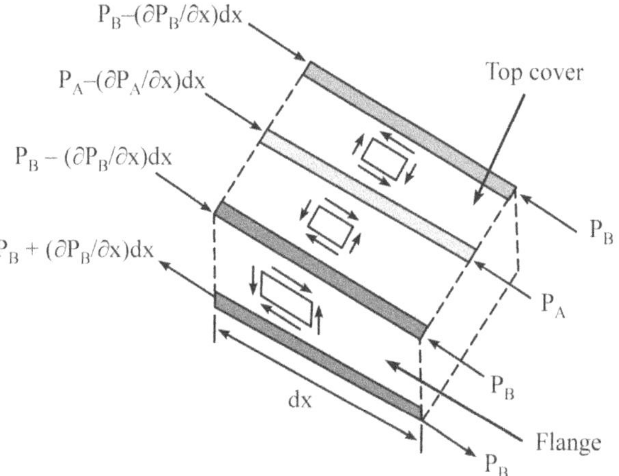

FIGURE 16.5 Shear flow in panels between booms

For equilibrium of bottom boom of the flange,

$$P_B + (\partial P_B/\partial x) \times dx - P_B - q \times dx + q_1 \times dx = 0$$

or $\qquad (\partial P_B/\partial x) - q + (P/2h) = 0 \qquad\qquad$ .....(16.2)

Similarly, for equilibrium of central boom, $(\partial P_A/\partial x) + 2q = 0 \qquad$ .....(16.3)

Considering overall equilibrium of the top or bottom planes,

$$2P_B + P_A + 2\,q_1 \times z = 2P_B + P_A + (P/h)\,z = 0 \qquad\qquad .....(16.4)$$

Solving eq. 16.2, 16.3 and 16.4 for the three unknown variables P_A, P_B and q, and applying displacement compatibility conditions between booms and adjacent panels for evaluating integration constants, we get

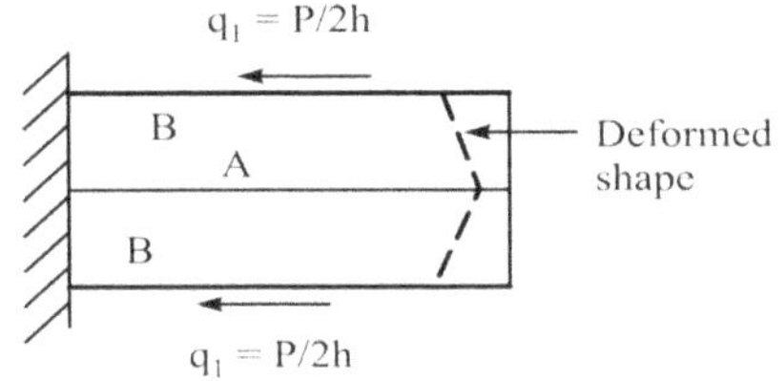

$$q = P \times A / [\, 2h \times (2B + A)\,]$$

$$P_A = [\, P \times A / \{h \times (2B + A)\}\,] \times [\, z - \{\, \sinh \lambda x / (\lambda \cosh \lambda L)\,\}\,]$$

and $\quad P_B = [\, P \times B / \{h\,(2B + A)\}\,] \times [\, z + \{\, A \sinh \lambda x / (2B\lambda \cosh \lambda L)\,\}\,]$

where, $\quad \lambda^2 = G \times t \times (2B + A) / (d \times E \times A \times B)$

Special Case

In many wing structures, only spars are connected to the fuselage and the intermediate stringers are not subjected to bending stresses. Then, boundary conditions for the central stringer are $P_A = 0$ at its two ends and the solution changes to

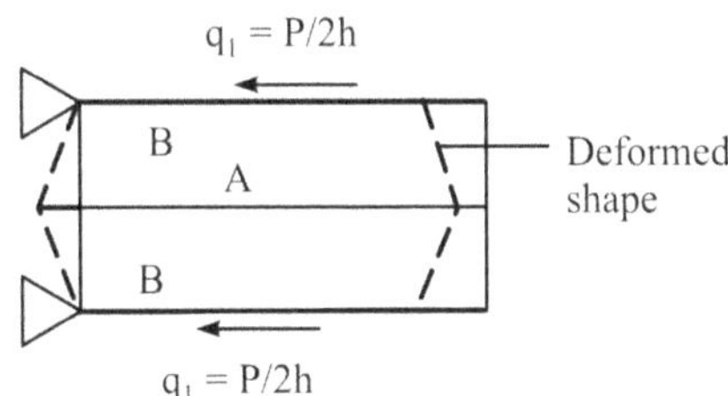

$$q \;= [P \times A / \{2h \times (2B + A)\}] \times [1 - \lambda \times L \times (\cosh \lambda x / \sinh \lambda L)]$$

$$P_A = [P \times A / \{h \times (2B + A)\}] \times [x - L \times (\sinh \lambda x / \sinh \lambda L)]$$

and $\quad P_B = [P \times B / \{h \times (2B + A)\}] \times [x + (AL/2B) \times (\sinh \lambda x / \sinh \lambda L)]$

16.3   ANALYSIS OF CLOSED SECTION BEAMS SUBJECTED TO BENDING

Wings are usually tapered both chord-wise and depth-wise along the wing span towards the tip, to provide a more efficient aerodynamic and structural shape. The effect of taper on the prediction of direct stresses in flanges / stringers produced by bending is minimal, if the taper is small and the section properties are calculated at the particular section. However, calculation of shear stresses in beam webs can be significantly affected by taper.

In a particular i^{th} boom, if P is the axial force with its components P_X, P_Y and P_Z along X, Y and Z directions respectively and A is the area of cross section, then

$$P_X = \sigma_X \times A; \quad P_Y = P_X \times (\delta y/\delta x); \quad P_Z = P_X \times (\delta z/\delta x)$$

and $\quad P = \sqrt{\left(P_X^{\,2} + P_Y^{\,2} + P_Z^{\,2}\right)} \quad$ or $\quad P = P_X \times \sqrt{\left(\delta x^2 + \delta y^2 + \delta z^2\right)}/\delta x$

The applied shear loads S_Y and S_Z are reacted by the resultants of the shear flows in the skin panels and webs, together with the components P_Y and P_Z of the axial loads in all the 'n' booms. If $S_{Y,w}$ and $S_{Z,w}$ are the resultants of skin and web shear flows, then

$$S_Y = S_{Y,w} + \sum (P_Y)_i \qquad \text{and} \qquad S_Z = S_{Z,w} + \sum (P_Z)_i \qquad .....(16.5)$$

Example 16.7

A cantilever beam is uniformly tapered along its length in both Y and Z directions and carries a load of 100 kN at its free end. Calculate the forces in the booms and shear flow distribution in the walls at a section 2 m from the built-in end, if the booms resist all the direct stresses while the walls are effective only in shear. Each corner boom has a cross sectional area of 900 mm² while both central booms have cross sectional areas of 1200 mm². Thickness of vertical panels is 3 mm and horizontal panels 2 mm.

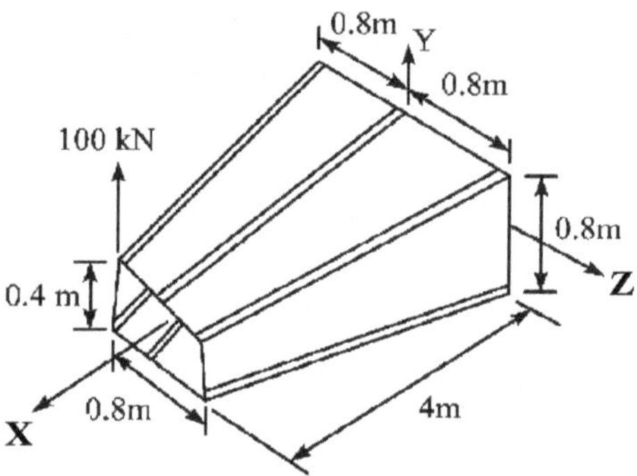

Solution

Section at a distance of 2m from built-in end of the beam is shown here. It is a doubly symmetrical section. Therefore, $I_{YZ} = 0$

$$I_{ZZ} = 4 \times 900 \times 300^2 + 2 \times 1200 \times 300^2$$

$$= 5.4 \times 10^8 \text{ mm}^4$$

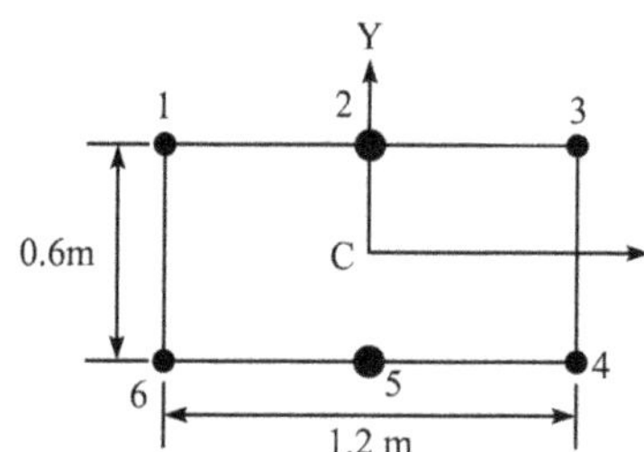

The force system on this section consists of

$$S_Y = 100 \text{kN}; \quad S_Z = 0$$

$$M_Z = -100 \times 2 \text{ kN m} = -200 \times 10^6 \text{ N mm} \quad M_Y = 0$$

$$(\sigma_X)_i = M_Z \times y_i / I_{ZZ} = -200 \times 10^6 \times y_i / (5.4 \times 10^8) = -0.37 \, y_i \ \text{N/mm}^2$$

and $(P_X)_i = (\sigma_X)_i \times A_i = -0.37 \times y_i \times A_i \ \text{N}$

For boom 1, $(P_X)_1 = -0.37 \times 300 \times 900 = -100 \text{ kN}$

$$(P_Y)_1 = (P_X)_1 \times (\delta y / \delta x) = (P_X)_1 \times [(400 - 300)/2000] = -5 \ \text{kN}$$

$$(P_Z)_1 = (P_X)_1 \times (\delta z / \delta x) = (P_X)_1 \times [(400 - 200)/2000] = -10 \text{ kN}$$

$$y_1 = 0.3 \text{ m} ; \ z_1 = -0.6 \text{ m} ; \ (P_Z)_1 \times y_1 = -3 \text{ kN m} ; \ (P_Y)_1 \times z_1 = 3 \text{ kN m}$$

Components of forces in others can be calculated in a similar way as

$$(P_X)_2 = -133 \text{ kN}; \qquad (P_Z)_2 = 0 \text{ kN}; \qquad (P_Y)_2 = 6.7 \text{kN}$$

$$(P_X)_3 = -100 \text{ kN}; \qquad (P_Z)_3 = 10 \text{ kN}; \qquad (P_Y)_3 = 5 \text{kN}$$

$$(P_X)_4 = 100 \text{ kN}; \qquad (P_Z)_4 = -10 \text{ kN}; \qquad (P_Y)_4 = 5 \text{kN}$$

$$(P_X)_5 = 133 \text{ kN}; \qquad (P_Z)_5 = 0 \text{ kN}; \qquad (P_Y)_5 = 6.7 \text{ kN}$$

$$(P_X)_6 = 100 \text{ kN}; \qquad (P_Z)_6 = 10 \text{ kN}; \quad (P_Y)_6 = 5 \text{ kN}$$

On summing up, we get $\sum (P_X)_i = 0 ; \qquad \sum (P_Y)_i = 0; \qquad \sum (P_Z)_i = 33.4 \text{ kN}$

$$\sum (P_Y)_i \times z_i = 0 ; \qquad \sum (P_Z)_i \times y_i = 0$$

Then, $S_{Z,w} = S_Z - \sum (P_Z)_i = 0 - 0 = 0$

and $S_{Y,w} = S_Y - \sum (P_Y)_i = 100 - 33.4 = 66.6 \text{ kN}$

Shear flow, for the particular case of $I_{YZ} = 0$ and $S_Z = 0$, from section 15.5, is given by $\qquad q_s = q_{s,0} - (s_{Y,w} / I_{ZZ}) \times (\sum y_i \times A_i) = q_{s,0} - 1.23 \times 10^{-4} \times (\sum y_i \times A_i)$

Taking $q_{b,1\text{-}6} = 0$, relative values of shear flow in other sections can be calculated as

$$q_{b,1\text{--}2} = 0 - 1.23 \times 10^{-4} \times (900 \times 300) \qquad\qquad = -33.2 \text{ N/mm}$$

$$q_{b,2\text{--}3} = 0 - 1.23 \times 10^{-4} \times (900 \times 300 + 1200 \times 300) \qquad = -77.5 \text{ N/mm}$$

$$q_{b,3\text{--}4} = 0 - 1.23 \times 10^{-4} \times (2 \times 900 \times 300 + 1200 \times 300) \quad = -110.7 \text{ N/mm}$$

By symmetry, $q_{b,2\text{--}3} = -77.5 \text{ N/mm}$ and $q_{b,1\text{--}2} = -33.2 \text{ N/mm}$

Taking moments about the center of symmetry,

$$S_Z \times y_0 - S_Y \times z_0 = q_{b,3\text{-}4} \times s_{3\text{-}4} \times x_{3\text{-}4} + 2 \times q_{b,4\text{-}5} \times s_{4\text{-}5} \times y_{4\text{-}5}$$

$$+ 2 \times q_{b,5\text{-}6} \times s_{5\text{-}6} \times y_{5\text{-}6} + 2 \times A \times q_{s,0} - \{\sum (P_Z)_i \times y_i\} - \{\sum (P_Y)_i \times z_i\}$$

$$\Rightarrow \quad 0 - 100\times10^3 \times 600 = 110.7 \times 600 \times 600 + 2 \times 33.2 \times 600 \times 300$$

$$+ 2 \times 77.5 \times 600 \times 300 + (2 \times 1200 \times 600)\, q_{s,0} - 0 - 0$$

$$\Rightarrow \quad q_{s,0} = -97 \text{ N/mm (clockwise)}$$

Actual shear flow distribution is found by adding $q_{s,0}$ to q_b in each panel, as shown.

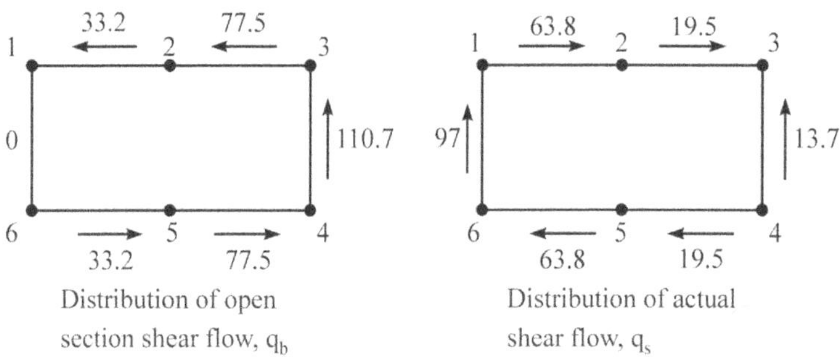

Distribution of open   Distribution of actual

section shear flow, q_b shear flow, q_s

16.4 BEAMS HAVING VARIABLE STRINGER AREAS

In many aircrafts, structural beams in wings have stringers whose cross-sectional areas vary in the span-wise direction. If the stringer stress is made constant by varying the area of cross section, there is no change in shear flow as the stringer is crossed. If stringer loads $P_{X,1}$ and $P_{X,2}$ are calculated at two sections x_1 and x_2 of the beam, a convenient distance apart, change in stringer load per unit length of beam assuming linear variation of load along its length is given by

$$\Delta P = (P_{X,1} - P_{X,2}) / (x_1 - x_2) \qquad\qquad .....(16.6)$$

Calculation of shear flow distribution is carried out as explained in the previous section.

16.5 ANALYSIS OF FUSELAGE

Aircraft fuselages consist of thin sheets of material stiffened by large numbers of longitudinal stringers together with transverse frames. The stringer-panel arrangement is idealised into one comprising axial load resisting booms and shear load resisting skin. Distance between adjacent stringers is usually so small that skin between adjacent stringers is assumed flat and variation in shear flow in the connecting panel will be small and, therefore, can be assumed constant between adjacent stringers. Stresses due to bending, shear and torsion are separately calculated as explained earlier and superimposed. The procedure is explained through the following example.

Example 16.8

Fuselage of a light passenger carrying aircraft has plate of 1.5 mm thickness in the shape of circular section of mean radius 2 m with 16 stringers, each of area 100 mm^2, distributed uniformly around the fuselage as shown. If the fuselage is subjected to a bending moment of 200 kNm applied in the vertical plane of symmetry, calculate direct stress distribution in the booms.

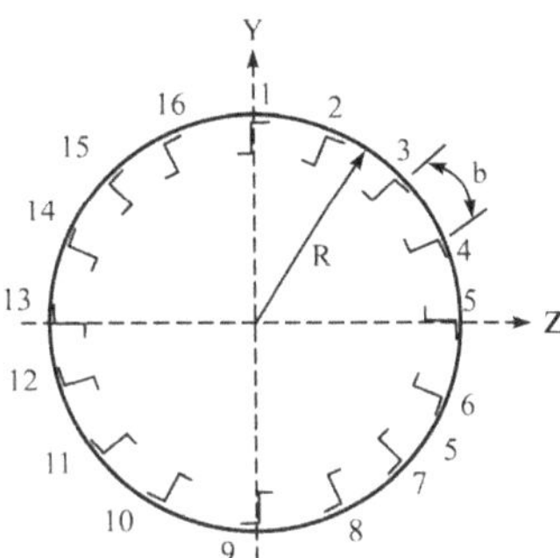

Solution

Distances of booms from Z-axis are

$$y_1 = -y_9 = R = 1000 \text{ mm} \; ; \; y_5 = -y_{13} = R \sin 0 = 0 \text{ mm}$$

$$y_2 = y_{16} = -y_8 = -y_{10} = R \sin 67.5 = 923.9 \text{ mm}$$

$$y_3 = y_{15} = -y_7 = -y_{11} = R \sin 45 \quad = 707.1 \text{ mm}$$

$$y_4 = y_{14} = -y_6 = -y_{12} = R \sin 22.5 = 382.7 \text{ mm}$$

Length of panel between any two booms, $b = (\pi \times 2000) / 16 = 392.9$ mm

(a) Due to bending

Due to symmetry,

$$\sigma_2 = \sigma_{16} = -\sigma_8 = -\sigma_{10} \; ; \quad \sigma_3 = \sigma_{15} = -\sigma_7 = -\sigma_{11}$$

$$\sigma_4 = \sigma_{14} = -\sigma_6 = -\sigma_{12} \; ; \quad \sigma_5 = \sigma_{13} = 0 \quad \text{and} \quad \sigma_1 = -\sigma_9$$

$$\sigma_2 / \sigma_1 = \sigma_{16} / \sigma_1 = (M \times y_2/I_{ZZ}) / (M \times y_1/I_{ZZ}) = y_2/y_1 = 0.9239$$

$$\sigma_3 / \sigma_2 = \sigma_{15} / \sigma_{16} = (M \times y_3/I_{ZZ}) / (M \times y_2/I_{ZZ}) = y_3/y_2 = 0.7653$$

$$\sigma_4 / \sigma_3 = \sigma_{14} / \sigma_{15} = (M \times y_4/I_{ZZ}) / (M \times y_3/I_{ZZ}) = y_4/y_3 = 0.5412$$

$$\sigma_5 / \sigma_4 = \sigma_{13} / \sigma_{14} = (M \times y_5/I_{ZZ}) / (M \times y_4/I_{ZZ}) = y_5/y_4 = 0$$

$$\sigma_1 / \sigma_2 = \sigma_1 / \sigma_{16} = y_1 / y_2 = 1.0824$$

$$\sigma_2 / \sigma_3 = \sigma_{16} / \sigma_{15} = y_2 / y_3 = 1.3067$$

$$\sigma_3 / \sigma_4 = \sigma_{15} / \sigma_{14} = y_3 / y_4 = 1.8477$$

Replacing circular sheet of fuselage section by zero thickness skin and booms at the locations of stringers, areas of booms at different locations are calculated using eq (16.1),

Total area of cross section of boom at 1 is given by

$$B_1 = B_9 = A_S + (t \times b/6) \times (2 + \sigma_2/\sigma_1) + (t \times b/6) \times (2 + \sigma_{16}/\sigma_1)$$

$$= 100 + 2 \times (1.5 \times 392.9 / 6) \times (2 + 0.9239) = 674.4 \text{ mm}^2$$

$$B_2 = B_8 = B_{10} = B_{16} = A_S + (t \times b/6) \times (2 + \sigma_1/\sigma_2)$$
$$+ (t \times b/6) \times (2 + \sigma_3/\sigma_2)$$
$$= 100 + (1.5 \times 392.9/6) \times (2 + 1.0824)$$
$$+ (1.5 \times 392.9/6) \times (2 + 0.7653)$$
$$= 674.4 \text{ mm}^2$$

$$B_3 = B_7 = B_{11} = B_{15} = A_S + (t \times b/6) \times (2 + \sigma_2/\sigma_3)$$
$$+ (t \times b/6) \times (2 + \sigma_4/\sigma_3)$$
$$= 100 + (1.5 \times 392.9/6) \times (2 + 1.3067)$$
$$+ (1.5 \times 392.9/6) \times (2 + 0.5412)$$
$$= 674.4 \text{ mm}^2$$

$$B_4 = B_6 = B_{12} = B_{14} = A_S + (t.b/6).(2 + \sigma_3/\sigma_4) + (t.b/6).(2 + \sigma_5/\sigma_4)$$
$$= 100 + (1.5 \times 392.9/6) \times (2 + 1.8477) + (1.5 \times 392.9/6) \times (2 + 0)$$
$$= 674.4 \text{ mm}^2$$

$$B_5 = B_{13} = A_S = 100$$

Then, $I_{ZZ} = 2 \times (B_1 \times y_1^2) + 4 \times (B_2 \times y_2^2) + 4 \times (B_3 \times y_3^2)$
$$+ 4 \times (B_4 \times y_4^2) + 2 \times (B_5 \times y_5^2)$$
$$= 674.4 \times (2 \times 1000^2 + 4 \times 923.9^2 + 4 \times 765.3^2 + 4 \times 541.2^2)$$
$$\text{since} \quad B_1 = B_2 = B_3 = B_4 = 674.4 \text{ mm}^2 \quad \text{and} \quad y_5 = 0$$
$$= 6021.5 \times 10^6 \text{ mm}^4$$

Axial stress in boom 'I' is given by $\sigma_i = M \times y_i / I_{ZZ}$

Thus, $\sigma_1 = M.y_1 / I_{ZZ} = 200 \times 10^6 \times 1000 / (6021.5 \times 10^6) = 33.21 \text{ N/mm}^2$

$\sigma_2 = M.y_2 / I_{ZZ} = 200 \times 10^6 \times 923.9 / (6021.5 \times 10^6) = 30.69 \text{ N/mm}^2$

$\sigma_3 = M.y_3 / I_{ZZ} = 200 \times 10^6 \times 765.3 / (6021.5 \times 10^6) = 25.42 \text{ N/mm}^2$

$\sigma_4 = M.y_4 / I_{ZZ} = 200 \times 10^6 \times 541.2 / (6021.5 \times 10^6) = 17.98 \text{ N/mm}^2$

$\sigma_5 = M.y_5 / I_{ZZ} = 200 \times 10^6 \times 0 / (6021.5 \times 10^6) \quad = 0 \quad \text{N/mm}^2$

(b) **Due to shear force:** Shear flow in the fuselage due to the shear force can be calculated as explained in section 15.6. In this example, shear force in Z-direction, S_Z, is zero. Since the circular section is symmetric, $I_{YZ} = 0$.

Shear flow in each section, between two booms is calculated using

$$q_s = -(S_Y/I_{ZZ}) \times (\Sigma B_i \times y_i) + q_{s,0}.$$

First term in this equation is the shear flow in an open section beam, with an imaginary cut between any two booms. Shear flow $q_{s,0}$ in the cut portion can be calculated from moment equilibrium, by taking moments of all forces about any point and added to the open section shear flow values.

The solution can be checked by calculating the resultant of shear flow distribution along Y-axis, which should be equal to the applied shear load S_Y.

Example 16.9

A fuselage, idealised as a zero-thickness shell of 380mm radius with 16 equi-spaced booms of area 216.4 mm^2, is subjected to a vertical shear load of 100 kN applied at a distance of 150 mm from the vertical axis of symmetry. Calculate distribution of shear flow.

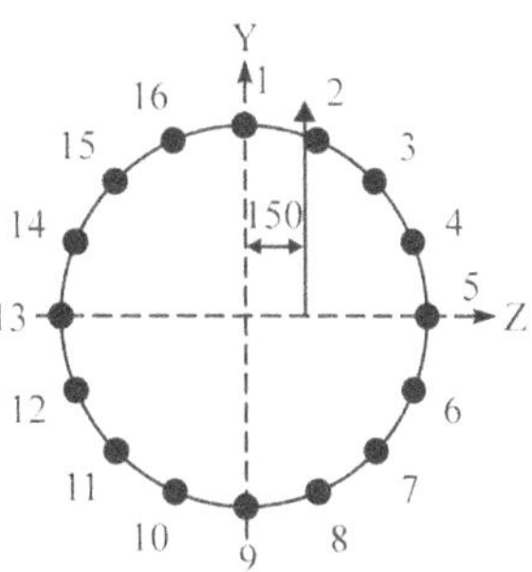

Solution

Since $S_Z = 0$ and $I_{YZ} = 0$,

$$q_s = -(S_Y / I_{ZZ}) \times \left[\sum_{r=1}^{n} B_r \times y_r \right] + q_{s,0} = -\frac{100 \times 10^3}{2.52 \times 10^8} \times \left[\sum_{r=1}^{n} B_r \times y_r \right] + q_{s,0}$$

$$= -3.97 \times 10^{-4} \times \left[\sum_{r=1}^{n} B_r \times y_r \right] + q_{s,0}$$

Due to symmetry of the section, we can expect symmetric shear flow about Z-axis. Imagining a cut in the shell between booms 1 and 2,

$$(q_s)_{1\text{-}2} = (q_s)_{8\text{-}9} = 0$$

$$(q_s)_{2\text{-}3} = (q_s)_{7\text{-}8} = -3.97 \times 10^{-4} \times (B_2 \times y_2) = -3.97 \times 10^{-4} \times (216.4 \times 351.1)$$

$$= -30.16$$

$$(q_s)_{3\text{-}4} = (q_s)_{6\text{-}7} = -30.16 - 3.97 \times 10^{-4} \times (216.4 \times 268.7) \qquad = -53.24$$

$$(q_s)_{4\text{-}5} = (q_s)_{5\text{-}6} = -53.24 - 3.97 \times 10^{-4} \times (216.4 \times 145.4) \qquad = -65.73$$

$$(q_s)_{9\text{-}10} = (q_s)_{16\text{-}1} = 0 - 3.97 \times 10^{-4} \times (216.4 \times 380) \qquad = +32.65$$

$$(q_s)_{10\text{-}11} = (q_s)_{15\text{-}16} = -32.65 - 3.97 \times 10^{-4} \times (216.4 \times 351.1) = +62.81$$

$$(q_s)_{11\text{-}12} = (q_s)_{14\text{-}15} = -62.81 - 3.97 \times 10^{-4} \times (216.4 \times 268.7) = +85.89$$

$$(q_s)_{12\text{-}13} = (q_s)_{13\text{-}14} = -85.89 - 3.97 \times 10^{-4} \times (216.4 \times 145.4) = +98.38$$

Shear flow $q_{s,0}$ in the panel 1-2 can now be found from

$$S_Y \times e = \int q_s \times p \times ds + 2A \times q_{s,0}$$

$$= (2A_{1\text{-}2} \times q_{s,1\text{-}2} + 2A_{2\text{-}3} \times q_{s,2\text{-}3} \dots + 2A_{16\text{-}1} \times q_{s,16\text{-}1}) + 2A \times q_{s,0}$$

Here, $A = \pi r^2 = \pi \times 380^2 = 45.4 \times 10^4$ mm^2

and $A_{1\text{-}2} = A_{2\text{-}3} = \dots = A_{16\text{-}1} = A/16 = 2.84 \times 10^4$ mm^2

Then, $100 \times 10^3 \times 50 = 2 \times 2.84 \times 10^4 \times (-261.2) + 2 \times 45.4 \times 10^4 \times q_{s,0}$

$\Rightarrow q_{s,0} = +32.86$ N/mm (anticlockwise)

Net shear flows in N/mm are :

$$q_{1\text{-}2} = q_{8\text{-}9} = q_{s,0} = -32.86$$
$$q_{2\text{-}3} = q_{7\text{-}8} = -30.16 - 32.86 = -63.02$$
$$q_{3\text{-}4} = q_{6\text{-}7} = -53.24 - 32.86 = -85.1$$
$$q_{4\text{-}5} = q_{5\text{-}6} = -65.73 - 32.86 = -98.59$$
$$q_{9\text{-}10} = q_{16\text{-}1} = +32.65 - 32.86 = -0.21$$
$$q_{10\text{-}11} = q_{15\text{-}16} = +62.81 - 32.86 = +29.95$$
$$q_{11\text{-}12} = q_{14\text{-}15} = +85.89 - 32.86 = +53.03$$
$$q_{12\text{-}13} = q_{13\text{-}14} = +98.38 - 32.86 = +65.52$$

These shear flow values are plotted with positive values representing counter-clockwise shear flows and negative values representing clockwise shear flows.

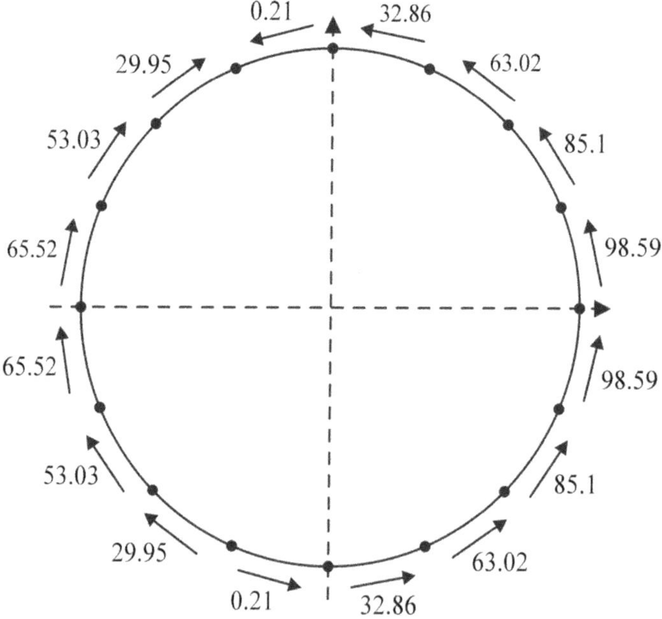

Check : Vertical components of shear flows,

$$\sum F_Y = 2 \times [(98.59 + 65.52) \times 145.4 + (85.1 + 53.03) \times 123.3$$
$$+ (63.02 + 29.95) \times 82.4 + (32.86 - 0.21) \times 28.9]$$
$$= 98.99 \text{ kN} \approx S_Y$$

Special case : If the same shear force is applied along the Y-axis, open section shear flows remain the same while the shear flow in the cut portion $q_{s,0}$ is obtained from

$$M_C = S_Y \times 0 = 0 = \int q_s \times p \times ds + 2A \times q_{s,0}$$

$$= (2A_{1\text{-}2} \times q_{s,1\text{-}2} + 2A_{2\text{-}3} \times q_{s,2\text{-}3} \ldots. + 2A_{16\text{-}1} \times q_{s,16\text{-}1}) + 2^a \times q_{s,0}$$

$$= 2 \times 2.84 \times 10^4 \times (-261.2) + 2 \times 45.4 \times 10^4 \times q_{s,0}$$

$\Rightarrow$ $q_{s,0} = +16.34$ N/mm (anticlockwise)

Net shear flows in N/mm are :

$$q_{1\text{-}2} = q_{8\text{-}9} = q_{s,0} = -16.34$$

$$q_{2\text{-}3} = q_{7\text{-}8} = -30.16 - 16.34 = -46.50$$

$$q_{3\text{-}4} = q_{6\text{-}7} = -53.24 - 16.34 = -69.38$$

$$q_{4\text{-}5} = q_{5\text{-}6} = -65.73 - 16.34 = -82.07$$

$$q_{9\text{-}10} = q_{16\text{-}1} = +32.65 - 16.34 = +16.31$$

$$q_{10\text{-}11} = q_{15\text{-}16} = +62.81 - 16.34 = +46.47$$

$$q_{11\text{-}12} = q_{14\text{-}15} = +85.89 - 16.34 = +69.55$$

$$q_{12\text{-}13} = q_{13\text{-}14} = +98.38 - 16.34 = +82.04$$

Check : Vertical components of shear flows,

$$\sum F_Y = 2 \times [(82.07 + 82.04) \times 145.4 + (69.38 + 69.55) \times 123.3$$

$$+ (46.50 + 46.47) \times 82.4 + (16.34 + 16.31) \times 28.9]$$

$$= 99.19 \text{ kN} \approx S_Y$$

(c)   Due to torsion

Fuselage is a single cell closed section and shear flow distribution remains constant in all the sections, since booms do not carry shear stresses. Torque load can be expected only when load S_Y does not act along the Y-axis but parallel to it at a normal distance of z_e.

Then, $T = S_Y \times z_e$ and   shear flow, $q = T/2A$

Total shear flow in the skin in each segment is the algebraic sum of shear flows due to bending load, shear load and torsion.

Example: Consider the previous example of shear load of 100kN acting at a distance of 150mm parallel to Y-axis as a combination of shear load of 100kN acting along Y-axis and a torque T of 100kN × 150mm acting about the center of fuselage as shown.

Uniform shear flow due to torque T, from Bredt-Batho theory, is

$$q = T/2A = 100 \times 10^3 \times 150/(2 \times 45.4 \times 10^4)$$

$$= 16.52 \text{ N/mm (counter-clockwise)}$$

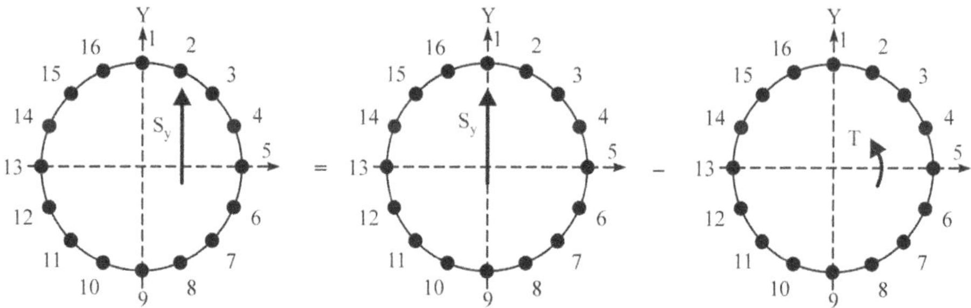

Net shear flows in N/mm can therefore be obtained by adding q to the shear flows obtained for shear load acting along Y-axis (special case of earlier example)

$$q_{1\text{-}2} = q_{8\text{-}9} = -16.34 - 16.52 = -32.86$$

$$q_{2\text{-}3} = q_{7\text{-}8} = -46.50 - 16.52 = -63.02$$

$$q_{3\text{-}4} = q_{6\text{-}7} = -69.38 - 16.52 = -85.9$$

$$q_{4\text{-}5} = q_{5\text{-}6} = -82.07 - 16.52 = -98.59$$

$$q_{9\text{-}10} = q_{16\text{-}1} = +16.31 - 16.52 = -0.21$$

$$q_{10\text{-}11} = q_{15\text{-}16} = +46.47 - 16.52 = +29.95$$

$$q_{11\text{-}12} = q_{14\text{-}15} = +69.55 - 16.52 = +53.03$$

$$q_{12\text{-}13} = q_{13\text{-}14} = +82.04 - 16.52 = +65.52$$

which are same as those obtained earlier for eccentric shear load.

16.6 ANALYSIS OF WINGS

Wing sections consist of thin skins stiffened by combination of stringers, spar webs, caps and ribs. The resulting structure frequently comprises one, two or more cells and is highly redundant. The large number of closely spaced stringers allows the assumption of a constant shear flow in the skin between adjacent stringers so that the direct stress carrying part of the skin can be taken as concentrated area booms at the locations of stringers. The wings are subjected to bending, shear and torsion loads.

Consider a simple wing section which is idealised into a combination of three direct stress carrying booms and skin panels carrying only shear stress (Ref Fig. 16.6). The part of wing section aft (left) of the vertical spar 3-1 performs aerodynamic function only and is unstressed. Lift force S_Y and drag force S_Z induce shear flows in the skin panels.

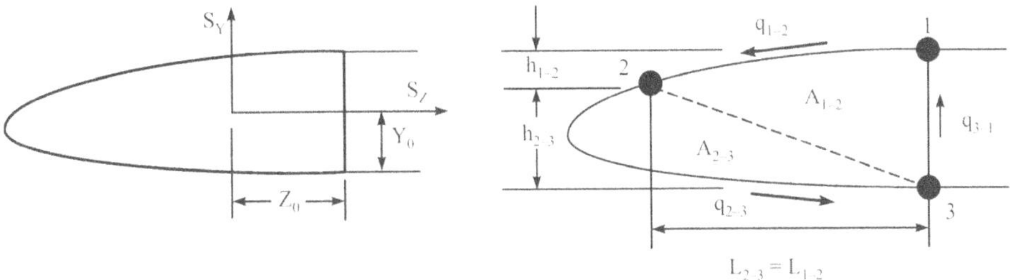

FIGURE 16.6 Idealisation of a single-cell wing section

Resolving the shear flows in Z-direction, $S_Z = - q_{1-2} \times L_{1-2} + q_{2-3} \times L_{2-3}$

Resolving them in Y-direction, $S_Y = q_{3-1} \times (h_{1-2} + h_{2-3}) - q_{1-2} \times h_{1-2} - q_{2-3} \times h_{2-3}$

If the aerodynamic center, where lift and drag forces are assumed to act, is located at z_0 and y_0 from boom–3, taking moments about boom-3,

$$S_Z \times y_0 + S_Y \times z_0 = -2A_{1-2} \times q_{1-2} - 2A_{2-3} \times q_{2-3} \qquad .....(16.7)$$

From these three equations of static equilibrium, the three unknown shear flows q_{1-2}, q_{2-3} and q_{3-1} can be evaluated.

16.6.1 SHEAR FLOW DUE TO TORSION IN MULTI-CELL WINGS

The chord-wise pressure distribution on an aerodynamic surface may be represented by shear loads (lift and drag) acting through the aerodynamic center together with a pitching moment M_0. This system of loads can be transferred to the shear center in the form of shear loads S_Y and S_Z together with a torque T. Axial load carrying booms have no role in resisting torque. Assuming that no axial constraints exist and that the shape of the section remains unchanged due to the applied load, the applied torque T produces individual but unknown

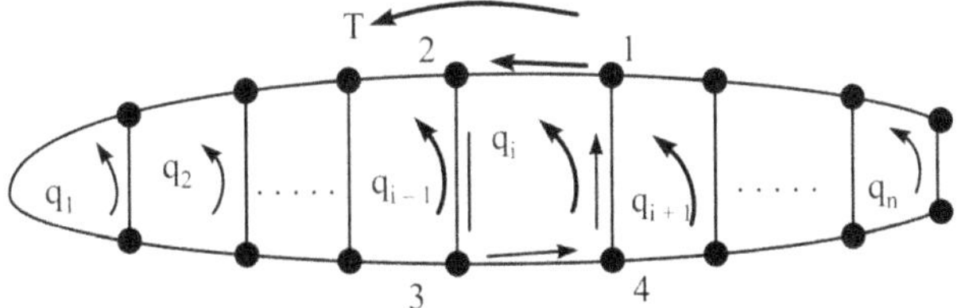

FIGURE 16.7 Multi-cell wing section subjected to torsion

torques in each of the 'n' cells forming the wing section. As shown in Fig 16.7, each cell, therefore, develops a constant shear flow q_1, q_2, ..q_i..q_n.

The total torque is then given by

$$T = \sum (2A_i \times q_i) \qquad .....(16.8)$$

Equation (16.8) can be used to calculate shear flow in the special case of a single–cell section, which is statically determinate. Solution for this equation, in the case of an n-cell section, requires additional equations by considering compatibility of displacements for the same rate of twist dθ/dx to ensure undistorted cross section.

The shear flow q is constant along each wall of a cell. Rate of twist in i^{th} cell, enclosed by walls 1–2, 2–3, 3–4 and 4–1 is given by,

$$d\theta/dx = [\,1/(2A_iG)\,] \times \oint q.(ds/t) \qquad \qquad .....(16.9)$$

Substituting $\oint (ds/t)$ for each wall with δ, the above equation can be written as

$$d\theta/dx = [1/(2A_iG)] \times [\, q_i \times \delta_{12} + (q_i - q_{i-1}) \times \delta_{23} + q_i \times \delta_{34} + (q_i - q_{i+1}) \times \delta_{41}\,]$$

$$= [1/(2A_iG)] \times [-q_{i-1} \times \delta_{23} + q_i \times (\delta_{12} + \delta_{23} + \delta_{34} + \delta_{41}) - q_{i+1} \times \delta_{41}\,]$$

$$= [1/(2A_iG)] \times [\,-q_{i-1} \times \delta_{i-1,i} + q_i \times \delta_i - q_{i+1} \times \delta_{i+1,i}\,] \qquad .....(16.10)$$

where, $\qquad \delta_{i-1,i} = \oint (ds/t)$ for the wall common to $(i–1)^{th}$ cell and i^{th} cell

$$\delta_i = \oint (ds/t) \text{ for all the walls enclosing } i^{th} \text{ cell}$$

and $\qquad \delta_{i+1,i} = \oint (ds/t)$ for the wall common to $(i+1)^{th}$ cell and i^{th} cell

Eq (16.10) is the set of 'n' compatibility conditions for the 'n' cells which, together with eq (16.8), give shear flows in all the walls of the entire section.

It is a common practice to fabricate skin panels and spar webs from materials possessing different shear modulus (G) values. In that case, G should be taken inside the integral in eq (16.10). For example, if shear modulus is G_2 for panel 2-3 and G_1 for all other panels, then eq (16.10) changes to

$$d\theta/dx = (1/2) \times A_i \times [q_i \times \delta_{12}/G_1 + (q_i - q_{i-1})\,\delta_{23}/G_2 + q_i\,\delta_{34}/G_1$$

$$+ (q_i - q_{i+1})\,\delta_{41}/G_1]$$

or $\quad d\theta/dx = (1/2) \times (A_i/G_1) \times [q_i\,\delta_{12} + (q_i - q_{i-1})\,f_G\,\delta_{23} + q_i\,\delta_{34} + (q_i - q_{i+1})\,\delta_{41}\,]$

$$\text{where, } \quad f_G = G_1/G_2$$

Example 16.10

Calculate shear stress distribution in the walls of three-cell wing section shown here, when it is subjected to an anticlockwise torque of 12 kNm. Let shear

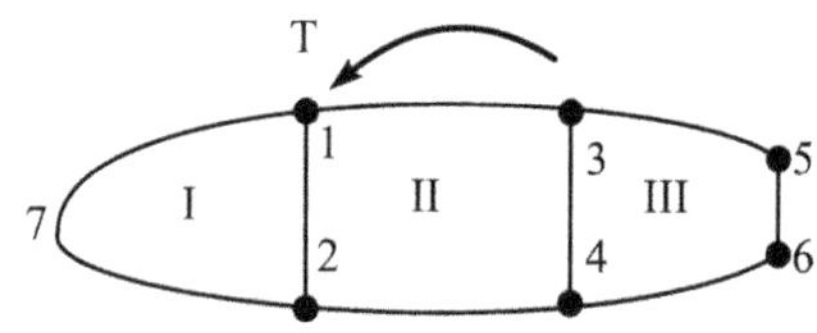

modulus be 24 GPa for all panels. Lengths and thicknesses of different panels in mm are given in Table.

Section	1-7-2	1-2	1-3	2-4	3-4	3-5	4-6	5-6
Length	1650	500	775	775	380	500	500	254
Thickness	1.2	2.0	1.2	1.2	1.6	1.0	1.0	1.0

Areas of cells are : $A_I = 258000$ mm^2; $A_{II} = 355000$ mm^2 and $A_{III} = 161000$ mm^2

Solution

$$\delta_{1-7-2} = \oint (ds / t) = L_{1-7-2} / t_{1-7-2} = 1650 / 12 = 1375$$

Similarly, $\delta_{1-2} = L_{1-2} / t_{1-2} = 500/2.0 = 250$; $\delta_{1-3} = L_{1-3} / t_{1-3} = 775/1.2 = 645.8$

$\delta_{2-4} = L_{2-4} / t_{2-4} = 775/1.2 = 645.8$; $\delta_{3-4} = L_{3-4} / t_{3-4} = 380/1.6 = 237.5$

$\delta_{3-5} = L_{3-5} / t_{3-5} = 500/1.0 = 500$; $\delta_{4-6} = L_{4-6} / t_{4-6} = 500/1.0 = 500$

$\delta_{5-6} = L_{5-6} / t_{5-6} = 254/1.0 = 254$

For cell-I, $(d\theta/dx)_I = [1/(2 \times A_I \times G)] \times [q_I \times (\delta_{1-7-2} + \delta_{1-2}) - q_{II} \times \delta_{1-2}]$

$$= [1/(2 \times 258000 \times 24200)] \times [q_I \times (1375 + 250) - q_{II} \times 250]$$

$$= 13 \times 10^{-8} \times q_I - 2.0 \times 10^{-8} \times q_{II} \qquad \ldots(a)$$

For cell-II, $(d\theta/dx)_{II} = [1/(2 \times A_{II} \times G)] \times [-q_I \times \delta_{1-2}$

$$+ q_{II} \times (\delta_{1-2} + \delta_{2-4} + \delta_{4-3} + \delta_{3-1}) - q_{III} \times \delta_{4-3}]$$

$$= [1/(2 \times 355000 \times 24200)] \times [-q_I \times 250$$

$$+ q_{II} \times (250 + 645.8 + 237.5 + 645.8) - q_{III} \times 237.5]$$

$$= -1.455 \times 10^{-8} \times q_I + 10.35 \times 10^{-8} \times q_{II} - 13.82 \times 10^{-8} \times q_{III}$$
$$\ldots(b)$$

For cell-III, $(d\theta/dx)_{III} = [1/(2 \times A_{II} \times G)] \times [-q_{II} \times \delta_{3-4}$

$$+ q_{III} \times (\delta_{3-4} + \delta_{4-6} + \delta_{6-5} + \delta_{5-3})]$$

$$= [1/(2 \times 161000 \times 24200)]$$

$$\times [- q_{II} \times 237.5 + q_{III} \times (237.5 + 500 + 254 + 500)]$$

$$= 3.0478 \times 10^{-8} \times q_{II} + 19.14 \times 10^{-8} \times q_{III} \qquad \ldots(c)$$

Total torque on the wing, $T = \sum (2A_i \times q_i)$

$$12 \times 10^6 \text{ Nmm} = 2 \times (258000 \times q_I + 355000 \times q_{II} + 161000 \times q_{III})$$

$\Rightarrow \qquad 2.58 \times q_I + 3.55 \times q_{II} + 1.61 \times q_{III} = 60 \qquad \ldots(d)$

Since rate of twist $d\theta/dx$ is same in all the cells, from (a) and (b),

$$13 \times 10^{-8} \times q_I - 2 \times 10^{-8} \times q_{II}$$

$$= -1.455 \times 10^{-8} \times q_I + 10.35 \times 10^{-8} \times q_{II} - 13.82 \times 10^{-8} \times q_{III}$$

$$\Rightarrow \quad 14.455 \times q_I - 12.35 \times q_{II} + 13.82 \times q_{III} = 0 \qquad \ldots(e)$$

and from (b) & (c),

$$(-1.455 \times q_I + 10.35 \times q_{II} - 13.82 \times q_{III}) \times 10^{-8}$$

$$= (3.0478 \times q_{II} + 19.14 \times q_{III}) \times 10^{-8}$$

$$\Rightarrow \quad 1.455 \times q_I - 7.3022 \times q_{II} + 32.96 \times q_{III} = 0 \qquad \ldots(f)$$

Eliminating q_I from (e) and (f),

$$(-12.35/14.455 + 7.3022/1.455) \times q_{II} + (13.82/14.455 - 32.96/1.455) \times q_{III} = 0$$

$$\Rightarrow \quad 4.1646 \times q_{II} - 21.7 \times q_{III} = 0 \qquad \ldots(g)$$

Eliminating q_I from (d) and (f),

$$(3.55/2.58 + 7.3022/1.455) \times q_{II} + (1.61/2.58 - 32.96/1.455) \times q_{III} = 60/2.58$$

$$\Rightarrow \quad 6.91 \times q_{II} - 22.03 \times q_{III} = 23.26 \qquad \ldots(h)$$

Eliminating q_{II} from (g) and (h),

$$(-21.7/4.1646 + 22.03/6.91) \times q_{III} = -23.26/6.91$$

$$\Rightarrow \quad q_{III} = 1.664$$

Then, from eq (g), $q_{II} = (21.7/4.1646) \times q_{III} = 8.672$

from eq (d), $q_I = (60 - 3.55 \times q_{II} - 1.61 \times q_{III}) / 2.58 = 10.285$

Thus, $q_I = 10.285$ N/mm ; $q_{II} = 8.672$ N/mm ; $q_{III} = 1.664$ N/mm

Shear stress in each wall is obtained by dividing shear flow in each wall by the wall thickness. Shear stress distribution is shown in the following figure.

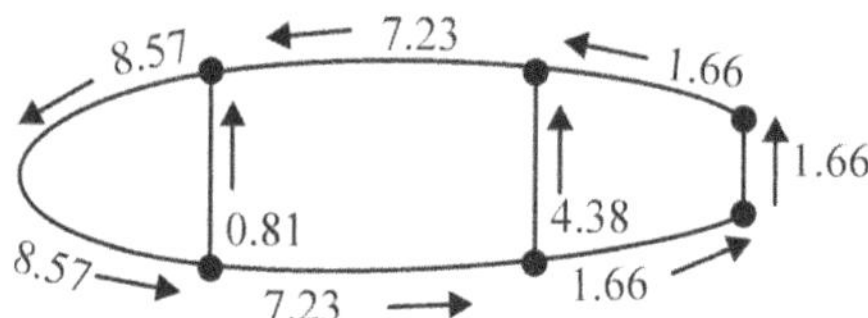

16.6.2 SHEAR FLOW DUE TO BENDING IN MULTI-CELL WINGS

Bending moments at any section of a multi-cell wing (Ref Fig 16.8) are usually produced by shear loads at other sections of the wing. Centroid of the direct stress carrying area is located at C. All the booms are used for calculating location of centroid as well as moments of inertia. Direct (axial) stress is calculated only in the booms 2, 4, 6, 8, 10 and 12 at the ends of webs from the following relation, in which coordinates y and z of each boom are referred to axes through centroid, as shown in Fig 16.8.

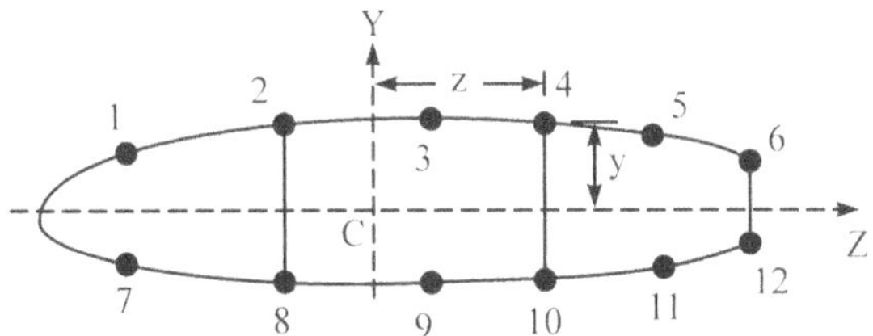

FIGURE 16.8 Multi-cell wing section subjected to bending

$$\sigma_X = \frac{(M_Z \times I_{YY} - M_Y \times I_{YZ}) \times y}{I_{YY} \times I_{ZZ} - I_{YZ}^{2}} + \frac{(M_Y \times I_{ZZ} - M_Z \times I_{YZ}) \times z}{I_{YY} \times I_{ZZ} - I_{YZ}^{2}} \qquad .....(16.11)$$

where, $I_{ZZ} = \sum B_i\, y_i^{2}$; $I_{YY} = \sum B_i\, z_i^{2}$; $I_{YZ} = \sum B_i\, y_i\, z_i$

16.6.3 SHEAR FLOW DUE TO SHEAR FORCE IN MULTI-CELL WINGS

A general n-cell wing section, as shown in Fig 16.9, is subjected to shear loads S_Y and S_Z whose lines of action do not necessarily pass through the shear center. The resulting shear flow distribution is therefore due to the combined effects of shear and torsion. The method of analysis for 'n' cell wing is an extension of the method for determining shear flow distribution and rate of twist in a single-cell box beam subjected to shear loads.

A single-cell beam is statically indeterminate, the single redundancy being the value of shear flow at an arbitrarily positioned 'cut'. Thus, the 'n' cell wing is made statically determinate by imagining a 'cut' in the top or bottom skin panel of each cell, as shown. In this method of successive approximations, 'cuts' are sometimes made in the spar webs although 'cutting' top or bottom skin panels produces a more rapid convergence in the numerical iteration process. In the special case of a wing section having a horizontal axis of symmetry, a 'cut' in the top skin panel is preferred as this will result in the 'open section' shear flow being zero in the bottom skin panel.

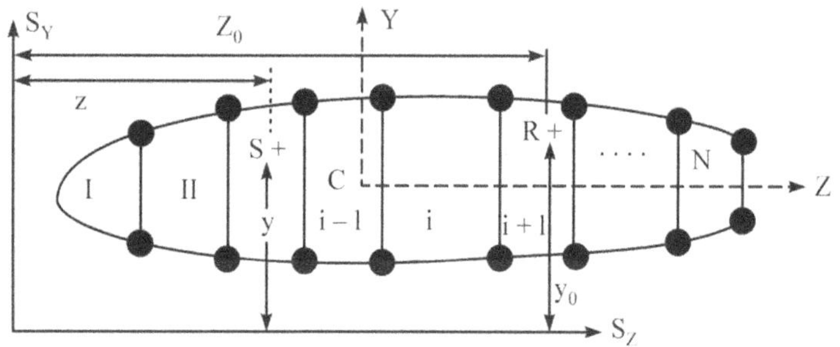

FIGURE 16.9 Multi-cell wing section subjected to shear loads

The open section shear flow q_b in i^{th} cell of the wing section is given by

$$q_b = -\left[\frac{S_Y \times I_{YY} - S_Z \times I_{YZ}}{I_{YY} \times I_{ZZ} - I_{YZ}^2}\right]\left(\int t \times y \times ds + \Sigma B_i \times y_i\right)$$

$$-\left[\frac{S_Z \times I_{ZZ} - S_Y \times I_{YZ}}{I_{YY} \times I_{ZZ} - I_{YZ}^2}\right]\left(\int t \times z \times ds + \Sigma B_i \times z_i\right) \qquad .....(16.12)$$

In this equation, t is the direct stress carrying thickness of the wall of the section and in this idealization, t = 0.

In the ith cell of closed section, there is an additional unknown shear flow $q_{s,0}$ at the imaginary cut. Actual shear flow around each cell is given by the summation of open section shear flow and the shear flow at the 'cut'. Rate of twist, from the pure torsion case, in each cell is given by

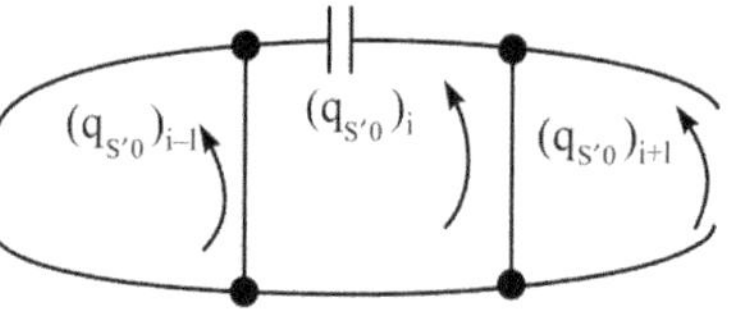

$$d\theta/dx = (1/2A_iG)\oint q_i \times (ds/t) = (1/2A_iG)\oint\left[q_b + (q_{s,0})_i\right] \times (ds/t)$$

$$= (1/2A_iG)\left[-(q_{s,o})_{i-1} \times \delta_{i-1-i} + (q_{s,o})_i \times \delta_i - (q_{s,o})_{i+1} \times \delta_{i+1,i} + \oint q_b \times (ds/t)\right]$$

$$.....(16.13)$$

Moment produced by the total shear flow of ith cell about any convenient moment center O is given by

$$M_{q,i} = \oint q_i \times p \times ds = \oint q_b \times p \times ds + (q_{s,o})_i \oint p \times ds = \oint q_b \times p \times ds + 2^a{}_i \times (q_{s,o})_i$$

Total moment due to shear forces,

$$M_q = S_Z \times y_0 - S_Y \times z_0 = \Sigma M_{q,i} = \Sigma \oint q_b \times p \times ds + \Sigma\left[2A_i \times (q_{s,o})_i\right]$$

If the moment center is chosen to coincide with the point of intersection of the lines of action of S_Y and S_Z, the total moment on the wing section due to shear is zero. Then,

$$\sum \oint q_b \times p.ds + \sum 2A_i \times (q_{s,o})_i = 0 \qquad .....(16.14)$$

Example 16.11

The wing section shown in the figure is subjected to a bending moment of 300kNm applied in a vertical plane and a vertical shear force of 86.8 kN in the plane of web 5-7-2. The section has been idealised such that booms carry all the direct stresses while the walls are effective only in shear. Calculate direct stresses in the booms, shear flow distribution and rate of twist. Distances (h_i) of different points from Z-axis and lengths between different points (L_i) in mm are: $L_{2-3} = 1275$; $L_{7-8} = 1270$; $L_{1-2} = L_{5-6} = 1020$

$h_7 = h_8 = 50$; $h_3 = h_4 = 200$; $h_1 = h_6 = 165$; $h_2 = h_5 = 230$

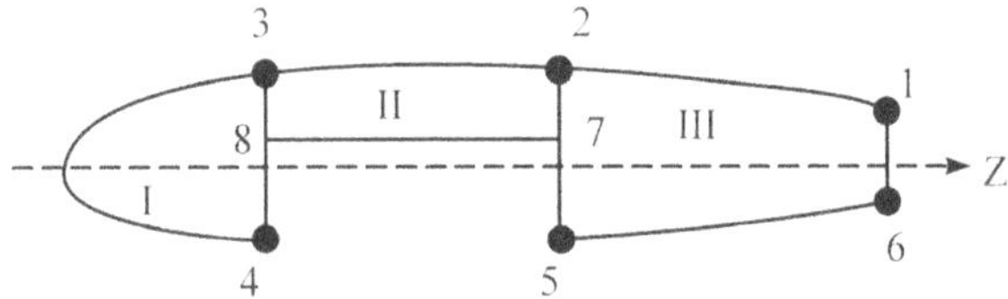

Areas of booms are : $B_1 = B_6 = 2580$ mm^2; $B_2 = B_5 = 3880$ mm^2; $B_3 = B_4 = 3230$ mm^2. Shear modulus 'G' for the panel 7-8 is 3 times that of all other panels, which is 27600 N/mm^2. Areas of the three cells are: $A_I = 265000$ mm^2; $A_{II} = 213000$ mm^2 and $A_{III} = 413000$ mm^2. Lengths and thicknesses of different panels are given in the table below.

Section	1-2	2-3	3-4	8-7	5-6	6-1	2-7-5	3-8-4
Length	1020	1275	2200	1270	1020	330	460	400
Thickness	1.2	1.6	2.0	1.2	1.2	1.6	2.6	2.6

Solution

(a) Due to bending

Since distribution of boom areas is symmetrical about X-axis, $I_{xy} = 0$

For $M_Z = 300$ kNm   and $M_Y = 0$, eq (16.8) reduces to

$$\sigma_X = \frac{(M_Z \times I_{YY} - M_Y \times I_{YZ}) \times y}{I_{YY} \times I_{ZZ} - I_{YZ}^2} + \frac{(M_Y \times I_{ZZ} - M_Z \times I_{YZ}) \times z}{I_{YY} \times I_{ZZ} - I_{YZ}^2} = \frac{M_Z \times y}{I_{ZZ}}$$

where, $I_{ZZ} = \sum (B_i \times y_i^2) = 2 (2580 \times 165^2 + 3880 \times 230^2 + 3230 \times 200^2)$

$$= 809 \times 10^6 \text{ mm}^4$$

Therefore, axial stress in i^{th} boom

$$(\sigma_X)_i = 300 \times 10^6 \times y_i \,/\, 809 \times 10^6 = 0.371 \times y_i$$

Then, $$(\sigma_X)_1 = -(\sigma_X)_6 = 61.2 \text{N/mm}^2$$

$$(\sigma_X)_2 = -(\sigma_X)_5 = 85.3 \text{N/mm}^2$$

$$(\sigma_X)_3 = -(\sigma_X)_4 = 74.2 \text{N/mm}^2$$

(b) Due to shear

With an imaginary cut in the top skin panel of each cell, open section shear flow from eq(16.9) is

$$q_b = -\left[\frac{S_Y \times I_{YY} - S_Z \times I_{YZ}}{I_{YY} \times I_{ZZ} - I_{YZ}^2}\right]\left(\int t \times y \times ds + \Sigma B_i \times y_i\right)$$

$$-\left[\frac{S_Z \times I_{ZZ} - S_Y \times I_{YZ}}{I_{YY} \times I_{ZZ} - I_{YZ}^2}\right]\left(\int t \times z \times ds + \Sigma B_i \times z_i\right)$$

For the particular case of direct stress carrying thickness of panel, $t = 0$; product of inertia for the section symmetric about Z-axis, $I_{YZ} = 0$ and $S_Z = 0$, this equation reduces to

$$q_b = -(S_Y/I_{ZZ}) \Sigma(B_i \times y_i) = [(86.8 \times 10^3)\,/\,(809 \times 10^6)] \times \Sigma(B_i \times y_i)$$

$$= -1.07 \times 10^{-4}\, \Sigma(B_i \times y_i)$$

Then, $(q_b)_{2-7} = -1.07 \times 10^{-4} \times 3880 \times 230 = -95.5$ N/mm

$(q_b)_{1-6} = -1.07 \times 10^{-4} \times 2580 \times 165 = -45.5$ N/mm

$(q_b)_{6-5} = -45.5 - 1.07 \times 10^{-4} \times 2580 \times (-165) = 0$

$(q_b)_{5-7} = -1.07 \times 10^{-4} \times 3880 \times (-230) = 95.5$ N/mm

$(q_b)_{3-8} = -1.07 \times 10^{-4} \times 3230 \times 200 = -69.0$ N/mm

$(q_b)_{4-8} = -1.07 \times 10^{-4} \times 3230 \times (-230) = 69.0$ N/mm

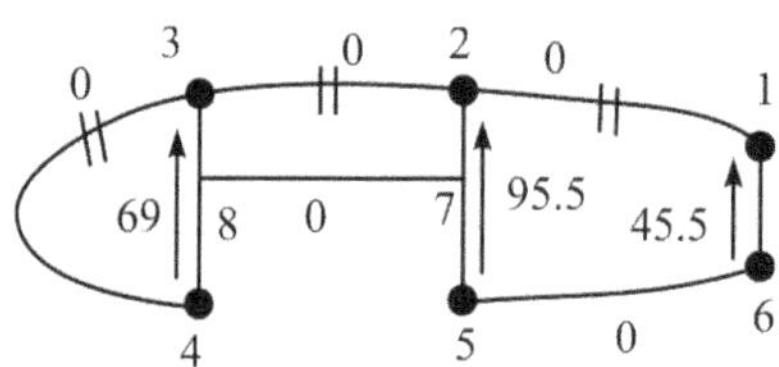

Since $(q_b)_{8-3} = (q_b)_{4-8}$ and $(q_b)_{7-2} = (q_b)_{5-7}$, $(q_b)_{7-8} = 0$

Open-section shear flow in different panels is shown in the figure.

Rate of twist for each cell is given by eq (16.10) as

$$d\theta/dx = (1/2A_iG)\,[-(q_{s,o})_{i-1}\,\delta_{i-1,i} + (q_{s,o})_i\,\delta_i - (q_{s,o})_{i+1}\,\delta_{i+1,i} + \oint q_b\,.(ds\,/\,t)$$

Here, $\delta_{7-8} = \oint(ds/t)_{7-8} = 1270/1.2 = 1058.3$

$$\delta_{2-3} = \oint(ds/t)_{2-3} = 1275/1.6 = 796.875$$

$$\delta_{1-2} = \delta_{5-6} = \oint(ds/t)_{1-2} = 1020/1.2 = 850$$

$$\delta_{3-4} = \oint(ds/t)_{3-4} = 2200/2.0 = 1100$$

$$\delta_{3-8} = \oint(ds/t)_{3-8} = 150/2.6 = 57.7$$

$$\delta_{8-4} = \oint(ds/t)_{8-4} = 250/2.6 = 96.15$$

$$\delta_{2-7} = \oint(ds/t)_{2-7} = 180/2.6 = 69.23$$

$$\delta_{7-5} = \oint(ds/t)_{7-5} = 280/2.6 = 107.7$$

$$\delta_{1-6} = \oint(ds/t)_{1-6} = 330/1.6 = 206.25$$

Then, from eq (16.10), with $f_G = G/G_{7-8} = 1/3$ applicable for the panel 7–8,

$$(d\theta/dx)_I = (1/2 \times 265000 \times 27600) \times [(q_{s,0})_I \times (\delta_{3-4} + \delta_{3-8} + \delta_{8-4}) - (q_{s,0})_{II} \times \delta_{3-8}$$
$$+ (q_b)_{3-8} \times \delta_{3-8} + (q_b)_{4-8} \times \delta_{8-4}]$$

$$= 0.684 \times 10^{-10} \times [(q_{s,0})_I \times (1100 + 57.7 + 96.15)$$
$$- (q_{s,0})_{II} \times 57.7 + 69 \times 57.7 + 69 \times 96.15]$$

$$= [8.57\,(q_{s,0})_I - 0.395\,(q_{s,0})_{II} + 72.55\,] \times 10^{-8} \qquad \ldots(a)$$

$$(d\theta/dx)_{II} = (1/2 \times 213000 \times 27600)$$
$$\times [-(q_{s,0})_I \times \delta_{3-8} + (q_{s,0})_{II} \times (\delta_{2-3} + \delta_{3-8} + f_G \times \delta_{8-7} + \delta_{7-2})$$
$$- (q_{s,0})_{III} \times \delta_{2-7} + (q_b)_{2-7} \times \delta_{2-7} + (q_b)_{3-8} \times \delta_{3-8}\,]$$

$$= 0.85 \times 10^{-10} \times [-(q_{s,0})_I \times 57.7$$
$$+ (q_{s,0})_{II} \times \{796.875 + 57.7 + (1/3) \times 1058.3 + 69.23\}$$
$$- (q_{s,0})_{III} \times 69.23 + 95.5 \times 69.23 - 69.0 \times 57.7\,]$$

$$= [-(q_{s,0})_I \times 0.49 + (q_{s,0})_{II} \times 10.85$$
$$- (q_{s,0})_{III} \times 0.588 + 22.36] \times 10^{-8} \qquad \ldots(b)$$

$$(d\theta/dx)_{III} = (1/2 \times 413000 \times 27600)$$
$$\times [-(q_{s,0})_{II} \times \delta_{2-7} + (q_{s,0})_{III} \times (\delta_{1-2} + \delta_{2-7} + \delta_{7-5} + \delta_{5-6} + \delta_{6-1})$$
$$+ (q_b)_{2-7} \times \delta_{2-7} + (q_b)_{7-5} \times \delta_{7-5} + (q_b)_{1-6} \times \delta_{1-6}\,]$$

$$= 0.439 \times 10^{-10} \times [- (q_{s,0})_{II} \times 69.23$$
$$+ (q_{s,0})_{III} \times (850 + 69.23 + 107.7 + 850 + 206.25)$$
$$- 95.5 \times 69.23 - 95.5 \times 107.7 + 45.5 \times 206.25]$$
$$= [- (q_{s,0})_{II} \times 0.3 + (q_{s,0})_{III} \times 1.04 - 3.75] \times 10^{-8} \qquad ...(c)$$

Taking moments of shear forces about the intersection of Z-axis and the web 5-7-2, we get from eq (16.14),

$$0 = - (q_b)_{4-8} \times 250 \times 1270$$
$$- (q_b)_{8-3} \times 150 \times 1270 + q_b)_{1-6} \times 330 \times 1020$$
$$+ 2(q_{s,0})_I \times 265000 + 2(q_{s,0})_{II} \times 213000 + 2(q_{s,0})_{III} \times 413000$$
$$\Rightarrow \quad 5.3(q_{s,0})_I + 4.26(q_{s,0})_{II} + 8.26(q_{s,0})_{III}$$
$$= 69 \times 3.175 + 69 \times 1.905 - 45.5 \times 3.366 = 197.367 \qquad ...(d)$$

Since rate of twist $d\theta/dx$ is same in all the cells,

from (a) and (b), $\quad [8.57 (q_{s,0})_I - 0.395 (q_{s,0})_{II} + 72.55] \times 10^{-8}$
$$= [- (q_{s,0})_I \times 0.49 + (q_{s,0})_{II} \times 10.85 - (q_{s,0})_{III} \times 0.588 + 22.36] \times 10^{-8}$$
$$\Rightarrow \quad 9.06 (q_{s,0})_I - 11.245 (q_{s,0})_{II} + 0.588 (q_{s,0})_{III} = 50.19 \qquad ...(e)$$

from (b) and (c), $[- (q_{s,0})_I \times 0.49 + (q_{s,0})_{II} \times 10.85 - (q_{s,0})_{III} \times 0.588 + 22.36] \times 10^{-8}$
$$= [- (q_{s,0})_{II} \times 0.3 + (q_{s,0})_{III} \times 1.04 - 3.75] \times 10^{-8}$$
$$\Rightarrow \quad 0.49 (q_{s,0})_I - 11.15 (q_{s,0})_{II} + 1.628 (q_{s,0})_{III} = 26.11 \qquad ...(f)$$

Eliminating $(q_{s,0})_I$ from (e) and (f),

$$[(-11.245/9.06) + (11.15/0.49)] \times (q_{s,0})_{II} + [(0.588/9.06) - (1.628/0.49)] \times (q_{s,0})_{III}$$
$$= - 50.19/9.06 - 26.11/0.49$$
$$\Rightarrow \quad 21.46 \times (q_{s,0})_{II} - 3.2575 \times (q_{s,0})_{III} = -53.86 \qquad ...(g)$$

Eliminating $(q_{s,0})_I$ from (d) and (f),

$$(4.26/5.3 + 11.15/0.49) \times (q_{s,0})_{II} + (8.26/5.3 - 1.628/0.49) \times (q_{s,0})_{III}$$
$$= 197.367/5.3 - 26.11/0.49$$
$$\Rightarrow \quad 23.56 \times (q_{s,0})_{II} - 1.76 \times (q_{s,0})_{III} = -16.05 \qquad .....(h)$$

Eliminating $(q_{s,0})_{II}$ from (g) and (h),

$$(- 3.2575/21.46 + 1.76/23.56) \times (q_{s,0})_{III}$$
$$= - 53.86/21.46 + 16.05/23.56$$

or $\qquad - 0.0771 \times (q_{s,0})_{III} = -1.8285$

$$\Rightarrow \quad (q_{s,0})_{III} = 23.72 \ N/mm$$

Substituting this value in eq (g), we get,

$$(q_{s,0})_{II} = (-53.86 + 3.2575 \times 23.72) / 21.46 = 1.09 \text{ N/mm}$$

Substituting these values in eq (f), we get

$$(q_{s,0})_I = [26.11 + 11.15 \times (q_{s,0})_{II} - 1.628 \times (q_{s,0})_{III}] / 0.49$$

$$= [26.11 + 11.15 \times 1.09 - 1.628 \times 23.72] / 0.49$$

$$= -0.702 \text{ N/mm}$$

Thus, shear flow at the three 'cuts' are

$$(q_{s,0})_I = -0.702 \text{ N/mm}$$

$$(q_{s,0})_{II} = 1.09 \text{ N/mm}$$

$$(q_{s,0})_{III} = 23.72 \text{ N/mm}$$

Actual shear flow in each skin panel is obtained by summing up open section shear flow and shear flow at the cut. Positive value of shear flow $q_{s,0}$ in each cell represents counter-clockwise direction. Thus, positive contribution of $(q_b)_{8-3}$ to $\oint q_b.(ds/t)$ in cell-I becomes a negative contribution to $\oint q_b.(ds/t)$ in cell-II.

16.7 TRANSVERSE FRAMES IN FUSELAGE AND WINGS

Aeroplane fuselages and wings are constructed primarily from thin metal sheets (panels) which are capable of resisting in-plane tension and shear loads but buckle under low values of in-plane compressive loads. The sheet panels are therefore stiffened by longitudinal stiffeners which resist in-plane compressive loads. These long stiffeners are also susceptible to buckling at low compressive loads. Effective length of these stiffeners is reduced by transverse frames - bulkheads in fuselages and ribs in wings. These transverse frames also resist concentrated loads in transverse planes and transmit them to the stringers and sheet panels. These frames are also fabricated from thin sheets and therefore require stiffeners to distribute concentrated loads to thin webs. The loads are applied along the stiffeners or, if applied at an angle, at the junctions of stiffeners so that each stiffener carries one component of applied load without undergoing any bending. Different axial loads in the adjacent stiffeners are balanced by shear flows in web panels.

16.7.1 FUSELAGE FRAMES

These frames take the form of open rings so that interior of the fuselage is not obstructed (Ref Fig 16.10). These are symmetrical about a vertical axis and are not necessarily circular in form. They are connected continuously around their periphery to the fuselage shell. If a vertical load 'W' produces shear flows q_1

and q_2 on the right and left sides, as shown, net shear flow 'q_f' transmitted to the periphery of the frame is $q_f = q_1 - q_2$, as shown in Fig 16.10.

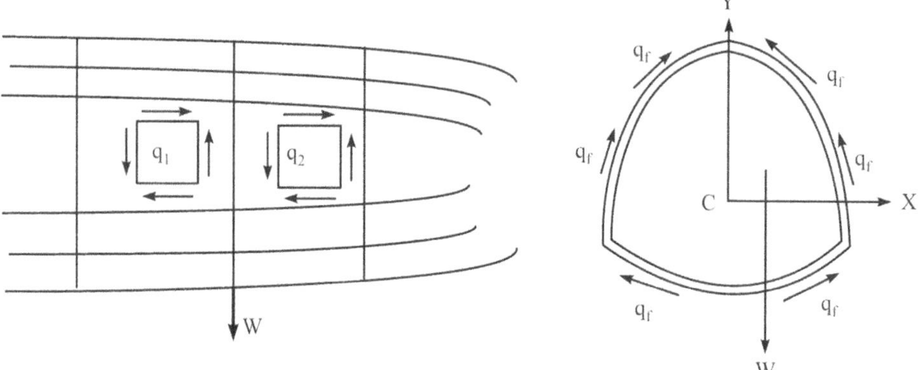

FIGURE 16.10 Loads on a fuselage frame

16.7.2 WING RIBS

The wing ribs (Ref Fig 16.11), similar to fuselage frames, have three main functions –

- maintain shape of wing

- transmit external applied loads to the wing skin and

- reduce column length of stringers

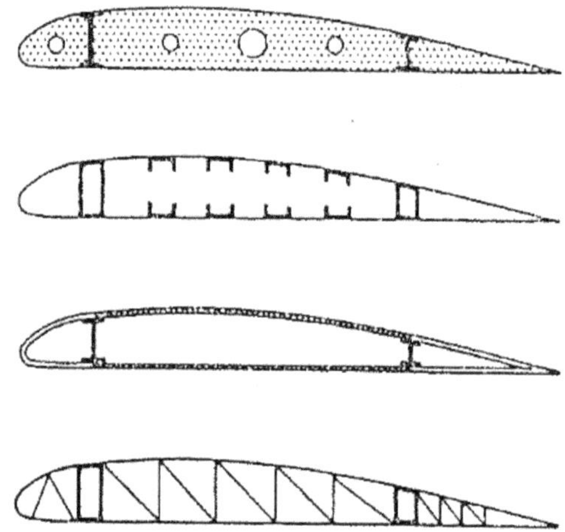

FIGURE 16.11 Different rib sections

The ribs are usually not symmetrical in shape and possess continuous webs except for holes to reduce weight and/or to run control cables. With three equations of equilibrium $\sum F_Z = 0$, $\sum F_Y = 0$ and $\sum M_X = 0$, it is possible to evaluate three shear flows in wing skin between three booms. Thus, *a three-boom wing structure is statically determinate*.

Example 16.12

Calculate shear flows in the web panels and axial loads in the flanges 1, 2 and 3 of a three-cell closed section wing, shown in figure, when subjected to horizontal load of 12 kN at 3 and vertical load of 15 kN at 6.

Areas of three cells and flanges and their distances are given below.

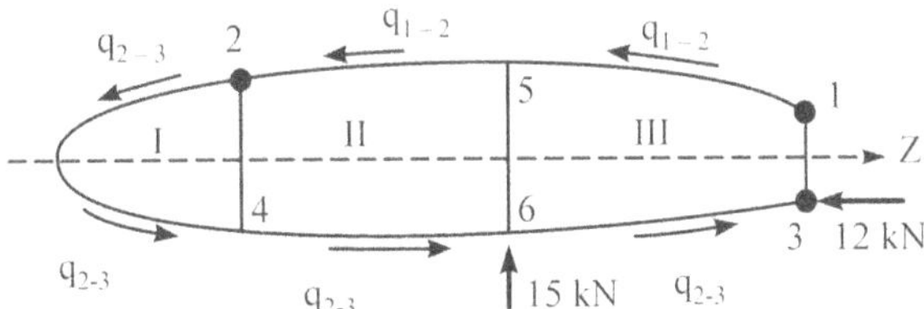

$z_{4-6} = z_{6-3} = 300\text{mm}$; $y_{1-3} = 300\text{mm}$; $y_{5-6} = 320\text{mm}$; $y_{2-4} = 300\text{mm}$

$A_I = 50{,}000\text{mm}^2$; $A_{II} = 95{,}000\text{mm}^2$; $A_{III} = 95{,}000\text{mm}^2$

Solution

Since the bending moments are fully resisted by the flanges 1, 2 and 3, the shear flows between flanges q_{1-2}, q_{2-3} and q_{3-1} are constant.

Resolving forces vertically,

$$q_{1-2} \times z_{1-2} - q_{2-3} \times z_{2-3} = 12000\text{N}$$

or $600\, q_{1-2} - 600\, q_{2-3} = 12000$

$\Rightarrow$ $q_{1-2} - q_{2-3} = 20$ $\hspace{4cm}$ ...(a)

Resolving forces horizontally,

$$q_{3-1} \times y_{3-1} - q_{2-3} \times y_{2-3} = 15000\text{N}$$

or $300\, q_{3-1} - 300\, q_{2-3} = 15000$

$\Rightarrow$ $q_{3-1} - q_{2-3} = 50$ $\hspace{4cm}$ ...(b)

Taking moments about flange-3, from $T = 2\, qA$,

$$2\,(50000 + 95000) \times q_{2-3} + 2 \times 95000 \times q_{1-2} = -15000 \times 300$$

$\Rightarrow$ $29\, q_{2-3} + 19\, q_{1-2} = -450$ $\hspace{3.5cm}$ ...(c)

Solving these three equations, we get

$$q_{1-2} = 13 \text{ N/mm} ; \quad q_{2-3} = -7 \text{ N/mm} ; \quad q_{3-1} = 43 \text{ N/mm}$$

Negative value for q_{2-3} indicates direction of the shear flow opposite to the assumed direction

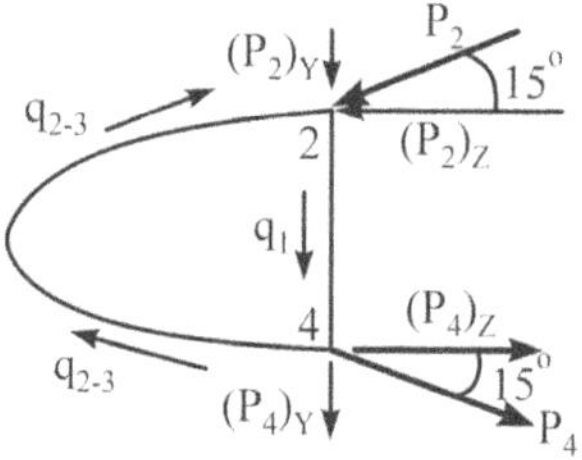

Shear flow q_1 in web of cell-I can be found, from the free body diagram, for equilibrium of vertical forces to the left of stiffener 2-4

$$q_1 \times L_{2-4} = [S_Y - (P_2)_Y - (P_4)_Y] \qquad \text{...(d)}$$

where, S_Y = Total vertical shear force at section 2-4

$$= q_{2-3} \times y_{2-4} = 7 \times 300 = 2100 \text{ N}$$

Taking moments about 2 or 4, $T - (P_4)_Z \times L_{2-4} = 0$,

$$(P_2)_Z = (P_4)_Z = T/ L_{2-4} = 2A \times q_{2-3}/L_{2-4} = 2 \times 50000 \times 7 / 300$$

$$= 7000/3$$

$$(P_2)_Y = (P_4)_Y = (P_4)_Z \times \tan \theta = (7000/3) \times \tan 15^0 = 625.2 \text{ N}$$

$$(\theta = 15^0, \text{ given for this wing})$$

Therefore, from eq (d),

$$q_1 = [2100 - 625.2 - 625.2] / 300 = 2.8 \text{ N/mm}$$

Shear flow to the right of stiffener 2-4 in cell-II can be found from

$$q_2' = [q_1 \times y_{2-4} + (P_2)_Y + (P_4)_Y] / y_{2-4} = 7 \text{ N/mm}$$

Similarly, shear flow q_2 in the web to the right of stiffener $5 - 6$ can be found by considering free body diagram as shown.

Let $A_2 = 46000 \text{ mm}^2$ and

$A_2' = 49000 \text{ mm}^2$; $A_2 + A_2' = A_{II}$

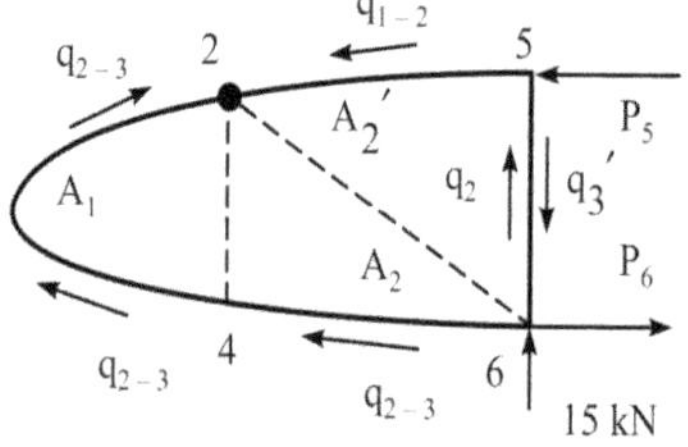

Shear force in the skin at 5 and 6 are horizontal and hence have no vertical components

Taking moments about 5 or 6,

$$P_5 = P_6 = 2\,[(A_1 + A_2) \times q_{2\text{-}3} - A_2' \times q_{1\text{-}2}]\,/\,y_{5\text{-}6}$$

$$= 2\,[(50000{+}46000) \times 7 - 49000 \times 13\,]\,/\,320 = 218.8\text{N}$$

Shear force at this section is resisted solely by the web. Assuming that the wing is symmetric about mid-plane, for equilibrium of vertical forces to the left of stiffener 5-6,

$$q_2 \times y_{5\text{-}6} = q_{2\text{-}3} \times [y_{2\text{-}4} + (y_{5\text{-}6} - y_{2\text{-}4})/2] - q_{1\text{-}2} \times [(y_{5\text{-}6} - y_{2\text{-}4})/2] - (P_5)_Y$$

$$q_2 = [7 \times (300 + 10) - 13 \times 10]\,/\,320 = 6.4\ \text{N/mm}\quad \text{since } (P_5)_Y = 0$$

Shear flow q_3' to the right of stiffener $5 - 6$ in web of cell-III is obtained from equilibrium of stiffener 5–6 as

$$q_3 \times y_{5\text{-}6} = q_2 \times y_{5\text{-}6} + 15000 \;\Rightarrow\; q_3 = 53.3\ \text{N/mm}$$

Shear flow to the left of stiffener 1–3 in cell-III is found in a similar way from the complete wing structure.

$$M_3 = 2[\{A_I + (A_{II} + A_{III})/2\} \times q_{2\text{-}3} + 15000 \times z_{3\text{-}6}] = 4.06 \times 10^6\ \text{Nmm}$$

$$(P_3)_Z = (P_1)_Z = M_3\,/\,y_{1\text{-}3} = 13533.3\ \text{N}$$

$$(P_3)_Y = (P_1)_Y = (P_1)_Z \times \tan 15 = 3626.2\ \text{N}$$

$$q_3' = [15000 + q_{2\text{-}3} \times y_{1\text{-}3} - (P_3)_Y - (P_1)_Y]\,/\,y_{1\text{-}3} = 32.8\ \text{N/mm}$$

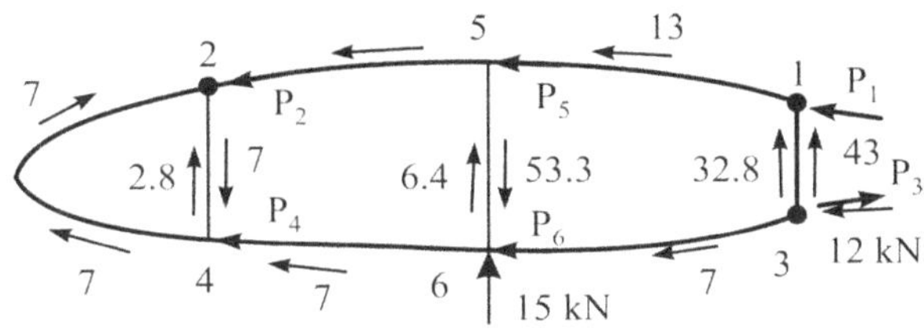

Shear flow values in N/mm and flange loads are plotted on the wing structure.

16.8   CUT-OUTS IN WINGS AND FUSELAGE

Wings and fuselages are so far treated as closed boxes stiffened with longitudinal stringers and transverse frames. In practice, these shells contain openings –

- in wings to accommodate retractable undercarriages, fuel tanks, engines,..

- in fuselage to accommodate doors, windows, cockpit, bomb bays,….

These cut-outs result in discontinuities in load distribution and these regions need to be heavily reinforced, thereby increasing weight of structure.

Doors are classified as Plugged (closes from inside to provide safety) or non-plugged and Stressed or non-stressed. In doors in pressurised area, hoop tension loads from pressure are carried through the door but, due to the quick-opening requirements, the major portion of tensile and shear loads are carried around the cutout structures (Ref Fig 16.12). Means of opening must be simple and obvious. It must be possible to open doors from either inside or outside and operated even in darkness. Jamming of doors due to deformation in a major crash must be avoided. A door may also be designed to move inwards and then slide upwards on rollers; Such a door is normally operated electrically, either from a panel inside the cabin or from an exterior control panel

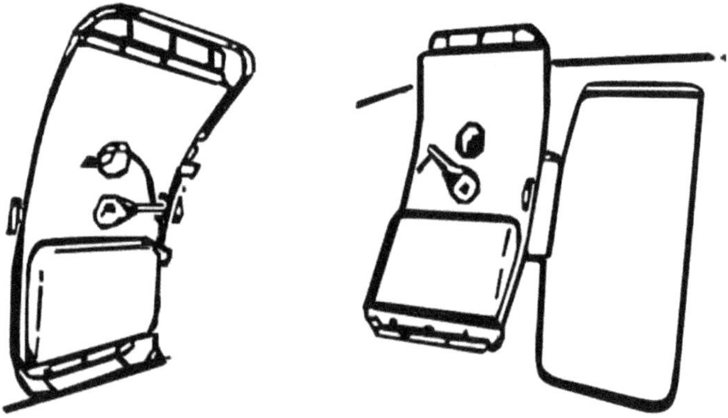

FIGURE 16.12 A typical passenger door in closed position and opened position

Window configuration is influenced by size, quantity and shape. The window cutouts fall in the area of highest skin shear from fuselage bending. A typical passenger window installation contains two panes as a fail-safe design - the outer pane carries the entire pressure load with a safety factor of about 8 and an inner pane capable of resisting full pressure load, if a failure should ever occur in the outer pane (Ref Fig 16.13). A third pane is usually provided for acoustic reasons.

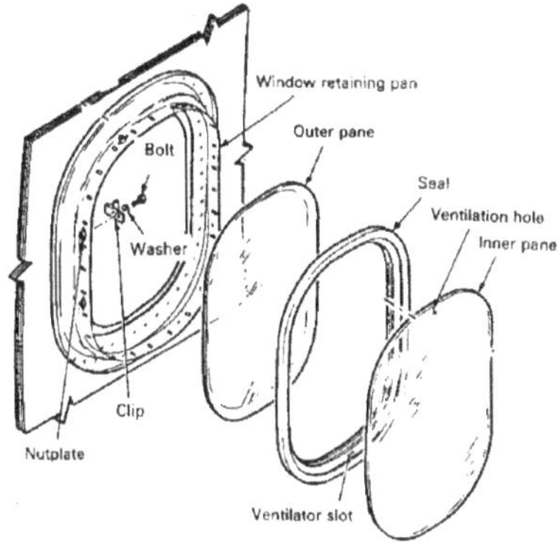

FIGURE 16.13 A typical window with two panes

In a military cargo aeroplane, big cargo such as tanks may be loaded from a opening provided in the cockpit side or aft section of fuselage. A swinging fuselage nose (including the flight deck) creates considerable difficulties in carrying through cables, wires and plumbing etc. Swing nose and swing tail impose a penalty on structure weight (Ref. Fig.16.14).

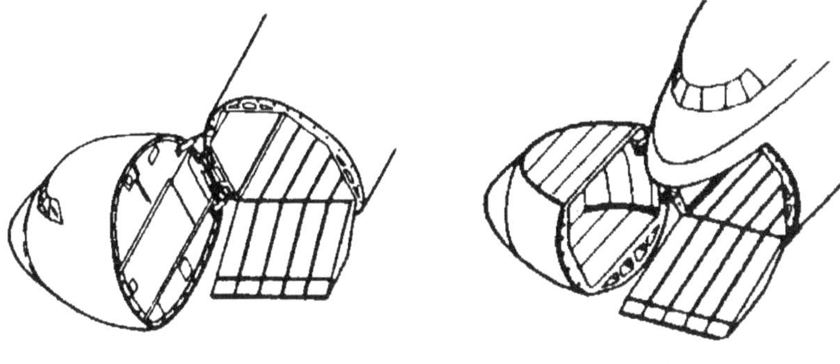

(a) Full swing nose   (b) Raised cockpit, swing nose

FIGURE 16.14 Typical loading arrangements from front end of fuselage

16.8.1 CUT-OUT IN WING

If the wing structure is continuous, shear flow due to torque in the skin panels will be q = T/2A and the flanges will be unloaded. In the box frame of a wing shown in Fig 16.15, the plane 1-4-8-5 is the fixed plane while 1-2-3-4 forms the front spar and 5-6-7-8 forms the rear spar. If the panel a-b-c-d on the lower surface of wing is cut-out, bay-2 becomes an open section beam with both ends fixed. It is torsionally weak and transmits torque by the differential bending of the front and rear spars. The applied torque results in a couple in bay-2 with shear loads 'S' in the front and rear spars (Ref Fig. 16.16) so that $T = S \times B$ and shear flow $q_1 = S/H = T/(B \times H)$

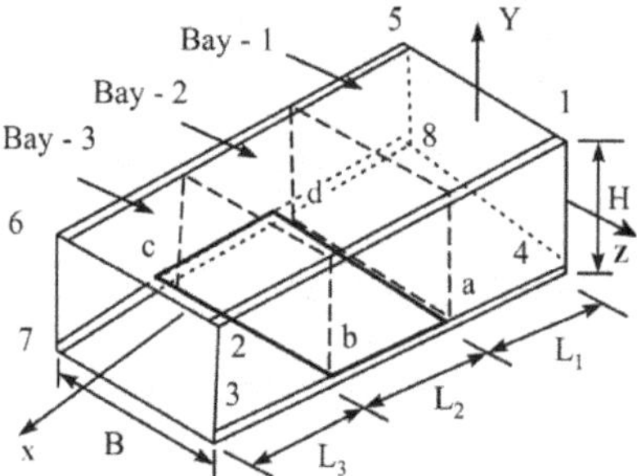

FIGURE 16.15 Box frame of a wing

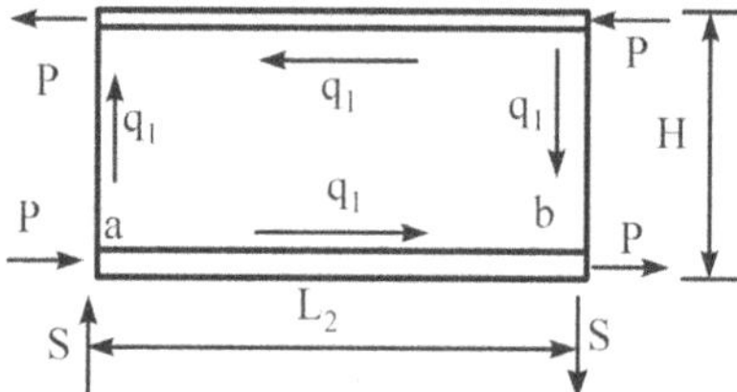

FIGURE 16.16 Flange loads in cut out bay of a wing

Midway in bay-2, a point of contra flexure occurs in the front and rear spars, so that axial load in the stringers 2-1, 3-4 of panel 1-2-3-4 or stringers 6-5, 7-8 of panel 5-6-7-8 is given by

$$M = S \times (L_2/2) - P \times H = 0 \quad \text{or} \quad P = S \times L_2/2H = q_1 \times L_2/2$$

Alternatively, P can be evaluated from equilibrium condition of spar flanges (Ref Fig. 16.17) as $\quad 2P = q_1 \times L_2 \quad\quad\quad$ or $\quad P = q_1 \times L_2/2$

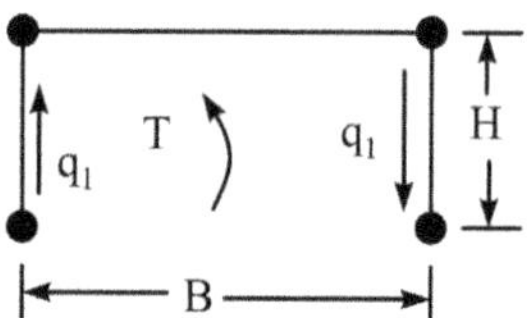

FIGURE 16.17 Torque and shear flow in a cut out bay

These flange loads in bay-1 and bay-3 are transmitted to the adjacent spar webs and skin panels as shown in Fig. 16.18. For equilibrium of flange 1-2 in bay-3, $\quad\quad L_3 \times (q_2 - q_3) = P \quad$ or $\quad q_2 - q_3 = P/L_3$

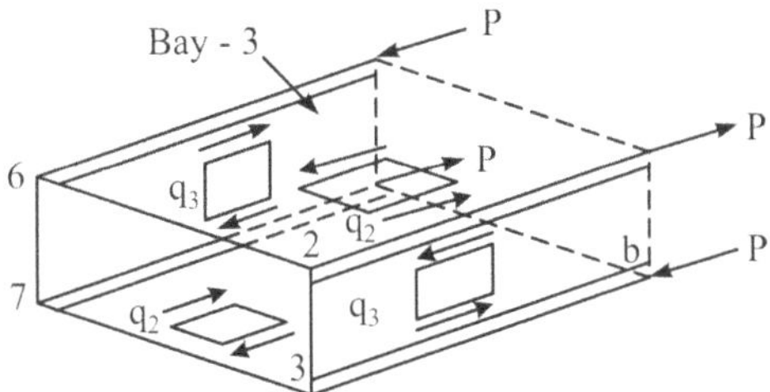

FIGURE 16.18 Shear flows in a bay adjacent to cut-out

For moment equilibrium, taking moments about 3,

$$T = (q_3 \times H) \times B + (q_2 \times B) \times H \quad \Rightarrow \quad q_2 + q_3 = T/(B \times H)$$

From the above two equations, we get,

$$q_2 = [(P/L_3) + \{T/(B \times H)\}] / 2 \quad \text{and} \quad q_3 = [-(P/L_3) + \{T/(B \times H)\}] / 2$$

The shear flow distribution on the section between bay-2 and bay-3 is as shown in Fig 16.19. Here, $q_2 = q_1 - q_3 = [(P/L_3) + \{T/(B \times H)\}] / 2$

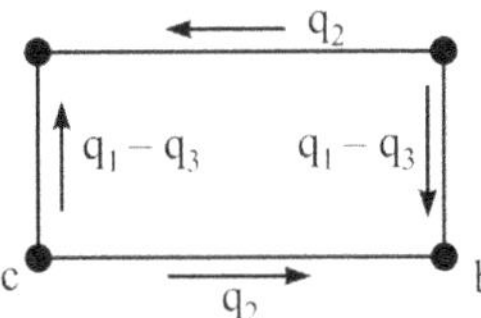

FIGURE 16.19 Shear flows on an intermediate section

Thus, the cut-out has the effect of increasing q_2 and decreasing q_3 by (P/L_3). Flange load varies along its length due to shear flows in the adjacent skin panels and spar webs. As an example, it is shown here for flange 1-2.

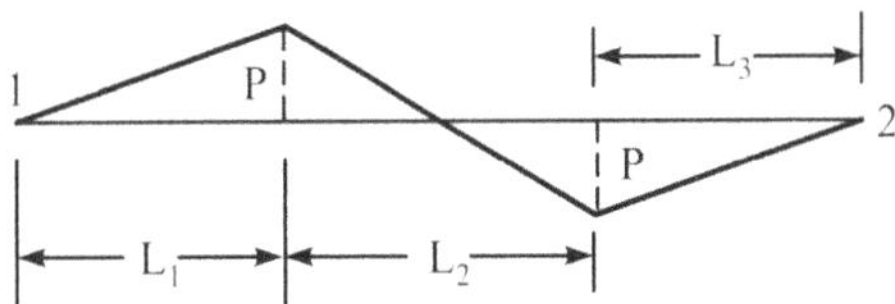

16.8.2 CUT-OUT IN FUSELAGE

Large cut-outs in fuselages required for cockpits and doors are treated as in the case of cut-outs in wings. The effects of smaller cut-outs, as those required for rows of windows of width L_w and height H_w placed L apart (Ref Fig. 16.20) are analysed as explained here. Taking q_{av} as the average shear flow in the area without cut-outs, shear flows in the regions near the cut-out can be evaluated from force equilibrium condition, using

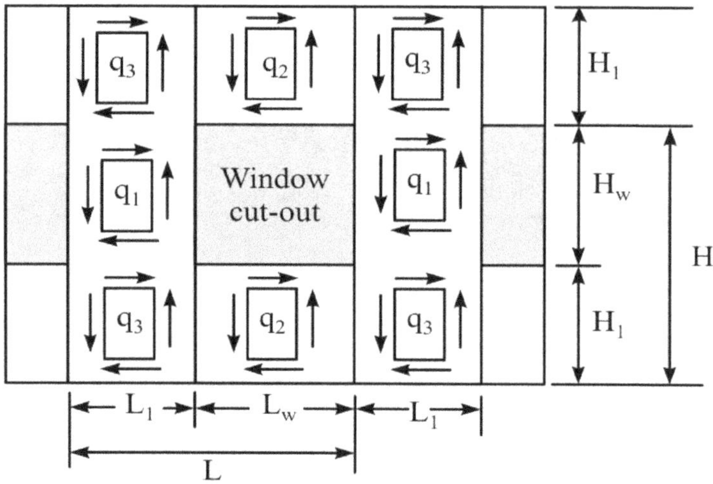

FIGURE 16.20 Shear flows around a window cut-out

$$q_1 \times L_1 = q_{av} \times L \qquad \Rightarrow \quad q_1 = q_{av} \times L / L_1$$

$$q_2 \times H_1 = q_{av} \times H \qquad \Rightarrow \quad q_2 = q_{av} \times H / H_1$$

$$q_3 \times L_1 + q_2 \times L_w = q_{av} \times L \quad \text{and} \quad q_3 \times H_1 + q_1 \times H_w = q_{av} \times H$$

where, $\quad L = L_1 + L_w \quad \text{and} \quad H = H_1 + H_w$

From either of the above equations, q_3 can be evaluated. From the first equation,

$$
\begin{aligned}
q_3 \ &= (q_{av} L - q_2 \times L_w) / L_1 \\
&= [q_{av} \times (L_1 + L_w) - q_{av} \times (H / H_1) \times L_w] / L_1 \\
\Rightarrow \quad &= [q_{av} \times (L_1 + L_w) - q_{av} \times \{(H_1 + H_w)/H_1\} \times L_w] / L_1 \\
\Rightarrow \quad &= [\, 1 + (L_w/ L_1) - \{1 + (H_w/H_1)\} \times (L_w/L_1)] \times q_{av} \\
\Rightarrow \quad &= [\, 1 - (H_w/H_1) \times (L_w/L_1) \,] \times q_{av}
\end{aligned}
$$

Thus, because of the cut-out, shear flow values q_1 and q_2 in the adjacent panels increased while q_3 value decreased in comparison with the average shear flow.

CHAPTER 17

FATIGUE ANALYSIS

Fatigue is defined as the progressive deterioration of the strength of a material or structural component during service such that failure can occur at much lower stress levels than the ultimate stress level. Fatigue is a dynamic phenomenon which initiates small (micro) cracks in the material or component and causes them to grow into large (macro) cracks which can cause, if not detected and corrected, catastrophic failure.

Many types of fatigue can occur in aircrafts – cyclic fatigue, caused by repeated fluctuating loads; fretting fatigue, due to small-scale rubbing movements and abrasion of adjacent parts; thermal fatigue, due to stress fluctuations induced by temperature changes (for example - on wing and fuselage surfaces at ground and high altitude) and sonic or acoustic noise, due to vibrations caused by jet or propeller.

Airworthiness requirement demands that the strength of an aircraft throughout its operational life under the action of repeated loads of variable magnitudes shall be such as to avoid disastrous fatigue failure (probability of failure $< 10^{-7}$).

17.1 FLUCTUATING STRESS

An aircraft suffers fatigue damage during all phases of the ground-air-ground cycle. For example, wing bends downwards due to self weight while on ground and its bottom surface experiences compressive stress while the same wing bends upwards due to lift during flight and the bottom surface experiences tensile stress. Such a fluctuation of stress can fail a component by fatigue after a certain number of stress cycles. The stress which a material can withstand for an unlimited number of stress cycles ($>10^6$ cycles) is called the ***endurance limit***.

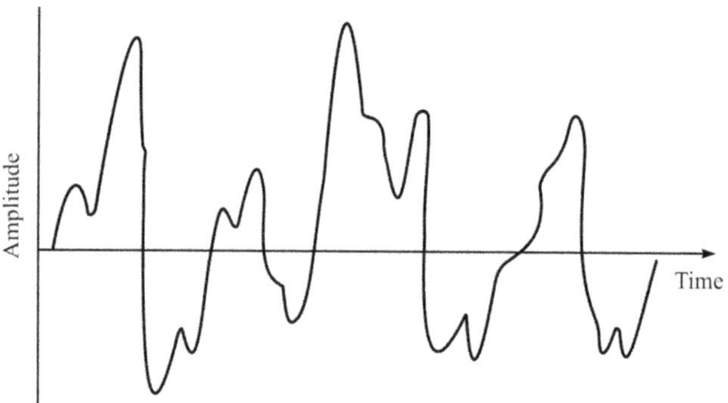

FIGURE 17.1 Typical time-varying stress amplitude

A typical time-varying stress amplitude is shown in fig 17.1. Actual stress variation can be considered as a combination of many individual cyclic stresses, as shown in Fig 17.2 (a), superimposed, while nomenclature of a stress cycle is shown in Fig 17.2 (b).

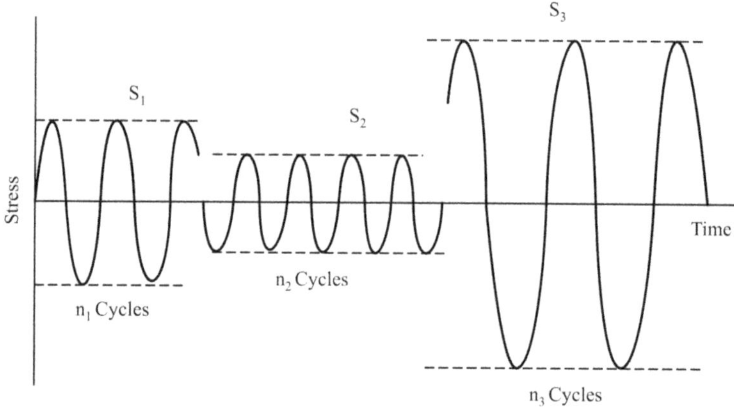

(a) Fluctuating stress as a combination of different stress cycles

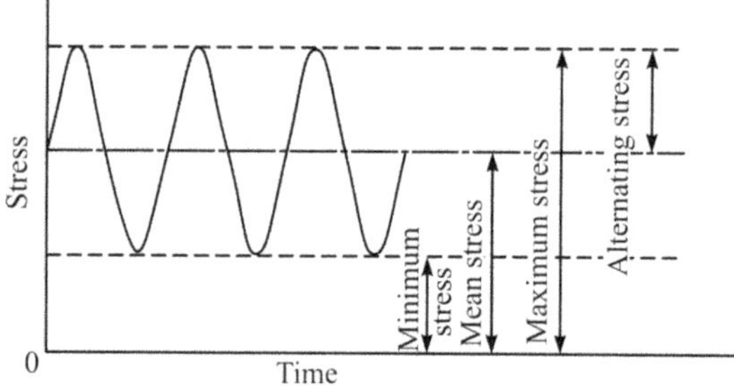

(b) Nomenclature of a stress cycle

FIGURE 17.2 Stress cycles

If σ_{max} and σ_{min} are the maximum and minimum stresses at a point due to a single type of stress cycle, mean stress and stress amplitude are defined as

$$\sigma_m = (\sigma_{max} + \sigma_{min}) / 2 \; ; \qquad \sigma_a = (\sigma_{max} - \sigma_{min}) / 2$$

Fatigue strength is a material property, which is significantly affected by the shape and size of the product, surface finish etc. These are accounted for by multiplying endurance limit stress with factors given below.

S_e' = Endurance limit stress for a rotating beam specimen subjected to reversed bending stress

= 0.5 S_{UT} for steels

= 0.4 S_{UT} for cast iron, cast steel,

S_e = Endurance limit stress for a particular component subjected to reversed bending stress. It is not just a material property like S_u or S_y, but depends on size, shape, surface finish and notch sensitivity

$$= (K_a \times K_b \times K_c \times K_d) \times S_e'$$

where, K_a is the surface finish factor

K_b is the size factor

(1.0 for d≤7.5,   0.85 for 7.5<d≤50,   0.75 for d>50)

K_c is the reliability factor

(1.0 for 50%,   0.897 for 90%,   0.868 for 95%,

0.814 for 99%   and   0.753 for 99.9%

K_d is the modifying factor for stress concentration = 1 / K_f

Endurance strength in shear,

S_{se} = 0.5 S_e ,   according to max shear stress theory

= $S_e/\sqrt{3}$,   according to distortion energy theory

$\sigma_a = S_e/f_s$; $\tau_a = S_{se}/f_s$

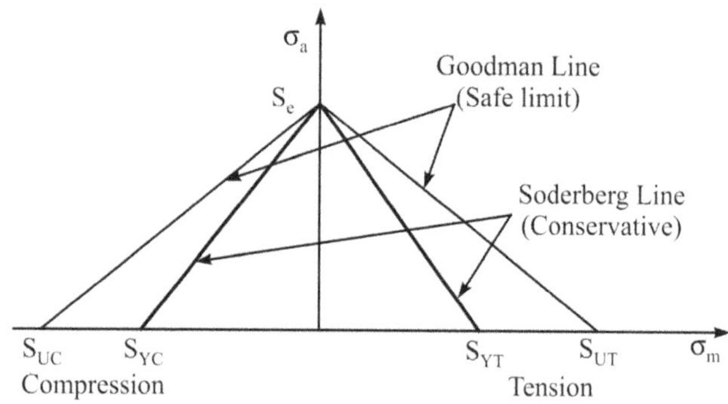

FIGURE 17.3 Goodman line and Soderberg line

Safe operating stresses for infinite life of a component are depicted in Fig 17.3. The plot shows mean stress component on X-axis and alternating stress component on Y-axis. For a single application of load ($\sigma_a = 0$), component is safe below the ultimate stress of the material (S_Y). In a completely reversed stress cycle ($\sigma_m = 0$), a component is safe for infinite cycles upto the endurance limit (S_e) of the material. Any particular stress cycle ($\sigma_a \neq 0$ and $\sigma_m \neq 0$) below the line joining ultimate stress on X-axis and endurance limit on the Y-axis is considered safe from fatigue point of view. This line is called **Goodman line.** Considering the tensile and compressive limits of ultimate stress, this line is plotted in the 1st quadrant and 2nd quadrant.

While the ultimate stress is considered as the limiting stress for brittle materials, yield stress (S_Y) is taken as the limiting stress for usefulness of a product, made of ductile materials. Thus, for ductile materials, any particular stress cycle ($\sigma_a \neq 0$ and $\sigma_m \neq 0$) below the line joining yield stress on X-axis and endurance limit on the Y-axis is considered safe from fatigue point of view. This line is called **Soderberg line.** Considering the tensile and compressive limits of yield stress, this line is also plotted in the 1st quadrant and 2nd quadrant. Ductile materials are used in most of the engineering applications and, hence, Soderberg line is more commonly used for design against fatigue failure.

17.2 FATIGUE LOADS

17.2.1 COMPLETELY REVERSED STRESSES

S-N curve, drawn from 0.9 S_{UT} at 10^3 cycles to S_e at 10^6 cycles on log-log graph, is shown in Fig 17.4, for completely reversed stress cycles ($S_m = 0$)

(i) **Design for infinite life** – For any alternating stress below the endurance limit, the component has an infinite life indicated by the horizontal line of S-N curve beyond B (10^6 cycles).

Criterion of failure is $\sigma_a \leq S_e/f_s$ for normal stress ($\tau_a \leq S_{se}/f_s$ in shear)

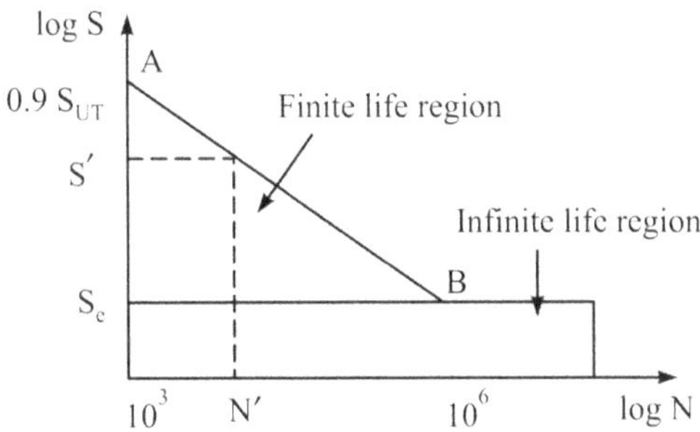

FIGURE 17.4 Log-log plot of alternating stress Vs No. of cycles (S-N curve)

(ii) **Design for finite life** – For stress (S') above endurance limit, the component has finite life (N') which can be read from the line AB in the S-N curve.

If multiple stress cycles of amplitudes 'S_i' act for 'n_i' cycles, the maximum number of cycles 'N_i', if this stress cycle alone is acting can be obtained from the fatigue (S-N) curve on log-log scale, as shown in Fig 17.5. Then according to *Miner's rule*, total life fraction used is given by,

$$U = \sum \frac{n_i}{N_i} \qquad \qquad .....(17.1)$$

If these stress cycles occur in a single flight, number of flights permitted for an aircraft with similar operational loads, $N_{max} = 1/U$. Even though $U < 1$ is an acceptable value, $U < 0.5$ is commonly used as design criterion, taking a factor of 2 for variations in loading conditions and material properties.

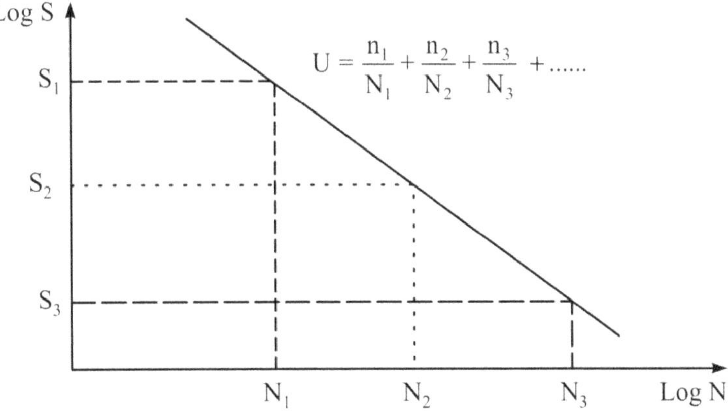

FIGURE 17.5 Miner's rule for multiple types of stress cycles

17.2.2 FLUCTUATING STRESSES

The stress may be purely tensile, purely compressive or mixed, depending on the magnitudes of σ_m and σ_a. When different types of loads are acting on a component, ***Modified Goodman diagram*** is used to obtain allowable S_a and S_m. Separate diagrams are used for axial and bending loads, based on yield stress in tension, S_{YT} (Ref Fig 17.6) and for torsion load based on yield stress in shear, S_{SY} (Ref Fig 17.7).

(i) For fluctuating normal stress due to axial and bending loads

Point of intersection of OE and AB is X having coordinates S_m and S_a

Then, $\sigma_a \leq S_a / f_s$; $\quad \sigma_m \leq S_m / f_s$ $\quad$ and $\quad \tan\theta = S_a/S_m = \sigma_a/\sigma_m$

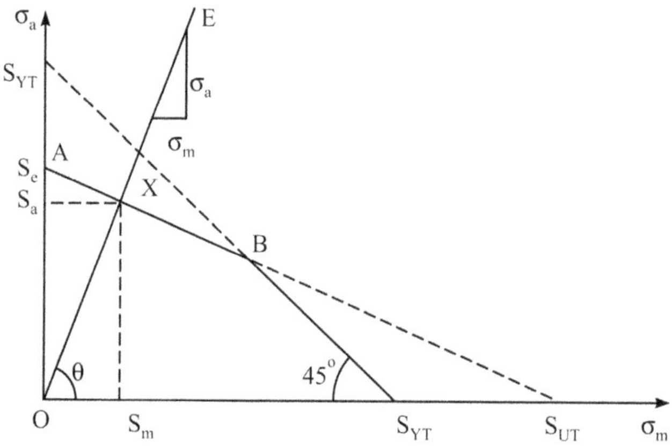

FIGURE 17.6 Modified Goodman diagram – for axial & bending stresses

(ii) For fluctuating shear stress due to torsion load

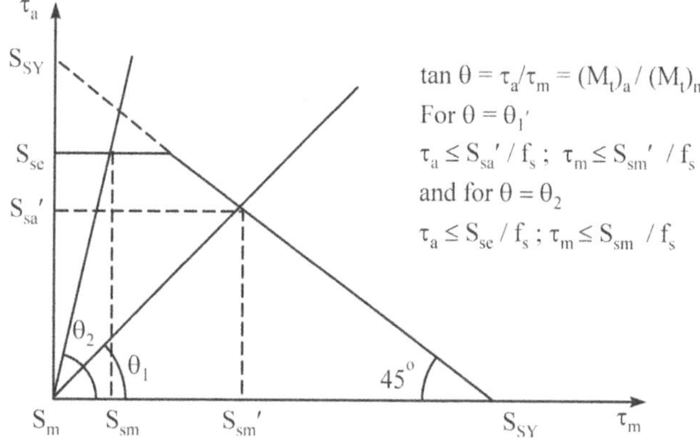

FIGURE 17.7 Modified Goodman diagram – for torsional stresses

17.3 FATIGUE DUE TO COMBINED NORMAL AND SHEAR STRESSES

Equivalent or von Mises stress is calculated from principal stresses or from the combination of normal and shear stresses at a point, using the equations,

$$\sigma_{eq} = \sqrt{\frac{\left[\left(\sigma_X - \sigma_Y\right)^2 + \left(\sigma_Y - \sigma_Z\right)^2 + \left(\sigma_Z - \sigma_X\right)^2 + 6\left(\tau_{XY}^2 + \tau_{YZ}^2 + \tau_{ZX}^2\right)\right]}{2}}$$

$$\ldots\ldots(17.2)$$

$$\text{or } \sigma_{eq} = \sqrt{\frac{\left[\left(\sigma_1 - \sigma_2\right)^2 + \left(\sigma_2 - \sigma_3\right)^2 + \left(\sigma_3 - \sigma_1\right)^2\right]}{2}}$$

In a 2-D stress case, $\sigma_Z = \tau_{XY} = \tau_{YZ} = \tau_{ZX} = 0$; $\sigma_{eq} = \sqrt{\sigma_X^2 - \sigma_X\sigma_Y + \sigma_Y^2}$

$$\ldots\ldots(17.3)$$

σ_X and σ_Y can have mean and alternating values

Then, $\left(\sigma_{eq}\right)_m = \sqrt{\sigma_{Xm}^2 - \sigma_{Xm}\sigma_{Ym} + \sigma_{Ym}^2}$

and $\left(\sigma_{eq}\right)_a = \sqrt{\sigma_{Xa}^2 - \sigma_{Xa}\sigma_{Ya} + \sigma_{Ya}^2}$

are calculated and used in modified Goodman diagram for fluctuating normal stresses

17.4 FATIGUE DUE TO COMBINED BENDING AND TORSION

If a component is subjected to normal stress σ_X due to bending and shear stress τ_{XY} due to torsion, equivalent stress is obtained from eq (17.2) by substituting

$$\sigma_Y = \sigma_Z = \tau_{YZ} = \tau_{ZX} = 0$$

Then, eq(17.2) simplifies to $\sigma_{eq} = \sqrt{\sigma_X^2 + 3\tau_{XY}^2}$ $\ldots\ldots(17.4)$

Therefore, $\left(\sigma_{eq}\right)_m = \sqrt{\sigma_{Xm}^2 + 3\tau_{XYm}^2}$ and $\left(\sigma_{eq}\right)_a = \sqrt{\sigma_{Xa}^2 + 3\tau_{XYa}^2}$ are

used in modified Goodman diagram for fluctuating shear stresses.

17.5   SOME DESIGN CONSIDERATIONS TO MINIMISE FATIGUE DAMAGE

Best detail design requires diligent effort in fabrication and installation so that the airframe requires minimum maintenance and minimum down-time for inspection and replacement of damaged parts, mainly due to fatigue failure.

(a)   Avoid designs with no radii, short fillet radii and sharp bend radii. Drawings should clearly spell out rounding off all corners.

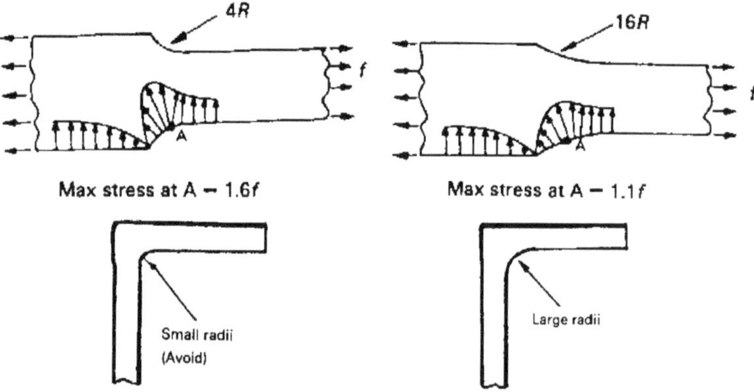

(b)   Avoid poor load distribution and superimposed stress concentrations in various forms

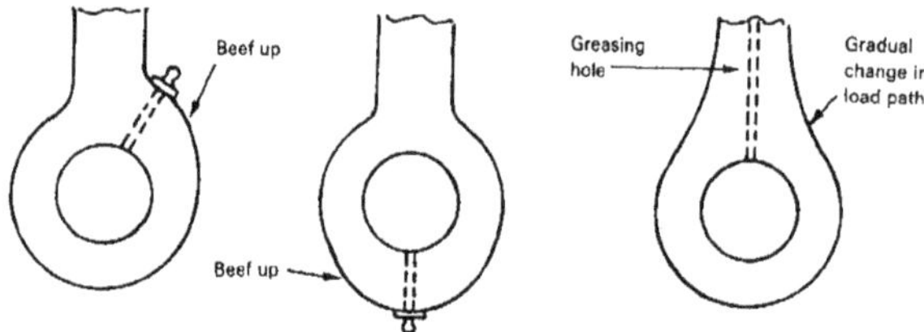

(c)   Radius for two changes of section should not be at the same location

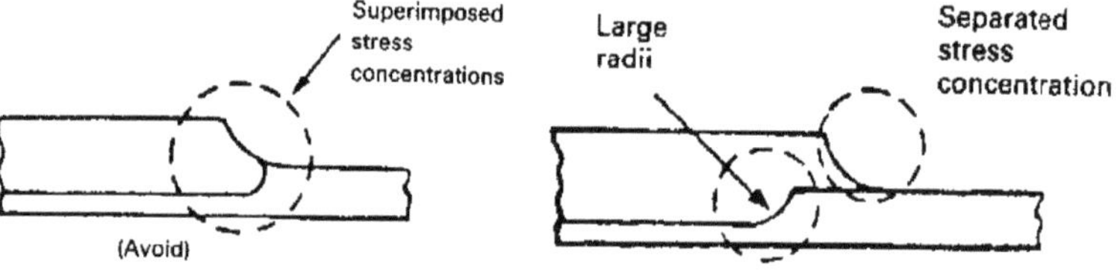

(d) Rough surface finish of a machined part, especially with tool marks normal to the direction of loading, and grain structure normal to the direction of loading should be avoided

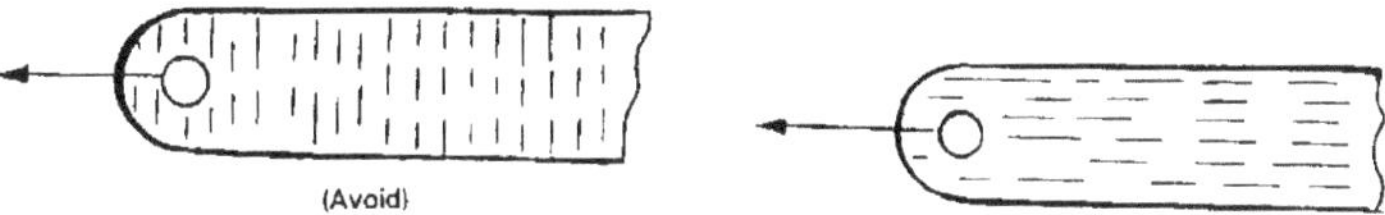

(e) Due to scatter in the experimental results, many regulations suggest use of a *scatter factor* of 3 on strength to account for fretting, clamped assembly stresses, size and surface effects (which are not accounted for, by the material S-N data)

17.6 DAMAGE PER FLIGHT

While rotating parts like motor shafts and turbine shafts are designed for infinite life (by limiting alternating stress to the endurance limit), structural components are designed for finite life since they experience limited number of stress cycles ($< 10^6$) during their operating life. If σ_m and σ_a are the mean and alternating stress values in one ground-air-ground cycle, and N is the number of cycles to failure for these stress levels from fatigue life curve, then safe life is N/f_s, where f_s is the life factor. Damage done during one cycle is f_s/N. With F_L as the factor for variability of loading, damage per flight from one ground-air-ground cycle is given by $(f_s \times f_L)/N$. Usual values of f_s and f_L are 3 and 1.5 respectively.

Part - III

Other Relevant Topics

MATERIALS OF AEROPLANE CONSTRUCTION

Properties like modulus of elasticity, fatigue strength, yield strength, resistance to corrosion, resistance to creep and corrosion, thermal conductivity over a large temperature range (from a maximum of about $60°C$ on the ground to a minimum of about $-55°C$ at high attitude) and compatibility with other materials as well as availability and ease of fabrication influence selection of materials for an aeroplane.

The materials used for the structure of an aeroplane must have a high ratio of strength to weight. The main groups of materials used for aeroplane structures are – wood, steel, light alloys of aluminium and magnesium, titanium and its alloys, plastics and composites. Each has its advantages and disadvantages and none of them can be considered ideal for the purpose.

As an illustration, let us take a simple bar subjected to uni-axial tension, uni-axial compression or buckling and ultimate bending. Using the applicable formulae, we can find relation between weights of two different materials for these three specific loads as

$$W_1/W_2 = (\rho_1/\rho_2) \times (\sigma_{t2}/\sigma_{t1})$$
$$W_1/W_2 = (\rho_1/\rho_2) \times (E_2/E_1)^{1/3}$$
$$W_1/W_2 = (\rho_1/\rho_2) \times (\sigma_{b2}/\sigma_{b1})^{1/2}$$

Similar relations can be found for other loads or other types of members. Since the allowable values of yield stress in tension, buckling stress in compression, shear stress in torsion, etc. of any two different materials do not have the same ratio, no single material is ideally suited for all types of loads.

Development of fracture mechanics by engineers concerned with airframe and turbine engineers helped choose a material based on its ability to withstand minor damage in service without endangering safety of the airframe. The residual strength after damage, described as toughness, is now uppermost in the designer's mind. Metallurgists have, therefore, changed compositions and treatment techniques to meet new toughness requirements. In addition, more refined techniques for damage detection have been developed.

The mechanical and physical properties of the material grain are not equal in all three directions (Ref Fig 18.1) – longitudinal (L), long transverse (LT) and short transverse (ST)

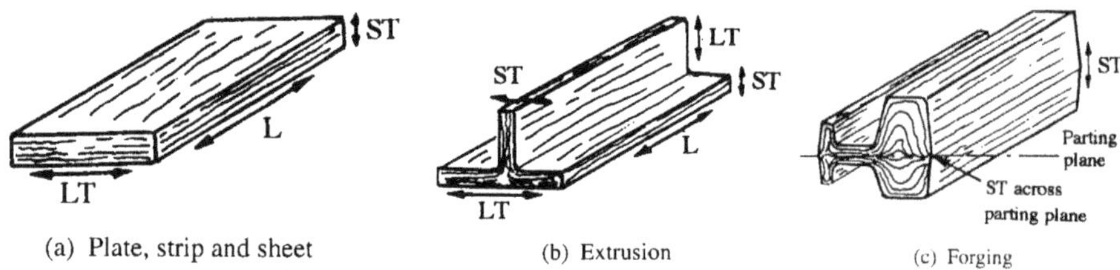

FIGURE 18.1 Material grain direction

18.1 WOOD

The first generation conventional powered aeroplane used wood and canvas. Spruce and birch were the most widely used timbers with tensile strengths of 70 MPa and 100 MPa; specific gravities of 0.4 and 0.63; and Young's modulii of 9 GPa and 14 GPa respectively. Even though these values compare favourably with heat treated aluminium alloys, natural wood has many disadvantages - absorbs moisture from atmospheric humidity and exhibits inconsistent properties; pronounced anisotropy caused by its grain structure and varying Young's modulus in the ratio of 150:1

Anisotropy was controlled with the introduction of plywood and development of synthetic resin adhesives. Non-availability of the right type of wood, during world war, forced engineers to consider metals for aeroplane structures. In addition, wood has some disadvantages of poor machinability for optimum shape, difficulty of fabrication (welding, riveting and forging not feasible) and larger volume for the same strength. With the increase in the payload and speed of aeroplanes, wood is no longer suitable for increased wing loadings and associated high stress concentration in most modern types of aeroplanes.

18.2 STEEL

The first all-metal aeroplane, using iron and steel, was constructed in 1915 in Germany. In spite of steel having advantages of high tensile strength and high modulus of elasticity, it has one serious disadvantage of high specific gravity - three times that of aluminium and ten times that of plywood. Since optimum weight is main criterion in any aeroplane, even at a higher cost, steel gave way to Aluminium alloys and composite materials in most of the aerospace applications. However, steel is specifically used where high strength, high stiffness and high wear resistance are required.

18.3 ALUMINIUM ALLOYS

Aluminium alloys (with different compositions of copper, magnesium, manganese, silicon, nickel and zinc) became popular for aeroplane structures, mainly because of their high strength to weight ratio. Other factors – introduction of extrusions in a wide range of sections and use of aluminium cladding to provide greater resistance to corrosion - helped in replacement of steel with aluminium alloys.

A patented material 'duralumin' containing 4% copper, 0.5% magnesium, 0.5% manganese, 0.3% silicon, 0.2% iron and the remainder aluminium was used for a long time in aeroplane structures. A second group of aluminium alloys have 1-2% Nickel, higher content of magnesium and variations in the amounts of copper, silicon, iron and the remainder aluminium was popular subsequently. Third and latest group, containing 2.5% copper, 5% zinc, 3% magnesium, and 1% nickel is significantly used in the present day aeroplanes. In modern versions, chromium is used in place of nickel. Frequently, special alloys are being developed for each particular aeroplane, such as Hiduminium (aluminium-copper-magnesium-nickel-iron alloy) for forged components in gas turbine aero engines of Concorde, for elevated temperature operation.

Use of aluminium-magnesium-silicon alloys has increased as they are cheaper than aluminium-copper alloys and reduce manufacturing costs, because of their weldability.

18.4 TITANIUM AND LITHIUM ALLOYS

The latest material to find general use in fuselage frames is titanium alloys and aluminium-lithium alloys, with high fracture toughness and high resistance to corrosion and crack propagation besides weldability. Main advantage of titanium alloys is in very high-speed aeroplane, in which the 'heat barrier' becomes an important problem. Titanium alloys are used in the early stages of compressor while nickel-based alloys are used in the hotter later stages. New fabrication processes, such as superplastic forming combined with diffusion bonding, enable large and complex components to be produced. Because of their high cost and higher density, they are not commonly used in structures.

18.5 COMPOSITE MATERIALS

The first important event was the development of glass fiber reinforced plastics in the early 1940s. A second important event was the discovery that short fibers or 'whiskers' are very strong (Ex. Carbon fibers have a tensile strength of about 2400 MPa and a modulus of elasticity of 400 GPa). Composite materials consist of laminas in which a stiff, high-strength filament (for example, glass or carbon fiber) is embedded in a matrix such as epoxy, polyester etc. Main advantages of composites are – saving in weight for the same strength and orientation of fibers in the direction of major loads at a particular point for optimum use of material.

The law of mixtures at the strain of composite fracture determines the strengths. The calculation is based on the conditions –

(i)   Load is transferred from the matrix to the filaments by shear stresses at their interface

(ii)   The filaments, matrix and the composite are elongated equally, according to Hooke's law

(iii)   All the filaments have the same strength, uniform size and shape and are fully bonded to the matrix

(iv)   The stress on the components is determined by the modulus and the strain.

For discontinuous fibers, since stresses are communicated between fibers through the matrix, the average stress is always less than that found in a continuous fiber. When the response of a composite is to be measured in terms of average stress and average strain, the material can be represented by an effective homogeneous but anisotropic material having the same average response Relationship between constituent properties and elastic modulii is reasonably well understood. However, when a fracture criterion is desired, an understanding of the average stress-strain response is no longer sufficient and consideration must be given to internal irregularities in the state of stress.

Since the components of the composite will of necessity possess different elastic and plastic properties, it is clear that there will be a rheological interaction at the interface when the composite is subjected to a loading system. This interaction will produce effects that may enhance the usefulness of the composite or may limit its usefulness.

18.5.1   PROPERTIES OF COMPOSITE MATERIALS

Composite materials may be described adequately as consisting of homogeneous and isotropic matrix in which particles of a second homogeneous and isotropic phase are dispersed. Assuming that the volume concentration of the particles is uniform and that the material may therefore be described as quasi-homogeneous and quasi-isotropic, it should be possible to calculate the properties of the composite material given the properties of composing elements, from the relation

$$\sigma_L = \sigma_f \times V_f + \sigma_m \times V_m$$

where, σ_f, V_f and σ_m, V_m are the tensile strength and volume fraction of fibers and matrix respectively, such that $V_f + V_m = 1$

and σ is the tensile strength of the composite material.

For a plate of composite material of thickness 't' and subjected to longitudinal load 'P_L', the above equation can also be expressed as

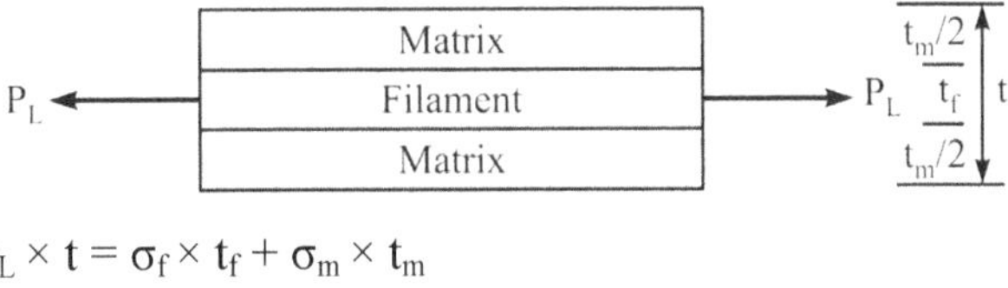

$$\sigma_L \times t = \sigma_f \times t_f + \sigma_m \times t_m$$

or $E_L = E_f \times (t_f/t) + E_m \times (t_m/t)$ since $\varepsilon_L = \varepsilon_f = \varepsilon_m$

This is called *law of mixtures*

For a plate of composite material of matrix thickness 't_m' and filament thickness 't_f' and subjected to transverse load 'P_T', the total elongation in the transverse direction is given by

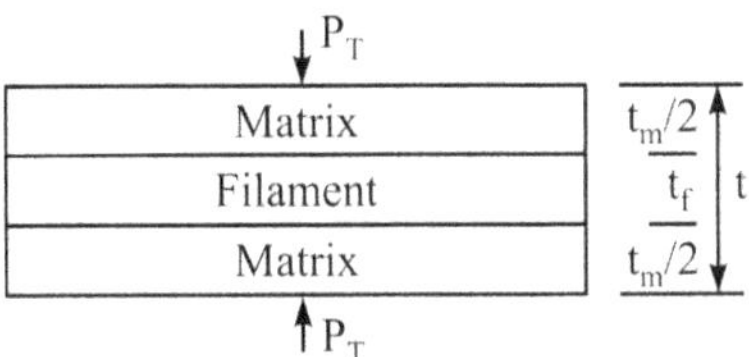

$$\varepsilon_T \times t = \varepsilon_f \times t_f + \varepsilon_m \times t_m$$

or $1/E_T = (1/E_f) \times (t_f/t) + (1/E_m) \times (t_m/t)$ since $\sigma_T = \sigma_f = \sigma_m$

Composite materials are anisotropic and so elastic modulii in longitudinal direction (E_L) and transverse (E_T) direction are different.

Major Poisson's ratio is given by $\nu_{LT} = \nu_f \times (t_f/t) + \nu_m \times (t_m/t)$

since $\delta_T = (\delta_f)_T + (\delta_m)_T$ due to P_L and $\delta_T = \nu_{LT} \times \varepsilon_L \times t$, ...

Minor Poisson's ratio is given by

$$\nu_{TL}/E_T = (\nu_f/E_f) \times (t_f/t) = (\nu_m/E_m) \times (t_m/t)$$

since $\delta_L = (\delta_f)_L = (\delta_m)_L$ due to P_T and $\delta_L = \nu_{TL} \times \varepsilon_T \times L$, ...

These two ratios are related by $\nu_{TL}/E_T = \nu_{LT}/E_L$

The shear modulus for the composite material is obtained from

$$t/G = t_m/G_m + t_f/G_f$$

18.6 SANDWICH CONSTRUCTION

In many aerospace vehicles, depending on the specific mission requirements of the vehicle, sandwich construction is employed. In this case, material of the inner and outer faces may be reinforced composites, titanium, aluminium etc.. while core materials of different properties (such as thermal insulators) may be used between the two faces, providing adequate stiffening, moment of inertia etc.

CHAPTER 19

FINITE ELEMENT METHOD

19.1 INTRODUCTION

Several methods, such as method of joints for trusses, simple theory of bending, simple theory of torsion, analyses of cylinders and spheres for axi-symmetric pressure load etc., are available for designing simple components of a structure. These methods try to obtain exact solutions of second order partial differential equations and are based on several assumptions on sizes of the components, loads, end conditions, material properties, likely deformation pattern etc. Also, these methods are not amenable for generalisation (for structures having different types of elements) and effective utilisation of the computer for repetitive jobs.

Strength of materials approach deals with a single beam member for different loads and end conditions (free, simply supported and fixed). In a space frame involving many such beam members, each member is analysed independently by an assumed distribution of loads and end conditions.

For example, in a 3-member structure (portal frame) shown in Fig. 19.1, the (horizontal) beam is analysed for deflection and bending stress by strength of materials approach considering its both ends simply supported or fixed. The load and moment reactions obtained at the ends are then used to calculate the deflections and stresses in the two columns separately.

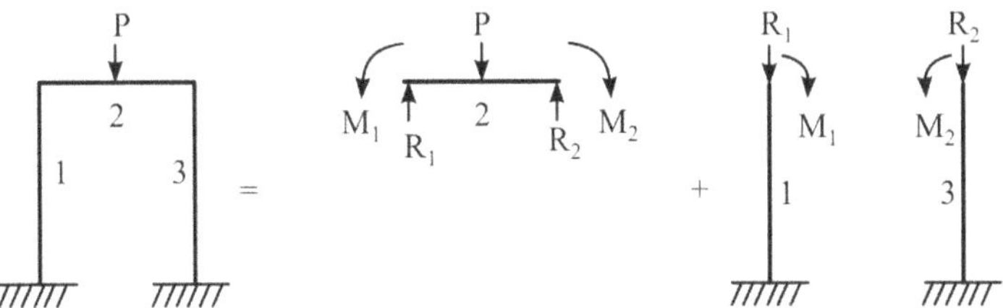

FIGURE 19.1 Analysis of a simple frame by strength of materials approach

Simple supports for the beam imply that the columns do not influence slope of the beam at its free ends (valid when bending stiffness of columns = 0 or the column is highly flexible). Fixed supports for the beam imply that the slope of the beam at its ends is zero (valid when bending stiffness of columns = ∞ or the column is extremely rigid). But, the ends of the horizontal beam are neither simply supported nor fixed. Supporting members at the two ends of the beam have finite, non-zero stiffness value and the maximum deflection of the beam depends upon the relative stiffness of the beam and the supporting columns.

For example, in a beam of length 'L', modulus of elasticity 'E', moment of inertia 'I' subjected to a uniformly distributed load of 'p' (Refer Fig. 19.2).

$$\text{Deflection,} \quad \delta = \frac{5\,pL^4}{384\,EI} \quad \text{with simple supports at its two ends (case - a)}$$

$$\delta' = \frac{pL^4}{384\,EI} \quad \text{with fixed supports at its two ends (case -b)}$$

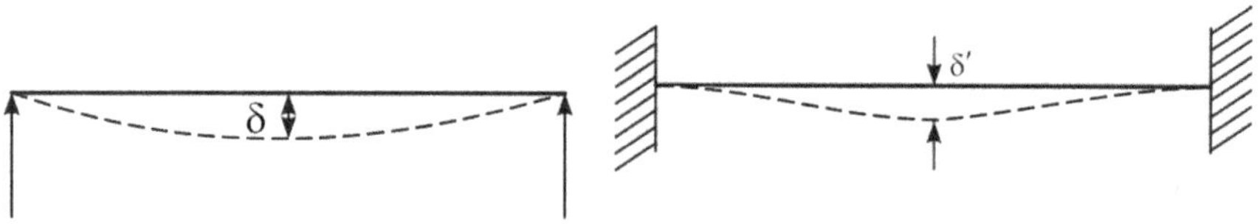

Case (a) Simple supports   Case (b) Fixed supports

FIGURE 19.2 Deflection of a beam with different end conditions

Deflection of the beam of a portal frame with I_B and I_C as Moments of inertia for beam & columns respectively, when analysed along with columns by FEM, gives

$$\delta_{max} = 0.5268\ \delta \text{ or } 2.634\ \delta', \text{ when } I_B = I_C$$

$$= 0.3011\ \delta \text{ or } 1.5055\ \delta', \text{ when } 5\ I_B = I_C$$

and $\qquad = 0.8224\ \delta \text{ or } 4.1118\ \delta', \text{ when } I_B = 5\ I_C$

All the three deflection values clearly indicate presence of columns with finite and non-zero stiffness and, hence, the deflection values are in between those of beam with free ends and beam with fixed ends $\qquad (\delta > \delta_{max} > \delta')$.

Thus, designing a single beam member of a frame leads to under-designing if fixed end conditions are assumed while it leads to over-designing if simple supports are assumed at its ends. Simply supported end conditions are, therefore, normally used for a conservative design in the conventional approach and, hence, results in higher factor of safety.

The individual member method was acceptable for civil and mechanical structures, where weight of the structure is not a serious constraint. ***Optimum*** *design is achieved by analysing the entire structure considering finite stiffness of supporting members.* Analysis of the complete structure was necessitated by the need for airplanes with minimum factor of safety (and, hence, minimum weight), during World War-II.

Finite element method, popular as FEM, was developed initially as Matrix method of structural analysis for discrete structures like trusses and frames. Subsequently, this method is extended for continuum structures to get better estimation of stresses and deflections even in components of variable cross-section as well as with non-homogeneous and non-isotropic materials.

19.2　APPROXIMATE METHOD VS EXACT METHOD

An analytical solution is a mathematical expression that gives the values of the desired unknown quantity at any location of a body and hence is valid for an infinite number of points in the component. However, it is not possible to obtain analytical mathematical solutions for many engineering problems. Numerical methods provide approximate but acceptable solutions (with reasonable accuracy) for the unknown quantities – only at discrete or finite number of points in the component. Approximation is carried out in two stages:

(a) In the formulation of the mathematical model, w.r.t. the physical behaviour of the component. Example : Approximation of joint with multiple rivets at the junction of any two members of a truss as a pin joint, assumption that the joint between a column and a beam behaves like a simple support for the beam,.... The results are reasonably accurate far away from the joint.

(b) In obtaining numerical solution to the simplified mathematical model. The methods usually involve approximation of a functional (such as Potential energy) in terms of unknown functions (such as displacements) at finite number of points. There are three broad categories:

　(i) **Weighted residual methods** such as Galerkin method, Collocation method, Least squares method, etc.

　(ii) **Variational methods** such as Rayleigh-Ritz method, FEM

　(iii) ***Principle of minimum potential energy -*** For a component in static equilibrium, this principle helps in the evaluation of unknown displacements of deformable solids (discrete and continuum structures).

The Rayleigh-Ritz method and minimum potential energy approach are now of only academic interest. For a big problem, it is difficult to deal with a polynomial having as many coefficients as the number of DOF or nodal displacements. FEM is a better generalization of these methods and extends beyond the discrete structures. Rayleigh-Ritz method of choosing a polynomial for displacement field and evaluating the coefficients for minimum potential energy is used in FEM, at the individual element level to obtain element stiffness matrix (representing load-displacement relations) and assembled to analyse the structure.

19.3   PRINCIPLE OF FEM

In FEM, actual component is replaced by a simplified model, identified by a finite number of *elements* connected at common points called *nodes*, with an assumed behaviour or response of each element to the set of applied nodal loads, and evaluating the unknown field variable (displacement, stress) at these finite number of points.

FEM approach, based on minimum potential energy theorem, *converges to the correct solution from a higher value* as the number of elements in the model increases. Thus, results obtained by FEM are useful (on a more conservative side) even with lesser number of elements due to limitations of computer memory and cost.

Finite Element Analysis (FEA) based on FEM is a simulation, not reality, applied to the mathematical model. The error in solution can result from three different sources.

Modeling error – associated with the approximations made to the real problem

Discretisation error – associated with type, size and shape of finite elements used to represent the mathematical model; can be reduced by modifying mesh

Numerical error – based on the algorithm used and the finite number of digits used to represent data in the computer; most software use double precision for data

It is entirely possible for an unprepared software user to misunderstand the problem, prepare the wrong mathematical model, discretise it inappropriately, fail to check computed output and yet accept nonsensical results. Thus, *the accuracy of FEA depends on the knowledge of the analyst in understanding the problem correctly*.

19.4   CLASSIFICATION OF FEM

The basic problem in any engineering design is to evaluate displacements, stresses and strains in any given structure under different loads and boundary conditions. Several approaches of Finite Element Analysis have been developed to meet the needs of specific applications. The common methods are:

- ***Displacement method***: Here the structure is subjected to applied loads and/or specified displacements. The primary unknowns are displacements, obtained by inversion of the stiffness matrix, and the derived unknowns are stresses and strains. Stiffness matrix for any element can be obtained by variational principle, based on minimum potential energy of any stable structure and, hence, this is the most commonly used method.

- ***Force method:*** Here the structure is subjected to applied loads and/or specified displacements. The primary unknowns are member forces, obtained by inversion of the flexibility matrix, and the derived unknowns are nodal displacements, stresses and strains. Calculation of flexibility matrix is possible only for discrete structural elements (such as trusses, beams and piping) and hence, this method is limited in the early analyses of discrete structures and in piping analysis

- ***Mixed method:*** Here the structure is subjected to applied loads and/or specified displacements. The method deals with large stiffness coefficients as well as very small flexibility coefficients in the same matrix. Analysis by this method leads to numerical errors and is not possible except in some very special cases.

- ***Hybrid method:*** Here the structure is subjected to applied loads and stress boundary conditions. This deals with special cases, such as airplane door frame which should be designed for stress-free boundary, so that the door can be opened during flight, in cases of emergencies.

Displacement method is the most common method and is suitable for solving most of the engineering problems. The discussion in the remaining chapters is confined to displacement method.

19.5 BASIC STEPS IN ANALYSIS BY FEM

Based on the relative dimensions of the element, the individual elements can be broadly classified as 1-D, 2-D and 3-D elements. The load-displacement relationships of these elements depend on the nature of loads (axial/in-plane loads, torsion or bending loads) and are calculated using variational principle. Some such elements and their degrees of freedom at each node in element (or local) coordinate system are given below.

	Axial/In-plane loads	**Bending (Normal loads and/or moments)**
1-D	Spar or Truss (1 DOF / node)	Beam (2 DOF/node for bending in 1-plane)
2-D	Plane stress/Plane strain/ Axisymmetric (2 DOF/node)	Plate bending (3 DOF/node) Thin shell (6 DOF/node)
3-D	Solid (3 DOF/node)	Thick shell (6 DOF/node)

Analysis by FEM involves the following steps.

- Calculation of element stiffness matrix in local (or element) coordinate system

- Transformation of the stiffness matrix into global coordinate system, common to all elements

- Assembling element stiffness matrices such that stiffness coefficients at any node in a particular DOF from different elements meeting at that node are added

- The assembled stiffness matrix is square, symmetric and singular. Application of boundary conditions (in the form of zero or specified displacements) reduces the size of the stiffness matrix and makes it non-singular.

- Solution for the primary unknowns (nodal displacements, in the displacement method) are evaluated using one of the many numerical techniques available

- Secondary unknowns (stresses, strains and reactions at supports, in displacement method) are evaluated using appropriate element matrices relating them with nodal displacements

Each of these steps is explained here in detail.

19.6   ANALYSIS OF DISCRETE STRUCTURES

A discrete structure is assembled from a number of easily identifiable 1-D elements like spars, beams whose cross sectional dimensions are very small compared to their lengths. So, the displacement is taken as a function of x, along the axis of the member. Nodes are chosen at the junctions of two or more discrete members, at junctions of two different materials, at points of change of cross section or at points of load application. The solution obtained in most of these cases exactly matches with closed form solution.

19.6.1   CALCULATION OF ELEMENT STIFFNESS MATRIX

1-D elements are broadly classified based on the load applied on them as spar or truss element for axial load, torsion element for torque load and beam element for bending in one or two planes through neutral axis/plane. For such discrete structural elements, stiffness matrix can be calculated by direct method (strength of materials approach) or by variational principle. Second approach is more general and common to all elements – discrete or continuum. This method is explained here.

Stiffness matrix of each element is calculated, using the principle of minimum potential energy which states that "Every component, subjected to some external loads, reaches a stable equilibrium condition when its potential energy is minimum". So, the problem lies in identifying the set of displacements at various points in the component which ensures that the potential energy of the component is minimum.

This is analogous to the problem in *variational calculus* of finding a stationary value y(x) such that the functional (function of functions)

$$I = \int_{x_1}^{x_2} F\left(x, y, \frac{dy}{dx}\right) dx \qquad .....(19.1)$$

is rendered stationary. Integral I is stationary when its first variation vanishes

i.e., $\qquad \delta I = 0$ $\qquad\qquad\qquad\qquad\qquad\qquad .....(19.2)$

In *Rayleigh-Ritz method*, which is the basis for FEM, a mathematical expression in the form of a power series in x is assumed for the unknown function y(x).

Then, eq. (19.2) becomes $\quad \dfrac{\partial I}{\partial a_i} = 0 \quad$ for $\ i = 0,1,...n$ $\qquad .....(19.3)$

where $a_o, a_i,...a_n$ are the coefficients of the assumed power series.

Finite element method is based on the variational principle where the functional I is the potential energy of the system with nodal displacements as the independent variables y(x) and strains as the functions (dy/dx) of the independent variables. This method leads to an approximate solution. Potential energy is an *extensive property* i.e. the energy of the entire component is the sum of the energy of its individual sub regions (or elements). Hence eq.(19.2) can be written as

$$\delta I = \Sigma \, \delta I_e = 0$$

Since the number and size of the elements are arbitrary, this relation is satisfied only when

$$\delta I_e = 0 \qquad\qquad\qquad .....(19.4)$$

In using this method, y(x) must be kinematically admissible i.e., y(x) must be selected so that it satisfies the displacement boundary conditions prescribed for the problem. Choice of a function for the entire component satisfying this condition becomes difficult for complex problems. Finite element method overcomes this difficulty by relating the primary unknown function to the individual element, rather than to the total problem. Hence, geometry of the overall component and the system boundary conditions are of no concern when choosing the function.

For the individual element,

$$\delta I_e = \delta W_{ext} - \delta U = \{\partial u_e\}^T \{P_e\} - \int_V \{\delta\varepsilon_e\}^T \{\sigma\} dV \qquad .....(19.5)$$

$$= \{\delta u_e\}^T (\{P_e\} - [K_e]\{u_e\}) = 0$$

where, $[K_e]$ is the stiffness matrix of the element

Since $\{\delta u_e\}$ represents the arbitrary nodal values of displacements, they can not be identically zero in a loaded component for a non-trivial solution.

$$\therefore \qquad \{P_e\} - [K_e]\{u_e\} = 0 \quad or \quad \{P_e\} = [K_e]\{u_e\} \qquad .....(19.6)$$

(a) *Conditions to be satisfied by the Displacement function:* In the displacement method, calculation of stiffness matrix for an element starts with an assumed displacement function over the element in each degree of freedom, usually in the form of a polynomial. By substituting nodal coordinate values in these polynomials, the unknown constants in the polynomial can be evaluated in terms of nodal displacements. This condition necessitates choosing displacement polynomial with as many coefficients as the number of nodes for that element.

The function shall be continuous over the entire element with no singularities and easily differentiable to obtain strains for calculation of potential energy. The polynomial should be symmetric in terms of the global coordinate axes, to ensure geometric isotropy, so that results are not influenced by the coordinate system chosen by the analyst.

Strains in the element are obtained as derivatives of the displacement polynomial, and are thus expressed in terms of the nodal displacements. Stresses are expressed in terms of strains, using the appropriate stress-strain relationship. By evaluating work done by the external forces and the change in internal strain energy of the element and applying variational principle, load-displacement relationships of the element in terms of stiffness coefficients are obtained. They represent a system of simultaneous equations in terms of nodal loads and nodal displacements.

(b) *Spar element Stiffness matrix:* A spar element is subjected to axial load only and therefore has one degree of freedom (axial displacement) per node, along the axis of the spar element. Variation of axial displacement u(x) between the two end nodes is represented by a linear relationship in the form of a polynomial with two constants.

Let $u(x) = a_1 + a_2 x = \begin{bmatrix} 1 & x \end{bmatrix} \begin{Bmatrix} a_1 \\ a_2 \end{Bmatrix} = \{f(x)\}^T \{a\}$ .....(19.7)

Choosing node i as the origin of local coordinate system for this element with X-axis along the axis of the element and substituting the values of x for the two end points of the spar element, (x = 0 at node i and x = L at node j), nodal displacement vector $\{u_e\}$ or $[u_i \ u_j]^T$ can be written as

$$\begin{Bmatrix} u_i \\ u_j \end{Bmatrix} = \begin{bmatrix} 1 & 0 \\ 1 & L \end{bmatrix} \begin{Bmatrix} a_1 \\ a_2 \end{Bmatrix}$$

or $\{u_e\} = [G] \{a\}$ and $\{a\} = [G]^{-1} \{u_e\}$ .....(19.8)

Solving for the coefficients $\{a\}$ from eq. (19.8) and substituting in eq. (19.7),

$$u(x) = \{f(x)^T\} [G]^{-1} \{u_e\}$$

$$= \begin{bmatrix} 1 & x \end{bmatrix} \begin{bmatrix} 1 & 0 \\ \dfrac{-1}{L} & \dfrac{1}{L} \end{bmatrix} \begin{Bmatrix} u_i \\ u_j \end{Bmatrix}$$

$$= \begin{bmatrix} 1 - \dfrac{x}{L} & \dfrac{x}{L} \end{bmatrix} \{u_e\} = [N]^T \{u_e\}$$ .....(19.9)

Strain, $\{\varepsilon\} = \dfrac{du}{dx} = \{f'(x)\}^T [G]^{-1} \{u_e\} = [B]\{u_e\}$ .....(19.10)

where, $[B] = \{f'(x)\}^T [G]^{-1} = \begin{bmatrix} 0 & 1 \end{bmatrix} \begin{bmatrix} 1 & 0 \\ \dfrac{-1}{L} & \dfrac{1}{L} \end{bmatrix} = \begin{bmatrix} -1/L & 1/L \end{bmatrix}$ .....(19.11)

and $\{\delta\varepsilon\} = [B] \{\delta u_e\}; \quad \{\delta\varepsilon\}^T = \{\delta u_e\}^T [B]^T$

Stress, $\{\sigma\} = [D] \{\varepsilon\} = [D] [B] \{u_e\}$

Here, $[D] = E$, since only one axial stress component is relevant for a 1-D spar element

From eq. (19.5),

$$\{\delta u_e\}^T \{P_e\} - \int_v \{\delta\varepsilon_e\}^T \{\sigma\} dv = 0$$

$$\{\delta u_e\}^T \{P_e\} - \int_v \{\delta u_e\}^T [B]^T [D][B]\{u_e\} dv = 0$$

$$\{\partial u_e\}^T \left[\{P_e\} - \left(\int_V [B]^T [D][B]dV \right)\{u_e\} \right] = 0$$

$$\{\delta u_e\}^T \left(\{P_e\} - [K_e]\{u_e\} \right) = 0$$

Since $\{\delta u_e\}$ cannot be zero for a non-trivial solution,

$$\{P_e\} = [K_e]\{u_e\} \qquad\qquad .....(19.12)$$

where, $\quad [K_e] = \int_V [B]^T [D][B]dv = \iiint [B]^T E[B]dx\ dy\ dz$

$$= AE\int_0^L [B]^T [B]dx$$

since, $\quad$ [B] is not a function of y or z $\quad$ and $\quad \iint dy\ dz = A$

Therefore, $\quad [K_e] = AE\int_0^L \begin{Bmatrix} -1/L \\ 1/L \end{Bmatrix} [-1/L \quad 1/L]dx = \dfrac{AE}{L}\begin{bmatrix} 1 & -1 \\ -1 & 1 \end{bmatrix}$ .....(19.13)

(c) ***Beam Element Stiffness Matrix:*** Let us consider bending of a typical beam of uniform cross section in a plane perpendicular to its axis, due to load and moment applied at its two ends. From strength of materials, it is well known that its deformed shape is a curve and can not be represented by a linear function. Considering 2nd order polynomial having three coefficients, with deflection alone as the nodal degree of freedom fails to express the three coefficients in terms of the two nodal deflections. Also, natural boundary conditions at the ends of the beam may include not only deflection but also the slope. Hence, a cubic polynomial is generally used and its four unknown coefficients are represented in terms of deflection and slope (first derivative of deflection) at each end.

If X-Y is the plane of bending, P_Y and M_Z are the loads applied while v and θ_z are the deflection and slope in the plane of bending,

Let $\qquad v(x) = a_1 + a_2 \times x + a_3 \times x^2 + a_4.x^3 = [\ 1\ \ x\ \ x^2\ \ x^3\ ]\ \{a\}$

$$.....(19.14)$$

and $\qquad \theta_z = \dfrac{dv}{dx} = [0\ \ 1\ \ 2x\ \ 3x^2]\{a\}$

or $\qquad \begin{Bmatrix} v \\ \theta_z \end{Bmatrix} = \{u\} = \{f(x)\}^T\ \{a\} \qquad\qquad .....(19.15)$

Choosing node i as the origin of local coordinate system for this element with X-axis along the axis of the element and substituting $x = 0$ at node i and $X = L$ at node j, we get the nodal displacement vector,

$$\begin{Bmatrix} v_i \\ (\theta_z)_i \\ v_j \\ (\theta_z)_j \end{Bmatrix} = \begin{bmatrix} 1 & 0 & 0 & 0 \\ 0 & 1 & 0 & 0 \\ 1 & L & L^2 & L^3 \\ 0 & 1 & 2L & 3L^2 \end{bmatrix} \begin{Bmatrix} a_1 \\ a_2 \\ a_3 \\ a_4 \end{Bmatrix} \qquad \text{or} \qquad \{u_e\} = [G]\{a\} \quad\text{.....(19.16)}$$

Solving for the coefficients $\{a\}$ and substituting in eq. (19.15)

$$u = \{f(x)\}^T \{a\} = \{f(x)\}^T [G]^{-1} \{u_e\} \qquad\qquad \text{.....(19.17)}$$

From theory of bending,

$$\text{strain,} \quad \varepsilon = \frac{\sigma}{E} = \frac{y}{R} = y \times \frac{d\theta}{dx} = y\left(\frac{d^2 v}{dx^2}\right) = [B]\{u_e\} \qquad \text{.....(19.18)}$$

where, $\dfrac{d^2 v}{dx^2} = \begin{bmatrix} 0 & 0 & 2 & 6x \end{bmatrix}\{a\} = \begin{bmatrix} 0 & 0 & 2 & 6x \end{bmatrix}[G]^{-1}\{u_e\}$

and so, $[B] = y\begin{bmatrix} 0 & 0 & 2 & 6x \end{bmatrix} \begin{bmatrix} 1 & 0 & 0 & 0 \\ 0 & 1 & 0 & 0 \\ -3/L^2 & -2/L & 3/L^2 & -1/L \\ 2/L^3 & 1/L^2 & -2/L^3 & 1/L^2 \end{bmatrix}$

$$= (y/L^3)\begin{bmatrix} 6(2x-L) & 2L(3x-2L) & -6(2x-L) & 2L(3x-L) \end{bmatrix} \quad\text{.....(19.19)}$$

Then, $\quad \{P_e\} = [K_e]\{u_e\} \qquad\qquad\qquad\qquad\qquad \text{.....(19.20)}$

where, $\quad [K_e] = \int\limits_V [B]^T E[B]\, dv = \iiint [B]^T E[B]\, dx\, dy\, dz$

$$[K_e] = \left(\frac{E}{L^6}\right)\int\limits_A y^2 \,(dy\,dz)\int\limits_L \begin{Bmatrix} 6(2x-L) \\ 2L(3x-2L) \\ -6(2x-L) \\ 2L(3x-L) \end{Bmatrix}\begin{bmatrix} 6(2x-L) & 2L(3x-2L) & -6(2x-L) & 2L(3x-L) \end{bmatrix} dx$$

$$= \frac{EI_Z}{L^3}\begin{bmatrix} 12 & 6L & -12 & 6L \\ 6L & 4L^2 & -6L & 2L^2 \\ -12 & -6L & 12 & -6L^2 \\ 6L & 2L^2 & -6L & 4L^2 \end{bmatrix} \qquad\qquad \text{.....(19.21)}$$

since $\displaystyle\int\limits_A y^2\,(dy.dz) = I_Z$

19.6.2 TRANSFORMATION MATRIX

Stiffness matrix and load vector of any element are initially derived in the *local coordinate system*, with its x-axis along the element, and can vary from one element to another. A *global coordinate system* is common to all the elements. If different elements have different local coordinate systems, stiffness coefficients relating nodal load vector and nodal displacement vector can not be numerically added together unless directions of load and displacements of different elements joining at a common node coincide i.e., sum of two vectors is equal to their algebraic sum only when the vectors are collinear. If the local coordinate system of an element is inclined to the global coordinate system at an angle θ, then transformation of load vector, displacement vector and the stiffness matrix are to be carried out before they are assembled with other elements.

Transformation of spar element stiffness matrix in 2-D plane

For a spar element, if P_i, P_j, u_i and u_j represent axial load and displacement values in the local or element coordinate system at nodes i and j and $[K_e]$ is the 2×2 stiffness matrix of the element and $\left(P_X'\right)_i$, $\left(P_Y'\right)_i$, u_i', v_i', $\left(P_X'\right)_j$, $\left(P_Y'\right)_j$, u_j'

and v_j' are the components of axial load and displacement along global x and y axes at nodes i and j then $[K'_e]$, stiffness matrix of the element in the global or structure coordinate system, is derived below.

$$P_i = (P_X')_i \cos\theta + (P_Y')_i \sin\theta \qquad u_i = u'_i \cos\theta + v'_i \sin\theta$$

$$P_j = (P_X')_j \cos\theta + (P_Y')_j \sin\theta \qquad u_j = u'_j \cos\theta + v'_j \sin\theta$$

These relations can be expressed in matrix form as

$$\begin{Bmatrix} P_i \\ P_j \end{Bmatrix} = \begin{bmatrix} \cos\theta & \sin\theta & 0 & 0 \\ 0 & 0 & \cos\theta & \sin\theta \end{bmatrix} \begin{Bmatrix} \left(P_X'\right)_i \\ \left(P_Y'\right)_i \\ \left(P_X'\right)_j \\ \left(P_X'\right)_j \end{Bmatrix}$$

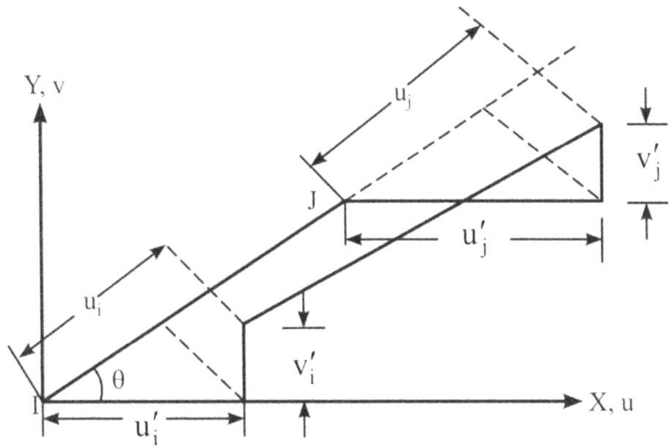

or $\{P_e\} = [T_e]\{P'_e\}$ and similarly, $\{u_e\} = [T_e]\{u'_e\}$ ...(19.22)

These can also be expressed as

$$\{P'_e\} = [T_e]^T \{P_e\} \text{ and } \{u'_e\} = [T_e]^T \{u_e\}$$

$\{P_e\} = [K_e]\{u_e\}$ in local coordinate system can now be written in global coordinate system as

$$[T_e]\{P'_e\} = [K_e][T_e]\{u'_e\}$$

or $$\{P'_e\} = [T_e]^T [K_e][T_e]\{u'_e\} = [K'_e]\{u'_e\}$$

where, $[K'_e] = [T_e]^T [K_e][T_e]$

or

$$[K'_e] = \frac{AE}{L}\begin{bmatrix} \cos^2\theta & \cos\theta\sin\theta & -\cos^2\theta & -\cos\theta\sin\theta \\ \cos\theta\sin\theta & \sin^2\theta & -\cos\theta\sin\theta & -\sin^2\theta \\ -\cos^2\theta & -\cos\theta\sin\theta & \cos^2\theta & \cos\theta\sin\theta \\ -\cos\theta\sin\theta & -\sin^2\theta & \cos\theta\sin\theta & \sin^2\theta \end{bmatrix}$$

$$.....(19.23)$$

By substituting $l = \dfrac{(x_2 - x_1)}{L} = \cos\theta$ and $m = \left(\dfrac{y_2 - y_1}{L}\right) = \sin\theta$

where, $L = \sqrt{(x_2 - x_1)^2 + (y_2 - y_1)^2}$

Eq.(19.23) can also be written in the form

$$[K'_e] = \frac{AE}{L} \begin{bmatrix} l^2 & lm & -l^2 & -lm \\ lm & m^2 & -lm & -m^2 \\ -l^2 & -lm & l^2 & lm \\ -lm & -m^2 & lm & m^2 \end{bmatrix} \qquad .....(19.24)$$

Transformation of spar element stiffness matrix in 3-D space

For a spar element arbitrarily oriented in 3-D space, a similar transformation matrix can also be derived and the stiffness matrix in 3-D space can be written using

$$l = \frac{(x_2 - x_1)}{L}; \quad m = \frac{(y_2 - y_1)}{L} \quad \text{and} \quad n = \frac{(z_2 - z_1)}{L}$$

where, $L = \sqrt{(x_2 - x_1)^2 + (y_2 - y_1)^2 + (z_2 - z_1)^2}$

in the form $[K'_e] = \dfrac{AE}{L} \begin{bmatrix} l^2 & lm & ln & -l^2 & -lm & -ln \\ lm & m^2 & mn & -lm & -m^2 & -mn \\ ln & mn & n^2 & -ln & -mn & -n^2 \\ -l^2 & -lm & -ln & l^2 & lm & ln \\ -lm & -m^2 & -mn & lm & m^2 & mn \\ -ln & -mn & -n^2 & ln & mn & n^2 \end{bmatrix}$ $\qquad .....(19.25)$

19.6.3 ASSEMBLING ELEMENT STIFFNESS MATRICES

In a truss having three spar elements connecting nodes 1-2, 2-3 and 3-1, the element stiffness matrices (each of 4 × 4) after transformation to global coordinate system can be obtained from eq. (19.13) and eq. (19.24) using appropriate values of l and m. Then,

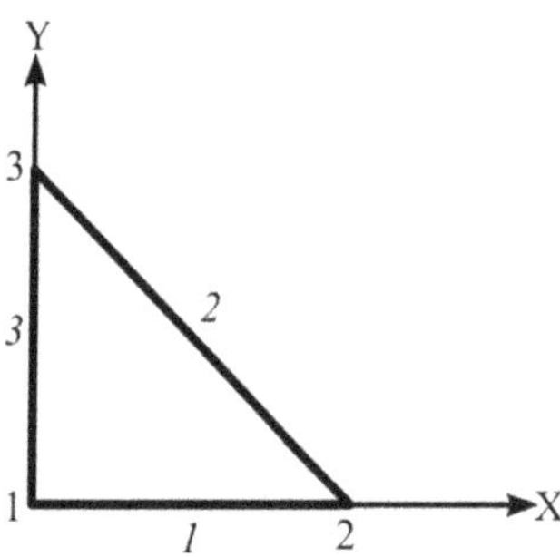

For element-1 joining nodes 1 and 2, the local coordinate system coincides with global coordinate system ($l = 1$, m = 0) and, so, displacement components v_1 and v_2 are zero.

Therefore,

$$\begin{Bmatrix} (P_X)_1 \\ (P_Y)_1 \\ (P_X)_2 \\ (P_Y)_2 \end{Bmatrix} = \begin{bmatrix} (k_{11})_1 & 0 & (k_{13})_1 & 0 \\ 0 & 0 & 0 & 0 \\ (k_{31})_1 & 0 & (k_{33})_1 & 0 \\ 0 & 0 & 0 & 0 \end{bmatrix} \begin{Bmatrix} u_1 \\ v_1 \\ u_2 \\ v_2 \end{Bmatrix}$$

For element-2 joining nodes 2 and 3 and inclined to the global axes, local coordinate system is different from the global coordinate system ($l \neq 0$, m $\neq 0$).

Therefore,

$$\begin{Bmatrix} (P_X)_2 \\ (P_Y)_2 \\ (P_X)_3 \\ (P_Y)_3 \end{Bmatrix} = \begin{bmatrix} (k_{11})_2 & (k_{12})_2 & (k_{13})_2 & (k_{14})_2 \\ (k_{21})_2 & (k_{22})_2 & (k_{23})_2 & (k_{24})_2 \\ (k_{31})_2 & (k_{32})_2 & (k_{33})_2 & (k_{34})_2 \\ (k_{41})_2 & (k_{42})_2 & (k_{43})_2 & (k_{44})_2 \end{bmatrix} \begin{Bmatrix} u_2 \\ v_2 \\ u_3 \\ v_3 \end{Bmatrix}$$

For element-3 joining nodes 1 and 3, the local coordinate system is perpendicular to the global coordinate system ($l = 0$, m $= 1$) and so displacement components u_1 and u_3 are zero.

$$\begin{Bmatrix} (P_X)_1 \\ (P_Y)_1 \\ (P_X)_3 \\ (P_Y)_3 \end{Bmatrix} = \begin{bmatrix} 0 & 0 & 0 & 0 \\ 0 & (k_{22})_3 & 0 & (k_{24})_3 \\ 0 & 0 & 0 & 0 \\ 0 & (k_{42})_3 & 0 & (k_{44})_3 \end{bmatrix} \begin{Bmatrix} u_1 \\ v_1 \\ u_3 \\ v_3 \end{Bmatrix}$$

Then, the process of assembling element stiffness matrices involves combining the nodal stiffness values of all the elements joining at every common node so that the order of the assembled stiffness matrix equals the total number of degrees of freedom of the structure.

$$\begin{Bmatrix} (P_X)_1 \\ (P_Y)_1 \\ (P_X)_2 \\ (P_Y)_2 \\ (P_X)_3 \\ (P_Y)_3 \end{Bmatrix} = \begin{bmatrix} (k_{11})_1 & 0 & (k_{13})_1 & 0 & 0 & 0 \\ 0 & (k_{22})_3 & 0 & 0 & 0 & (k_{24})_3 \\ (k_{31})_1 & 0 & (k_{33})_1 + (k_{11})_2 & (k_{12})_2 & (k_{13})_2 & (k_{14})_2 \\ 0 & 0 & (k_{21})_2 & (k_{22})_2 & (k_{23})_2 & (k_{24})_2 \\ 0 & 0 & (k_{31})_2 & (k_{32})_2 & (k_{33})_2 & (k_{34})_2 \\ 0 & (k_{24})_3 & (k_{41})_2 & (k_{42})_2 & (k_{43})_2 & (k_{44})_2 + (k_{44})_3 \end{bmatrix} \begin{Bmatrix} u_1 \\ v_1 \\ u_2 \\ v_2 \\ u_3 \\ v_3 \end{Bmatrix}$$

or $\quad \{P\} = [K]\{u\}$

which represents a set of n simultaneous equations, where n is the total number of degrees of freedom in the structure. In the case of structures, the assembled stiffness matrix is symmetric, singular and positive definite. Therefore, solution for primary unknowns by inversion of the stiffness matrix is not possible at this stage.

19.6.4 BOUNDARY CONDITIONS

The singularity of the matrix indicates possibility of rigid body movement of the structure in different directions and hence the possibility of infinite solutions for the unknown nodal displacements. For example, in a 2-node spar element, displacement at node j of $u_j = u_i + \delta$, where u_i is displacement at node i which can have any arbitrary value. Strain, stress and, hence, potential energy in this element are functions of δ and are independent of u_i. Boundary conditions, in terms of fixed degrees of freedom or known values of displacements at some points of the structure, are therefore applied. In some structures, where no part of the structure is fixed, it is possible to apply different boundary conditions. Each solution gives displacements at other points in the structure, with reference to the chosen fixed points.

(a) *Elimination Method*

The columns and rows of the stiffness matrix, displacement vector and load vector are rearranged so that the set of equations can be written as

$$\begin{Bmatrix} P_1 \\ P_2 \end{Bmatrix} = \begin{bmatrix} K_{11} & K_{12} \\ K_{21} & K_{22} \end{bmatrix} \begin{Bmatrix} q_1 \\ q_2 \end{Bmatrix} \qquad \ldots\ldots(19.26)$$

where q_1 is the set of unknown displacements and q_2 is the set of specified displacements.

From static equilibrium considerations, terms in the load vector $\{P_2\}$ corresponding to the fixed or specified values of degrees of freedom $\{q_2\}$ represent reactions at those degrees of freedom to balance the applied loads.

Taking all known values to the left side, first set of these equations can be rewritten as

$$\{P_1\} - [K_{12}]\{q_2\} = [K_{11}]\{q_1\} \qquad \ldots\ldots(19.27)$$

The reduced stiffness matrix of the structure $[K_{11}]$ is usually non-singular and can be inverted so that unknown displacements $\{q_1\}$ and the reactions $\{R\}$ can be evaluated from

$$\{q_1\} = [K_{11}]^{-1}(\{P_1\} - [K_{12}]\{q_2\}) \qquad \ldots\ldots(19.28)$$

$$\{R\} = \{P_2\} = [K_{21}\ K_{22}]\begin{Bmatrix} q_1 \\ q_2 \end{Bmatrix} = [K_{21}]\{q_1\} + [K_{22}]\{q_2\} \qquad \ldots\ldots(19.29)$$

In case of specified zero displacements $\{q_2\}$, eq.(19.29) reduces to

$$\{P_1\} = [K_{11}]\{q_1\} \quad \text{or} \quad \{P_r\} = [K_r]\{u_r\} \qquad \ldots\ldots(19.30)$$

This procedure is equivalent to deleting rows and columns corresponding to the fixed degrees of freedom from the assembled stiffness matrix, displacement vector and load vector. The reduced stiffness matrix $[K_r]$ is a non-singular matrix.

The unknown displacements are now obtained by using a suitable matrix inversion algorithm like Gauss elimination method or Gauss-Jordan method or Cholesky method. Eq. (19.28) and (19.29) thus simplify to

$$\{q_1\} = [K_{11}]^{-1}\{P_1\} \quad \text{or} \quad \{u_r\} = [K_r]^{-1}\{P_r\} \qquad \dots(19.31)$$

$$\text{and} \quad \{R\} = \{P_2\} = [K_{21}]\{q_1\} \qquad \dots(19.32)$$

In the displacement formulation, displacements are calculated in the global coordinate system for the entire structure while the stresses are calculated in each element, in the local or element coordinate system, from the nodal displacements of that element using

$$\{\sigma_e\} = [D]\{\varepsilon\} = [D][B_e]\{u_e\}$$

In a structure with 'm' fixed degrees of freedom, assembling the complete stiffness matrix and then deleting some rows and columns will involve more computer memory as well as more time. It is therefore a common practice to ignore the rows and columns of element stiffness matrices corresponding to the fixed degrees of freedom during the assembly process, thus storing $[K_r]$ of order $(n–m) \times (n–m)$ only. In that case, calculation of reaction values corresponding to the fixed degrees of freedom requires storing appropriate terms $[K_{21}]$ in a different matrix.

(b) *Multi-point Constraints*

There are many situations in trusses where the end supports are on inclined plane and do not coincide with the coordinate system used to describe the truss. In such cases, the displacement and force components along the coordinate axes have to be resolved along and perpendicular to the inclined plane and necessary conditions specified on them.

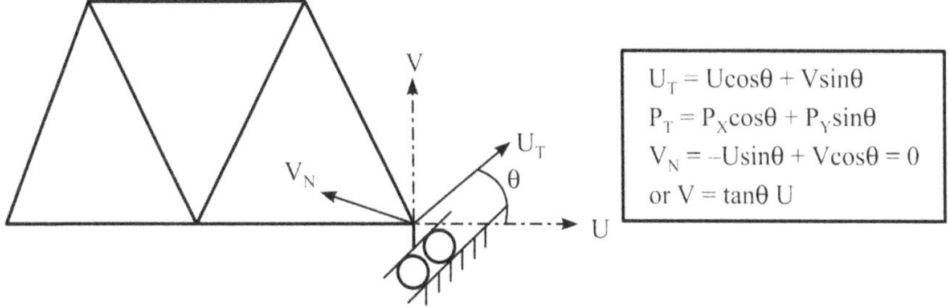

Some other types of multi-point constraints are those linking displacement of one node with that of another. A few of them are shown below, with node 1 as the fixed point and node 2 as the point of load application. If 'δ' is the gap, $U_3 = U_2 - \delta$ can be substituted in the load-displacement relations to reduce the number of unknowns by 1 and corresponding columns of stiffness matrix are modified. Accordingly, order of the stiffness matrix also reduces by 1.

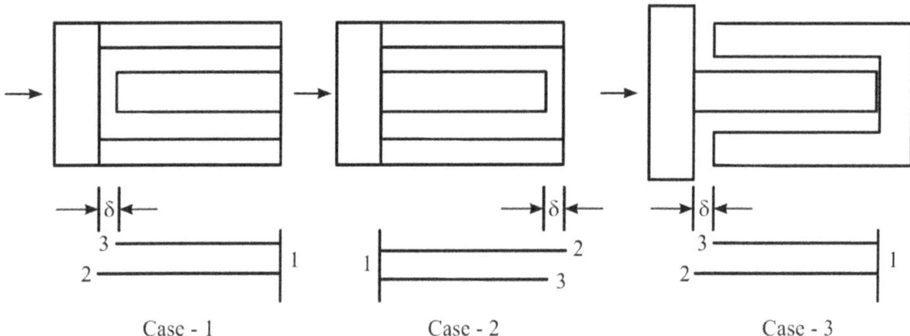

Case - 1   Case - 2   Case - 3

19.7 GENERAL BEAM ELEMENT STIFFNESS MATRIX

A beam in a space frame is generally subjected to axial load, torsion load and bending loads in two planes, due to the combined effect of loads acting at different locations of the space frame and in different directions, as shown in the figure.

If a single beam AB in a space frame with concentrated loads P_1, P_2, P_3 and P_4 acting on some members is considered, load P_1 contributes to axial load giving rise to displacement u; load P_2 contributes to bending in X-Y plane giving rise to deflection v and slope θ_z, load P_3 contributes to bending in X-Z plane giving rise to deflection w and slope θ_y and load P_4 contributes to torsion in AB giving rise to θ_X. Therefore, general stiffness matrix for beam AB in its local coordinate system should include response of the beam for these four types of deformations, which are mutually independent. Such DOFs are called ***uncoupled degrees of freedom*** *i.e.* torsion of a beam does not result in axial elongation or compression of the beam; deflection of the beam in X-Y plane does not cause any displacement of the beam in X-Z plane etc. Hence, stiffness contribution of a beam in these 6 DOFs can be placed directly, without any modifications, in the appropriate positions of the general stiffness matrix of order 12 (2 nodes × 6 DOF/node).

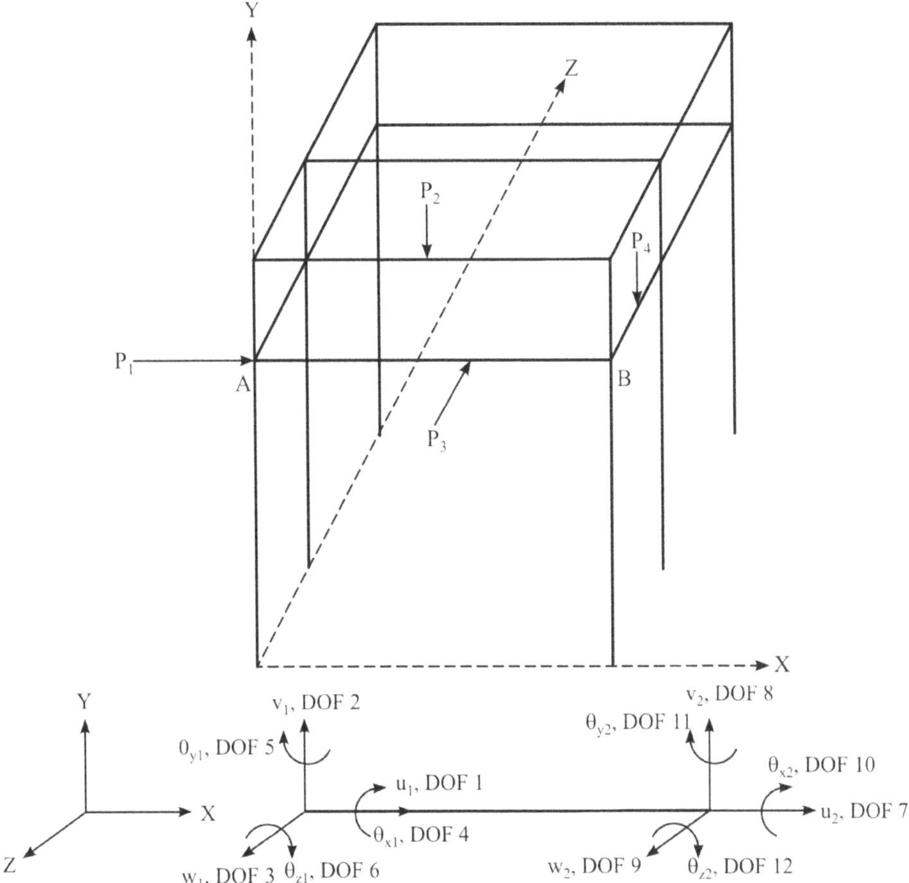

The combined stiffness matrix of a general beam element thus includes stiffness coefficients linking loads along the three coordinate axes and moments about the three axes at each end of the beam to the corresponding displacements and rotations, and is obtained by a simple addition of the coefficients of the four matrices, corresponding to the four uncoupled modes of deformation.

$$
\begin{Bmatrix} P_{x1} \\ P_{y1} \\ P_{z1} \\ M_{x1} \\ M_{y1} \\ M_{z1} \\ P_{x2} \\ P_{y2} \\ P_{z2} \\ M_{x2} \\ M_{y2} \\ M_{z2} \end{Bmatrix} = \frac{1}{L}
\begin{bmatrix}
EA & 0 & 0 & 0 & 0 & 0 & -EA & 0 & 0 & 0 & 0 & 0 \\
0 & 12EI_z/L^2 & 0 & 0 & 0 & 6EI_z/L & 0 & -12EI_z/L^2 & 0 & 0 & 0 & 6EI_z/L \\
0 & 0 & 12EI_y/L^2 & 0 & 6EI_y/L & 0 & 0 & 0 & -12EI_y/L^2 & 0 & 6EI_y/L & 0 \\
0 & 0 & 0 & GJ & 0 & 0 & 0 & 0 & 0 & GJ & 0 & 0 \\
0 & 0 & 6EI_y/L & 0 & 4EI_y & 0 & 0 & 0 & -6EI_y/L & 0 & 2EI_y & 0 \\
0 & 6EI_z/L & 0 & 0 & 0 & 4EI_z & 0 & -6E_z/L & 0 & 0 & 0 & 2EI_z \\
-EA & 0 & 0 & 0 & 0 & 0 & EA & 0 & 0 & 0 & 0 & 0 \\
0 & -12EI_z/L^2 & 0 & 0 & 0 & -6EI_z/L & 0 & 12EI_z/L^2 & 0 & 0 & 0 & -6EI_z/L \\
0 & 0 & -12EI_y/L^2 & 0 & -6EI_y/L & 0 & 0 & 0 & 12EI_y/L^2 & 0 & -6EI_y/L & 0 \\
0 & 0 & 0 & -GJ & 0 & 0 & 0 & 0 & 0 & GJ & 0 & 0 \\
0 & 0 & 6EI_y/L & 0 & 2EI_y & 0 & 0 & 0 & -6EI_y/L & 0 & 4EI_y & 0 \\
0 & 6EI_z/L & 0 & 0 & 0 & 12EI_z/L^2 & 0 & -6EI_z/L & 0 & 0 & 0 & 4EI_z
\end{bmatrix}
\begin{Bmatrix} u_1 \\ v_1 \\ w_1 \\ \theta_{x1} \\ \theta_{y1} \\ \theta_{z1} \\ u_2 \\ v_2 \\ w_2 \\ \theta_{x2} \\ \theta_{y2} \\ \theta_{z2} \end{Bmatrix}
$$

$$.....(19.33)$$

SOLVED PROBLEMS

Example 19.1

Determine the stiffness matrix, stresses and reactions in the truss structure shown below, assuming points 1 and 3 are fixed. Use E = 200 GPa and A = 1000 mm^2.

Solution

Stiffness matrix of any truss element is given by

$$[K] = \frac{AE}{L} \begin{bmatrix} l^2 & lm & -l^2 & -lm \\ lm & m^2 & -lm & -m^2 \\ -l^2 & -lm & l^2 & lm \\ -lm & -m^2 & lm & m^2 \end{bmatrix}$$

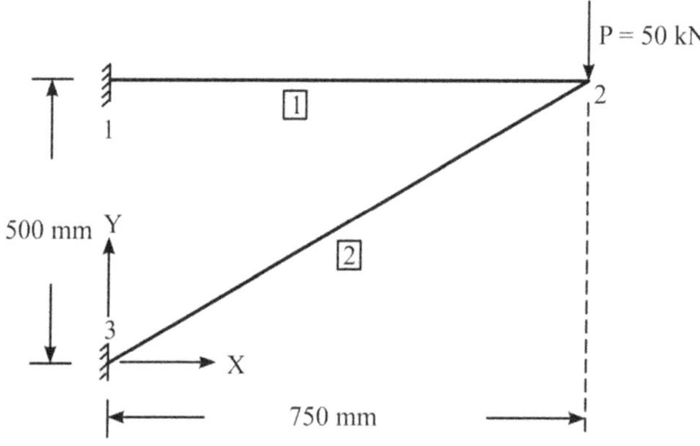

In the given problem, L$_1$ = 750 mm; $L_2 = \sqrt{[750^2 + 500^2]} = 250\sqrt{13}$

For element-1, $l = \dfrac{(x_2 - x_1)}{L_1} = 1$ and $m = \dfrac{y_2 - y_1}{L_1} = 0$

$$[K]_1 = \frac{AE}{750} \begin{bmatrix} 1 & 0 & -1 & 0 \\ 0 & 0 & 0 & 0 \\ -1 & 0 & 1 & 0 \\ 0 & 0 & 0 & 0 \end{bmatrix}$$

$$\frac{AE}{750} = 266.67 \times 10^3$$

For element-2, $l = \dfrac{(x_3 - x_2)}{L_2} = \dfrac{3}{\sqrt{13}}$ and $m = \dfrac{y_3 - y_2}{L_2} = \dfrac{2}{\sqrt{13}}$

$$[K]_2 = \frac{AE}{250 \times 13 \times \sqrt{13}} \begin{bmatrix} 9 & 6 & -9 & -6 \\ 6 & 4 & -6 & -4 \\ -9 & -6 & 9 & 6 \\ -6 & -4 & 6 & 4 \end{bmatrix}$$

$$\frac{AE}{250 \times 13 \sqrt{13}} = 17.07 \times 10^3$$

The assembled stiffness matrix is given by appropriate addition of stiffness coefficients of the two elements,

$$[K] = 10^3 \begin{bmatrix} 266.67 & 0 & -266.67 & 0 & 0 & 0 \\ 0 & 0 & 0 & 0 & 0 & 0 \\ -266.67 & 0 & 266.67-153.63 & 102.42 & -153.63 & -102.42 \\ 0 & 0 & -102.42 & 68.28 & -102.42 & -68.28 \\ 0 & 0 & -153.63 & -102.42 & 153.63 & 102.42 \\ 0 & 0 & -102.42 & -68.28 & 102.42 & 68.28 \end{bmatrix}$$

After applying boundary conditions that $u_1 = v_1 = u_3 = v_3 = 0$, the load-displacement relationships reduce to $\{P\}_R = [K]_R \{u\}_R$

$$\left\{ \begin{array}{c} 0 \\ -50000 \end{array} \right\} = 10^3 \begin{bmatrix} -266.67+153.63 & 102.42 \\ 102.42 & 68.28 \end{bmatrix} \left\{ \begin{array}{c} u_2 \\ v_2 \end{array} \right\}$$

Solving these two simultaneous equations gives

$$u_2 = 0.2813 \text{ mm} \qquad \text{and} \qquad v_2 = -1.154 \text{ mm}$$

Displacements of element-1 in local coordinate system are given by

$$\{q_1'\} = \begin{bmatrix} 1 & 0 & 0 & 0 \\ 0 & 0 & 1 & 0 \end{bmatrix} \left\{ \begin{array}{c} 0 \\ 0 \\ 0.2813 \\ -1.154 \end{array} \right\} = \left\{ \begin{array}{c} 0 \\ 0.2813 \end{array} \right\}$$

Stress in element-1, $\sigma_1 = E\, \varepsilon_1 = E[-1/L \quad 1/L]\{q_1\}$

$$= 200 \times 10^3 \times 0.2813 / 750 = 75 \text{ N/mm}^2$$

Displacements of element-2 in local coordinate system are given by

$$\{q_2'\} = \begin{bmatrix} 3/\sqrt{13} & 2/\sqrt{13} & 0 & 0 \\ 0 & 0 & 3/\sqrt{13} & 2/\sqrt{13} \end{bmatrix} \left\{ \begin{array}{c} 0.28313 \\ -1.154 \\ 0 \\ 0 \end{array} \right\} = \left\{ \begin{array}{c} -0.406 \\ 0 \end{array} \right\}$$

Stress in element-2,

$$\sigma_2 \;=\; E\,\varepsilon_2 \;=\; E\left[\dfrac{-1}{L}\quad \dfrac{1}{L}\right]\{q_2'\} \;=200\times10^3\times\dfrac{(-0.406)}{250\,\sqrt{13}}=90.08\ \text{N/mm}^2$$

Reactions at the two fixed ends are obtained from the equations of the assembled stiffness matrix corresponding to the specified zero displacements

$$\begin{Bmatrix} R_{1-X} \\ R_{1-Y} \\ R_{3-X} \\ R_{3-Y} \end{Bmatrix} =10^3 \begin{bmatrix} 266.67 & 0 & -266.67 & 0 & 0 & 0 \\ 0 & 0 & 0 & 0 & 0 & 0 \\ 0 & 0 & -153.63 & -102.42 & 153.63 & 102.42 \\ 0 & 0 & -102.42 & -68.28 & 102.42 & 68.28 \end{bmatrix} \begin{Bmatrix} 0 \\ 0 \\ 0.2813 \\ -1.154 \\ 0 \\ 0 \end{Bmatrix}$$

$$= \begin{Bmatrix} -75014.3 \\ 0 \\ 74976.6 \\ 49984.4 \end{Bmatrix}$$

The exact solution can be obtained from the equilibrium conditions applied at the three nodes as follows -

At node 2, $\Sigma F_Y = (F_2)_Y + P = 0$ or $(F_2)_Y = -P = 50\ \text{kN}$

Horizontal component of force in element-2 is given by

$$(F_2)_X = (F_2)_Y \times (750/500) = 75\ \text{kN}$$

At node 2, $\Sigma F_X = (F_2)_X + F_1 = 0$ or $F_1 = -(F_2)_X = -75\ \text{kN}$

At node 1, $\Sigma F_X = R_{1-X} - F_1 = 0$ or $R_{1-X} = +F_1 = -75\ \text{kN}$

At node 3, $\Sigma F_X = R_{3-X} - (F_2)_X = 0$ or $R_{3-X} = +(F_2)_X = +75\ \text{kN}$

At node 3, $\Sigma F_Y = R_{3-Y} - (F_2)_Y = 0$ or $R_{3-Y} = +(F_2)_Y = +50\ \text{kN}$

These results can also be cross checked for the overall equilibrium of the structure.

$$\Sigma F_X = R_{3-X} + R_{1-X} = -75 + 75 = 0$$

and $\Sigma F_Y = R_{3-Y} + P = 50 - 50 = 0$

It can be seen that the approximate solution obtained by FEM is in close agreement with the exact solution obtained from equilibrium consideration. The small variation is due to numerical rounding off at different stages.

Example 19.2

A concentrated load P = 50 kN is applied at the center of a fixed beam of length 3m, depth 200 mm and width 120 mm. Calculate the deflection and slope at the mid point. Assume $E = 2 \times 10^5$ N/mm^2.

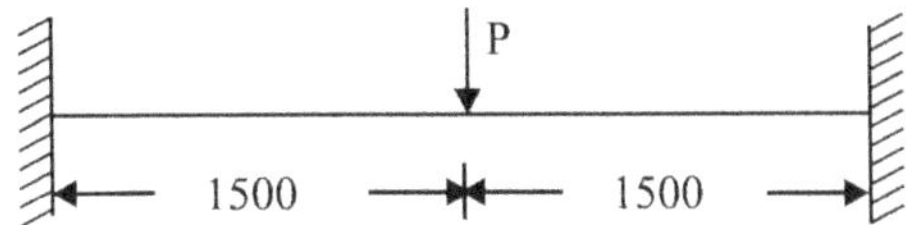

Solution

The finite element model consists of 2 beam elements, as shown here, with nodes 1 and 3 at the two fixed supports and node 2 at the location where load P is applied.

Stiffness matrices of elements 1 and 2 (connected by nodes 1 and 2 ; 2 and 3 respectively, each with L = 1500 mm) are given by,

$$[K] = \frac{E\,I_z}{L^3}\begin{bmatrix} 12 & 6L & -12 & 6L \\ 6L & 4L^2 & -6L & 2L^2 \\ -12 & -6L & 12 & -6L \\ 6L & 2L^2 & -6L & 4L^2 \end{bmatrix} = \frac{2\times10^5 \times \dfrac{\left(120\times200^3\right)}{12}}{L^3}\begin{bmatrix} 12 & 6L & -12 & 6L \\ 6L & 4L^2 & -6L & 2L^2 \\ -12 & -6L & 12 & -6L \\ 6L & 2L^2 & -6L & 4L^2 \end{bmatrix}$$

Assembling the element stiffness matrices, we get

$$\begin{Bmatrix} P_1 \\ M_1 \\ P_2 \\ M_2 \\ P_3 \\ M_3 \end{Bmatrix} = \frac{2\times10^5 \times \dfrac{120\times200^3}{12}}{1500^3}\begin{bmatrix} 12 & 6L & -12 & 6L & 0 & 0 \\ 6L & 4L^2 & -6L & 2L^2 & 0 & 0 \\ -12 & -6L & 12+12 & -6L+6L & -12 & 6L \\ 6L & 2L^2 & -6L+6L & 4L^2+4L^2 & -6L & 2L^2 \\ 0 & 0 & -12 & -6L & 12 & -6L \\ 0 & 0 & 6L & 2L^2 & -6L & 4L^2 \end{bmatrix}\begin{Bmatrix} w_1 \\ (\theta_z)_1 \\ w_2 \\ (\theta_z)_2 \\ w_3 \\ (\theta_z)_3 \end{Bmatrix}$$

After applying boundary conditions $v_1 = v_3 = 0$ and $(\theta_z)_1 = (\theta_z)_3 = 0$, the equations reduce to

$$\begin{Bmatrix} P_2 \\ M_2 \end{Bmatrix} = \frac{2\times10^5 \times \dfrac{\left(120\times200^3\right)}{12}}{1500^3}\begin{bmatrix} 12+12 & -6L+6L \\ -6L+6L & 4L^2+4L^2 \end{bmatrix}\begin{Bmatrix} v_2 \\ (\theta_z)_2 \end{Bmatrix}$$

The applied loads are $P_2 = -50000$ N   and $M_2 = 0$

Therefore, $v_2 = \dfrac{-50000 \times 1500^3}{\left[2 \times 10^5 \times \dfrac{\left(120 \times 200^3\right)}{12} \times 24\right]} = -0.4395 \text{ mm}$

and $(\theta_z)_2 = 0$

Check: From strength of materials approach,

$$v_3 = \frac{-PL^3}{24\,EI} \text{ or } \frac{P(2L)^3}{192\,EI} = -0.4395 \text{ mm}$$

and the deflection being symmetric, slope at the center $(\theta_z)_2 = 0$.

19.8 ANALYSIS OF THIN PLATES (2-D ELEMENTS)

Unlike discrete elements, there is no unique way of discretising a 2-D plate into a certain number of elements (smaller or larger size elements) or into a particular shape (triangular, quadrilateral,..). Also, these elements may be lower order elements (with only corner nodes representing an element) or higher order elements (with additional nodes along the boundary or inside), based on the nature of assumed displacement polynomial over the element. For the same accuracy of solution, one may use a large number of lower order elements or a small number of higher order elements. Similar to 1-D truss and beam elements, 2-D elements or thin plates are broadly categorized as elements subjected to in-plane loads and elements subjected to normal (or bending) loads.

19.8.1 ANALYSIS OF PLATES FOR IN-PLANE LOADS

When two dimensions of a solid (length and breadth) are very large compared to the other dimension (thickness), the solid is treated as a 2-D plate and discretised using 2-D elements. Let us consider the element in the X-Y plane while dimension in the Z-direction represents the thickness of the element. The load is assumed to be acting in the plane of the element, along X-direction and/or Y-direction. Such a plane element has two DOF per node, displacements along X and Y directions.

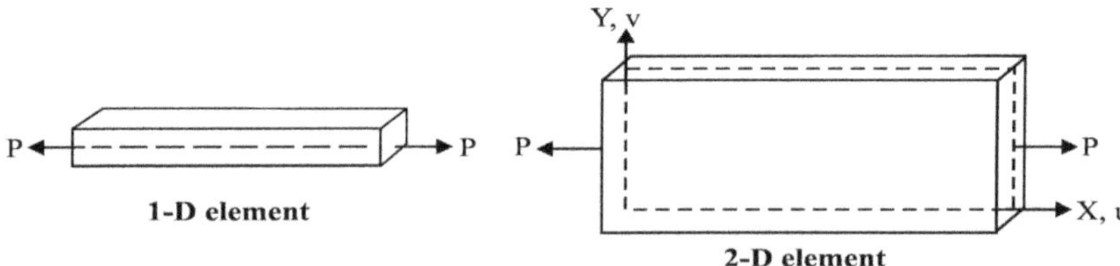

If a concentrated load is applied at a point on the width of the plate, load can not be considered as uniformly distributed over the width and hence the displacement 'u' at any point is a function of its x and y coordinates. Load along X-direction produces lateral strain and, hence, a displacement 'v' in the y-direction (because of Poisson's effect). Thus, displacements u and v are functions of x and y coordinates of the point.

In the case of discrete structures, with each member treated as a 1-D element, nodes are chosen at junctions of two discrete members, junctions of two different materials, at points of change of cross section or at points of load application. However, in the case of continuum, which is modeled by 2-D or 3-D elements, there is no unique finite element model for analysis. ***Each engineer may use a particular number of nodes and a particular orientation of elements. Hence, the results obtained by different engineers may vary.*** Increasing number of elements or changing type of elements, but at a higher computational cost, may improve accuracy. A judicial compromise has to be made between better accuracy of results and computational cost. This aspect is further discussed under 'modelling techniques'

Calculation of stiffness matrix for a triangular element is first considered since triangular elements are the simplest and can be used to define arbitrary boundaries of a component more conveniently, by approximating curved boundary with a large number of elements having straight edges.

(a) STIFFNESS MATRIX OF A CST ELEMENT

Let $u(x,y) = a_1 + a_2 \times x + a_3 \times y$ and $v(x,y) = a_4 + a_5 \times x + a_6 \times y$ or $\{u\} = [f(x,y)] \{a\}$ be the displacements in the element. Displacement function $f(x,y)$, representing u or v, is graphically shown in Fig. 19.3. In general, 1-1′, 2-2′ and 3-3′ are not equal.

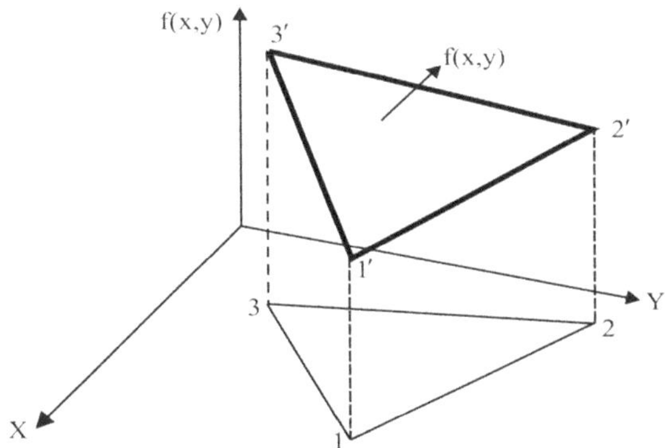

FIGURE 19.3 Graphical representation of displacement function on a triangle

Substituting nodal coordinates, while the element 1-2-3 displaces to 1'-2'-3' on application of load, we get nodal displacement vector as

$$\begin{Bmatrix} u_1 \\ u_2 \\ u_3 \\ v_1 \\ v_2 \\ v_3 \end{Bmatrix} = \begin{bmatrix} 1 & x_1 & y_1 & 0 & 0 & 0 \\ 1 & x_2 & y_2 & 0 & 0 & 0 \\ 1 & x_3 & y_3 & 0 & 0 & 0 \\ 0 & 0 & 0 & 1 & x_1 & y_1 \\ 0 & 0 & 0 & 1 & x_2 & y_2 \\ 0 & 0 & 0 & 1 & x_3 & y_3 \end{bmatrix} \begin{Bmatrix} a_1 \\ a_2 \\ a_3 \\ a_4 \\ a_5 \\ a_6 \end{Bmatrix}$$

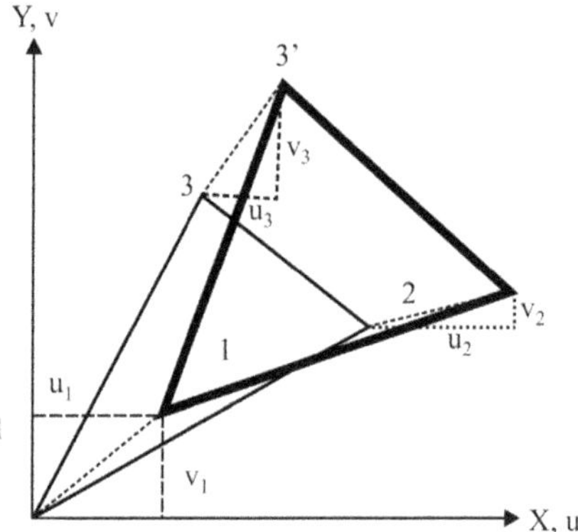

$$\{u_e\} = [G]\{a\} = \begin{bmatrix} [A] & [O] \\ [O] & [A] \end{bmatrix} \{a\}$$

$$\text{where,} \quad [A] = \begin{bmatrix} 1 & x_1 & y_1 \\ 1 & x_2 & y_2 \\ 1 & x_3 & y_3 \end{bmatrix} \quad \text{and} \quad [O] = \begin{bmatrix} 0 & 0 & 0 \\ 0 & 0 & 0 \\ 0 & 0 & 0 \end{bmatrix}$$

$$\text{or} \quad \{a\} = [G]^{-1}\{u_e\} = \begin{bmatrix} [A]^{-1} & [O] \\ [O] & [A]^{-1} \end{bmatrix} \{u_e\}$$

$$\{u\} = [f(x,y)]\{a\} = [f(x,y)][G]^{-1}\{u_e\}$$

$$\begin{Bmatrix} \varepsilon_X \\ \varepsilon_Y \\ \gamma_{XY} \end{Bmatrix} = \begin{Bmatrix} \partial u / \partial x \\ \partial v / \partial y \\ \partial u / \partial y + \partial v / \partial x \end{Bmatrix} = [f'(x,y)] [G]^{-1}\{u_e\}$$

$$\{\varepsilon\} = \begin{bmatrix} 0 & 1 & 0 & 0 & 0 & 0 \\ 0 & 0 & 0 & 0 & 0 & 1 \\ 0 & 0 & 1 & 0 & 1 & 0 \end{bmatrix} [G]^{-1}\{u_e\} = [B]\{u_e\}$$

where, strain-displacement matrix,

$$[B] = \frac{1}{\text{Det J}} \begin{bmatrix} y_{23} & 0 & y_{31} & 0 & y_{12} & 0 \\ 0 & x_{32} & 0 & x_{13} & 0 & x_{21} \\ x_{32} & y_{23} & x_{13} & y_{31} & x_{21} & y_{12} \end{bmatrix}$$

This triangular element, with 3 nodes and 2 DOF per node chooses linear displacement functions for u and v and hence gives constant strain terms over the entire element as seen from $[f'(x,y)]$ or [B] and hence is popularly known as 'Constant Strain Triangle (CST)' element.

Here, [J] is Jacobian of nodal coordinates

$$\text{and} \qquad \text{Det J} = \begin{vmatrix} 1 & x_1 & y_1 \\ 1 & x_2 & y_2 \\ 1 & x_3 & y_3 \end{vmatrix} = \begin{vmatrix} 0 & x_1 - x_3 & y_1 - y_3 \\ 0 & x_2 - x_3 & y_2 - y_3 \\ 1 & x_3 & y_3 \end{vmatrix} = \begin{vmatrix} 0 & x_{13} & y_{13} \\ 0 & x_{23} & y_{23} \\ 1 & x_3 & y_3 \end{vmatrix}$$

$$= x_{13} \times y_{23} - y_{13} \times x_{23}$$

It can be seen that area of the triangle, $A = \dfrac{1}{2} \text{Det J} = \left(\dfrac{1}{2}\right) \begin{vmatrix} 0 & x_{13} & y_{13} \\ 0 & x_{23} & y_{23} \\ 1 & x_3 & y_3 \end{vmatrix}$

If nodes are numbered counter-clockwise, in right-handed coordinate system, Det J is +ve

Stiffness matrix of the 3-noded triangular element can now be obtained from

$$[K]_v = \int [B]^T [D][B] dV = t \iint [B]^T [D][B] dx\, dy$$

where, $\quad t = \int dz$ is the thickness of the element

and $\qquad$ [B] is a function of x and y only

For a thin plate subjected to in-plane loads (plane stress case, since $\sigma_z = 0$), stress-strain relationship is given by

$$[D] = \frac{E}{(1 - v^2)} \begin{bmatrix} 1 & v & 0 \\ v & 1 & 0 \\ 0 & 0 & (1-v)/2 \end{bmatrix}$$

An explicit evaluation of the stiffness matrix is not generally feasible (except for a few special cases).

(b)   CONVERGENCE CONDITIONS

While choosing the function to represent u and v displacements at any point in the element, care should be taken to ensure that the following conditions are satisfied.

- The function should be continuous and differentiable (to obtain strains) within the element. This is automatically satisfied with polynomial functions.

- The displacement polynomial should include constant term, representing rigid body displacement, which any unrestrained portion of a component should experience when subjected to external loads.

- The polynomial should include linear terms, which on differentiation give constant strain terms. Constant strain is the logical condition as the element size reduces to a point in the limit.

- Compatibility of displacement and its derivatives, up to the required order, must be satisfied across inter-element boundaries. Otherwise the displacement solution may result in separated or overlapped inter-element boundaries when the displacement patterns of deformed elements with a common boundary are plotted separately.

- The polynomial shall satisfy geometric isotropy (terms symmetric in terms of coordinate axes x, y and z). Otherwise, different users analysing the same component may get different results by following different node number sequence to define the elements which result in different local coordinate systems.

The displacement function $u(x) = a_1 + a_2 \times x + a_3 \times y$ of a triangular element is complete and isotropic while $u(x) = a_1 + a_2 \times x + a_3 \times y + a_4 \times xy$ of a quadrilateral element is incomplete (since other 2nd order terms x^2 and y^2 are missing) but isotropic.

(c)   ASPECT RATIO

In 2-D elements, the displacement function is symmetric in x and y, whether it is complete or not, in terms of coefficients of a particular order Hence, the shape of the finite element in the idealised structure should also be oriented equally to all the relevant axes. For this purpose, certain conditions are generally specified in the standard packages on the sizes and included angles for various elements. Aspect ratio is defined for this purpose as the ratio of the longest side to the shortest side. It is usually limited to 5, while the included angle is usually limited to 45^0 to 135^0 for a triangular element and to 60^0 to 120^0 for a quadrilateral or 3-D element. A few 2-D elements with valid and invalid shapes are shown in Fig. 19.4.

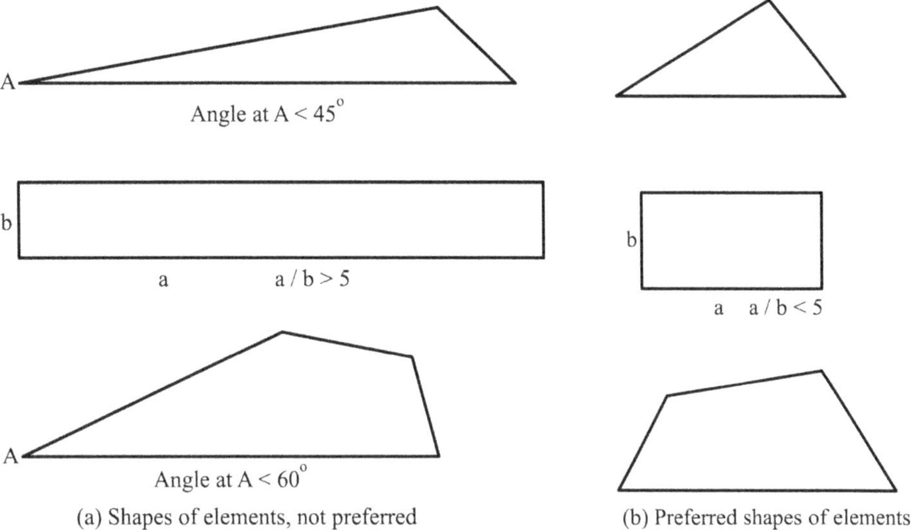

FIGURE 19.4 Aspect ratio of 2-D elements

19.8.2 2-D ELEMENTS SUBJECTED TO BENDING LOADS

Plate bending element: It is a plate element in X-Y plane subjected to bending load P_Z and/or bending moments M_x , M_y. A thin plate (span > 10 x thickness) with small deflection (< thickness/10) follows Kirchhoff's theory and is an extension of 1-D beam element into two dimensions. It will have three degrees of freedom at each node, displacement normal to the plate (w) and rotations about the two major axes of the element represented by derivatives θ_x and θ_y of w about x and y.

For a triangular Plate bending element, normal deflection is assumed by the polynomial,

$$w = a_1 + a_2 \times x + a_3 \times y + a_4 \times x^2 + a_5 \times y^2$$
$$+ a_6 \times x^2 y + a_7 \times xy^2 + a_8 \times x^3 + a_9 \times y^3$$

Rotations or slopes of the plate at any point are defined by,

$$\theta_x = \frac{\partial w}{\partial x} = a_2 + 2a_4 \times x + 2a_6 \times xy + a_7 \times y^2 + 3a_8 \times x^2$$

$$\theta_y = \frac{\partial w}{\partial y} = a_3 + 2a_5 \times y + a_6 \times x^2 + 2a_7 \times xy + 3a_9 \times y^2$$

Displacement function of triangular plate bending element is incomplete but isotropic.

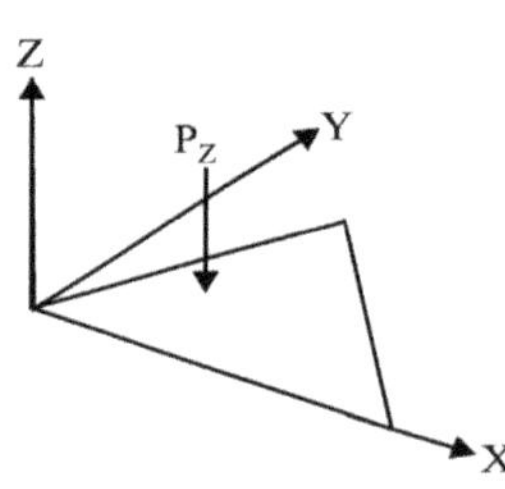

Plate bending element

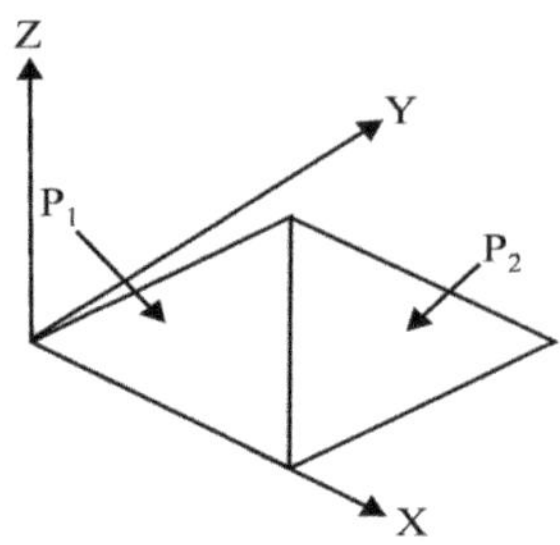

Thin shell element

Thin plate with large deflection is characterised by large tensile or compressive stresses in the middle plane. The corresponding u and v displacements are given by

$$u = -z\left(\frac{\partial w}{\partial x}\right) ; \qquad v = -z\left(\frac{\partial w}{\partial y}\right)$$

$$\varepsilon_X = \left(\frac{\partial u}{\partial x}\right) = -z\left[\frac{\partial^2 w}{\partial x^2}\right]$$

$$\varepsilon_Y = \left(\frac{\partial v}{\partial y}\right) = -z\left[\frac{\partial^2 w}{\partial y^2}\right]$$

$$\gamma_{XY} = \left(\frac{\partial u}{\partial y}\right) + \left(\frac{\partial y}{\partial x}\right) = -2z\left(\frac{\partial^2 w}{\partial x\,\partial y}\right)$$

The stress-strain matrix is given by $\{\sigma\} = [D]\,\{\varepsilon\}$

or

$$\begin{Bmatrix} \sigma_X \\ \sigma_Y \\ \tau_{XY} \end{Bmatrix} = \frac{E}{1-v^2} \begin{bmatrix} 1 & v & 0 \\ v & 1 & 0 \\ 0 & 0 & (1-v)/2 \end{bmatrix} \begin{Bmatrix} \varepsilon_X \\ \varepsilon_Y \\ \gamma_{XY} \end{Bmatrix}$$

This is more commonly expressed in terms of moments per unit width (b=1), also called stress resultants, using

$$M_X = \left(\frac{I}{z}\right) \times \sigma_x = \left(\frac{bh^3/12}{z}\right) \times \sigma_x = \left(\frac{h^3/12}{z}\right) \times \left\{ E\left(\varepsilon_x + v\varepsilon_y\right)/\left(1-v^2\right) \right\}$$

Thus,
$$\begin{Bmatrix} M_X \\ M_Y \\ M_{XY} \end{Bmatrix} = \frac{-Eh^3}{12(1-v^2)} \begin{bmatrix} 1 & v & 0 \\ v & 1 & 0 \\ 0 & 0 & (1-v)/2 \end{bmatrix} \begin{Bmatrix} \partial^2 w/\partial x^2 \\ \partial^2 w/\partial y^2 \\ \partial^2 w/\partial x\,\partial y \end{Bmatrix}$$

19.8.3 THIN SHELL ELEMENT

It is a 2-D element subjected to in-plane loads as well as bending loads. As in the case of a general beam, these two behaviours represent uncoupled degrees of freedom. It can therefore be considered as a combination of plane stress element and plate bending element. In the local coordinate system, this element will have five degrees of freedom since moment about normal to the plate is not included. However, if different elements are inclined to each other, transformation of the combined stiffness matrix of each element with five degrees of freedom per node in local coordinate system results in six degrees of freedom per node in the global coordinate system.

Example 19.3

Calculate displacements and stress in a triangular plate (with sides a = 30mm and b = 20mm), fixed along one edge (1-2) and subjected to concentrated load at its free end (3). Assume E = 70 GPa, t = 1 mm and v = 0.3.

Solution

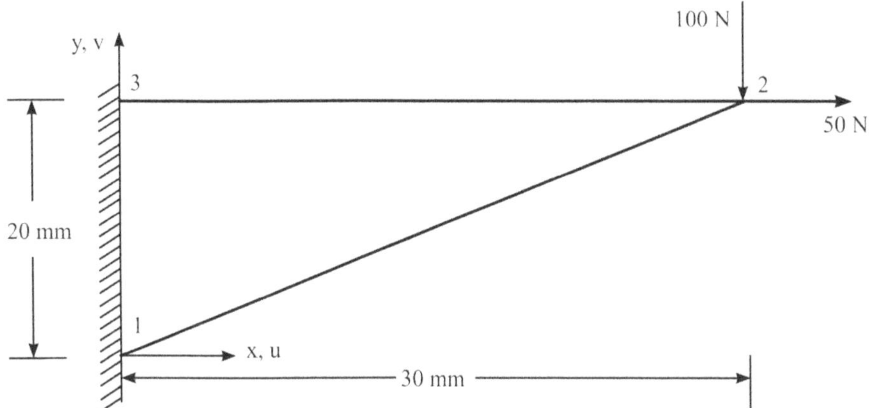

For the particular case of a right angled triangle with coordinates 1(0,0), 2(a, b) and 3(0, b), let $u = a_1 + a_2 \times x + a_3 \times y$ and $v = a_4 + a_5 \times x + a_6 \times y$ represent the displacement field. Substituting nodal coordinates, vector of nodal displacements can be written as

$$\begin{Bmatrix} u_1 \\ u_2 \\ u_3 \\ v_1 \\ v_2 \\ v_3 \end{Bmatrix} = \begin{bmatrix} 1 & 0 & 0 & 0 & 0 & 0 \\ 1 & a & b & 0 & 0 & 0 \\ 1 & 0 & b & 0 & 0 & 0 \\ 0 & 0 & 0 & 1 & 0 & 0 \\ 0 & 0 & 0 & 1 & a & b \\ 0 & 0 & 0 & 1 & 0 & b \end{bmatrix} \begin{Bmatrix} a_1 \\ a_2 \\ a_3 \\ a_4 \\ a_5 \\ a_6 \end{Bmatrix} \quad \text{or} \quad \{u_e\} = \begin{bmatrix} [A] & [O] \\ [O] & [A] \end{bmatrix} \{a\} = [G]\,\{a\}$$

Evaluating coefficients a_1 to a_6 in terms of nodal displacements,

$$\begin{Bmatrix} a_1 \\ a_2 \\ a_3 \\ a_4 \\ a_5 \\ a_6 \end{Bmatrix} = \begin{bmatrix} 1 & 0 & 0 & 0 & 0 & 0 \\ 0 & 1/a & -1/a & 0 & 0 & 0 \\ -1/b & 0 & 1/b & 0 & 0 & 0 \\ 0 & 0 & 0 & 1 & 0 & 0 \\ 0 & 0 & 0 & 0 & 1/a & -1/a \\ 0 & 0 & 0 & -1/b & 0 & 1/b \end{bmatrix} \begin{Bmatrix} u_1 \\ u_2 \\ u_3 \\ v_1 \\ v_2 \\ v_3 \end{Bmatrix}$$

or $\quad \{a\} = \{G\}^{-1} \{u_e\} = \{a\} = \begin{bmatrix} [A]^{-1} & [O] \\ [O] & [A]^{-1} \end{bmatrix} \{u_e\}$

Strain, $\{\varepsilon\} = \begin{Bmatrix} \partial u/\partial x \\ \partial u/\partial y \\ \partial u/\partial y + \partial v/\partial x \end{Bmatrix} = \begin{bmatrix} 0 & 1 & 0 & 0 & 0 & 0 \\ 0 & 0 & 0 & 0 & 0 & 1 \\ 0 & 0 & 1 & 0 & 1 & 0 \end{bmatrix} \{a\} = [B]\,\{u_e\}$

where, $[B] = \left(\dfrac{1}{ab}\right) \begin{bmatrix} 0 & b & -b & 0 & 0 & 0 \\ 0 & 0 & 0 & -a & 0 & a \\ -a & 0 & a & 0 & b & -b \end{bmatrix}$

$$[K] = \int_V [B]^T\,[D]\,[B]\,dv = t \int_0^a \int_0^b [B]^T\,[D]\,[B]\,dx\,dy = tA\,[B]^T\,[D]\,[B]$$

since elements of matrices [B] and [D] are not functions of x or y

For a plane stress case, stress-strain relationship is given by

$$[D] = \frac{E}{(1-v^2)} \begin{bmatrix} 1 & v & 0 \\ v & 1 & 0 \\ 0 & 0 & (1-v)/2 \end{bmatrix} \qquad \text{Let} \quad \beta = \frac{(1-v)}{2}$$

Then, with element DOFs arranged in the sequence of $[u_1\ u_2\ u_3\ v_1\ v_2\ v_3]^T$

$$[K] = \frac{EtA}{a^2 b^2 (1-v^2)} \begin{bmatrix} \beta\,a^2 & & & & & \\ 0 & b^2 & & & & \\ -\beta\,a^2 & -b^2 & b^2+\beta\,a^2 & & \text{Symmetric} & \\ 0 & -v\,ab & v\,ab & a^2 & & \\ -\beta\,ab & 0 & \beta\,ab & 0 & \beta\,b^2 & \\ \beta\,ab & v\,ab & -ab\,(v+\beta) & -a^2 & -\beta\,b^2 & a^2+\beta\,b^2 \end{bmatrix}$$

If the element DOFs are arranged in the sequence of $[u_1\ v_1\ u_2\ v_2\ u_3\ v_3]^T$, the elements of stiffness matrix are rearranged as

$$[K] = \frac{EtA}{a^2 b^2 (1-v^2)} \begin{bmatrix} \beta\,a^2 & & & & & \\ 0 & a^2 & & & & \\ 0 & -v\,ab & b^2 & & \text{Symmetric} & \\ -\beta\,ab & 0 & 0 & \beta\,b^2 & & \\ -\beta\,a^2 & v\,ab & -b^2 & \beta\,ab & b^2+\beta\,a^2 & \\ \beta\,ab & -a^2 & v\,ab & -\beta\,b^2 & -ab(v+\beta) & a^2+\beta\,b^2 \end{bmatrix}$$

The stiffness matrix calculated using one CST element to model the plate, if the element DOFs are arranged in the sequence of $[u_1\ v_1\ u_2\ v_2\ u_3\ v_3]^T$, is obtained after substituting the given dimensions and material properties, as

$$[K] = \frac{70000 \times 1 \times \left(\dfrac{30 \times 20}{2}\right)}{30^2 \times 20^2 \times \left(1-0.3^2\right)} \begin{bmatrix} 315 & & & & & \\ 0 & 900 & & & \textit{symmetric} & \\ 0 & -180 & 400 & & & \\ -210 & 0 & 0 & 140 & & \\ -315 & 180 & -400 & 210 & 715 & \\ 210 & -900 & 180 & -140 & -390 & 1040 \end{bmatrix}$$

After applying boundary conditions, $u_1 = v_1 = u_3 = v_3 = 0$, these equations reduce to

$$\begin{Bmatrix} P_{X2} \\ P_{Y2} \end{Bmatrix} = \begin{Bmatrix} 50 \\ -100 \end{Bmatrix} = 64.1026 \begin{bmatrix} 400 & 0 \\ 0 & 140 \end{bmatrix} \begin{Bmatrix} u_2 \\ v_2 \end{Bmatrix}$$

Therefore, $u_2 = 0.00195$ mm and $v_2 = -0.01114$ mm

19.9 ISOPARAMETRIC ELEMENTS

The derivation of stiffness matrix by the method described so far involves integration of the strain energy over the surface or volume. For straight boundaries, this integration can be carried out by numerical techniques (Ref Fig. 19.5a). Higher order elements are developed with better displacement functions so that accurate results can be obtained with lesser number of elements (Ref Fig.

19.5b). Inherent disadvantage with these elements is that as the size of the element increases, accuracy of boundary representation reduces since edges of element boundary are always assumed as straight lines. Also, as the size of the polynomial increases, inversion of G matrix, linking nodal displacements to the coefficients of the polynomial, takes more time.

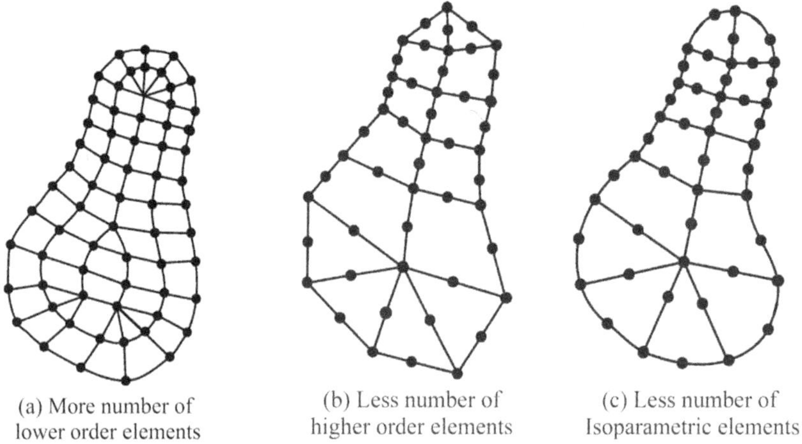

<table>
<tr><td>(a) More number of
lower order elements</td><td>(b) Less number of
higher order elements</td><td>(c) Less number of
Isoparametric elements</td></tr>
</table>

FIGURE 19.5 Representation of curved boundaries

A need was therefore felt to improve the method, in order to idealise the given structure with curved boundaries more accurately. In this method, element geometry as well as displacements are interpolated over the element **using shape functions or interpolation functions N_i in terms of natural or intrinsic or non-dimensional coordinates** (Ref Fig. 19.5c).

Two types of shape functions are commonly used.

- **Lagrange interpolation function**, which matches the function value (displacement) at specified points or nodes

- **Hermite interpolation function**, which matches the function value (displacement) as well as its derivatives (slopes) at the specified nodes

Curvilinear orthogonal coordinates ξ, η and ζ, whose magnitudes vary from -1 to $+1$ in any element, are used in place of cartesian coordinates x, y and z or cylindrical coordinates R, θ and Z. The shape functions are then defined in terms of the natural coordinates, which link displacement at any point in the element to the nodal displacements thus avoiding the need for inversion of G matrix. It is obvious that there will be as many shape functions as the number of nodes in the element. Each shape function will have a value equal to unity at one node and a value equal to zero at all other nodes

Three types of elements are possible:

- If a higher order function is used to represent displacement and a lower order function is used to represent geometry, it is called a **sub-parametric element**.

- If a lower order function is used to represent displacement and a higher order function is used to represent geometry, it is called a **super-parametric element**.

- If functions of same order are used to represent displacement as well as geometry, then the element is called an **iso-parametric element**.

  Of these, iso-parametric elements are most commonly used.

19.9.1 STIFFNESS MATRIX CALCULATION BY SHAPE FUNCTIONS APPROACH

(a) 1-D linear Interpolation for a Truss Element

If the origin is taken at the left end of the truss element, as shown in Fig. 19.6, then the non-dimensional coordinate ξ is given by

$$\xi = (x - x_1) / (x_2 - x_1)$$

At node 1, $x = x_1$ and $\xi = 0$ while at node 2, $x = x_2$ and $\xi = 1$

The shape functions are now defined by $N_1(\xi) = 1 - \xi$; $N_2(\xi) = \xi$

which give the values $N_1 = 1$ for $\xi = 0$ at node 1 and $N_1 = 0$ for $\xi = 1$ at node 2 while $N_2 = 0$ at node 1 and $N_2 = 1$ at node 2.

The Cartesian coordinate x and displacement u are again defined by

$$x = N_1 \times x_1 + N_2 \times x_2 = [N]\{x\}$$

and $u = N_1 \times u_1 + N_2 \times u_2 = [N]\{q\}$

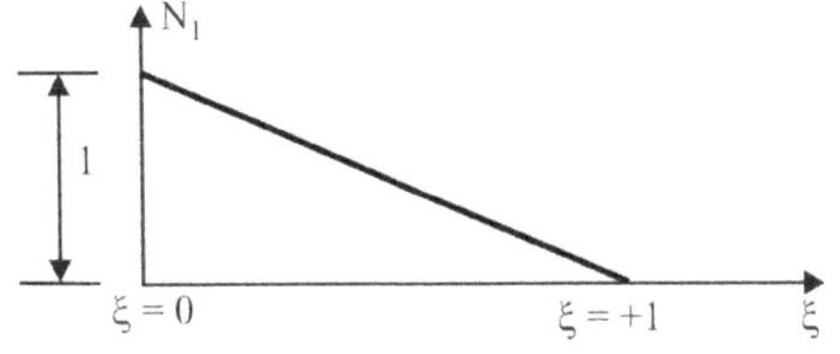
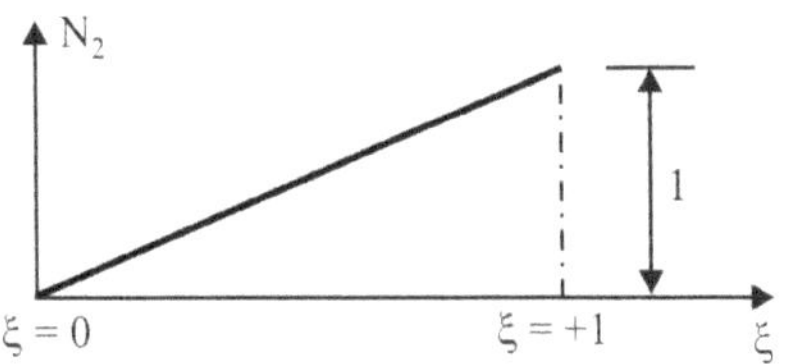

FIGURE 19.6 Shape functions of a 2-noded truss element (Case-2)

$$\varepsilon = \frac{du}{d\xi}\frac{d\xi}{dx} = \left(-u_1 + u_2\right).\frac{1}{x_2 - x_1} = \frac{-u_1 + u_2}{L} = [B]\{q\}$$

where, $[B] = \left(\dfrac{1}{L}\right) [-1\ 1] \; ; \; L = x_2 - x_1$

$$[K_e] = \int_v [B]^T [D][B]\,dv = \int_0^L [B]^T\ E\,[B]\,A\,dx = \dfrac{AE}{L}\begin{bmatrix} 1 & -1 \\ -1 & 1 \end{bmatrix}$$

This stiffness matrix is also identical to the one obtained by polynomial method.

(b) 2-D Linear Interpolation for a Triangular Element

The non-dimensional coordinates ξ and η and the shape functions N_1, N_2 and N_3 at any point P are given by

$$N_1 = \frac{A_1}{A} = \xi \; ; \qquad N_2 = \frac{A_2}{A} = \eta \; ; \qquad N_3 = \frac{A_3}{A} = \zeta$$

where A is the area of the triangle; A_1 is the area of the triangle formed by points P, 2 and 3; A_2 is the area of the triangle formed by points P, 3 and 1; and A_3 is the area of the triangle formed by points P, 1 and 2, as shown in Fig. 19.7. Hence, the shape functions N_1, N_2 and N_3 are also called **area coordinates**.

It can be seen that for any point P, $A = A_1 + A_2 + A_3$

and so $N_1 + N_2 + N_3 = 1$ or $N_3 = 1 - N_1 - N_2 = 1 - \xi - \eta$

The local Cartesian coordinates x and y and displacements u and v along these Cartesian coordinates are given by

$$x = N_1 \times x_1 + N_2 \times x_2 + N_3 \times x_3 = x_1 \times \xi + x_2 \times \eta + x_3 \times (1 - \xi - \eta)$$
$$= x_{13}\times\xi + x_{23}\times\eta + x_3 \quad \text{where,}\quad x_{13} = x_1 - x_3 \quad \text{and} \quad x_{23} = x_2 - x_3$$

Similarly,

$$y = y_{13}\times\xi + y_{23}\times\eta + y_3 \hspace{4cm} .....(19.34)$$

$$\text{where,}\quad y_{13} = y_1 - y_3 \quad \text{and}\quad y_{23} = y_2 - y_3$$

The displacements u and v can be represented in terms of the same non-dimensional coordinates as

$$u = N_1 \times u_1 + N_2 \times u_2 + N_3 \times u_3 = u_1 \times \xi + u_2 \times \eta + u_3 \times (1 - \xi - \eta)$$

$$= (u_1 - u_3) \times \xi + (u_2 - u_3) \times \eta + u_3$$

$$v = N_1 \times v_1 + N_2 \times v_2 + N_3 \times v_3 = v_1 \times \xi + v_2 \times \eta + v_3 \times (1 - \xi - \eta)$$

$$= (v_1 - v_3) \times \xi + (v_2 - v_3) \times \eta + v_3$$

These equations can also be represented in matrix form

$$\begin{Bmatrix} u \\ v \end{Bmatrix} = \begin{bmatrix} N_1 & 0 & N_2 & 0 & N_3 & 0 \\ 0 & N_1 & 0 & N_2 & 0 & N_3 \end{bmatrix} \begin{Bmatrix} u_1 \\ v_1 \\ u_2 \\ v_2 \\ u_3 \\ v_3 \end{Bmatrix} \quad \text{or} \quad \{u\} = [N]\,\{q\}$$

Shape functions are plotted in Fig. 19.7 with non-dimensional coordinates indicated for each node. The three shape functions have a value of unity at one node and a value of zero at all other nodes. In the figure, A_1, A_2 and A_3 indicate areas used for the calculation of non-dimensional coordinates ξ and η of any point P.

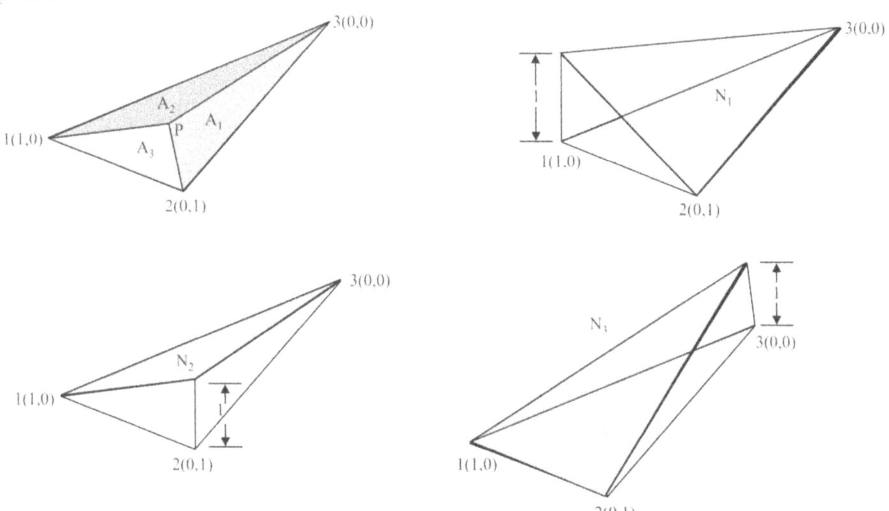

FIGURE 19.7 Shape functions of a 3-noded triangular element

(c) 2-D linear Interpolation for a Quadrilateral Element

Unlike in the case of triangular element, which is identified by three non-dimensional coordinates each having values in the range 0 to 1, a quadrilateral element is identified by two non-dimensional coordinates each having values in the range -1 to $+1$, as shown in Fig. 19.8.

$$x = N_1 \times x_1 + N_2 \times x_2 + N_3 \times x_3 + N_4 \times x_4$$

$$y = N_1 \times y_1 + N_2 \times y_2 + N_3 \times y_3 + N_4 \times y_4$$

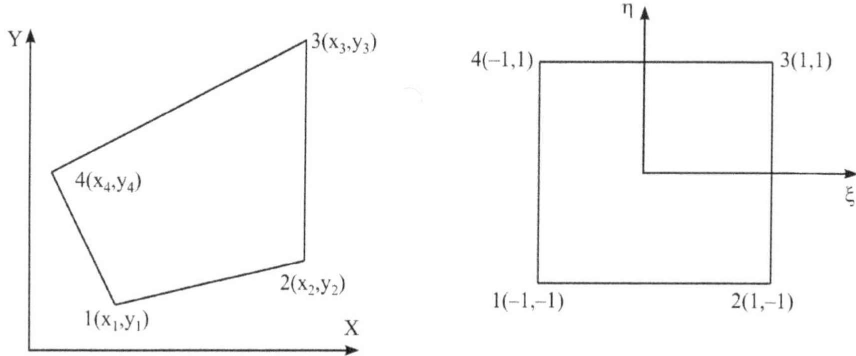

FIGURE 19.8 4-noded quadrilateral element in $\xi - \eta$ coordinate system

The coordinates and displacements at every point in the element are expressed in terms of nodal values, using shape functions N_1, N_2, N_3 and N_4 as

$$u = N_1 \times u_1 + N_2 \times u_2 + N_3 \times u_3 + N_4 \times u_4$$

and $\quad v = N_1 \times v_1 + N_2 \times v_2 + N_3 \times v_3 + N_4 \times v_4$

Shape functions at each node of a 2-D element can be derived as the product of shape functions along ξ direction and η direction passing through the particular node.

Thus, $\quad N_1 = N_{1\xi} \times N_{1\eta}$

where, $\quad N_{1\xi}$ and $N_{1\eta}$ are the shape functions of 1-D elements with

$$\xi = -1 \text{ to } +1 \quad \text{and} \quad \eta = -1 \text{ to } + 1$$

$$\therefore \quad N_1 = \frac{(1-\xi)}{2} \times \frac{(1-\eta)}{2} = \frac{(1-\xi)(1-\eta)}{4}$$

Similarly, we can get $\quad N_2 = \frac{(1+\xi)(1-\eta)}{4} ; \quad N_3 = \frac{(1+\xi)(1+\eta)}{4}$

and $\quad N_4 = \frac{(1-\xi)(1+\eta)}{4}$

Their values are graphically represented in Fig. 19.9.

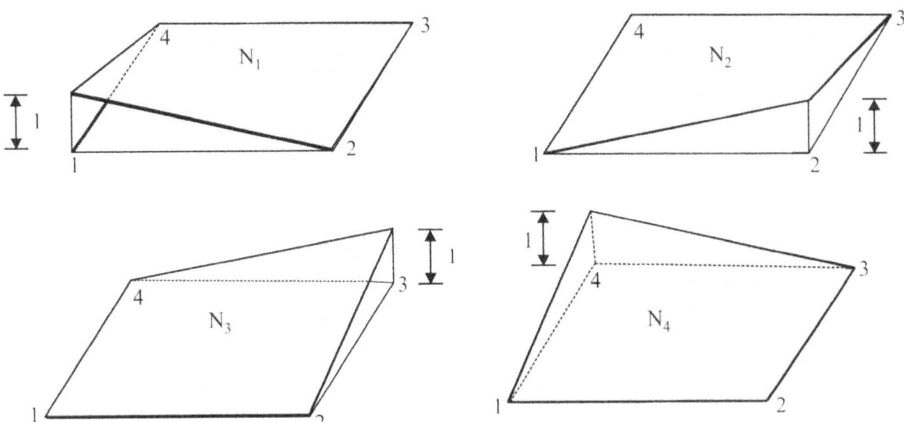

FIGURE 19.9 Shape functions of a 4-noded quadrilateral element

(d) 2-D Quadratic Interpolation for a Quadrilateral Element

This quadrilateral element, with 4 corner nodes and 4 mid-side nodes, is also identified by two non-dimensional coordinates each having values in the range -1 to +1. But, the coordinates and displacement of any point in the element are expressed by using 8 shape functions (Ref. Fig. 19.10), as

$$x = N_1\,x_1 + N_2\,x_2 + N_3\,x_3 + N_4\,x_4 + N_5\,x_5 + N_6\,x_6 + N_7\,x_7 + N_8\,x_8$$

$$y = N_1\,y_1 + N_2\,y_2 + N_3\,y_3 + N_4\,y_4 + N_5\,y_5 + N_6\,y_6 + N_7\,y_7 + N_8\,y_8$$

$$u = N_1\,u_1 + N_2\,u_2 + N_3\,u_3 + N_4\,u_4 + N_5\,u_5 + N_6\,u_6 + N_7\,u_7 + N_8\,u_8$$

$$v = N_1\,v_1 + N_2\,v_2 + N_3\,v_3 + N_4\,v_4 + N_5\,v_5 + N_6\,v_6 + N_7\,v_7 + N_8\,v_8$$

where,

$$N_1 = \frac{-(1-\xi)(1-\eta)(1+\xi+\eta)}{4}; \qquad N_5 = \frac{(1-\xi^2)(1-\eta)}{2}$$

$$N_2 = \frac{-(1+\xi)(1-\eta)(1-\xi+\eta)}{4}; \qquad N_6 = \frac{(1+\xi)(1-\eta^2)}{2}$$

$$N_3 = \frac{-(1+\xi)(1+\eta)(1-\xi-\eta)}{4}; \qquad N_7 = \frac{(1-\xi^2)(1+\eta)}{2}$$

$$N_4 = \frac{-(1-\xi)(1+\eta)(1+\xi-\eta)}{4}; \qquad N_8 = \frac{(1-\xi)(1-\eta^2)}{2}$$

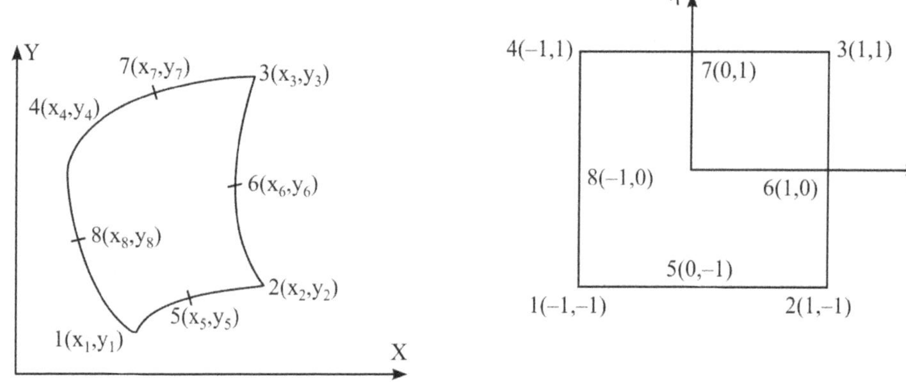

FIGURE 19.10 Mapping of 8-noded quadrilateral in ξ–η coordinate system

19.10 DISTRIBUTED LOADS

In FEM, analysis is confined to structures subjected to specified nodal loads. But, many engineering problems include distributed loads like

- Loads along the length of 1-D elements such as wind load on columns, self weight of beams

- Loads along the edges of 2-D elements such as in-plane pressure on edges of plates; pressure (bending) load normal to the surface of the plate

Such loads are usually represented by *equivalent loads, based on force equilibrium*, in strength of materials.

For example, a uniformly distributed load 'p' on a beam AB of length 'L', as shown in case-1 of Fig. 19.11, is approximated by two equal parts of the beam as shown in case-2. The distributed load on both the parts is transferred to the ends of the beam as point load of pL/2 and moment due to the distributed load represented by the resultant load of pL/2 acting at a distance of L/4 from beam end, as shown in case-3. Thus,

$$P_1 = P_2 = \frac{p\,L}{2} \quad \text{and} \quad M_1 = -M_2 = \left(\frac{p\,L}{2}\right)\left(\frac{L}{4}\right) = \frac{p\,L^2}{8}$$

These statically equivalent loads are shown in case-4

If beam AB is simply supported at its two ends, then the reactions based on the static force equilibrium conditions $\Sigma F = 0$ and $\Sigma M = 0$ will be equal and opposite to these equivalent loads. Statically equivalent loads satisfy force and moment equilibrium, but do not give the same nodal displacements as the actual loads.

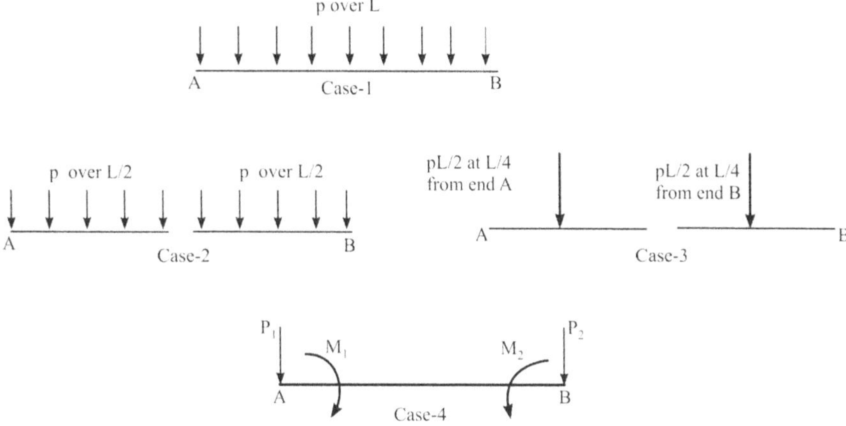

FIGURE 19.11 Statically equivalent loads

19.10.1 CONSISTENT LOADS BASED ON ENERGY

Finite element method is based on minimum potential energy theory for the calculation of stiffness matrix or load-displacement relations. It will, therefore, be *consistent* if the equivalent *loads* are also *based on energy*. Consistent loads, based on energy equivalence, give the same displacements as obtained with the actual loads, in addition to satisfying force and moment equilibrium. For this reason, consistent loads are used in FEM.

The shape functions used to define displacement, in natural coordinate system, over a finite element can also be used to calculate nodal loads vector consistent with the loads distributed over an edge or a surface of an element. These consistent loads are calculated for different types of distributed loads (along a beam; along an edge or area of a plate), as explained below.

(a) For a beam element with uniformly distributed self weight,

$$[N] = [L_1^2 \times (3 - 2L_1) \quad L \times L_1^2 \times L_2 \quad L_2^2 \times (3 - 2L_2) \quad -L \times L_1 \times L_2^2]$$

where $L_1 = 1 - (x/L)$ and $L_2 = x/L$

$$\{P_B\} = b\,h \int_L [N]^T \{\rho\}\,dL = \left(\frac{b\,h\,\rho\,L}{12}\right) \begin{Bmatrix} 6 \\ L \\ 6 \\ -L \end{Bmatrix}$$

(b) Consistent nodal loads corresponding to distributed surface or traction forces are given by $\{P_S\} = \int_S [N]^T \{T\}\,ds$

where $\{T\}$ indicates distributed load over the surface of the body

(c) For a beam element with uniformly distributed pressure load $\{p\}$, in units of force per unit length (w), assuming that the load is uniformly spread across the width 'b' of the cross-section,

$$\int_S p\,ds = b\int_L p\,dL = \int_L w\,dL$$

Thus, $\{P_S\} = b\int_L [N]^T \{p\}dL$

$$= \frac{bLp}{12}\begin{Bmatrix} 6 \\ L \\ 6 \\ -L \end{Bmatrix} = \frac{wL}{12}\begin{Bmatrix} 6 \\ L \\ 6 \\ -L \end{Bmatrix}$$

19.11   OTHER ELEMENTS FOR CONTINUUM ANALYSIS

Treatment in this chapter is limited to basics of modeling and analysis using Finite Element Method. The subject is limited here to 1-D truss and beam elements as well as 2-D plate, which are more common in aircraft structures. For other elements such as 2-D plane strain and axisymmetric elements as well as 3-D elements, which are more common in mechanical engineering applications, the user can refer to any book on FEM.

Also, analysis of shear panels and stiffened plates etc., can be carried out following similar procedure, by suitably modeling the behaviour of the element and evaluating its stiffness matrix. Analysis of continuum for thermal loads is deliberately excluded from the scope of this chapter.

19.12   DYNAMIC ANALYSIS

Dynamics is a special branch of mechanics where inertia of accelerating masses must be considered in the force-deflection relationships. In order to describe motion of the mass system, a component with distributed mass is approximated by a finite number of mass points.

For a simple spring of stiffness 'k' and a lumped mass 'm' under steady state undamped condition of oscillation without external force, the force equilibrium condition of the system using D'Alembert's principle is given by

$$k \times u(t) + m \times \ddot{u}(t) = 0,$$

where, $F_i = -k \times u(t)$ is the reactive elastic force applied to the mass

Displacement in vibration is a simple harmonic motion and can be represented by a sinusoidal function of time as $u(t) = u \times \sin \omega t$

where, ω is the frequency of vibration in radians/sec

Then, acceleration $\ddot{u}(t) = -\omega^2 \times u \times \sin \omega t = -\omega^2 \times u(t)$

$\therefore$ $k \times u(t) + m \times \ddot{u}(t) = (k - \omega^2 \times m) \times u(t) = 0$

In general, for a system with 'n' degrees of freedom, stiffness 'k' and mass 'm' are represented by stiffness matrix [K] and mass matrix [M] respectively.

Then, $([K] - \omega^2 [M]) \{u\} = \{0\}$

Here, [M] is the mass matrix of the entire structure and is of the same order, say n × n, as the stiffness matrix [K]. This is also obtained by assembling element mass matrices in a manner exactly identical to assembling element stiffness matrices. The mass matrix is obtained by two different approaches, as explained subsequently.

19.12.1 NORMALISATION OF MODE SHAPES

The equation of motion of free vibrations $([K] - \omega^2 [M]) \{u\} = \{0\}$ is a system of homogeneous equations (right side vector zero) and hence does not give unique numerical solution. *Mode shape is a set of relative displacements* in various degrees of freedom, while the structure is vibrating in a particular frequency and is usually expressed in normalised form, by following one of the three normalisation methods explained here.

(a) The maximum value of any one component of the eigenvector is equated to '1' and, so, all other components will have a value less than or equal to '1'.

(b) The length of the vector is equated to '1' and values of all components are divided by the length of this vector so that each component will have a value less than or equal to '1'.

(c) The eigenvectors are usually normalised so that

$\{u\}_i^T [M] \{u\}_i = 1$ and $\{u\}_i^T [K] \{u\}_i = \lambda_i$

For a positive definite symmetric stiffness matrix of size n × n, the eigen values are all real and eigenvectors are **orthogonal**

i.e., $\{u\}_i^T [M] \{u\}_j = 0$ and $\{u\}_i^T [K] \{u\}_j = 0$ for $i \neq j$

Solution for any dynamic analysis is an iterative process and, hence, is time consuming. Geometric model of the structure for dynamic analysis can be significantly simplified, giving higher priority for proper representation of distributed mass.

19.12.2 MASS MATRIX

Mass matrix [M] differs from the stiffness matrix in many ways:

(i)   The mass of each element is equally distributed at all the nodes of that element

(ii)   Mass, being a scalar quantity, has **same** effect along the three translational degrees of freedom (u, v and w) and is **not** shared

(iii)   Mass, being a scalar quantity, is not influenced by the local or global coordinate system. Hence, no transformation matrix is used for converting mass matrix from element (or local) coordinate system to structural (or global) coordinate system.

Two different approaches of evaluating mass matrix [M] are commonly considered.

(a)   Lumped Mass Matrix

Total mass of the element is assumed equally distributed at all the nodes of the element in each of the translational degrees of freedom. Lumped mass is not used for rotational degrees of freedom. Off-diagonal elements of this matrix are all zero. This assumption *excludes dynamic coupling* that exists between different nodal displacements. Lumped mass matrices [M] of some elements are given here.

(i) *Lumped mass matrix of truss element* with 1 translational DOF per node along its local X-axis

$$[M] = \frac{\rho AL}{2} \begin{bmatrix} 1 & 0 \\ 0 & 1 \end{bmatrix}$$

(ii) *Lumped mass matrix of plane truss element* in a 2-D plane with 2 translational DOF per node (Displacements along X and Y coordinate axes)

$$[M] = \frac{\rho AL}{2} \begin{bmatrix} 1 & 0 & 0 & 0 \\ 0 & 1 & 0 & 0 \\ 0 & 0 & 1 & 0 \\ 0 & 0 & 0 & 1 \end{bmatrix}$$

Please note that the same lumped mass is considered in each translational degree of freedom (without proportional sharing of mass between them) at each node.

(iii) *Lumped mass matrix of a beam element* in X-Y plane, with its axis along x-axis and with two DOF per node (deflection along Y axis and slope about Z axis) is given below. Lumped mass is not considered in the rotational degrees of freedom.

$$[M] = \frac{\rho AL}{2} \begin{bmatrix} 1 & 0 & 0 & 0 \\ 0 & 0 & 0 & 0 \\ 0 & 0 & 1 & 0 \\ 0 & 0 & 0 & 0 \end{bmatrix}$$

Note that lumped mass terms are not included in 2nd and 4th rows, as well as columns corresponding to rotational degrees of freedom.

(iv) *Lumped mass matrix of a CST element* with 2 DOF per node. In this case, irrespective of the shape of the element, mass is assumed equally distributed at the three nodes. It is distributed equally in all DOF at each node, without any sharing of mass between different DOF

$$[M] = \frac{\rho AL}{3} \begin{bmatrix} 1 & 0 & 0 & 0 & 0 & 0 \\ 0 & 1 & 0 & 0 & 0 & 0 \\ 0 & 0 & 1 & 0 & 0 & 0 \\ 0 & 0 & 0 & 1 & 0 & 0 \\ 0 & 0 & 0 & 0 & 1 & 0 \\ 0 & 0 & 0 & 0 & 0 & 1 \end{bmatrix}$$

(b) Consistent Mass Matrix

Element mass matrix is calculated here, *consistent* with the assumed displacement field or element stiffness matrix. [M] is a banded matrix of the same order as the stiffness matrix. This is evaluated using the same interpolating functions which are used for approximating displacement field over the element. It yields more accurate results but with more computational cost. Consistent mass matrices of some elements are given here.

(i) ***Consistent mass matrix of a Truss element*** along its axis (in local coordinate system)

$$\{u\}^T = [u \quad v]$$

$$[N]^T = [N_1 \quad N_2]$$

where, $N_1 = \dfrac{(1-\xi)}{2}$ and $N_2 = \dfrac{(1+\xi)}{2}$

$$[M] = \int_V [N]\rho[N]^T \, dV = \int_0^L A[N]\rho[N]^T \, dx = \int_{-1}^{+1} A \times \rho \times [N][N]^T \times (dx/d\xi) \times d\xi$$

Here, $x = N_1 x_1 + N_2 x_2 = \dfrac{1-\xi}{2}x_1 + \dfrac{1+\xi}{2}x_2 = \dfrac{(x_2+x_1)}{2} + \dfrac{(x_2-x_1)\xi}{2}$

and $dx = \dfrac{dx}{d\xi} \times d\xi = \det J \, d\xi = \left(\dfrac{L}{2}\right) d\xi$

Using the values of integration in natural coordinate system,

$$[M] = \rho A \left(\frac{L}{2}\right) \int_{-1}^{+1} \begin{bmatrix} (1-\xi)/2 \\ (1+\xi)/2 \end{bmatrix} \begin{bmatrix} (1-\xi)/2 & (1+\xi)/2 \end{bmatrix} d\xi$$

$$= \frac{\rho AL}{8} \begin{bmatrix} \int (1-\xi)^2 d\xi & \int (1-\xi^2) d\xi \\ \int (1-\xi^2) d\xi & \int (1+\xi)^2 d\xi \end{bmatrix}$$

$$= \frac{\rho AL}{8} \begin{bmatrix} (\xi - \xi^2 + \xi^3/3) & (\xi - \xi^3/3) \\ (\xi - \xi^3/3) & (\xi + \xi^2 + \xi^3/3) \end{bmatrix}$$

$$= \frac{\rho AL}{8} \begin{bmatrix} 8/3 & 4/3 \\ 4/3 & 8/3 \end{bmatrix} = \frac{\rho AL}{6} \begin{bmatrix} 2 & 1 \\ 1 & 2 \end{bmatrix}$$

(ii) ***Consistent mass matrix of a*** **Plane *Truss* element**, inclined to global X-axis -Same elements of 1-D mass matrix are repeated in two dimensions (along X and Y directions) without sharing mass between them. Mass terms in X and Y directions are uncoupled.

$$[M] = \frac{\rho AL}{6} \begin{bmatrix} 2 & 0 & 1 & 0 \\ 0 & 2 & 0 & 1 \\ 1 & 0 & 2 & 0 \\ 0 & 1 & 0 & 2 \end{bmatrix}$$

(iii) *Consistent mass matrix of a Beam element*

$$[M] = \rho A \left(\frac{L}{2}\right) \int \{H\}^T \{H\} d\xi$$

with Hermite shape functions $\{H\}$ as used in a beam element

$$= \frac{\rho AL}{128} \int \begin{bmatrix} 2(2 - 3\xi + \xi^3) \\ L(1 - \xi + \xi^2 + \xi^3) \\ 2(2 + 3\xi - \xi^3) \\ L(-1 - \xi + \xi^2 + \xi^3) \end{bmatrix} \times$$

$$\begin{bmatrix} 2(2 - 3\xi + \xi^3) & L(1 - \xi - \xi^2 + \xi^3) & 2(2 + 3\xi - \xi^3) & L(-1 - \xi + \xi^2 + \xi^3) \end{bmatrix} d\xi$$

$$= \frac{\rho AL}{420} \begin{bmatrix} 156 & 22L & 54 & -13L \\ 22L & 4L^2 & 13L & -3L^2 \\ 54 & 13L & 156 & -22L \\ -13L & -3L^2 & -22L & 4L^2 \end{bmatrix}$$

(iv) *Consistent mass matrix of a CST element in a 2-D plane*

$$[N]^T = \begin{bmatrix} N_1 & 0 & N_2 & 0 & N_3 & 0 \\ 0 & N_1 & 0 & N_2 & 0 & N_3 \end{bmatrix}$$

$$[M] = \int [N] \rho [N]^T \, dV = t \int [N] \rho [N]^T \, dA$$

$$= \frac{\rho tA}{12} \begin{bmatrix} 2 & 0 & 1 & 0 & 1 & 0 \\ & 2 & 0 & 1 & 0 & 1 \\ & & 2 & 0 & 1 & 0 \\ & & & 2 & 0 & 1 \\ & & & & 2 & 0 \\ \text{symmetric} & & & & & 2 \end{bmatrix}$$

Note: Natural frequencies obtained using lumped mass matrix are LOWER than exact values.

Example 19.4

Find the natural frequencies of longitudinal vibrations of the constrained stepped shaft of areas A and 2A and of equal lengths (L), as shown. Compare the results obtained using lumped mass matrix approach and consistent mass matrix approach.

Solution

Let the finite element model of the shaft be represented by 3 nodes and 2 truss elements (as only longitudinal vibrations are being considered) as shown,

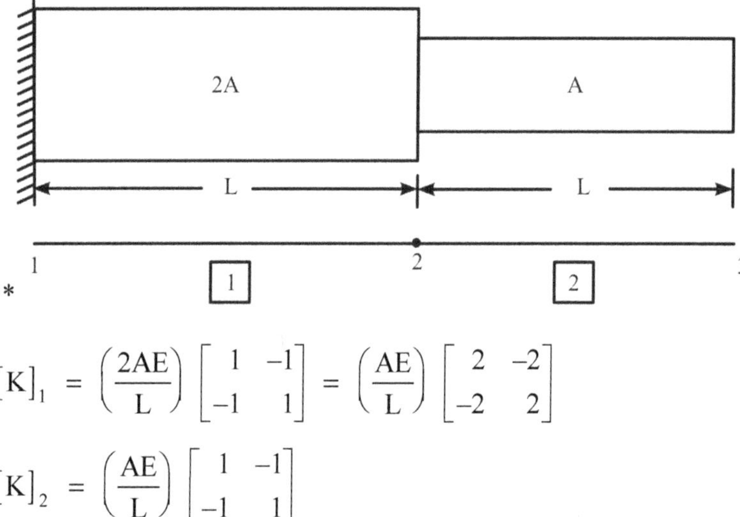

$$[K]_1 = \left(\frac{2AE}{L}\right)\begin{bmatrix} 1 & -1 \\ -1 & 1 \end{bmatrix} = \left(\frac{AE}{L}\right)\begin{bmatrix} 2 & -2 \\ -2 & 2 \end{bmatrix}$$

$$[K]_2 = \left(\frac{AE}{L}\right)\begin{bmatrix} 1 & -1 \\ -1 & 1 \end{bmatrix}$$

(a) *Using lumped mass matrix approach*

$$[M]_1 = \frac{\rho(2A)L}{2}\begin{bmatrix} 1 & 0 \\ 0 & 1 \end{bmatrix} = \frac{\rho AL}{2}\begin{bmatrix} 2 & 0 \\ 0 & 2 \end{bmatrix}; \quad [M]_2 = \frac{\rho AL}{2}\begin{bmatrix} 1 & 0 \\ 0 & 1 \end{bmatrix}$$

Assembling the element stiffness and mass matrices,

$$[K] = \frac{AE}{L}\begin{bmatrix} 2 & -2 & 0 \\ -2 & 3 & -1 \\ 0 & -1 & 1 \end{bmatrix}; \quad [M] = \frac{\rho AL}{2}\begin{bmatrix} 2 & 0 & 0 \\ 0 & 3 & 0 \\ 0 & 0 & 1 \end{bmatrix}$$

Application of boundary condition (node 1 constrained) eliminates row 1 and column 1, thus reducing the size of stiffness and mass matrices to 2 ×

2. Eigen values of the equation $([K] - \omega^2 [M]) \{u\} = \{0\}$ are the roots of the characteristic equation represented by

$$\begin{vmatrix} 3AE/L - \omega^2 3\rho AL/2 & -AE/L \\ -AE/L & AE/L - \omega^2 \rho AL/2 \end{vmatrix} = 0$$

Multiplying all the terms by (L/AE) and substituting $\beta = \dfrac{\rho L^2 \omega^2}{2E}$

$$\begin{vmatrix} 3(1-\beta) & -1 \\ -1 & (1-\beta) \end{vmatrix} = 0$$

or $\quad 3\beta^2 - 6\beta + 2 = 0$

The roots of this equation are $\quad \beta = \dfrac{(3 \pm \sqrt{3})}{3} \quad$ or $\quad 0.423, \ 1.577$

Corresponding eigenvectors are $\quad [0 \quad -0.57734 \quad 1]^T \quad$ for $\quad \beta = 1.577$

$$\text{and} \quad [0 \quad 0.57734 \quad 1]^T \quad \text{for} \quad \beta = 0.423$$

(b) *Using consistent mass matrix approach*

$$[M]_1 = \frac{\rho(2A)L}{6}\begin{bmatrix} 2 & 1 \\ 1 & 2 \end{bmatrix} = \frac{\rho AL}{6}\begin{bmatrix} 4 & 2 \\ 2 & 4 \end{bmatrix}$$

$$[M]_2 = \frac{\rho AL}{6}\begin{bmatrix} 2 & 1 \\ 1 & 2 \end{bmatrix}$$

Assembling the element stiffness and mass matrices,

$$[K] = \frac{AE}{L}\begin{bmatrix} 2 & -2 & 0 \\ -2 & 3 & -1 \\ 0 & -1 & 1 \end{bmatrix}; \quad [M] = \frac{\rho AL}{6}\begin{bmatrix} 4 & 2 & 0 \\ 2 & 6 & 1 \\ 0 & 1 & 2 \end{bmatrix}$$

Application of boundary condition (node 1 constrained) eliminates row 1 and column 1, thus reducing the size of stiffness and mass matrices to 2×2. Eigen values of the equation $([K] - \omega^2 [M]) \{u\} = \{0\}$ are the roots of the characteristic equation represented by

$$\begin{vmatrix} 3AE/L - \omega^2 6\rho AL/6 & -AE/L - \omega^2 \rho AL/6 \\ -AE/L - \omega^2 \rho AL/6 & AE/L - \omega^2 2\rho AL/6 \end{vmatrix} = 0$$

Multiplying all the terms by (L/AE) and substituting $\beta = \dfrac{\rho L^2 \omega^2}{6E}$

$$\begin{vmatrix} 3(1-2\beta) & -(1+\beta) \\ -(1+\beta) & (1-2\beta) \end{vmatrix} = 0$$

or $\quad 11\,\beta^2 - 14\,\beta + 2 = 0$

The roots of this equation are $\beta = \dfrac{\left(7 \pm 3\sqrt{3}\right)}{11}$ $\quad$ or $\quad$ 1.10874, 0.16399

Corresponding eigenvectors are $[0 \quad -0.57734 \quad 1]^{\mathrm{T}}$ $\quad$ for $\quad \beta = 1.577$

$\qquad\qquad\qquad$ and $\qquad [0 \quad 0.57734 \quad 1]^{\mathrm{T}}$ $\quad$ for $\quad \beta = 0.423$

Note: ***Natural frequencies obtained with lumped mass matrices are LOWER*** than those obtained with consistent mass matrices, while the mode shapes are practically same.

C H A P T E R 20

AEROELASTICITY

An aeroplane is borne due to aerodynamic forces generated during flight. These forces mainly depend on the shape and orientation of wing based on the position of flap, aileron, etc. The aerodynamic forces are transferred to the main aeroplane structure, which cause elastic deformation of the structural components. These structural deformations change the shape and orientation of aerodynamic surfaces resulting in change of aerodynamic forces. The interaction of aerodynamic forces and elastic behaviour of aeroplane structure is known as *aeroelasticity*.

Such interactions in an adequately stiff structure, in a static stability condition, reach a stable equilibrium condition. Dynamic loading systems, of which gusts are of primary importance, induce oscillations of structural components. If the natural or resonant frequency of the component is in the region of the frequency of applied loads, then the amplitude of the oscillations may diverge, causing failure. The problem of dynamic stability includes flutter, buffeting and dynamic response.

Wing distortion produces significant changes in lift distribution from that calculated on the assumption of a rigid wing. Actual lift distribution depends on the angle of incidence of the wing at all stations along its span. Obviously, this is effected by twisting of the wing, due to non-coinciding aerodynamic center (or center of pressure) where resultant lift force at that section is considered and the shear center (or center of twist). This twist changes angle of incidence further increasing lift force and, hence, the twist and so on.

At speeds below a critical value, called the *divergence speed*, a condition of static equilibrium is reached in which the torsional moment of aerodynamic forces about the shear center is balanced by the torsional rigidity of the wing.

481

For a straight wing, the redistribution of lift usually causes an outward spanwise movement of the center of pressure resulting in greater bending moment at the wing root. In the case of a swept wing, a reduction in angle of incidence of outboard sections due to bending deflection causes a movement of the center of pressure towards the wing root.

20.1 INFINITE WING TORSIONAL DIVERGENCE – 2-D CASE

Let us first consider the case of a wing of area 'S' without ailerons and in a two-dimensional flow. Representing torsional stiffness of the wing by a spring of stiffness 'K', for moment equilibrium of the wing section for an angle of twist 'θ' of the wing (Ref Fig. 20.1),

$$M_0 + L \times e \times c = K \times \theta$$

where, e×c is the normal distance between the aerodynamic center and the shear center as a function of wing chord c

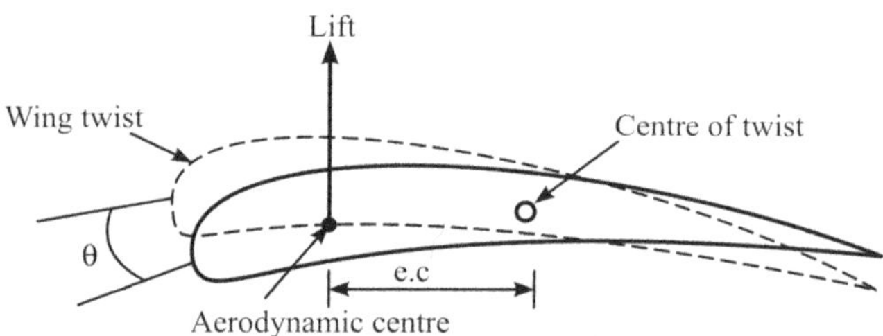

FIGURE 20.1 Increase of wing incidence due to wing twist

Substituting for M_0 and L from aerodynamics, we get

$$[(1/2) \rho \times V^2 \times S \times c \times C_{M,0}] + [(1/2) \rho \times V^2 \times S \times C_L] \times e \times c = K \times \theta$$

where, V   is the free stream velocity of air

$C_{M,0}$ is the coefficient of moment at the aerodynamic center

and C_L is the coefficient of lift

or $(1/2) \rho \times V^2 \times S \times c \times [C_{M,0} + e \times C_L] = K \times \theta$

Substituting, $C_L = C_{L,0} + (\partial C_L / \partial \alpha) \times (\alpha + \theta)$,   where $\partial C_L / \partial \alpha$ is the wing lift curve slope,

$$\theta = \frac{\left(\dfrac{1}{2}\right) \times \rho \times V^2 \times S \times c \times \left[C_{M,0} + e \times C_{L,0} + e \times \left(\dfrac{\partial C_L}{\partial \alpha}\right) \times \alpha \right]}{K - \left(\dfrac{1}{2}\right) \times \rho \times V^2 \times S \times c \times e \times \left(\dfrac{\partial C_L}{\partial \alpha}\right)}$$

Divergence or instability occurs (i.e. θ becomes infinite) when

$$K = (1/2)\,\rho \times V^2 \times S \times c \times e \times (\partial C_L/\partial\alpha)$$

or when the aeroplane moves forward at divergence speed,

$$V_d = \sqrt{2k}\ /\ [\rho \times S \times c \times e \times (\partial C_L/\partial\alpha)]$$

The wing is stable at speeds below V_d. It can be seen from the above equation that V_d can be increased either by stiffening the wing (increasing K) or by reducing the distance 'e × c' between aerodynamic center and shear center. The latter is a more economical solution. If the aerodynamic center coincides or aft of shear center (behind, w.r.t leading edge), then the wing is stable at all speeds.

20.2   FINITE WING TORSIONAL DIVERGENCE – 3-D CASE

20.2.1   STRAIGHT WING WITH ITS FLEXURAL AXIS PERPENDICULAR TO AEROPLANE'S PLANE OF SYMMETRY

Considering a small strip of wing of chord 'c' and width 'dz' along its span (Ref Fig. 20.2a), let 'dL' be the additional lift and 'dM_0' be the additional moment at the aerodynamic center developed by this strip. Let 'T' be the applied torque at the starting section of the strip and 'T + (dT/dz) × dz' is torque at the end section of the strip (Ref Fig. 20.2b). Then, for moment equilibrium of the strip, neglecting weight of the wing

$$[T + (dT/dz)dz] - T + dL \times e \times c + dM_0 = 0$$

Substituting $dL\ = (1/2)\,\rho \times V^2 \times c \times dz \times (\partial C_L/\partial\alpha) \times (\alpha + \theta)$

and $dM_0 = (1/2)\,\rho \times V^2 \times c^2 \times dz \times C_{M,0}$

from aerodynamics where $\partial C_L/\partial\alpha$ is the local two-dimensional curve slope

and $T = G \times J \times (d\theta/dz)$ from torsion equation

and dividing throughout by 'dz', we get a second-order differential equation in θ as

$$G \times J \times (d^2\theta/dz^2) - (1/2)\,\rho \times V^2 \times e \times c^2 \times (\partial C_L/\partial\alpha) \times \theta$$

$$= (1/2)\,\rho \times V^2 \times e \times c^2 \times (\partial C_L/\partial\alpha) \times \alpha + (1/2)\,\rho \times V^2 \times c^2 \times C_{M,0}$$

Its solution for θ is in the form

$$\theta = A \times \sin \lambda z + B \times \cos \lambda z - [\, C_{M,0}\ /\ \{e \times (\partial C_L/\partial\alpha)\} + \alpha\,]$$

where $\lambda^2 = (1/2)\,\rho \times V^2 \times e \times c^2 \times (\partial C_L/\partial\alpha)\ /\ (G \times J)$

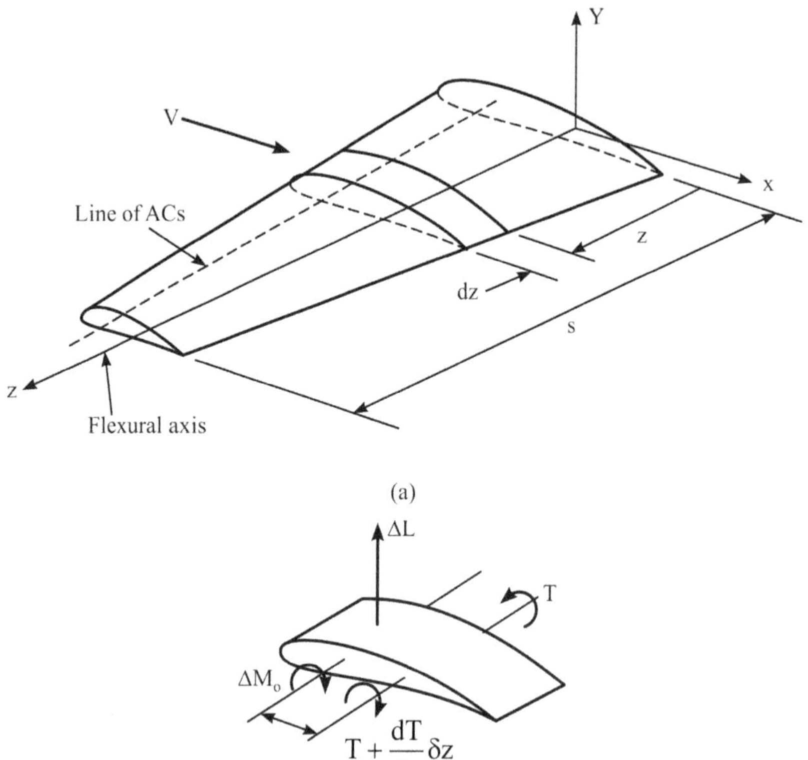

FIGURE 20.2 Wing divergence in a finite wing

A and B are unknown constants to be derived from the boundary conditions $\theta = 0$ when $z = 0$ at the wing root and $d\theta/dz = 0$ when $z = s$ at the wing tip ($T = 0$)

Then,

$$\theta = [\, C_{M,0} / \{e \times (\partial C_L/\partial\alpha)\} + \alpha \,] \times [\, \cos \lambda(s - z) / \cos \lambda s - 1 \,]$$

At divergence, when elastic twist θ becomes infinite, $\cos \lambda s = 0$

$$\text{or} \qquad \lambda s = (2n + 1) \times \pi/2 \quad \text{for} \ n = 0, 1, 2, \ldots \infty$$

The smallest value of divergence speed V_d corresponds to $n = 0$ or $\lambda s = \pi/2$ from which we get

$$V_d = \sqrt{\frac{\pi^2 G \times J}{2\rho \times e \times c^2 \times S^2 \times (\partial C_L / \partial\alpha)}}$$

This equation leads to conservative estimates of V_d when 2-D (infinite wing approximation) lift curve slope ($\partial C_L / \partial \alpha$) is used and gives reasonably accurate value when 3-D (actual wing) lift curve slope ($\partial C_L / \partial \alpha$) is used.

20.2.2 Swept Wing Divergence

When the flexural axis is not perpendicular to the aircraft's plane of symmetry, bending of the wing influences divergence speed V_d. Points A and B on a line perpendicular to the reference axis will deflect by approximately same amount, greater than the deflection of A$'$ along the streamline through B (Ref Fig. 20.3). This reduces stream-wise incidence of the wing and corresponding lift, thereby opposing elastic twist and reducing the possibility of wing divergence. It has been proved that wings with moderate to large sweepback can not diverge.

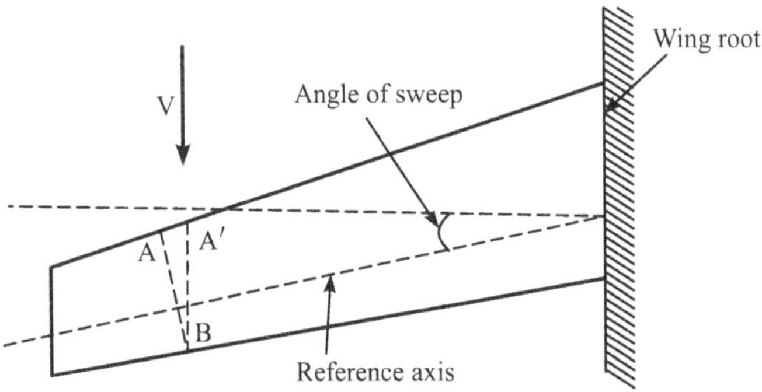

FIGURE 20.3 Divergence on swept wing

20.3   Aileron Effectiveness and Reversal

Downward deflection of an aileron causes a nose down twisting of a flexible wing which reduces the aileron incidence and the lift compared to a rigid wing. The aerodynamic twisting moment on the wing due to aileron deflection increases as the square of the speed but the elastic restoring moment due to torsional stiffness of the wing structure is constant. Therefore, ailerons become less effective as the speed increases until at a particular speed aileron deflection does not produce any rolling moment. This particular speed is called ***aileron reversal speed***. At higher speeds, reversed aileron movements are necessary i.e., a positive increment of wing lift requires an upward aileron deflection and vice versa.

Similar problems occur in the loss of effectiveness and reversal of the rudder and elevator controls. They are complicated by the additional deformation of the fuselage and tail plane. These are not discussed in detail here.

20.3.1 2-D CASE (INFINITE WING)

In a two-dimensional flow over a wing-aileron combination, an aileron deflection 'ξ' produces changes 'dL' in the wing lift 'L' and 'dM_0' in the pitching moment 'M_0' (Ref Fig. 20.4).

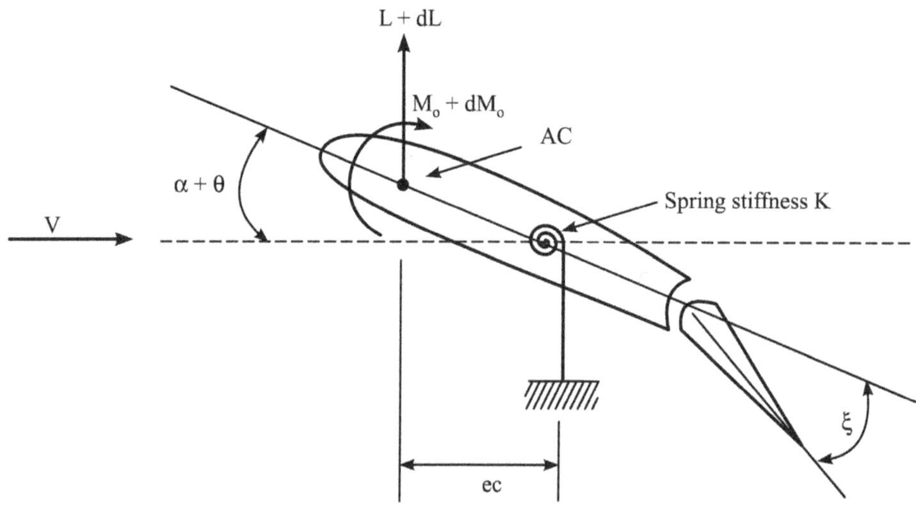

FIGURE 20.4 Aileron effectiveness and reversal speed (2-D wing)

These changes cause an elastic twist 'θ' of the wing. Thus,

$$dL = [(\partial C_L/\partial\alpha) \times \theta + (\partial C_L/\partial\xi) \times \xi] \times [(1/2)\,\rho \times V^2 \times S]$$

and

$$dM_0 = (\partial C_{M,0}/\partial\xi)\times\xi \times [(1/2)\,\rho\times V^2\times S] \times c$$

where, $\partial C_L/\partial\alpha$ is the wing lift curve slope

$\partial C_L/\partial\xi$ is the rate of change of lift coefficient (C_L) with aileron angle (ξ)

and $\partial C_{M,0}/\partial\xi$ is the rate of change of wing pitching moment coefficient ($C_{M,0}$) with aileron angle (ξ)

The moment produced by these increments in lift and pitching moment is balanced by an increment of torque 'dT' about the flexural axis given by

$$dT = K \times \theta = [\{(\partial C_L/\partial\alpha) \times \theta + (\partial C_L/\partial\xi) \times \xi\} \times e$$
$$+ (\partial C_{M,0}/\partial\xi) \times \xi] \times [(1/2)\,\rho \times V^2 \times S] \times c$$

$$\Rightarrow \theta = \frac{\left(\frac{1}{2}\right) \times \rho \times V^2 \times S \times c \times \left[\left(\frac{\partial C_{M,0}}{\partial \xi}\right) \times \xi + e \times C_{L,0} + e \times \left(\frac{\partial C_L}{\partial \xi}\right) \times \xi\right]}{K - \left(\frac{1}{2}\right) \times \rho \times V^2 \times S \times c \times e \times \left(\frac{\partial C_L}{\partial \alpha}\right)}$$

Substituting this value of θ in the expression for dL and simplifying, we get

$$dL = (1/2) \times \rho \times V^2 \times S \frac{\left[\left(\frac{1}{2}\right) \times \rho \times V^2 \times S \times c\right] \times \left(\frac{\partial C_{M,0}}{\partial \xi}\right) \times \left(\frac{\partial C_L}{\partial \alpha}\right) + K \times \left(\frac{\partial C_L}{\partial \xi}\right)}{K - \left(\frac{\partial C_L}{\partial \alpha}\right) \times \left[\left(\frac{1}{2}\right) \times \rho \times V^2 \times S \times c \times e\right]} \times \xi$$

The increment of wing lift is therefore a linear function of aileron angle ξ and becomes zero (or aileron reversal occurs) when

$$[(1/2) \rho \times V^2 \times S \times c] \times (\partial C_{M,0}/\partial \xi) \times (\partial C_L/\partial \alpha) + K \times (\partial C_L/\partial \xi) = 0$$

Aileron reversal speed V_R, corresponding to this condition, is

$$V_R = \sqrt{\frac{-K \times \left(\frac{\partial C_L}{\partial \xi}\right)}{\left(\frac{1}{2}\right) \times \rho \times S \times c \times \left(\frac{\partial C_{M,0}}{\partial \xi}\right) \times \left(\frac{\partial C_L}{\partial \alpha}\right)}}$$

If the wing is considered rigid, i.e. $\partial C_L/\partial \alpha = 0$, the corresponding lift increment

$$dL_R = (\partial C_L/\partial \xi) \times \xi \times [(1/2) \rho \times V^2 \times S]$$

Aileron effectiveness is defined as dL/dL_R or in terms of wing divergence speed V_d and aileron reversal speed V_R as $[1 - (V^2/V_R^2)] / [1 - (V^2/V_d^2)]$

It is seen from the above equation that when $V_R = V_d$ the aileron is 100% effective at all speeds. This condition occurs when $\partial C_L/\partial \xi = - (\partial C_{M,0}/\partial \xi) / e$ i.e., nose-down wing twist caused by aileron deflection is balanced completely by the nose-up twist produced by the increase in wing lift.

20.3.2 3-D CASE (FINITE WING)

Consider a small strip of the wing of width 'dz' at a distance 'z' from the wing root in the aileron region. If a small deflection of the aileron by 'ξ' produces a rolling velocity 'p' rad/sec, then the incidence at the section is reduced by pz/V (Ref Fig. 20.5). Since the ailerons on the starboard wing and port wing operate in opposite directions, this roll results in increase of incidence on the section of the other wing by the same magnitude.

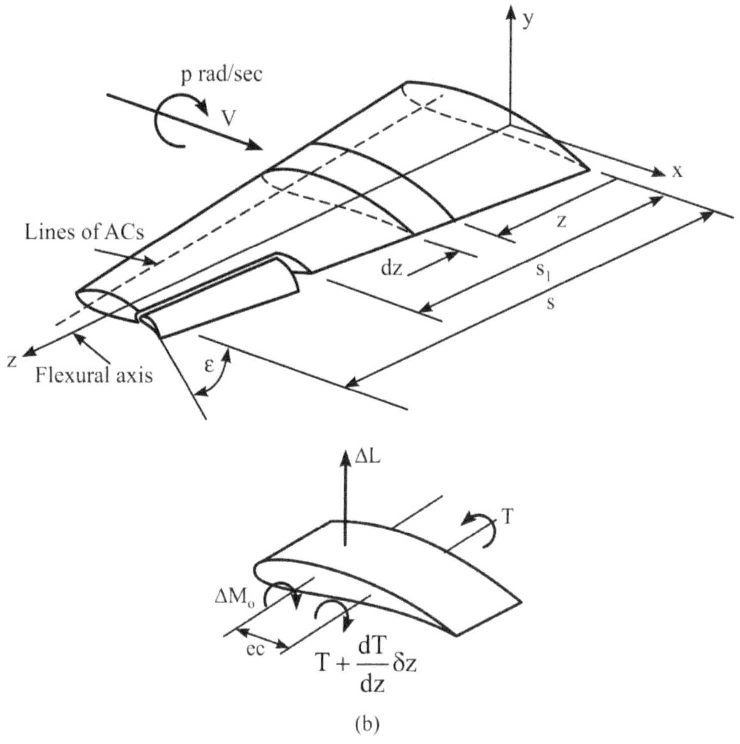

FIGURE 20.5 Aileron effectiveness and reversal speed (Finite wing)

The change in lift 'dL' from level flight condition on any one wing is given by

$$dL = [(1/2)\,\rho \times V^2 \times c \times dz] \times [(\partial C_L/\partial\alpha) \times (\theta - p \times z/V)$$

$$+ (\partial C_L/\partial\xi) \times f_a(z) \times \xi]$$

and $$dM_0 = [(1/2)\,\rho \times V^2 \times c^2 \times dz] \times [(\partial C_{M.0}/\partial\xi) \times f_a(z) \times \xi]$$

Here, function $f_a(z)$ represents aileron forces and moments along the span such that

$$f_a(z) = 0 \quad \text{for} \ \ 0 \le z \le s_1 \quad \text{and} \quad\quad f_a(z) = 1 \quad \text{for} \ \ s_1 \le z \le s$$

where, s_1 and s are the distances of start and end of aileron from the wing root.

For the moment equilibrium of the elemental strip,

$$(dT/dz) \times dz + dL \times e \times c + dM_0 = 0$$

Substituting for dL, dM$_0$ and T = G×J×(dθ/dz), the above equation can be written as

$$G \times J \times (d^2\theta/dz^2) \times dz + [(1/2) \rho \times V^2 \times c^2 \times dz] \times [\{(\partial C_L/\partial\alpha) \times (\theta - p \times z/V)$$
$$+ (\partial C_L/\partial\xi) \times f_a(z) \xi\} \times e + \{(\partial C_{M.0}/\partial\xi) \times f_a(z) \times \xi\}] = 0$$

Substituting $[(1/2) \rho \times V^2 \times e \times c^2 \times (\partial C_L/\partial\alpha)] / (G \times J) = \lambda^2$, the above equation can be simplified as

$$d^2\theta/dz^2 + \lambda^2 \theta = \lambda^2 \times (p \times z/V) - [\lambda^2 / (\partial C_L/\partial\alpha)]$$
$$\times [(\partial C_L/\partial\xi) + (1/e) \times (\partial C_{M.0}/\partial\xi)] \times f_a(z) \times \xi$$

Solution for 'θ' of this 2nd order partial differential equation satisfying the boundary conditions, $\theta = 0$ at $z = 0$ and   dθ/dz $= 0$ at $z = s$ is given by

$$\theta = (p/V) \times [z - (\sin \lambda z/\lambda) \times \cos \lambda s] - [1/(\partial C_L/\partial\alpha)] \times [(\partial C_L/\partial\xi)$$
$$+ (1/e) \times (\partial C_{M.0}/\partial\xi)] \times [f_a(z) \{1 - \cos \lambda(z - s_1)\}$$
$$- \{\sin \lambda(s - s_1)/\cos \lambda s\} \times \sin \lambda s] \xi$$

where, $\cos \lambda(z - s_1) = 0$ when $z < s_1$

Since ailerons on opposite wings deflect in opposite directions, jointly contributing to the rolling motion of the aeroplane, we have

$$2 \int dL \times z = 0 \text{ in the limits } 0 \leq z \leq s$$
$$\Rightarrow \quad \int (\partial C_L/\partial\alpha) \times (\theta - p \times z/V) \times z \times dz = - \xi \int (\partial C_L/\partial\xi) \times f_a(z) \times z \times dz$$

The aileron reversal speed occurs when aileron effectiveness (ps/V) / ξ is zero

20.4   FLUTTER

It is defined as the dynamic instability of an elastic body in an airstream. It is found most frequently in aircraft structures subjected to large aerodynamic loads such as wings, tail units and control surfaces. Critical or flutter speed V$_f$ is the lowest airspeed at which a given structure with sustained simple harmonic motion. Flight at speeds below and above the flutter speed represents conditions of stable and unstable (or divergent) structural oscillation respectively.

Flutter speed is a function of natural modes of vibration of the elastic body and its determination for the continuous structure of an aircraft is a complex process. Simplifying assumptions such as breaking down the structure into a number of concentrated or lumped masses connected by weightless elastic beams (springs) are made. Generally, an elastic lumped mass system having just one degree of freedom can not be unstable.

20.4.1 NATURAL MODES OF A LUMPED MASS SYSTEM

The equation of motion of a simple spring-mass system of spring stiffness 'k' and mass 'm' without any damping force, displaced by a small amount x_0 and suddenly released, is given by

$$m \times (d^2x/dt^2) + k \times x = 0 \ldots$$

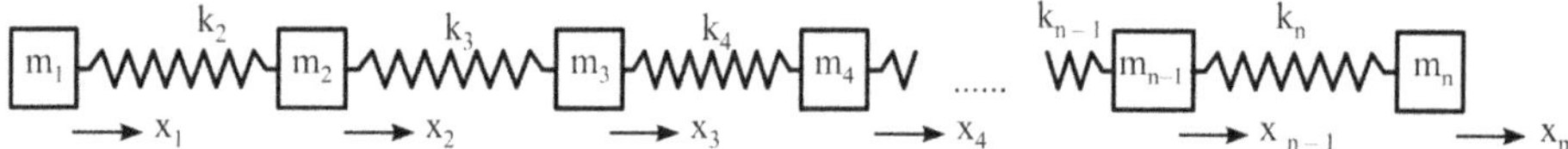

The solution to the above 2nd order differential equation gives simple harmonic motion of the form $x = x_0 \times \sin(\omega t + \varepsilon)$ where $\omega = \sqrt{(K/m)}$ and ε is the phase angle. The natural frequency of oscillation is $\omega/2\pi$ cycles per sec (Hertz) and its amplitude is x_0. It can be seen that amplitude is a function of initial disturbance while the frequency of oscillation is independent of it.

A system of 'n' masses connected by 'n − 1' springs, whose motion takes place only along the spring axes, has 'n' degrees of freedom and 'n' modes of vibration. They can be determined by solving a set of 'n' simultaneous 2nd order differential equations. In each of these modes, the masses oscillate in phase so that they all attain maximum amplitude at the same time and pass through zero displacement at the same time. Each mode has a particular set of amplitude and frequency values. The natural modes of vibration are orthogonal, which can be proved by the fact that the product of inertia forces in one mode and the displacements in the other results in zero work done. It also means that the natural modes are independent of one another and the response of each mode to an externally applied force can be found without reference to the other modes. Complete response of the system to the applied loads can be found by calculating the response of the system in the individual modes and summing up these responses.

Just as equation of motion for a single degree of freedom undamped spring-mass system was represented by displacement or stiffness approach as
$$m \times (d^2x/dt^2) + k \times x = 0$$
It can also be expressed in terms of force or flexibility approach as
$$m \times (d^2x/dt^2) + x / f = 0$$

where f is the displacement per unit applied force

Let us consider a 'n' degrees of freedom spring-mass system having masses 'm_i' and displacements 'x_i' and flexibility coefficients 'f_{ij}' representing displacement at 'i' due to unit force at 'j', the equations of motion can be written as

$$\sum m_i \times (d^2x_i/dt^2) \times f_{ij} + x_i = 0 \quad \text{for } i,j = 1,\ldots n$$

With x as a sine or cosine function $d^2x_i/dt^2 = \omega^2 \times x_i$ and for a non-trivial solution, determinant of the matrix of coefficients should be zero. i.e.,

$$\begin{vmatrix} \omega^2 m_1 f_{11} - 1 & \omega^2 m_2 f_{12} & \cdots & \omega^2 m_i f_{1i} & \cdots & \omega^2 m_n f_{1n} \\ \omega^2 m_1 f_{21} & \omega^2 m_2 f_{22} - 1 & \cdots & \omega^2 m_i f_{2i} & \cdots & \omega^2 m_n f_{2n} \\ \cdots & \cdots & \cdots & \cdots & \cdots & \cdots \\ \omega^2 m_1 f_{i1} & \omega^2 m_2 f_{i2} & \cdots & \omega^2 m_i f_{ii} - 1 & \cdots & \omega^2 m_n f_{in} \\ \cdots & \cdots & \cdots & \cdots & \cdots & \cdots \\ \omega^2 m_1 f_{n1} & \omega^2 m_2 f_{n2} & \cdots & \omega^2 m_i f_{ni} & \cdots & \omega^2 m_n f_{nn} - 1 \end{vmatrix} = 0$$

This equation gives a nth order polynomial in ω^2 and therefore will have 'n' roots. Some of them may be repeated and some others may be complex roots in pairs. A zero root indicates unconstrained rigid body motion.

Normal mode of vibration, corresponding to each natural frequency, gives proportional displacements and absolute displacements depend on amplitude or applied load at each mass.

Example 20.1

Determine normal modes of vibration of a weightless horizontal cantilever supporting masses m/3 and m at points 1 and 2 at distances of L and L/2 from the fixed end. Assume flexural rigidity of the cantilever as EI. Also calculate ratio of the displacements at 1 and 2 in the 2 modes of vibration

Solution

Since the cantilever moves normal to its length, let 'v_1' and 'v_2' be the displacements at 1 and 2. Then, the equations of motion can be written as

$$(m/3) \times (d^2v_1/dt^2) \times f_{11} + (m) \times (d^2v_2/dt^2) \times f_{12} + v_1 = 0$$

$$(m/3) \times (d^2v_1/dt^2) \times f_{21} + (m) \times (d^2v_2/dt^2) \times f_{22} + v_2 = 0$$

Flexibility coefficients are calculated from the relation $f_{ij} = \int (M_i \times M_j / EI) \, dz$

where, M_i and M_j are the bending moments at a distance 'z' from the fixed end due to unit loads at I and J respectively

Here, $M_1 = 1 \times (L - z)$ for $0 \le z \le L$

$M_2 = 1 \times (L/2 - z)$ for $0 \le z \le L/2$ and $M_2 = 0$ for $L/2 \le z \le L$

Then, $f_{11} = \int [M_1 \times M_1 / (E \times I)]\, dz = [1/(E \times I)] \times \int (L - z)^2\, dz$ for $0 \le z \le L$

$\qquad = L^3/(3E \times I)$

$f_{22} = \int [M_2 \times M_2 / (E \times I)]\, dz = [1/(E \times I)] \times \int (L/2 - z)^2\, dz$ for $0 \le z \le L/2$

$\qquad = L^3/(24E \times I)$

$f_{12} = \int [M_1 \times M_2 / (E \times I)]\, dz = [1/(E \times I)] \times \int (L - z) \times (L/2 - z)\, dz$

$\qquad\qquad\qquad\qquad\qquad\qquad\qquad\qquad\qquad$ for $0 \le z \le L/2$

$\qquad = L^3/(48E \times I)$

From Maxwell's reciprocal theorem, $f_{ij} = f_{ji}$ Therefore, $f_{21} = f_{12} = L^3/(48E \times I)$

For a non-trivial solution to exist,

$$\begin{vmatrix} 1 - 16\lambda\omega^2 & -15\lambda\omega^2 \\ 5\lambda\omega^2 & -(1-6)\lambda\omega^2 \end{vmatrix} = 0$$

where, $\lambda = m \times L^3 / (3 \times 48 \times E \times I)$

Solving this equation, we get $\lambda \times \omega^2 = 1/21$ or 1

$\qquad => \quad \omega^2 = 3 \times 48 \times E \times I / (21m \times L^3)$ or $3 \times 48 \times E \times I / (m \times L^3)$

Substituting in any one of the two equations of motion

$\qquad (1-16\lambda \times \omega^2) \times v_1 - 15\lambda \times \omega^2 \times v_2 = 0$

or $\quad 5\lambda \times \omega^2 \times v_1 - (1 - 6) \times \lambda \times \omega^2 \times v^2 = 0,$

the first frequency value, we get

$\quad v_1/v_2 = 15\lambda \times \omega^2 / (1 - 16\lambda \times \omega^2) = 15 \times (1/21) /[1 - 16 \times (1/21)] = 3,$ a +ve quantity

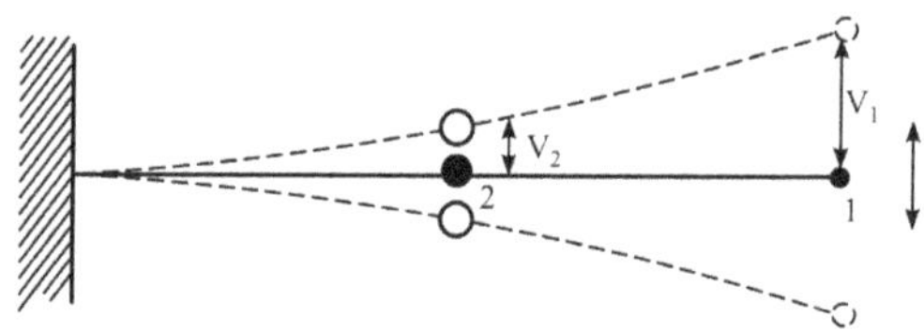

First mode of vibration

while, substituting the second frequency value gives

$v_1/v_2 = 15\lambda \times \omega^2 / (1 - 16\lambda \times \omega^2) = 15 \times (1) /[1 - 16 \times (1)] = -1,$ a –ve quantity

which means the two masses move in opposite directions, as shown

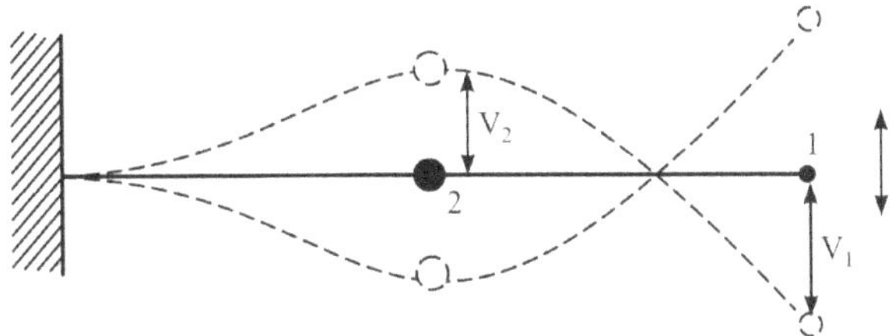

Second mode of vibration

Additional Practice

1. Find the two natural frequencies of the weightless beam-mass system shown in figure, if the two segments are perpendicular to each other. For the beam $G \times J = (2/3) \times E \times I$

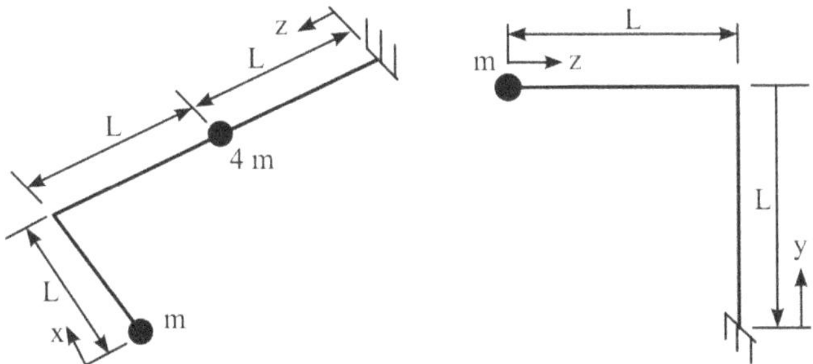

Figure for Problem 1   Figure for Problem 2

2. Determine the natural frequencies of the system shown in figure. Take $E \times I = 1.44 \times 10^6$ N m^2; L = 0.76 m; Radius of gyration of the mass = 0.152 m and its weight is 1435 N

20.4.2 NATURAL MODES OF A DISTRIBUTED MASS SYSTEM

In general, all structural members have distributed masses and, therefore, distributed inertia. The equations of motion of beams with distributed masses are derived on the assumption that vibration occurs in one of the principal planes of bending and that the effects of rotary inertia and shear displacements are neglected.

Consider a small segment (of length 'dx' at a distance 'x') of uniform beam of cross section 'A' vibrating in a principal plane about some axis OX. Let 'v' be the displacement of the element at an instant of time 't' and the shear force and bending moment at the two ends of the segment be S_Y and M_Z on its left end with increments of $(\partial S_Y/\partial x) \times dx$ and $(\partial M_Z/\partial x) \times dx$ over the segment. In addition, there is an inertia force acting opposite to the direction of motion of magnitude $\rho \times A \times dx \times (\partial^2 v/\partial t^2)$ acting at the mid-section. We know, shear force at any section, $S_Y = (\partial M_Z/\partial x)$.

Considering force equilibrium in Y-direction,

$$[S_Y + (\partial S_Y/\partial x) \times dx] - S_Y - \rho \times A \times dx \times (\partial^2 v/\partial t^2) = 0$$

$$\Rightarrow \quad \partial S_Y/\partial x = \partial^2 M_Z/\partial x^2 = \rho \times A \times (\partial^2 v/\partial t^2)$$

Substituting $M_Z = - E \times I \times (\partial^2 v/\partial x^2)$ from simple bending theory,

$$E \times I \times (\partial^4 v/\partial x^4) + \rho \times A \times (\partial^2 v/\partial t^2) = 0$$

In the normal modes of vibration, each element of the beam describes simple harmonic motion given by $\quad v(x,t) = v(x) \times \sin(\omega t + \varphi)$

where, $v(x)$ is the amplitude of vibration at any section x

and $\quad \varphi$ is the phase angle

With this understanding, we use v to mean $v(x)$ in the subsequent equations.

$$\ddot{v} = \frac{d^2 v}{dt^2} = -w^2 v$$

Substituting this in the above equation, we get

$$d^4 v/dx^4 - [\rho \times A \times \omega^2/(E \times I)] \times v = 0 \quad \text{or} \quad d^4 v/dx^4 - \lambda^4 \times v = 0$$

where, $\quad \lambda^4 = \rho \times A \times \omega^2/(E \times I)$

The general solution for this 4th order differential equation is of the form

$$v = C_1 \times \sin \lambda x + C_2 \times \cos \lambda x + C_3 \times \sinh \lambda x + C_4 \times \cosh \lambda x$$

where, C_1, C_2, C_3 and C_4 are constants, which depend on the end conditions of the beam

(i) $v = 0$ and $M_Z/(E \times I) = d^2 v/dx^2 = 0$

if the end is simply supported or pinned

(ii) $v = 0$ and slope, $dv/dx = 0$ if the end is fixed

(iii) $M_Z/(E \times I) = d^2 v/dx^2 = 0$ and $S_Y = \partial M_Z/\partial x = d^3 v/dx^3 = 0$

if the end is free

Example 20.2

Determine the first three natural modes of vibration and the corresponding natural frequencies of the uniform beam, with one end simply supported and the other end on roller support

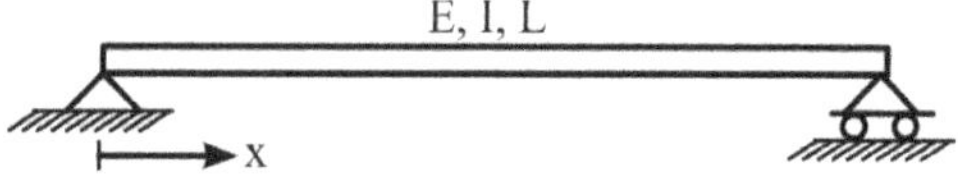

Solution

Since both ends of the beam are simply supported, $v = 0$ and $d^2 v/dx^2 = 0$ at $x = 0$ and at $x = L$

Applying these conditions at x = 0 in the equations

$$v = C_1 \times \sin \lambda x + C_2 \times \cos \lambda x + C_3 \times \sinh \lambda x + C_4 \times \cosh \lambda x$$

$$d^2v/dx^2 = -\lambda^2 \times C_1 \times \sin \lambda x - \lambda^2 \times C_2 \times \cos \lambda x + \lambda^2 \times C_3 \times \sinh \lambda x$$
$$+ \lambda^2 \times C_4 \times \cosh \lambda x$$

give $(v)_{X=0} = 0 = C_2 + C_4$ and $(d^2v/dx^2)_{X=0} = 0 = -\lambda^2 \times C_2 + \lambda^2 \times C_4$

These two conditions are simultaneously satisfied only if $C_2 = C_4 = 0$

Applying the end conditions at x = L give

$$(v)_{X=L} = 0 = C_1 \sin \lambda L + C_3 \sinh \lambda L \qquad \text{since, } \cos \lambda L = \cosh \lambda L = 0$$

and $(d^2v/dx^2)_{X=L} = 0 = -\lambda^2 \times C_1 \times \sin \lambda L + \lambda^2 \times C_3 \times \sinh \lambda L$

The only non-trivial solution ($\lambda L \neq 0$) exists when $C_3 = 0$ and $\sin \lambda L = 0$

$$\Rightarrow \lambda L = n \pi \qquad \text{for } n = 1,2,3,\ldots$$

$$\Rightarrow \omega^2 = \lambda^4 \times [E \times I/(\rho \times A)] = (n \times \pi/L)^4 \times [E \times I/(\rho \times A)]$$
$$\text{for } n = 1,2,3,\ldots$$

$$\Rightarrow \text{frequencies, } f = \omega_n/(2\pi) = (1/2\pi) \times (n \times \pi/L)^2 \sqrt{[E \times I/(\rho \times A)]}$$
$$\text{for } n = 1,2,3,\ldots$$

First two mode shapes are shown in Fig 20.6.

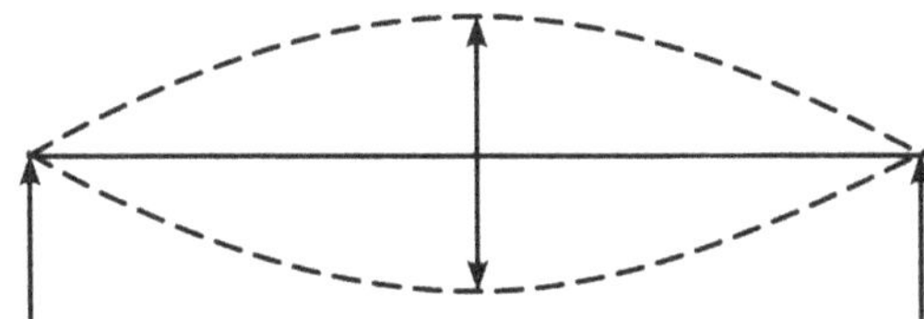

(a) First mode shape

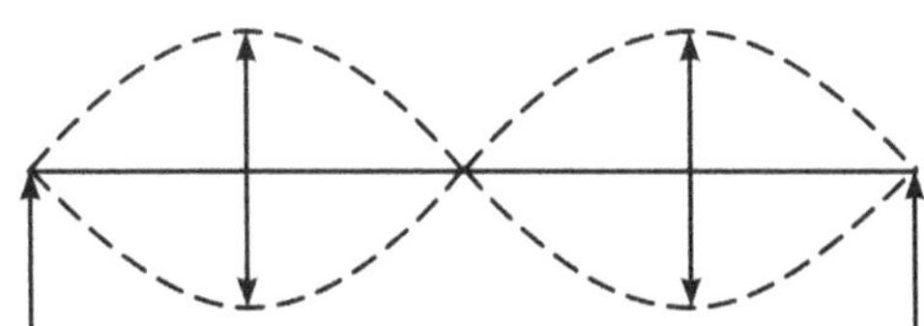

(b) Second mode shape

FIGURE 20.6   Simply supported beam - Natural modes of vibration

Example 20.3

Find the first three natural modes of vibration and corresponding frequencies for a uniform cantilever beam of length L and flexural rigidity E × I

Solution

The fixed end boundary conditions $v = 0$ and $dv/dx = 0$ at $x = 0$ and free end conditions of zero bending moment or $d^2v/dx^2 = 0$ and zero shear force or $d^3v/dx^3 = 0$ give $C_1 = -C_4$ and $C_2 = -C_3$. Substituting these conditions yield the frequency equation

$$\text{Cos } \lambda L \times \cosh \lambda L + 1 = 0$$

Solution of the first three roots of this equation graphically or by Newton's method gives

$$\lambda_1 L = 1.875 \; ; \quad \lambda_2 L = 4.694 \quad \text{and} \quad \lambda_3 L = 7.855$$

Corresponding mode shapes are shown in Fig 20.7.

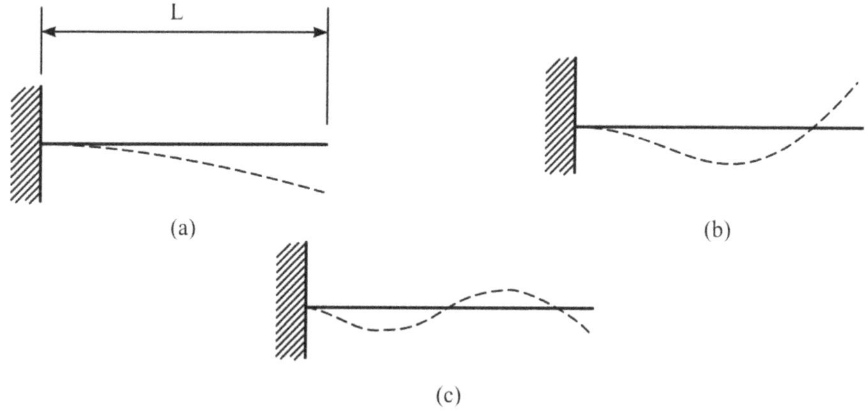

FIGURE 20.7 Cantilever - Natural modes of vibration

20.4.3 DYNAMIC INSTABILITY OR FLUTTER

It is possible for systems having two or more degrees of freedom to be unstable. The forces associated with each individual degree of freedom can interact, causing divergent oscillations for certain phase differences. Flutter of a wing in which flexural and torsional modes are coupled is an important example of this type of instability. In pure bending or pure torsional oscillation, the aerodynamic forces produced by the effective wing incidence oppose the motion. This is not the case in the combined oscillation when the maximum twist occurs at zero bending and vice versa, being 90^0 out of phase.

At the position of zero bending, twisting of the wing causes a positive geometric incidence and, therefore, an aerodynamic force in the same direction as the motion of the wing. A similar but reversed situation exists as the wing moves in a downward direction; the negative geometric incidence due to wing twist causes a downward aerodynamic force. Thus, although the effective wing incidence produces aerodynamic forces which opposes the motion at all stages, the aerodynamic forces associated with geometric incidence have a destabilizing effect. At a certain speed, the destabilization action becomes greater than the stabilizing forces and the oscillations diverge (or flutter occurs).

The phenomenon involving two distinctly different types of oscillating motions (such as bending and twist) interacting with each other resulting in divergent motion is known as ***classical flutter***. The phenomenon of divergence involving only one type of oscillating motions is known as ***non-classical flutter***. Many of them are possible, such as

- ***Stalling flutter***, which is the diverging self-excited oscillations at high incidence above a critical speed, for particular positions of the spanwise axis of twist

- ***Aileron buzz***, which occurs at high subsonic speeds associated with the shock wave on the wing forward of the aileron. If the aileron oscillates downwards, flow over the upper surface of the wing accelerates intensifying the shock and resulting in a reduction in pressure in the boundary layer behind the shock tending to suck back the aileron to its neutral position. Reverse phenomenon takes place when the aileron oscillates upwards

- ***Buffeting***, which is a resonant oscillation produced most commonly in a tail plane due to eddies caused by poor airflow in the wing wake

20.4.4 CLASSICAL FLUTTER

Two different phenomenon, which interact and cause flutter, are said to be *coupled*. Different forms of coupling occur – inertial, aerodynamic and elastic. Consider a small length of wing of mass 'm'. Let 'gc' be the normal distance between its centroid and the flexural axis (Ref Fig 20.8).

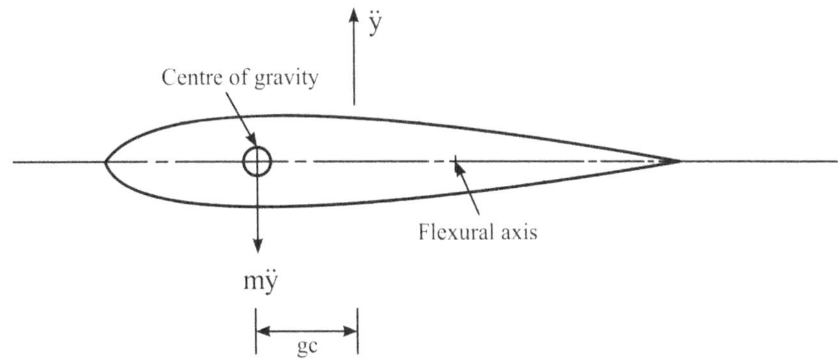

FIGURE 20.8 Inertial coupling of a wing

- **Inertial coupling** – It is *associated with vertical motion and angular motion* of the wing. An upward acceleration ÿ induces a downward inertia force m × ÿ through its centroid, which produces a torsional moment of (m × ÿ) × g × c about its flexural axis causing the wing to twist. Conversely, an angular acceleration $\ddot{\alpha}$ about the flexural axis causes a linear acceleration of g × c × $\ddot{\alpha}$ at the center of gravity with a corresponding inertia force m × (g × c × $\ddot{\alpha}$). It is seen that inertia torque due to unit linear acceleration (m × g × α) is equal to inertia force due to unit angular acceleration (m × g × c) – possess symmetry. Inertial coupling can be prevented if the center of gravity coincides with the flexural axis.

- **Aerodynamic coupling** – It is *associated with changes of lift produced by wing rotation or translation*. A change of wing incidence due to rotation of the wing induces a change of lift, which causes vertical movement. Conversely, a vertical motion with velocity ẏ results in an effective change in incidence and, hence, lift which causes rotation. These aerodynamic forces, which oscillate in flutter condition, act through a specific point known as the ***center of independence***.

- *Elastic coupling* – It is coupling of stiffness associated with translation and rotation. Let us represent stiffness of the wing by a spring of stiffness 'k' at the flexural axis. Vertical displacement of the wing 'y' of an arbitrary point 'O' produces spring force 'k×y' and, hence, torque 'k × y × d'. Conversely, rotation of the wing 'α' produces spring force 'k × α × d' and, hence, vertical displacement. Here also, moment due to unit displacement (k × d) is equal to the force due to unit rotation (k × d) – possess symmetry

Aerodynamic coupling and elastic coupling can be eliminated if the center of independence coincides with the flexural axis. However, in practice, center of independence is usually forward of the flexural axis while the center of gravity is behind it, which promote flutter.

20.4.5 CRITICAL FLUTTER SPEED

Consider a wing section of chord 'c' oscillating harmonically in an airflow of velocity 'V' and density 'ρ' having instantaneous displacements, velocities and accelerations of translation (Y, Y' and Y'') and rotation (α, α' and α''). The oscillation causes a change in lift from steady state lift. Total aerodynamic lift on the wing section due to the oscillating motion is given by

$$L = L_y \times y + L_{\dot{y}} \times \dot{y} + L_{\ddot{y}} \times \ddot{y} + L_\alpha \times \alpha + L_{\dot{\alpha}} \times \dot{\alpha} + L_{\ddot{\alpha}} \times \ddot{\alpha}$$

$$= \ell_y \times \rho \times c \times V^2 \times Y/c + \ell_y' \times \rho \times c^2 \times V \times Y'/c + \ell_y'' \times \rho \times c^3 \times Y''/c$$
$$+ \ell_\alpha \times \rho \times c \times V^2 \times \alpha + \ell_\alpha' \times \rho \times c^2 \times V \times \alpha' + \ell_\alpha'' \times \rho \times c^3 \times \alpha'' \quad .....(20.1)$$

where, ℓ_y, ℓ_y', ℓ_y'', ℓ_α, ℓ_α' and ℓ_α'' are non-dimensional coefficients,
analogous to lift curve slopes in steady motion

Similarly, total nose-up or nose-down moment on the wing section is

$$M = M_y \times y + M_{\dot{y}} \times \dot{y}' + M_{\ddot{y}} \times \ddot{y} + M_\alpha \times \alpha + M_{\ddot{\alpha}} \times \ddot{\alpha} + M_{\dot{\alpha}} \times \dot{\alpha}$$
$$= \ell_y \times \rho \times c^2 \times V^2 \times y/c + \ell_y' \times \rho \times c^3 \times V \times y'/c + \ell_y'' \times \rho \times c^4 \times y''/c$$
$$+ m_\alpha \times \rho \times c \times V^2 \times \alpha + m_\alpha' \times \rho \times c^2 \times V \times \alpha' + m_\alpha'' \times \rho \times c^3 \times \alpha''$$

$$.....(20.2)$$

where, m_α, m_α' and m_α'' are non-dimensional coefficients, analogous to local pitching moment coefficients in steady motion.

If the flexural and torsional stiffnesses are represented by k and k_θ, instantaneous equations of equilibrium for forces at aerodynamic center O and moments about O are

$$L - m \times y'' + m \times g \times c \times \alpha'' - k \times y = 0 \qquad .....(20.3)$$

$$M - I_O \times \alpha'' + m \times g \times c \times y'' - k_\theta \times \alpha = 0 \qquad .....(20.4)$$

where, I_O is the moment of inertia about O

Substituting for L and M from eq (20.1) and (20.2) in equations (20.3) and (20.4), we get

$$(m - L y'') \times y'' - L y' \times y' + (k - L_y) \times y - (mgc + M_y'') \times \alpha''$$

$$- L_\alpha' \times \alpha' - L_\alpha \times \alpha = 0 \qquad .... (20.5)$$

$$- (mgc + M_y'') \times y'' - M_y' \times y' - M_Y \times y + (I_O - M_\alpha'') \times \alpha''$$

$$- M_\alpha' \times \alpha' + (k_\theta - M_\alpha) - \alpha = 0 \qquad ..... (20.6)$$

In these equations, terms involving y in the force equation and α in the moment equation are known as *direct terms* while those containing α in the force equation and y in the moment equation are known as *coupling terms*. The critical or flutter speed V_f is contained in eq (20.3) and (20.4) in the terms L_Y, L_y', L_α, L_α', M_Y, M_Y', M_α and M_α'.

Solution for the oscillating motion from the above equation, can be represented by

$$y = y_0 \times e^{(\delta + i\omega t)} \qquad \text{and} \qquad \alpha = \alpha_0 \times e^{(\delta + i\omega t)}$$

Substituting these expressions in eq (20.5) and (20.6) and rearranging in matrix form, we get

$$\begin{vmatrix} -\omega^2\left(m - L_Y''\right) - i\omega \times L_Y' + k - L_Y & \omega^2\left(m \times g \times c + L_\alpha''\right) - i\omega \times L_\alpha' - L_\alpha \\ \omega^2\left(m \times g \times c + M_Y''\right) - i\omega \times M_Y' - M_Y & -\omega^2\left(I_O - M_\alpha''\right) - i\omega \times M_\alpha' + K_\theta - M_\alpha \end{vmatrix} \begin{Bmatrix} y_0 \\ \alpha_0 \end{Bmatrix} = \{0\}$$

For any speed V, the imaginary part ω gives frequency of the system while δ represents the exponential growth rate. At lower speeds, the equations represent *damped oscillatory* motion (δ is –ve). At the critical velocity V_f, these equations represent ***simple harmonic motion*** (zero growth rate or $\delta = 0$). At higher speeds, these equations represent ***divergent oscillatory*** motion (δ is +ve).

20.4.6 PREVENTION OF FLUTTER

- By preventing inertial, aerodynamic and elastic coupling - Designing the wing such that its center of gravity and center of independence coincide with its flexure axis (i.e., $g \times c = 0$ and $M_Y' \times y' = 0$). In such a design, In eq. (20.5) and (20.6), the terms M_y, $M_y \times y$ and $L_\alpha' \times \alpha'$ are very small and can be neglected. In such a design, then, eq (20.5) and (20.6) reduce to

$$(m - L_Y'') \times y'' - L_Y' \times y' + (k - L_Y) \times y - L_\alpha.\alpha = 0 \qquad \ldots\ldots(20.7)$$

$$(I_O - M_\alpha'') \times \alpha'' - M_\alpha' \times \alpha' + (k_\theta - M_\alpha) \times \alpha = 0 \qquad \ldots\ldots(20.8)$$

  The lone coupling term in Eq (20.7) can not be eliminated, since the vertical force required to maintain the flight is produced by wing incidence. Eq (20.8) has no coupling terms and so, any torsional oscillations produced (for example, by gust) will decay.

  However, it is not always possible to prevent flutter by eliminating coupling terms.

- Increasing structural stiffness, although carries the penalty of increased weight, can raise the value of V_f above the operating speed range.

REFERENCES

1. Theory and Analysis of structures–Robert M *Rivello* – McGraw Hill, 1969

2. Aircraft Structures – T H G *Megson* – Butterworth-Heinemann, 2005

3. Stresses in Aircraft and Shell Structures – **Paul *Kuhn*** - McGraw Hill, 1956

4. Airplane Structural Analysis and Design – Ernest E *Sechler* & Louis G *Dunn* – Dover Publications, 1963

5. Aircraft Structures – David J *Peery* - McGraw Hill, 1950

6. The Aeroplane Structure – A.C.*Kermode* – Sir Isaac Pitman & Sons, 1964

7. Aircraft Structures – David J *Peery* and J J **Azar** - McGraw Hill, 1982

8. Fundamentals of Aircraft Structural Analysis – Howard D *Curtis* – McGraw Hill, 1997

9. A Textbook of Machine Design – R S *Khurmi*, J K *Gupta* – S Chand & Co Ltd, New Delhi, 2001

10. Engineering Mechanics – *Pakirappa* – Durga Publishing house, Hyderabad, 2009

11. Engineering Mechanics – R K *Rajput* – Dhanpat Rai Publications, New Delhi, 2009

12. Strength of Materials – A R *Basu* – Dhanapat Rai & Co, New Delhi, 2003

13. Strength of Materials – R S *Khurmi* - S Chand & Co Ltd, New Delhi, 2001

14. Finite Element Analysis, 2nd Edition – G *Lakshmi Narasaiah* – B S Publications, 2019

Index